INFORMATION TECHNOLOGY IN THEORY

Pelin Aksoy, Ph.D.

Laura DeNardis, Ph.D.

THOMSON

COURSE TECHNOLOGY

Information Technology in Theory
by Pelin Aksoy and Laura DeNardis

Vice President, Technology and Trades ABU:
Dave Garza

Director of Learning Solutions:
Sandy Clark

Product Manager:
Kate Hennessy

Development Editor:
Dan Seiter

Content Project Manager:
Jennifer Goguen McGrail and Marisa Taylor

Marketing Manager:
Bryant Chrzan

Marketing Specialist:
Victoria Ortiz

Editorial Assistant:
Patrick Frank

Photo Researcher:
Abby Reip

Compositor:
GEX Publishing Services

Manufacturing Coordinator:
Justin Palmeiro

Disclaimer
Course Technology reserves the right to revise this publication and make changes from time to time in its content without notice.

ISBN-13: 978-1-4239-0140-2
ISBN-10: 1-4239-0140-1

To Miray, my beloved husband with a beautiful soul—PA

With love and gratitude to Deborah R. Smith, Len and Gail DeNardis, and in loving memory of Harry E. Hansen.—LD

BRIEF CONTENTS

TABLE OF CONTENTS

TABLE OF CONTENTS

Part 2 Fundamentals of Computers

TABLE OF CONTENTS

TABLE OF CONTENTS

TABLE OF CONTENTS

The information technology (IT) world in which we live and work is characterized by wireless mobility, multimedia devices, interoperability, peer-produced information creation, and an expectation of ubiquitous access to powerful computing resources. *Information Technology in Theory* seeks to fill an important gap in today's marketplace of introductory IT books in three ways. First, it takes as its starting point the multimedia and converged nature of today's global flow of information, in which integrated text, audio, images, and video are peer-produced, processed, stored, and exchanged in distributed architectures. Second, it challenges readers with questions about the social and economic contexts and controversies with which information technologies intersect, with the goal of making technical explanations more relevant. Finally, while recognizing the inevitability of unforeseen technological advancements, *Information Technology in Theory* introduces cutting-edge technologies such as Bluetooth, Wireless Fidelity (Wi-Fi), Voice over Internet Protocol (VoIP), audio compression standards, multimedia wireless devices, and advanced network security techniques, including cryptography and biometric identification.

This book is a response to the encouragement and positive feedback of hundreds of IT students who the authors have taught, using the material in this book. The book's intended audience is primarily students pursuing an undergraduate degree in information technology, management information systems, or other programs of study that require a technical and contextual grounding necessary for higher-level courses. People entering fields that are highly dependent on information technologies can also benefit from this introductory text. Other intended readers of this text are introductory electrical and computer engineering students who would like to be introduced to general IT concepts before taking classes that relate to the engineering of such information technologies. Finally, this text could be used in business programs to introduce students to technical concepts before taking classes that relate to financial, managerial, and administrative aspects of information technologies.

OBJECTIVES AND APPROACHES

Information Technology in Theory has three objectives: 1) to explain mathematical principles underlying multimedia information technologies, 2) to provide questions about the social, economic, and political contexts in which IT exists, and 3) to introduce cutting-edge technologies and trends such as those in the areas of wireless multimedia, computer security, digital audio, and high-performance computing.

MATHEMATICAL PRINCIPLES OF MULTIMEDIA IT

First, this book seeks to provide a solid grounding in the principles that enable IT. Students will learn how to represent images, text, numbers, video, and audio information in binary code. For example, the book explains binary coded decimal, hexadecimal shorthand notation, and 2's complement notation for representing negative integers in binary. It also explains important topics in digital logic, including Boolean logic, logic gates, and fundamentals of integrated circuits. Students will become grounded in the fundamentals of electronic signaling and radio wave communications. Furthermore, explanations of digital audio include descriptions of the amplitude, phase, and frequency properties of sound and pulse code modulation techniques for digitizing audio, including sampling, quantizing, and encoding. The book also explains the Nyquist sampling theorem, which mathematically calculates the minimum sampling rate necessary for digitizing an analog signal. After studying this book, students will understand error detection and correction techniques such as rectangular coding and convolutional coding and will be able to calculate the refractive index of an optical medium. The book also provides mathematical explanations of compression techniques for digital audio, image, and video. These mathematical underpinnings, reinforced with examples and end-of-chapter review exercises, explain how information technologies work and provide necessary background for higher-level courses.

SOCIOECONOMIC QUESTIONS

Understanding IT requires knowing the business, social, and political contexts that shape and are shaped by IT. One objective of the book is to challenge students to contemplate where they stand on contemporary questions and issues at the intersection of IT and society. The discussion questions at the end of each chapter serve as a vehicle for this exploration. For example, what do students think about such issues as illegal music sharing, electronic voting standards, Net neutrality, the outsourcing and off-shoring of call centers, and the economic and social implications of open source software? What are some ethical questions about biological computing or biometric identifiers? What are the economic implications of multiple audio and video formats or the rapid obsolescence of various types of storage media? What should businesses do to prepare for possible Internet security attacks or more widespread critical infrastructure attacks or cyberterrorism? While there are no singular answers to these questions, an important part of being in the IT field is to understand the issues and what is at stake in technical design and adoption.

Information Technology in Theory asks students to contemplate a broad range of social issues, from the importance of fiber-optic technologies in medical diagnostic procedures to the implications of video-sharing sites such as YouTube and the economics and politics of municipal broadband wireless projects. The book explains the economic implications of network protocols and addresses issues related to the control of centralized Internet functions such as the administration of Internet domain names and IP addresses. It explains the business drivers of VoIP along with more complex issues related to security, public safety, regulatory jurisdiction, and taxation. Readers also will be exposed to the current state of IT careers, ranging from computer forensics to management consulting to software development.

CUTTING-EDGE TECHNOLOGIES

Another important objective of *Information Technology in Theory* is to provide a comprehensive overview of the cutting-edge technologies critical to modern business and social life. The book explains how VoIP works and gives more granular attention to the many protocols, such as the Session Initiation Protocol (SIP), H.323, and Real-Time Transport Protocol (RTP), that actually comprise what is commonly called VoIP. Another topic of great pertinence to modern business and social contexts is wireless mobility, in particular how multimedia devices communicate over systems such as Wi-Fi and Bluetooth, and advanced cellular standards and approaches such as Global System for Mobile communications (GSM) and Code Division Multiple Access (CDMA). Multimedia information exchange is not possible without advanced compression techniques such as Moving Picture Experts Group 4 Audio (M4A), Advanced Audio Coding (AAC), Moving Picture Experts Group Audio Layer-3 (MP3), Waveform Audio

Format (WAV), Joint Photographic Experts Group (JPEG), and Moving Picture Experts Group (MPEG). Therefore, this text explains compression in various chapters.

Students will become familiar with how the Internet's Domain Name System (DNS), backbone routers, and exchange points work, and will understand the new Internet protocol standard, Internet Protocol Version 6 (IPv6), and how it globally expands the Internet address space. One of the most important functions of IT professionals is securing information, computing resources, and networks, so great attention is given to explaining security threats such as distributed denial-of-service attacks and identity theft, and security solutions such as public key encryption, firewalls, and digital signatures. The book explains advanced optical storage devices and electronic storage technologies that store bits by trapping electricity within an array of devices arranged on a chip. Other advanced topics include plasma devices, open source software, and standards for multiplexing over optical fiber telecommunications lines. The objective is for students to have a working knowledge of this broad selection of critical multimedia IT topics. The book uses real-world examples to explain these concepts and uses end-of-chapter case studies to promote further exploration. For example, one case study asks students to create, post, and share a digital video on a video-sharing site.

ORGANIZATIONAL STRUCTURE

Information Technology in Theory is divided into six parts and 18 chapters.

PART 1, INTRODUCTION TO INFORMATION TECHNOLOGY

Part 1 provides a basic understanding of the concept of IT and is divided into the following three chapters:

Chapter 1, Introduction to Information Technology—This chapter defines terms such as "information" and "information technology" and explains how information can be conveyed via four integrated methods: image, text, audio, and numbers. The chapter addresses the significance of IT in modern business, political, and social contexts and describes some of the most advanced current trends in IT. Finally, it explains what topics are included in the IT discipline and exposes readers to the current categories of available IT jobs.

Chapter 2, Understanding the Digital Domain—What does it mean that we live in a "digital age"? Chapter 2 explains how to represent any form of information in analog and digital formats, provides historical background on the development of digital technologies, and explains techniques for transmitting and storing digital information. The chapter concludes by explaining the advantages of the digital approach, including resistance to noise, higher transmission speeds, ease of reproduction, greater reliability in transmission and storage, more efficient security, capacity for error control, and the ability to use advanced compression techniques.

Chapter 3, Representing Numbers and Text in Binary—Binary representation is an important mathematical foundation in IT and is the basis for digital systems. Combinations of binary symbols called bits can be used to represent any form of information. To introduce this topic, the chapter explains mathematical techniques for converting between the decimal and binary numbering systems, including converting negative and positive noninteger numbers to binary. The chapter also introduces other numbering systems such as octal and hexadecimal, explaining how these systems are used as shorthand notations for important real-world IT functions such as Internet and hardware addresses. The chapter also introduces how standard codes such as American Standard Code for Information Interchange (ASCII), Unicode, and Extended Binary Coded Decimal Interchange Code (EBCDIC) can represent alphanumeric characters in binary.

PART 2, FUNDAMENTALS OF COMPUTERS

Part 2 is divided into two chapters that address essential concepts in computer hardware and software.

Chapter 4, Computer Hardware—After providing a brief history of computing, this chapter explains how transistors are the essential components of digital devices and how they can be combined to create logic gates, which can then be used as the basic building blocks for integrated circuits. The chapter explains the fundamental components of a computer and how computer performance depends on factors such as processor speed, chip set, bus width, bus speed, and number of processor cores and processor chips. Chapter 4 also examines future trends in computing, such as nanotechnology, quantum computing, and biological computing, and raises social questions related to these advances. Finally, the chapter explores storage technologies, including magnetic, optical, and electronic storage, and describes future possibilities for meeting ever-increasing demands for digital storage, including holographic and molecular storage.

Chapter 5, Software—Chapter 5 introduces the concepts of high-level and low-level programming languages. It describes different types of software, including system software and application programs, and examines the stages of the software development process, some of which include definition, design, programming, testing, releasing, training, maintenance, and support. The chapter also discusses the concept of open source software and related issues.

PART 3, CREATING DIGITAL MULTIMEDIA

Part 3 addresses issues related to digital audio, images, and video.

Chapter 6, Digital Audio Technology—This chapter explains the physical and mathematical basis of sound waves and how they can be captured electrically in analog format for digitization by electrical circuits. The chapter describes the three-step audio digitization process of sampling, quantization, and encoding and introduces the Nyquist sampling theorem. A major theme of this chapter is compression, including the need for digital audio compression, a technical explanation for how music files can be compressed without degrading sound quality, and a distinction between lossy and lossless compression techniques. The chapter concludes by introducing readers to popular digital audio formats such as MP3, AAC, WMA, WAV, and Audio Interchange File Format (AIFF).

Chapter 7, Digital Images and Video—This chapter introduces readers to digital imagery, including how images can be directly captured by a digital camera or indirectly captured by a scanner. The chapter explains the underlying technical process of producing digital video, the limitations of human vision, and how these limitations are exploited by image and video digitization. This chapter also helps readers to identify factors that affect the quality of digital images and discusses popular digital image and video file formats and state-of-the-art display technologies, including

Liquid Crystal Display (LCD), plasma, and Digital Light Processing (DLP) projection systems. Finally, the chapter describes the properties and technical features that allow calculation of the bit size of digital images and video.

PART 4, TRANSMISSION OF INFORMATION

Part 4 is divided into the following three chapters:

Chapter 8, Fundamentals of Communications—Bits can logically represent any type of information, but a physical entity such as electrical energy, optical energy, or electromagnetic energy is essential for carrying these bits over transmission channels. This chapter describes how bit streams are physically generated and how carriers are modulated to carry the binary streams. It discusses the basics of electrical, electromagnetic, and optical signaling for transmitting analog and digital forms of information over various transmission media, with an emphasis on digital electrical signaling. This chapter also introduces important transmission concepts such as modulation, multiplexing, attenuation, and bandwidth, and concludes by examining error detection and correction techniques for digital transmission systems.

Chapter 9, Introduction to Fiber Optics—This chapter introduces the optical communication systems that have facilitated many IT innovations ranging from the Internet's core transmission backbone to cutting-edge medical imaging. The chapter describes the basic components of fiber-optic cable, the various types of cables, and the physical and mathematical principles of transmitting light through fiber-optic cable, including refraction, total internal reflection, and wavelength-division multiplexing. The chapter also describes the advantages and drawbacks of using fiber as a transmission medium, and introduces some novel technologies being used in state-of-the-art optical communication systems.

Chapter 10, Wireless Communications—Chapter 10 introduces radio-based wireless communication systems and provides an overview of some important radio applications, including the Global Positioning System, satellite Internet, and radio frequency identification systems. The chapter also discusses the effects of the atmospheric channel on radio waves and techniques used to counteract some of these effects.

PART 5, INTRODUCTION TO COMPUTER NETWORKING

Part 5 is divided into five topics: local area networks, wide area networks, communication protocols, the Internet's technical architecture, and network security.

Chapter 11, Local Area Networks—This chapter explains the technology of underlying networks, such as Wi-Fi networks, that span a limited geographical area such as an office building or house. This chapter describes the most popular types of local area networks (LANs), with a focus on Ethernet, and explains LAN design characteristics, including topology, access mechanisms, physical transmission media, and equipment. This chapter emphasizes the technical architecture and standards underlying the wireless LAN technology that we commonly call Wi-Fi.

Chapter 12, Wide Area Networks—This chapter describes the distinguishing technical characteristics of wide area networks (WANs), including packet switching, virtual private networking, WAN equipment, and WAN protocols. The chapter also presents an overview of popular WAN services, including Internet services, frame relay, Asynchronous Transfer

Mode (ATM), Multiprotocol Label Switching (MPLS), and private networks. It also describes alternatives for accessing WANs, including dedicated lines, Digital Subscriber Line (DSL), and cable modem access. The chapter concludes with an overview of network management functions that are vital to enterprise networking.

Chapter 13, Communication Protocols—This chapter describes the important role of network protocols, also called communication protocols or standards, in enabling digital information exchange. It describes the Open Systems Interconnection (OSI) reference model and gives examples of important protocols such as the Internet Protocol (IP). The chapter also describes some of the organizations that establish these standards and discusses how these protocols are not only technological design decisions but have economic and political consequences.

Chapter 14, Internet Architecture—After providing some Internet history, Chapter 14 examines the key technologies required for the Internet to operate, such as Transmission Control Protocol/Internet Protocol (TCP/IP), Internet backbone routers, packet switching, Internet exchange points (IXPs), and the Domain Name System (DNS). It examines some important Internet applications that improve business productivity and individual communications, and some central administrative aspects of the Internet such as the administration of domain names and IP addresses. The chapter also discusses unique economic and social issues that have accompanied the Internet's evolution.

Chapter 15, Network Security—Network security is one of the most critical technical areas within IT. This chapter describes major security challenges such as denial-of-service attacks, viruses, worms, identity theft, spam, piracy, and threats to critical infrastructure. It also explains privacy, authentication, and access control security measures to counteract these threats, including private and public key cryptography, digital signatures, firewalls, and biometric authentication.

PART 6, TELEPHONY AND WIRELESS MULTIMEDIA

Part 6 focuses on the three primary network alternatives for transmitting multimedia information that includes voice. These alternatives include the traditional telephone system architecture, VoIP, and wireless multimedia systems such as WiMAX and cellular telephone systems.

Chapter 16, The Telephone System—Chapter 16 provides historical background on the traditional telephone system. It describes the system's physical components, including transmission systems, switching systems, and customer premises equipment such as Private Branch Exchanges (PBXs). The chapter also addresses underlying techniques for transmitting voice information over the telephone system, including multiplexing, pulse code modulation (PCM), and signaling.

Chapter 17, Voice over IP—This chapter describes the concept of VoIP and how it enables the transmission of voice over the Internet. It explains some of the many VoIP protocols, such as SIP, H.323, and RTP, and presents various implementation options for Internet telephony. The chapter describes the business drivers for using VoIP and its unique security and performance requirements. Finally, the chapter identifies interesting policy and regulatory challenges that have emerged with Internet telephony.

Chapter 18, Wireless Multimedia—This chapter describes how multimedia access devices such as cellular telephones have integrated previously separate applications such as voice, Internet applications, electronic commerce, and entertainment. The chapter describes important enabling wireless technologies such as spread spectrum and Bluetooth, and important cellular standards such as GSM. The chapter also examines emerging broadband metropolitan wireless technologies and how they might transform information access.

ACKNOWLEDGEMENTS

First and foremost, we thank the vast number of information technology students who have served as the inspiration for this project. Throughout the course of developing these materials, the students have been a source of motivation, new ideas, encouragement, and invaluable feedback.

We especially want to thank the Thomson Course Technology team for making this project possible. In particular, we thank Product Manager Kate Hennessy for skillfully shepherding the project to completion. Development Editor Dan Seiter deserves special recognition for his enormous contributions and tireless efforts. We also acknowledge the important contributions of Acquisitions Editor Maureen Martin, Senior Field Representative Mark Mochary, and Content Project Manager Marisa Taylor.

This book benefited greatly from the constructive suggestions of the following reviewers:

- Dr. Robert Bonometti, Shenandoah University
- Dr. Robert Friedman, New Jersey Institute of Technology
- Dr. Barry Lunt, Brigham Young University
- Dr. Anne Marchant, George Mason University
- Dr. Charles Snow, George Mason University
- Mark Stockman, University of Cincinnati
- Dr. Diana Wang, George Mason University

We thank them for their candor and thoughtful contributions to this project. Dr. Anne Marchant deserves special recognition, as her suggestions and helpful comments have contributed greatly to the development of this book.

We also want to thank Dr. Andre Manitius, the chair of the Electrical and Computer Engineering Department at George Mason University, for providing us with the encouragement and enthusiasm to embark upon this project. Pelin Aksoy would also like to thank the following faculty of George Mason University for their encouragement and support: Dr. Lloyd Griffiths, the Dean of the Volgenau School of Information Technology and Engineering; Dr. Donald Gantz, the Chair of the Department of Applied Information Technology; Dr. Sharon Caraballo, the Assistant Dean for Academic Affairs; and Ms. Kamaljeet Sanghera, Faculty Member of the Applied Information Technology Department. Pelin also thanks Dr. Michael Haney, her dissertation advisor, for his encouragement and support.

Finally, we want to extend our greatest appreciation to our families: Miray Kurtay; Deborah Smith; Ihsan, Lerzan, and Levent Aksoy; Filiz, Arican, Mine, and Melisa Kurtay; Len, Gail, Len Jr., Denise, Marissa, Abby, and Natalie DeNardis; and David and Elva Smith for their invaluable support in this major endeavor. We would not have been able to complete this book without their unconditional love.

PART

1

INTRODUCTION TO INFORMATION TECHNOLOGY

INTRODUCTION TO INFORMATION TECHNOLOGY

In this chapter you will:

- Define "information" and understand the four methods of representing and conveying information

- Describe some important historical milestones in recording and exchanging information

- Explain what technical topics are included in the field of information technology

- Describe the role information technologies play in modern society

- Articulate some cutting-edge trends in information technology

- Gain familiarity with current career specializations in information technology

INTRODUCTION

Information technology (IT) seems to inhabit its own time scale. As recently as the year 2000, few people had heard of Wi-Fi or MP3 file sharing, and YouTube did not exist. In 1995, eBay and Google had not yet been founded. As recently as 1990, there was no World Wide Web or widespread home Internet access. Thirty years ago, there were few cell phones or personal computers. One hundred and thirty years ago, there were no telephones, phonographs, movies, radios, or televisions. Given such exponential advancements in information technology, it is mind-bending to envision what the future might hold for the field.

Before exploring the cutting edge of IT, it is helpful to define the terms *information* and *information technology*. After all, there was no such thing as an information technology major until fairly recently. To place modern information technologies in perspective, we need to understand some important historical milestones and reflect on the relationship between IT advancements and societal conditions. This chapter introduces the subject of IT, places the field within its historical context, and describes the role of IT in today's economic and social climate. It also provides a snapshot of cutting-edge trends and careers in IT.

DEFINING INFORMATION

Modern society has graduated from the industrial era to the information era, but it would be easy to argue that information exchange has been critical in every historical era for several thousand years. Before exploring the technologies that enable the exchange of information, what exactly is meant by **information**?

The definition of information is not straightforward; it differs depending on one's perspective. For example, an engineer and a social scientist might have different perspectives on the same information. However, information is widely accepted to be a fact or series of facts that carry meaning. The value of these facts also varies and strongly depends on the context in which the information is provided. A lot of electronically exchanged information is hardly vital. For example, instant messaging conversations count as information, as do downloaded music files, e-mail, or interactive video games. Approximately half of all electronic mail is **spam**—unsolicited, unwanted e-mails that flood Internet mailboxes. These e-mails technically count as "information," albeit unwanted and hardly vital. In many contexts, however, information is vital: dialing 911, monitoring a patient's vital statistics during surgery, and conveying air traffic control conditions. Economically critical exchanges of information keep the stock market running, enable businesses to electronically exchange transactions, and allow consumers to order products. Instructions for building a nuclear bomb and yesterday's sports scores both count as information. Information sharing enables people to communicate, transact business, convey "how to" information, exchange news, and share opinions about entertainment (see Figure 1-1).

Figure 1-1

What counts as information

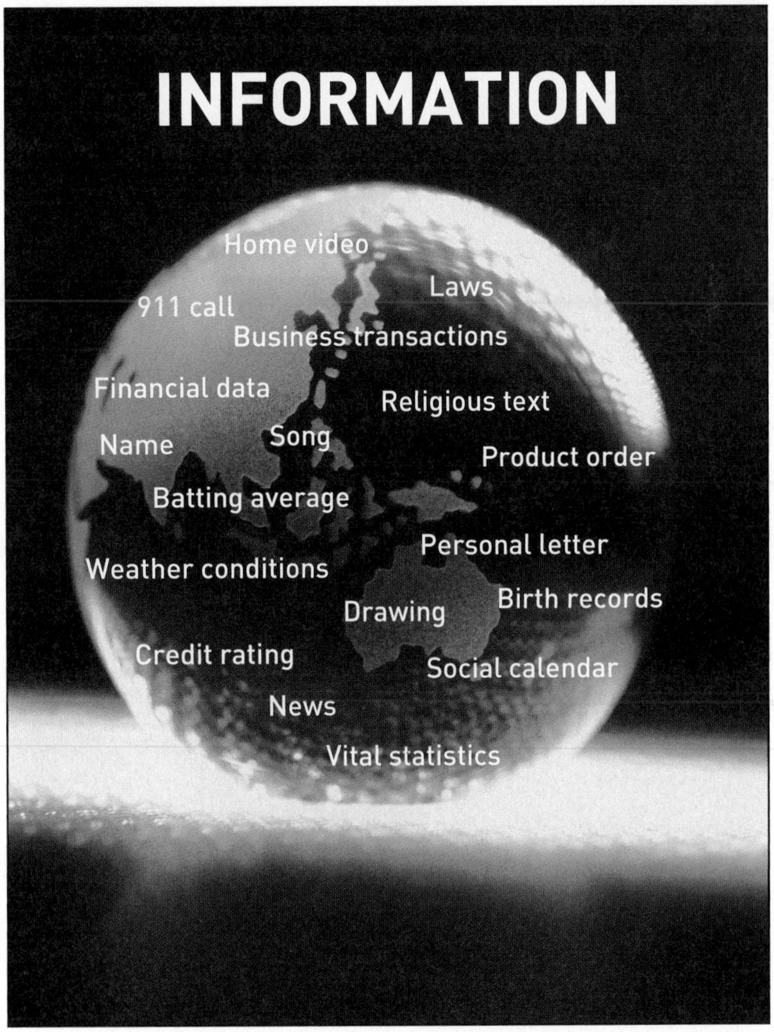

CHAPTER 1   Introduction to Information Technology   ◆   3

THE VARIOUS FORMS OF REPRESENTING INFORMATION

How do people represent and interpret information? News Web sites normally combine text, images (including video), numbers, and sound. Cell phones convey alphanumeric information via text messaging, sound in the form of voice conversations, and images. These routes to representing information have not changed for centuries (see Figure 1-2).

One of the first forms that people systematically employed to represent information was the spoken word. Before the advent of written languages, writing tools, and recording media such as parchment, historical events and important facts were transmitted orally. People shared information via oral histories, songs, poems, and stories that were passed from group to group and generation to generation. Historically, sound has perhaps been the dominant approach for representing and conveying almost any kind of information; to some extent, the importance of sound in representing information has not changed.

The *means* of exchanging and capturing audible information has changed, however, especially with the introduction of the telephone and phonograph in 1876. Prior to the invention of the telephone, spoken conversations obviously required two people to be in each other's presence. Before the phonograph, listening to music required attendance at live performances. The new sound technologies, especially the telephone, effectively expanded space—people could live far apart and still have real-time conversations. Newer technologies (see Figure 1-3) have improved on these groundbreaking innovations and added the dimension of **mobility** to distance. A person traveling across the country can listen to radio broadcasts, CDs, or music files and can talk to anyone almost anywhere in the world.

Figure 1-2

Four methods for conveying information

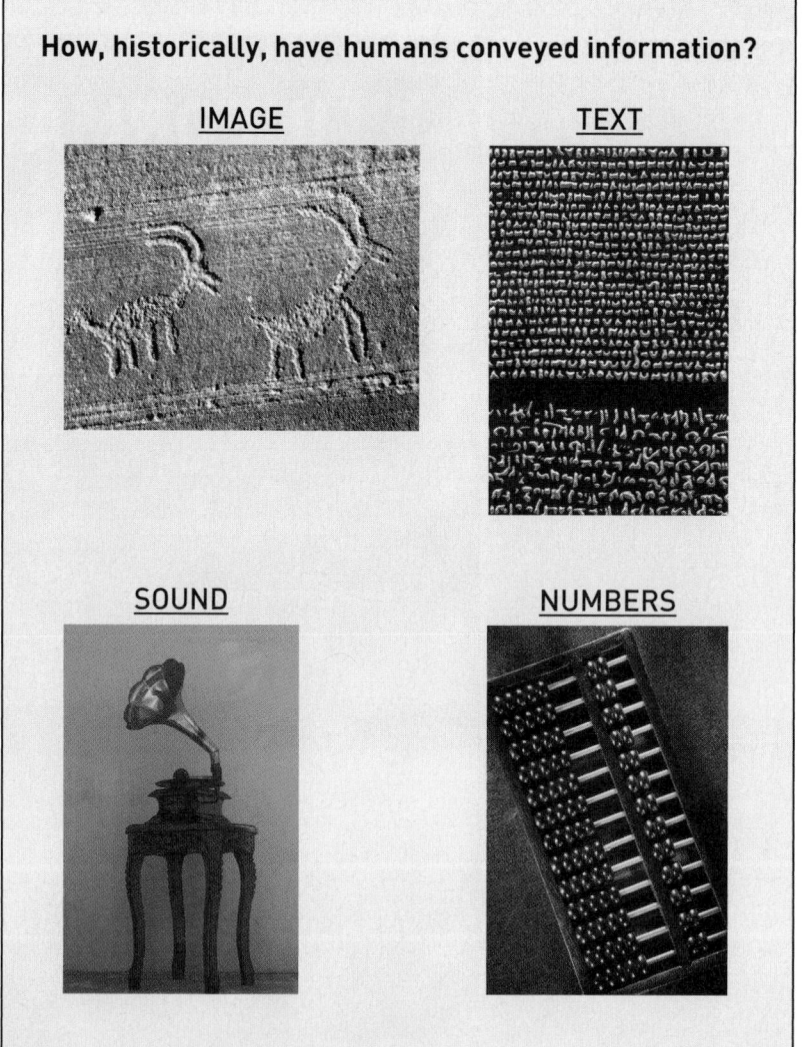

How, historically, have humans conveyed information?

IMAGE

TEXT

SOUND

NUMBERS

Figure 1-3

Sound technology
milestones

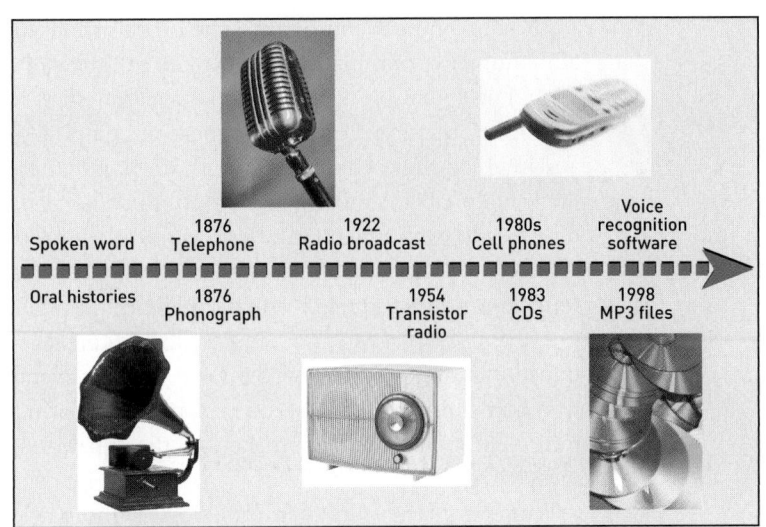

Figure 1-3

Sound technology
milestones

From the inception of recorded human history, images have also offered an important form for recording and exchanging information. Ancient cave drawings and rock carvings indicate the human penchant for information exchange, whether for pragmatic or artistic reasons, even in very early civilizations. For millennia, artistic renderings remained the exclusive means of capturing and conveying visual information. It took the nineteenth-century innovations of photography and silent movies and the twentieth-century technologies of television, cable TV, and digital photography to revolutionize approaches to capturing visual images (see Figure 1-4).

Many modern information technologies address the recording and conveying of imaging information. Healthcare imaging technologies capture ever more detailed medical images. Satellite surveillance technologies record images from space. Radar systems capture not only weather conditions but the locations of airplanes, ships, and submarines. Handheld digital video phones instantly capture images from war zones and transmit them thousands of miles for broadcast on news programs. Video repositories such as YouTube allow anyone with Internet access to provide or access a video.

Figure 1-4

Image technology
milestones

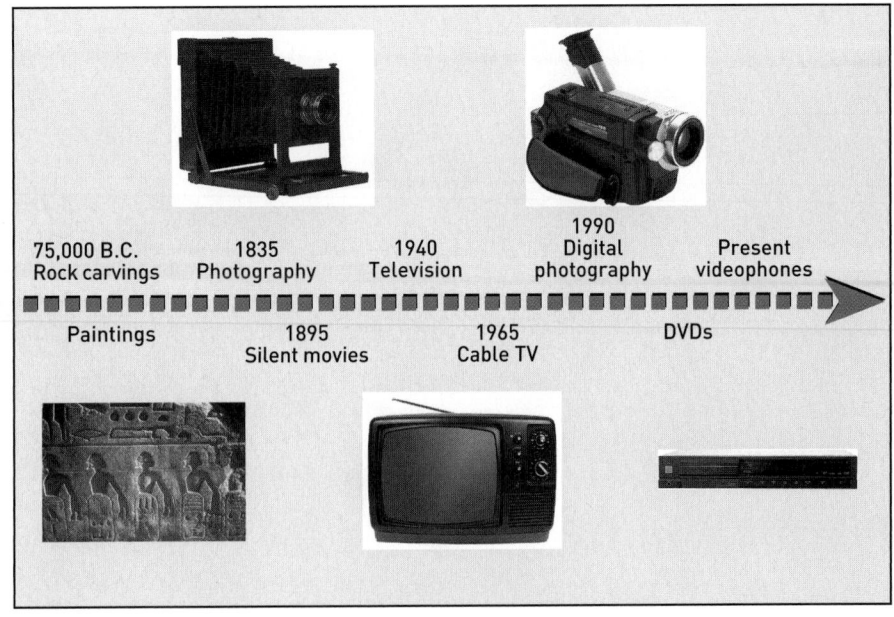

Besides sounds and images, the other primary methods of conveying information are alphanumeric combinations of text and numbers. Numeral systems, whether Arabic (0, 1, 2, 3, etc.) or Roman (I, II, III, IV, etc.), have long represented financial, computational, and statistical information. Textual alphabets and graphic systems have historically recorded events and enabled the exchange of information, whether through Egyptian hieroglyphics or the English alphabet. Until the fifteenth century, textual information was captured by hand—the only way to copy a manuscript was to rewrite it. Around 1450 AD, Gutenberg's printing press changed all this by enabling the mass production of the written word. Despite this new ability to mass-produce textual information, there still was no expeditious way to transmit information from one place to another. People had to exchange messages using the fastest available transportation, whether by foot, horse, boat, or eventually train. After centuries of transportation constraints that dictated the speed of information exchange, the nineteenth-century introduction of the telegraph changed the equation forever. Suddenly, people could transmit messages over long distances in a matter of seconds. The telegraph provided the first of many real-time or near real-time information technologies. Modern IT systems, most notably electronic mail and instant text messaging, continue this tradition of recording and transmitting textual information (see Figure 1-5).

The technological systems and computational approaches presented in this textbook address the capture, storage, and exchange of all four forms of information: sound, images, numbers, and text. These forms of information, as we will explain in detail in later chapters, can all be represented using a **binary code** that consists of the values 1 and 0. Any form of information can be represented as a series of ones and zeros, which are called **bits**. A piece of text, for example, may be represented in binary code as 1100100101010110000. Similarly, a sample of music or an image can be represented as a sequence of bits:

00101111010101010101010101010101000010101101011111110.

Technologies that capture, store, and exchange information historically have two important characteristics in common: **obsolescence** and **convergence**. Think about some of the technologies that have captured images and sound over the years. How many of these technologies are completely or partially obsolete? No one uses the telegraph anymore. Fax machines are increasingly unnecessary. Transistor radios, phonographs, and typewriters are quaint novelties rather than necessities. Almost without exception, the "next new thing" ultimately lands on the scrap heap of obsolescence. Interestingly, the exceptions to this obsolescence are the original means of information exchange, because people still write letters, share conversations, and paint pictures.

Figure 1-5

Alphanumeric technology milestones

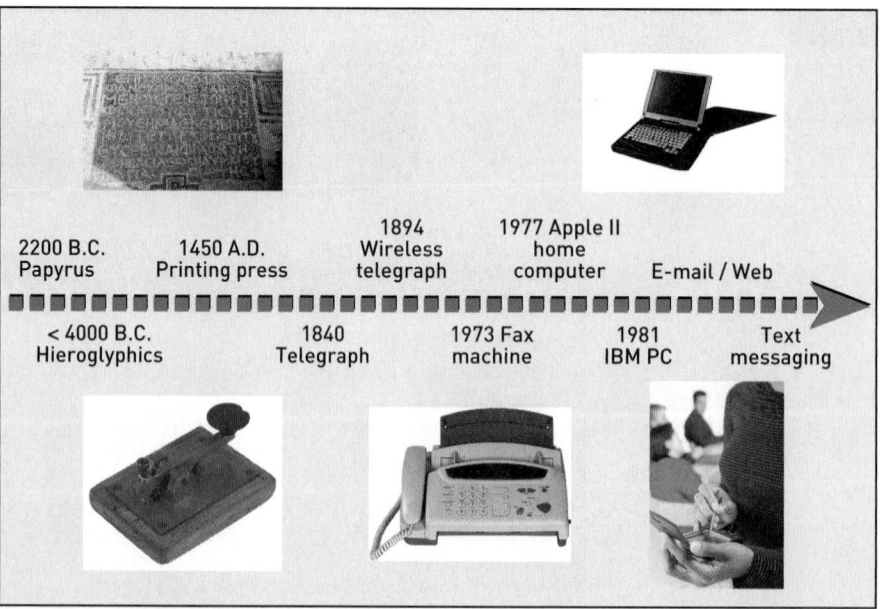

The second commonality is convergence. Most present-day technologies are multifunctional, as they capture or exchange not a single type of information but integrate *multiple* forms of information for different purposes. The term **multimedia** describes this integration. For example, cell phones are classified as multimedia devices because they can transmit information in the form of text, numbers, sound, images, and video. The term *convergence* describes the unification of technologies to offer different information services through a single device or system. A cellular telephone is an example of convergence because it can be used for a multitude of purposes, including telephony, accessing the Internet, and watching television. Convergence is now the norm rather than the exception, which means we must simultaneously consider and study all types of information exchange under a single umbrella.

THE SPEED AND SCOPE OF INFORMATION

While the basic forms of representing information have not changed for centuries, the speed and scope of the technologies that enable people to capture, store, and exchange information have changed dramatically. Today, information exchange permeates almost every day-to-day activity. Enormous amounts of information change hands in very brief periods of time. Consider a college student with an hour to expend before class. In the course of only an hour, the student could easily conduct the following information exchanges:

- Check out the day's most viewed videos on YouTube
- Make two cell phone calls
- Download a homework assignment for class
- Exchange several text messages with friends
- Book an airline reservation for a spring break trip
- Consult one news source
- Download a new music file
- Use the electronic library system to reserve a book
- Check the day's sports scores
- Play an interactive online game
- Spend five minutes posting a blog
- Send three e-mail messages

Wireless access and mobile devices allow the student to exchange all this information from anywhere on campus. This round-the-clock ability to instantly communicate with others, and the ability to exchange information at will from anywhere, exemplify the present information era.

DEFINING INFORMATION TECHNOLOGY

The ability of people to exchange information transcends millennia, but what does the term *information technology* specifically refer to in the modern sense? Consider the following commonly adapted description of information technology:

Information technologies are devices or systems that capture, process, exchange, or store information.

By this definition, the following devices and systems are considered information technologies:

- The postal system, because it exchanges information
- A filing cabinet, because it stores information
- A library, because it stores information
- A mathematical algorithm, because it processes information

These systems do not conventionally fall under the modern title of IT. Therefore, a more narrow definition is necessary to exclude certain systems, such as those that involve paper storage. This textbook recommends the following definition of IT:

Information technologies are systems of hardware and/or software that capture, process, exchange, store, and/or present information, using electrical, magnetic, and/or electromagnetic energy.

Even this more restrictive definition of IT cuts a wide swath across technical topics ranging from digital cameras to Internet radio to corporate computer networks. The following section discusses two examples of information technologies—cellular telephony and wireless Internet access—and describes how they meet our description of IT.

TWO IT EXAMPLES

Consider a call made between two cell phone users. A cell phone *captures* the sound of the voices and converts it into an electrical signal. Network equipment *processes* the call and determines how to route the signal to its destination. The information is *exchanged* from the cell phone to a cellular antenna via radio waves, using electromagnetic energy, and is received at the switching center. The call is then typically routed across the traditional telephone system, to the switching center, and then to the cellular antenna closest to the recipient's cell phone location. If the recipient is unavailable, the system *stores* the information in voice mail. If the recipient is available, the cell phone translates information from electrical signals into sound waves that are *presented* to the user and that replicate the original voice message of the sender. This cellular telephony system captures, processes, exchanges, stores, and presents information. Like most modern IT systems, it can also accommodate multimedia by transmitting alphanumeric characters and images as well as sound (see Figure 1-6).

Another IT example is wireless Internet access, as shown in Figure 1-7. A Web designer *captures* multimedia information by programming Web files in a standard Web format such as Hypertext Markup Language (HTML) for inclusion on a Web server. The Web server might act as a transaction site that *processes* information, such as merchandise purchases or dinner

Figure 1-6

Cellular telephony as an example of an IT system

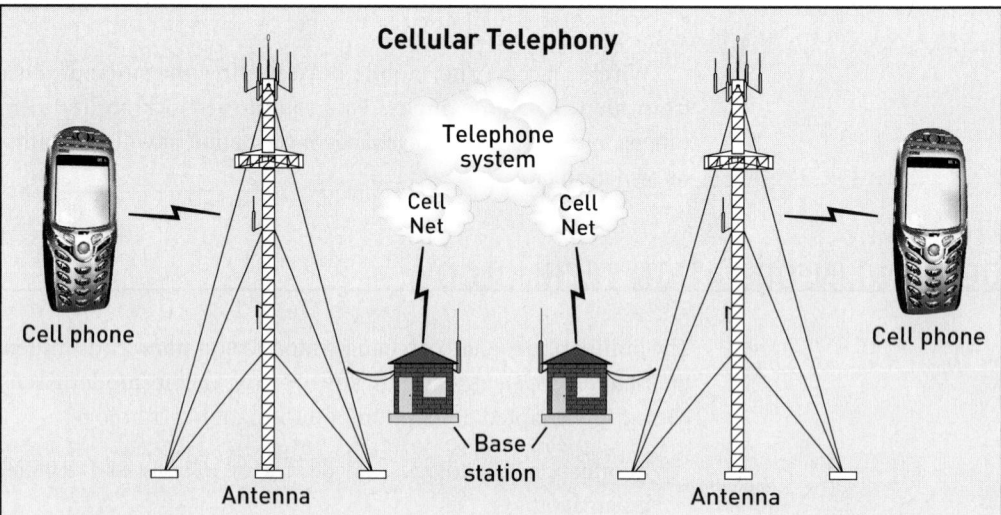

Cellular Telephony

Telephone system

Cell Net   Cell Net

Cell phone   Cell phone

Base station

Antenna   Antenna

Capture: A cell phone captures the sound of a human voice and converts it into electrical signals.
Process: Network equipment determines where to route the call.
Exchange: A network routes the call from origination to destination.
Store: A voice mail system stores information for later use.
Present: A cell phone translates information from electrical signals to sound waves the recipient can understand.

reservations by a Web user. The Web server *stores* information on a medium such as a magnetic disk. Information is *exchanged* over the Internet between the Web site and a user to a wireless access point and then to the user's wireless laptop or other computing device. Information is *presented* to a user via a Web browser on a laptop. Like the cell phone example, this IT system involves the transmission of multimedia information. The Web site may house a combination of text, numbers, images, video, and sound.

The preceding examples described information technologies that capture, process, exchange, store, *and* present information. Other examples also fall under the definition of an information technology, and accomplish one or more of the same tasks:

- A digital camera, because it can capture, process, store, and present information
- A hard disk, because it can store information
- A fiber-optic link, because it can exchange information
- A computer monitor, because it can present information

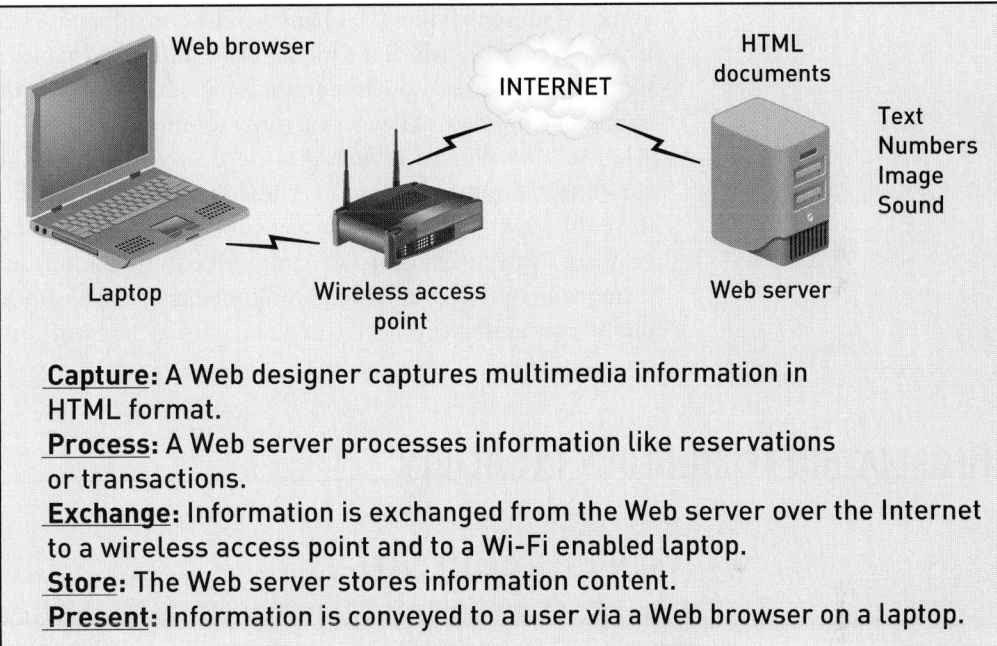

Figure 1-7

Wireless Internet access as an example of an IT system

Capture: A Web designer captures multimedia information in HTML format.
Process: A Web server processes information like reservations or transactions.
Exchange: Information is exchanged from the Web server over the Internet to a wireless access point and to a Wi-Fi enabled laptop.
Store: The Web server stores information content.
Present: Information is conveyed to a user via a Web browser on a laptop.

ORGANIZATION OF THIS TEXTBOOK

What else is included in IT? Because it represents such a vast knowledge area, the field can be separated into subtopics in various ways. To provide a comprehensive overview of information technologies, this textbook is organized into the following six sections:

- *Introduction to Information Technology*—Section I introduces concepts in information representation, defines IT, and discusses some cutting-edge IT topics. All advanced information technologies are digital, so the first task of understanding IT is to distinguish the digital approach from its analog predecessors and explain why digital is more advantageous for modern computing. Digital devices use binary code, a system with two values (1 and 0); this section introduces the concept of binary representation and explains how combinations of ones and zeros can encode textual and numerical information.
- *Fundamentals of Computers*—Section II introduces the underlying principles of computer hardware and software. It explains digital logic and introduces important computing topics such as Boolean logic, logic gates, and fundamentals of integrated circuits. It discusses the issue of human-computer interaction, and it introduces the essential components of computers, including the processor, types of memory, storage technology, and input/output devices. This section also presents the fundamentals of computer software, including operating systems, applications, and the hierarchy of programming languages.

- *Creating Digital Multimedia*—Section III addresses digital sound and digital imaging technologies. It describes the mathematical and physical properties of sound waves and how devices can be used to digitize and compress these waves. This section addresses similar issues for images, explains how images and video can be digitized by various devices, and describes how digital image and video files can be compressed.
- *Transmission of Information*—Section IV begins with an overview of electrical, optical, and radio-wave communications, explaining related concepts such as modulation, multiplexing, bandwidth, and attenuation. This section presents an overview of copper-based transmission media and discusses the techniques that are used for managing errors in digital communication systems. Fiber-optic systems and wireless communication technologies are also covered in depth in this section.
- *Introduction to Computer Networking*—Section V provides an overview of the field of computer networking, including local area networks (LANs), wide area networks (WANs), and the Internet's technical architecture. It explains different types of network access, such as digital subscriber line (DSL) and wireless broadband. The section introduces communication protocols, which are the standard rules that enable networks to interoperate. The section also addresses the important topic of network security, covering types of security attacks and a variety of network security solutions.
- *Telephony and Wireless Multimedia*—Section VI discusses the many platforms for voice communications. In particular, it describes the traditional telephone system, known as the Public Switched Telephone Network (PSTN), and explains how Voice over Internet Protocol (VoIP) technologies enable voice conversations over the Internet. The section includes an overview of wireless multimedia systems, with an emphasis on cellular telephony concepts such as cells, frequency reuse, hand-off, roaming, and current cellular standards.

INFORMATION TECHNOLOGY IN SOCIETY

IT-ENABLED ACTIVITIES

It is hard to imagine modern society without information technologies. We are surrounded by technologies that enable constant communication and information exchange between people. The nature of technology is such that it becomes almost transparent once it is ingrained in everyday life. Consider the kitchen sink, for example. Is it considered a *technology*? Its function is to deliver a controlled supply of hot and cold water into a kitchen. It has a critical function that was viewed as cutting-edge technology a hundred years ago. Now it is so taken for granted that people may not even consider the sink a technology. Many information technologies are taken for granted in a similar manner. For example, vacationers traveling in the United States assume ubiquitous availability of phones if they need to call home, ATMs to withdraw money, wireless LANs in airports to access the Internet, and TVs in hotels to watch 24-hour news networks. Without these technologies, travelers would be completely incommunicado. In fact, people who want to "get away from it all" and "unplug" must make a concerted effort, such as going camping in the middle of nowhere. In other words, being plugged in via information technology is the norm in modern society, and being unplugged is an exception that requires some effort.

People rely on information technology for entertainment, communication, and a variety of day-to-day functions (see Figure 1-8). For better or worse, IT supplies much of modern society's entertainment. This reality is most stark during power outages, when children suddenly lose access to television, online chatting, Web surfing, and video games. Entertainment still counts as information, and since the advent of the phonograph in 1876, technologies have certainly changed the nature of human pastimes.

Figure 1-8

IT-enabled activities for people

IT PERMEATES SOCIETY	
IT User	**IT-Enabled Activities**
I N D I V I D U A L S	**Personal Communications** ☐ Conversations via phones and cell phones ☐ Messaging via e-mail and text messages ☐ Video communications **Entertainment** ☐ Surfing the Web ☐ Listening to the radio ☐ Downloading MP3 files ☐ Watching television ☐ Playing interactive games **Day-to-Day Living** ☐ Buying an airline ticket ☐ Ordering books ☐ Checking the weather forecast or news ☐ Electronic banking ☐ Investing in the stock market

Today, almost anyone is instantly reachable via a telephone, cell phone, or text messenger. On any given day, people have the option of using the Internet or telephone to purchase merchandise or services. Online banking and direct deposit make a trip to the bank a rare occasion.

Most national economies are also enmeshed in information technologies. IT software and hardware manufacturers and service providers make up an important sector of the U.S. economy and other national economies (see Figure 1-9). Even more economically significant is the reality that almost every business and government sector depends on computers and networks to function. IT advancements have engendered productivity improvements in industries such as retail, finance, and manufacturing, and some economically important activities are completely reliant on computer networks. Examples of network-enabled business activities include airline reservation systems and the NASDAQ stock market network. Without computer systems and networks, entire national economies would be dramatically affected.

Figure 1-9

IT-enabled activities for businesses

IT PERMEATES SOCIETY	
IT User	**IT-Enabled Activities**
B U S I N E S S E S	**Internal Communications** ☐ Internal computer networks ☐ Internal corporate Web sites ☐ Video teleconferencing ☐ Phone systems, cell phones, voice mail ☐ Messaging via e-mail and text messaging **Electronic Commerce** ☐ Call centers ☐ Electronic transactions with suppliers ☐ Online sales ☐ Point of sale devices and networks ☐ Customer transaction servers **Business Operations** ☐ Factory floor systems ☐ Inventory tracking systems ☐ Customer databases ☐ Payroll and human resources

Uses of IT in academia, politics, and government also continue to escalate, as shown in Figures 1-10 and 1-11. Computer networks have improved communications and information sharing within the federal government, and have streamlined interaction between citizens and government. In-person interaction with government is increasingly unnecessary because people can file taxes online, find a host of information on government Web sites, and even register their car with the Department of Motor Vehicles over the Internet. Political uses of information technology have become an interesting development in democratic societies. U.S. presidential campaigns have used their Web sites as important avenues for political fundraising, information dissemination, and grassroots organizing.

Every new application of information technology in society comes with advantages and disadvantages. For example, electronic voting represents the gray area that can materialize around emerging areas of information technology. Should citizens have the option of electronic or Internet voting? Placing electronic voting machines in polling places streamlines the voting. process, but the possible absence of a paper trail for accountability and recounting is a disadvantage. On the surface, Internet voting sounds like an efficient use of information technology, as it would be convenient for millions of citizens, but it presents complex problems, such as how to authenticate that citizens are who they claim to be, how to prevent hackers from disrupting the process, and how to maintain voting equality, given that some citizens have Internet access and some do not.

Figure 1-10

IT-enabled activities for universities

IT PERMEATES SOCIETY	
IT User	**IT-Enabled Activities**
U N I V E R S I T I E S	**University Communications** ❑ University phone systems ❑ Messaging via e-mail and text messages ❑ University Web site ❑ Electronic course schedules ❑ Online academic calendar **Day-to-Day University Business** ❑ Electronic classrooms ❑ Library information systems ❑ Online registration ❑ Online applications system ❑ Payroll and human resources

Figure 1-11

IT-enabled activities for governments

IT PERMEATES SOCIETY	
IT User	**IT-Enabled Activities**
G O V E R N M E N T	**Government/Citizen Interactions** ❑ Electronic voting systems ❑ Motor vehicle registration ❑ Electronic tax filing ❑ Electronic voting systems ❑ Social Security transactions ❑ Information dissemination **Government Functions** ❑ Military information systems ❑ Electronic surveillance ❑ Intelligence networks ❑ Air traffic control systems ❑ IRS databases ❑ Internal information networks and phone systems ❑ Library of Congress catalogs

No matter what the application, every advancement in IT is driven by some social, economic, military, or political requirement, and carries both advantages and complex challenges. This textbook will present some of these socioeconomic complexities in conjunction with the more technical mechanics of information technology.

THE STATE OF IT

What are the top trends in IT? By observing life in a corporate office building, a coffee shop, or a university center, you can begin to answer this question. Cutting-edge IT products are unplugged (wireless), small, and fast. They offer access to multimedia information systems and are interoperable with other devices and networks. A top concern for any IT user or organization is how to secure information in such open and exposed computing environments, so advanced network security devices are in hot demand. This section introduces some of the IT trends that will dominate discussions in this textbook (see Figure 1-12).

Wireless

Only a few years ago, the dominant trend in network transmission media was widespread installation of fiber-optic cable, a high-speed glass (or plastic) medium over which light is transmitted for communication. Investments in "fiber to the curb" and "fiber to the home" were plentiful. Telephone companies and their competitors, known as alternative access providers, dug up city streets to lay fiber-optic cable in conduits. Some universities upgraded dormitories with high-speed fiber links to every room. However, this all occurred before the proliferation of wireless devices. University students now expect wireless Internet access in their dorm rooms.

Wireless connectivity in corporations is often easier to set up than fiber and often cheaper than conventional networks. Due to their high speed and other advantages, fiber-optic cable and other wired approaches are still important, especially as the backbone of networks, but much installed fiber remains unused. The IT industry calls this phenomenon "dark fiber." Wireless is definitely the trend, whether it is used with a laptop, a cell phone, or a handheld device. Wireless access accommodates all types of information exchange, especially voice

Figure 1-12

Top IT trends

What's Hot		What's Not	
☐ Being "Unplugged" ☐ Wireless LANs (Wi-Fi) ☐ Mobile Telephony	**WIRELESS**	☐ Being "Plugged" ☐ "Dark" Fiber ☐ Traditional Telephony	
☐ Critical Infrastructure Protection ☐ Firewalls ☐ Biometric Authentication	**SECURE**	☐ Wi-Fi Spoofing ☐ Unencrypted Transmission ☐ Worms and Viruses	
☐ Distributed File Sharing ☐ Video Downloads ☐ MP3 Files	**MEDIA FREE**	☐ CDs ☐ DVDs ☐ Floppy Drives	
☐ Open Source Code ☐ Interoperable Approaches ☐ Open Standards	**OPEN**	☐ Closed Source Code ☐ Proprietary Technology ☐ IM Incompatibility	
☐ Mobile Devices ☐ Nanotechnology ☐ Distributed Processing	**SMALL TECH**	☐ Personal Computers ☐ Macrotechnology ☐ Centralized Processing	
☐ Voice over Internet ☐ Video Phones ☐ Multimedia	**CONVERGED**	☐ Traditional Telephone Service ☐ Voice Only Cell Phones ☐ Data Only LANs	
☐ Compressed Formats ☐ Broadband ☐ Gbps+	**FAST**	☐ Uncompressed Formats ☐ Dial-up ☐ Mbps	

calls, text messaging, and all Internet applications. These technologies are cheap, easy to install, and most importantly, they provide the advantage of mobility. Unfortunately, the downside of the wireless explosion is that it presents its own set of security, performance, and network management challenges. Wireless technologies and challenges are extensively addressed throughout this textbook.

Security

A confluence of circumstances places **network security** technologies on top of the list of critical IT trends. Computer networks, the Internet, and voice communication technologies are so essential to modern society and world economies that a major disruption or outage would have dramatic consequences. Securing these networks and their underlying infrastructures from possible terrorist attacks, natural disasters, hackers, and technological vulnerabilities is a paramount concern. National governments are playing more of a role in critical infrastructure protection because national economies depend on the smooth operation of networks. Governments also require secure, private communications networks and protected electronic information to carry out day-to-day operations. Corporations consider the security of their computing resources and networks to be a routine but critical part of their daily operations.

Electronically stored and exchanged corporate information and business transactions are targets for corporate espionage or hackers. Individual users are increasingly concerned about security as wireless access makes them more vulnerable to data interception by hackers, or bandwidth theft by people who tap into wireless access devices. Personal information such as Social Security numbers, credit card numbers, healthcare information, and financial accounts are attractive targets for hackers. Viruses and worms pose a constant threat on the Internet. The possibility of identity and password theft plagues Internet users. Critical infrastructure attacks are possible not only on the Internet but on the phone system, other computing networks, and the enabler of all these systems: the power grid.

The explosion of software and products designed to improve network and computing security is a major IT trend that will not diminish any time soon. This textbook presents state-of-the-art techniques that help reduce security vulnerabilities. Various approaches, especially private key and public key encryption, seek to enhance and protect the privacy of information as it is transmitted over a network. Access control mechanisms, such as firewalls, password systems, and physical security, control who can use computing resources. A variety of authentication technologies ensure that people requesting access to resources are who they say they are. The most cutting-edge and most controversial authentication systems use biometric identification, which can include retinal eye scans, fingerprint scanners, facial recognition, or even DNA authentication. Other security approaches assess the integrity of information to ensure that it has not been modified in transmission.

Media-Free Information

An ongoing phenomenon in IT is that today's leading medium will be on tomorrow's scrap heap. Consider the physical media that have supported recorded music in the past few decades. Depending on their age, some music aficionados still have stacks of vinyl albums next to 8-track tape collections, cassette tapes, and stockpiles of CDs. They may also own an iPod Mini, a diminutive handheld device that holds 2000 songs from all the earlier media combined. Believe it or not, there are still stockpiles of paper "punch cards" that contain a variety of inaccessible textual information. Many computer users still have floppy disks containing information that current computers cannot directly read. What's the current state of media technologies?

The trend is toward **media-free information** exchange. For some time, information exchange has occurred without the need for paper: examples include instant messaging, electronic mail, bar codes for groceries, wireless handheld rental-car checkouts, direct deposit of paychecks, and so on. The use of "hard" media such as CDs, DVDs, and floppy disks is also diminishing. Rather than buying CDs, consumers purchase and download music files directly from the Internet onto an iPod or other MP3 player; others illegally download or share copyrighted files. Rather than renting or purchasing DVDs, consumers can download movies on demand from satellite TV or other systems. Instead of downloading a presentation onto a CD or floppy disk, many people use flash media devices that act as mobile, miniature hard drives or access their presentation from a server over a network. Some types of hard media are still important, but they are becoming increasingly unnecessary because of ubiquitous, instant, high-speed communications.

Interoperability and Openness

Two decades ago, different types of computers on different networks could not easily communicate. At one point, America Online subscribers could not send e-mail to subscribers of the online services Prodigy or CompuServe. With network protocols, the global Internet, and software portability, almost every system is now **interoperable**, or able to communicate with any other system anywhere in the world. Electronic mail standards allow any user on any system to communicate with any other user. Standard Web formats allow anyone with a Web browser to access any Web site. Proprietary systems that won't allow information exchange with other systems are completely out of vogue.

Several high-profile efforts are under way to advance international standards, such as the movement to upgrade the Internet to a new standard called IPv6. This upgrade is designed to dramatically increase the number of devices that can connect to the Internet. Another part of standards-based, collaborative, interoperable, and open technologies is the open source software movement. **Open source code** is software that anyone can access, view, and modify. The most popular open source software is Linux, an operating system developed by Linus Torvalds in 1991. Linux is freely available over the Internet and has posed a challenge to the hegemony of Microsoft's Windows operating system family. You will read more about IPv6 and open source code later in this book.

"Small" Technologies

Nanotechnologies, distributed computing, and small mobile devices are definitely the trend today. Microprocessors are not only getting smaller and more powerful, they are becoming cheap enough to place everywhere. Small computing technologies are critical because they help fulfill the need for mobile, **ubiquitous computing**. Cell phones and laptops are decreasing in size, and other multimedia devices are smaller than ever. People don't want to lug around a large laptop if they can accomplish their goals with a smaller, handheld device.

Rather than processing information on a large, centralized computer, some applications are better suited to distributed computing systems that spread storage space and processing power over numerous smaller systems connected by a network. Distributed file sharing is a major, user-driven phenomenon that falls into the category of small technology. Research developments in **nanotechnology** hold the potential for the next radical advancements in computing. Nanotechnology, named after the measurement for one billionth of a meter, refers to technologies so small that they are measured in nanometers. Research in nanotechnology seeks to assemble machines made of a group of molecules. Nanotechnologies hold promise to exponentially increase computer processing speeds and have great potential in medical fields. Devices that are smaller than human cells hold great promise for diagnoses and treatments from within a person's body.

Convergence

A significant theme of this textbook is that computing platforms, communications devices, and networks no longer support the transmission of a single information type. Think about all the systems in a typical house. There's probably a cable television line made of coaxial cable. A digital satellite system might sit on the roof. Twisted pair cable supports the telephone system. Several members of the household probably have their own cell phones. A wireless LAN might support Internet access for users in the household. The house probably has only one driveway, yet it might have as many as six information conduits concurrently serving the family. Each of these conduits delivered a single type of information at one point, but today converged technologies simultaneously support multiple information types and applications. As mentioned earlier, cell phones support images, text, and voice. Cable systems deliver television and Internet access for applications, including the Web and VoIP. Traditional telephone companies provide Internet access and voice services. This phenomenon of convergence has created an industry war for control of information in residences and businesses. Who will dominate this "last mile" information-access market? Will it be cable companies, cell phone providers, telephone companies, satellite service providers, or some new industry innovation such as high-speed metropolitan area wireless? Regardless, information technologies that adapt to requirements to support multimedia information exchange hold the most promise in today's telecommunications and computing environments.

Speed

Modern information exchange requires instant access, real-time communications, and high bandwidth. Common use of the term *bandwidth* simply means transmission rate, measured in bits per second. (This term will be further clarified later in the book.) The speed of information is faster than ever, with transmission rates of millions or billions of bits per second becoming the norm. By the time the morning paper arrives, much of the information in it has already been broadcast over the Internet, radio, or cable news channels. Many types of news and information place high demands on computer networks. As we will describe in mathematical terms later in this textbook, music files, digital images, and video have extremely high storage and bandwidth requirements. In other words, they consume a lot of space. Increasing the speed of transmission lines is a vital component of enabling fast downloads. This textbook will describe the fastest transmission techniques, sometimes referred to by the misnomer *broadband*. It will also address cutting-edge storage technologies, which are becoming cheaper and dramatically more efficient.

Another important technological approach has increased the speed of information exchange and reduced storage demands. Rather than increasing the size of transmission lines and storage media, why not reduce the size of information so there is less to store or transmit? **Compression** formats and technologies reduce file sizes dramatically without significantly degrading the quality of images, video, or sound. Mathematical algorithms exploit characteristics of digital information, such as redundancy and repetition, to decrease file sizes by 50% or more. Other techniques eliminate information that people can't detect anyway, such as color resolution that is invisible to the human eye or sound frequencies that are inaudible to human ears. This textbook thoroughly explains the mathematical and analytical ability of compression techniques to reduce the size of digital information.

THE STATE OF IT CAREERS

Given the variety of IT trends, what exactly falls under the umbrella of IT jobs? Because information technologies are a critical underpinning in all areas of modern society, IT experts work in every sector of the economy: entertainment, law enforcement, government, military, education, finance, manufacturing, retail, healthcare, and so on. The following sections provide brief descriptions of options for specializing in the IT field.

INFORMATION SECURITY JOBS

All institutions must address information security, and all information infrastructures increasingly face threats such as denial-of-service attacks, spyware, viruses, worms, unauthorized access, and data modification. The outgrowth of these requirements is a market for workers with expertise in IT security. Positions in this field have many names, including network security engineer, information security analyst, network security consultant, and cybersecurity analyst. Jobs in information security exist within institutions that seek to secure their information infrastructures and in companies that specialize in security products and consulting services. For example, security consulting firms often seek graduates with an IT specialization to conduct vulnerability assessments and to test applications, networks, and systems for risk. As we will explain later in this book, information security professionals implement and test intrusion detection systems (IDS), configure firewalls, implement encryption technologies, and perform security audits.

COMPUTER/IT ANALYST

In 2006, *Money* magazine rated computer/IT analyst positions among the 10 best jobs in America. The magazine's analysis included a number of jobs within the IT analyst category, including desktop support technicians and IT managers, and reported that entry-level IT analysts earn an average yearly salary of $60,000. Network operations directors, who are responsible for overseeing internal networks, can earn $250,000 and more.

One such job is called "IT business analyst"—someone who translates business requirements into technical specifications. These analysts generally work with business units to help determine and document business and technical requirements. They also serve as liaisons between business units and the technical developers who design systems to meet business needs. Technical IT analysts, especially in entry-level positions, focus more on supporting end-user desktop applications, hardware, and shared systems. The analysts also train employees who use these systems.

NETWORK ADMINISTRATION

Section V of this book discusses computer networking technologies. Every organization that has computer networks needs people to install, manage, and troubleshoot these networks. Network administration positions are hands-on jobs that deal with technology and usually require interacting directly with technology users. Network administrators configure and operate computer networks on a company-wide, division-wide, or local department basis, and often provide direct end-user support. In other words, if a user's network connection fails, the network administrator works directly with the user and fixes the problem. Network administrators have to communicate technical information with nontechnical people, so excellent communication skills are helpful. The network administrator not only resolves problems but proactively manages networks, performs hardware and software upgrades, maintains servers, ensures that information is adequately backed up, and reconfigures networks to add new users. Because most enterprises require networks to operate 24 hours a day and 7 days a week, including holidays, network administrators usually have to be available outside regular business hours to address any problems that arise. These positions require interaction with local area network equipment such as routers, switches, hubs, and wireless LAN technology. They also require interaction with operating systems such as Linux, Solaris, and Windows XP. All of these technologies will be discussed in later chapters. Certification in specific product environments is sometimes a requirement for these positions, although companies might arrange for this education after hiring the employee.

MANAGEMENT CONSULTING

Many large management consulting firms hire students directly out of college. IT consultants work as part of a team on limited-term engagements for large corporations, governments, and other institutions. Consulting is a stimulating job because it involves working on different engagements for different enterprises with unique business needs and technical environments. It also involves a great deal of travel, nationally and often internationally, because consulting teams spend much of their time at client locations. Some IT consultants specialize in helping clients in specific industries such as government, healthcare, retail, financial services, manufacturing, or entertainment. Other IT consultants gain expertise in certain areas of IT.

DATABASE ADMINISTRATION

Reams of information exist in electronic **databases**. Almost every business has a database management system that tracks thousands, millions, or billions of records. For example, a midsized company might operate a database with hundreds of thousands of records, including customer names, addresses, account information, purchase information, and so on. The organization, storage, and retrieval of data are critical parts of a successful IT infrastructure. Database administrators design, create, and maintain these computer systems. An important component of this job is the ability to work with business units to create reports that are essential to a corporation's strategic and financial success. Database administrators are also called database analysts, database engineers, applications database administrators, and database architects. These positions usually require familiarity with numerous computer platforms discussed later in this book, such as mainframes, UNIX servers, and Windows systems. The positions also require programming skills such as SQL, and familiarity with specific database products such as those created by Oracle, Microsoft, or Sybase.

COMPUTER FORENSICS

The productive and entertaining aspects of computers and networks come with a dark side of criminal activity, cyberterrorism, child pornography, and fraud. Traditional law enforcement services have to keep up with the times and develop expertise in detecting and prosecuting crimes committed with information technologies. **Computer forensics** experts extract computer evidence for detecting, preventing, and prosecuting crime. People who have an interest in law enforcement and homeland security and who have IT backgrounds can enter the field of computer forensics, although this career path often involves supplemental course work in law enforcement, law, or a graduate degree in computer forensics itself. Computer forensics experts can determine whether information has been deleted or modified, recover information that has been deleted from a computer, trace and locate the source of transmitted or stored material, search enormous stores of data for specific evidence, and serve as expert witnesses in judicial proceedings. For example, if a person is reported missing but is known to have sent an e-mail message at a particular time, a computer forensics specialist can trace the source of the e-mail and provide clues to the missing person's whereabouts.

IT SALES

Selling IT products and services can be a lucrative and exciting path for IT majors who have excellent interpersonal skills and decent powers of persuasion. Every company that makes IT software, applications, systems, hardware, or network equipment has a fleet of IT-savvy salespeople on the front lines to sell its products. IT salespeople often travel extensively. They must understand the business requirements and technical characteristics of the products or services they sell.

SOFTWARE DEVELOPMENT

Many IT jobs involve software development. Programming opportunities exist in every industry and touch on every type of information exchange: music, video, images, voice, and so on. The most well-known programming jobs for recent graduates are in Web development. However, development opportunities exist in numerous functional areas. Software developers work for the government, IT users and IT providers, the military, universities, and many other types of institutions.

On the more entertaining side, graduates who have a creative streak and skills in a programming language such as C++ or Java can find a career in game programming. Whether they are computer-based, online, on a cell phone, or on an arcade video console, games ultimately are built on computer code, and someone has to develop it. Breaking into software development for games is competitive, and having a portfolio of ideas or game demos can help programmers gain entry into the field. Some companies offer internships to students who are interested in game programming. For people who want to pursue this career option, the game programming site *http://www.gpwiki.org* provides more information and access to selected game source code. The International Game Developers Association (IGDA), a non-profit professional society, also provides considerable information about gaming careers. For more information, see *www.igda.org*.

IT ADVOCACY AND MARKETING

Many IT jobs require a solid technical understanding of IT but are not necessarily "technical jobs." One such job falls into the category of "IT activist": people who work for nonprofit groups, think tanks, or advocacy groups, and who advocate for certain technologies or uses of technology. Others work as lobbyists or for corporations as liaisons with governments. Others join large corporations as technical marketing engineers for IT vendors, interacting with customers and helping to deliver products and services. These positions usually involve market research in the IT industry, staying abreast of competitor offerings, and awareness of strategic industry directions.

WHAT AFFECTS JOB PROSPECTS?

The U.S. Department of Labor's Bureau of Labor Statistics has forecasted that hiring of computer support specialists will rise "faster than the average" relative to all occupations through the year 2014. It also forecasts that employment of systems and network administrators will increase "much faster than the average" relative to other occupations. Among the many variables that affect IT employment prospects are regional differences, overall economic conditions, geopolitical circumstances, and industry trends. These variables make it difficult to assess job prospects at any given moment. Often, one credible source will claim that IT hiring is at an all-time high, while another will predict that IT worker demand is declining and that outsourcing, or "offshoring," will continue to send thousands of IT jobs to other countries where workers receive lower wages for the same work. Those who predict increases in IT jobs cite the continuing need for people with any type of expertise in information security. They also see high demand for people with a combination of technical expertise and business acumen, because many IT jobs are integrated with business units.

It is not too early to begin assessing which skills and experiences would bolster credentials for securing employment in an IT career. One way to prepare for applying for IT jobs and internships is to start building a resume. Students who work on campuses in IT support roles, such as systems administrators, database administrators, Web designers, and IT help desk personnel, gain valuable experience and on-the-job training. Another way to develop technology leadership skills and hands-on experience is to volunteer in local communities. Options for volunteers include providing computer training to community members in schools, church programs, libraries, town centers, and nursing homes. Volunteers can also provide IT support or Web design for nonprofit organizations. Some students also benefit from joining professional societies such as the Institute of Electrical and Electronics Engineers (IEEE).

With so many variables affecting IT employment trends at any given moment, the safest bet is to determine which IT career sounds the most interesting and then go for it. This book will expose students to aspects of IT and help them discover which area of specialization they find most interesting.

CHAPTER SUMMARY

- This book introduces students to the most important concepts in modern IT, explains the computational methods that underlie all of IT, and explores the relationship between technological developments and societal phenomena.
- Information is represented in four ways: through text, numbers, images (including video), and sound. Rapid technology advances in each of these areas has made previously cutting-edge technologies obsolete. For example, fax machines and CDs were once considered groundbreaking, but now they are not really necessary. Some of today's emerging information technologies might become unnecessary in 5 or 10 years.
- IT hardware and software can capture, process, exchange, store, and present various types of information using electrical, magnetic, or electromagnetic energy. Cellular telephony systems and wireless Internet access systems are two examples of IT.
- Many current information technologies provide multimedia capture and exchange of multiple forms of information. For example, a cell phone can provide voice communications, image capture and exchange, streaming video, text messaging, and e-mail. Modern IT systems permeate all areas of society, and the most cutting-edge IT trends include technologies that are secure, high-speed, interoperable, mobile, small, and user controlled.

KEY TERMS

binary code

bits

compression

computer forensics

convergence

databases

information

information technology

interoperable

media-free information

mobility

multimedia

nanotechnology

network security

obsolescence

open source code

spam

ubiquitous computing

REVIEW QUESTIONS

1. Name several popular information technologies that did not exist 10 years ago.
2. In your own words, define *information*.
3. What examples of critical information have you exchanged electronically in the past week?
4. What completely unimportant information have you recently exchanged electronically?
5. Besides text and numbers, what are the other two approaches for representing information?
6. Name a sound-related technology and an image-related technology that are now obsolete.
7. In your own words, define *multimedia*.
8. Explain why a cellular telephone system is an example of convergence.
9. Give one example of an information technology that stores information using magnetic energy.
10. Name several forms of entertainment that are not electronically dependent.
11. In what ways do national economies depend on information technologies?
12. What is the downside of the proliferation of wireless network access?
13. Why are network security technologies so critical?
14. What are some examples of biometric security technologies?
15. List some reasons for the decline in CD sales.
16. What is the purpose of compression technologies?

DISCUSSION QUESTIONS

1. Discuss which information technologies you use every day will most likely be obsolete in 10 years. Do these technologies have any characteristics in common that suggest impending obsolescence? What current information technology do you think will look the same in 10 years? Why?

2. In your opinion, should it be legal or illegal to download and share free music files? What are the legal, political, social, and technical enablers and consequences of file sharing?

3. Discuss the advantages and disadvantages of electronic voting. Make a political and technical argument for and against electronic voting. Will Internet voting ever be possible from the standpoints of security and access? Describe the possible political consequences.

4. What information technology, if it suddenly became unavailable, would significantly change your daily routine? How? How would the absence of information technologies change how universities and businesses currently operate?

5. What do you believe are the most important trends in information technology?

CASE PROJECT

Search for entry-level IT jobs in a major newspaper or an online employment site such as Monster.com or Careerbuilder.com. Select the 10 most interesting IT job advertisements. While reading these job postings, note the required qualifications and experience. Also note any unfamiliar terms or acronyms and research them on the Internet.

Now build a sample IT resume that you might have upon graduation. Include the following components:

Name and contact information—Include your formal name, home address, phone number, e-mail address, and Web site (if applicable).

Employment objective—Tailor your employment objective to the type of job you would find interesting. Describe the type of job you seek and whether you prefer summer or full-time employment.

Educational background—Include your school, major, and expected graduation date. Consider listing key IT courses that you expect to take.

Professional experience—In reverse chronological order, describe any professional experience, campus jobs, and summer employment you have had or might have in the next few years.

Professional skills—Drawing from the required skills in the job advertisements you selected, invent a list of professional skills that you would like to develop over the next few years. Include professional certifications, programming skills, foreign languages, and familiarity with computing platforms and software applications.

Leadership skills—Include a final resume category such as "activities" or "leadership skills." This section can include such activities as community service, volunteer work, athletics, music, fraternity or sorority leadership, and student organizations.

2

UNDERSTANDING THE DIGITAL DOMAIN

In this chapter you will:

- Understand the difference between analog and digital representations of information

- Learn about techniques for transmitting and storing digital information

- Understand the use of multipliers for representing, transmitting, and storing large amounts of digital information

- Discuss the advantages of representing information in digital format and using digital devices for processing, exchanging, and storing information

INTRODUCTION

Society is immersed in digital technology. We often hear the term *digital information*, we talk about living in the *digital age*, and we discuss using *digital devices* such as computers, cell phones, and personal digital assistants (PDAs). However, we often do not stop to consider exactly what is meant by the term *digital* and what technologies have enabled this digital age. We also frequently encounter the term *analog*, but might not understand its exact meaning or how it differs from digital. This chapter describes some historical developments in digital technology, explains how to represent information in analog and digital formats, and discusses the numerous advantages of operating in a digital format rather than analog.

The term **digital information** refers to representations of numbers, text, sound, and images as a combination of two fundamental logical symbols: one and zero. These symbols are also called **binary symbols**, **binary digits**, or **bits**. A piece of text such as "Hello How Are You?" may be represented by a sequence of 144 bits:

010010000011001010110110001101100011011111001011000010000001001000011011110111
011100100000001000001011100100110010100100000010110010110111101110101

Similarly, a small section of music stored on an audio CD may be represented by the following sequence:

11100000000000000000000000000000000000000000000000000000000000000000
00000001111111111111111111111111111111111111111111111111111100000010101010
100000011001110010100011010011001110000111100000010100101001001100001101101
11110111010

Again, this is a combination of bits. Any form of information may be represented in this manner.

Digital devices process information in the form of ones and zeros; in other words, they speak a **binary language**. In order to process digital information, these devices contain basic switches that switch on and off to represent bits. Digital devices are built by combining millions or even billions of these switches. By connecting the switches to each other in a particular way and appropriately turning them on and off, the devices can manipulate digital information. An electronic digital device such as a computer contains millions of these switches, and processes digital information such as audio files, text files, and video files by electronically turning the switches on or off. The processing power of a digital device is essentially governed by the sheer number of these switches and by how fast they can be turned on and off.

Several fundamental advantages of representing information using only ones and zeros are discussed in detail later in this chapter. Once people recognized the possibilities of digital technology, they began to develop digital devices that could "speak" the binary language. Unless an efficient switch had been invented, powerful digital devices would not have been possible, and the true benefits of representing and processing information in digital form could not have been realized. The first of these switches was the **vacuum tube**, a device resembling a light bulb that could be turned on or off using electronic controls. The vacuum tube was used in the world's first electronic digital computers in the 1940s. Vacuum tubes were very large, consumed significant amounts of power, and had slow switching speeds, so the devices that used them were also large, expensive, and slow.

Developers understood the need for a smaller and more efficient switch. The invention of another type of switch, the **transistor**, put digital devices at the forefront of technology. The transistor is the elementary component in all digital devices today. Invented in 1947, a transistor is an electronic switch that can be controlled electrically to turn on or off, just like a mechanical switch that is controlled by hand. The transistor had the same functionality as the vacuum tube, but it was significantly smaller and more reliable. It could be switched at a faster rate and consumed a very small amount of power. Figure 2-1 illustrates a vacuum tube and a transistor.

Today, the typical size of a vacuum tube is a few centimeters (one tenth of a meter), and transistors are just fractions of a micrometer (one millionth of a meter).

With the evolution of the transistor, vacuum tubes disappeared from computers and other digital devices, dramatically reducing their size and making them more powerful. Vacuum tubes are no longer used to construct computers, but have different applications, such as in audio amplifiers.

Figure 2-1

Examples of a vacuum
tube and transistor

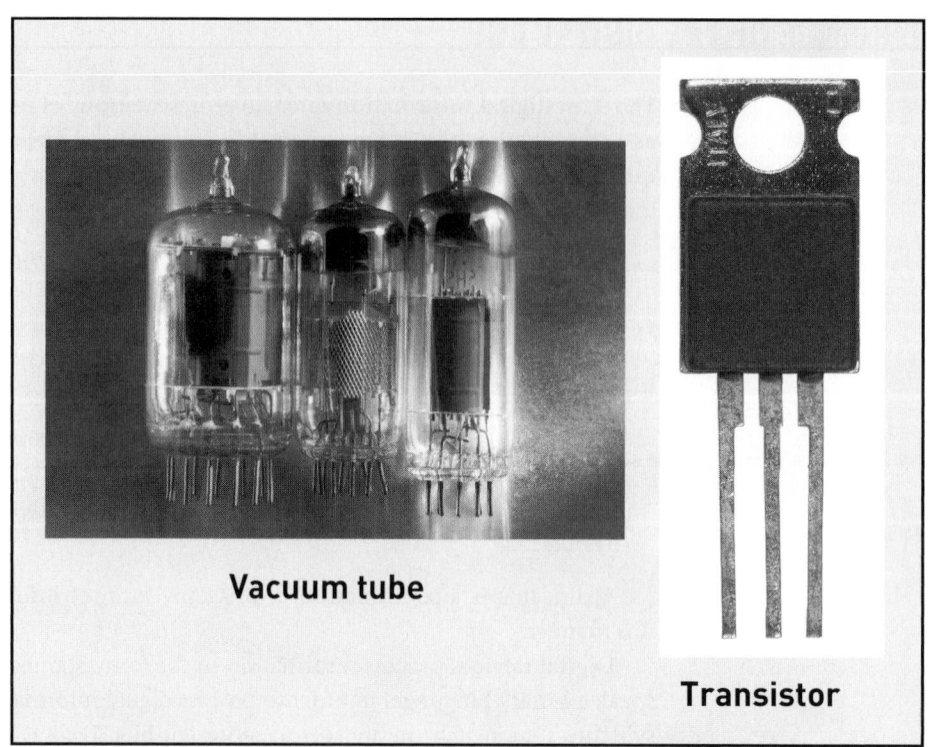

Vacuum tube

Transistor

The next major digital advancement was the invention of the **integrated circuit (IC)**, also commonly called a chip. Developed in the 1950s, the chip was composed of a thin layer of semiconductor material; several transistors on this layer were interconnected to perform digital processing. The invention of the IC enabled more efficient manufacturing and packaging of transistors, further reducing the size of digital devices. Today, the size and capability of digital devices is governed by the number of transistors that can be incorporated on the thin semiconductor layer of an IC made of silicon. As companies develop smaller transistors and fit more of them on ICs, digital devices that include ICs become smaller and more powerful. The exterior of an IC, which roughly has the dimensions of a quarter, is shown in Figure 2-2. The silicon layer that carries the several transistors, other components, and their interconnections is inside the black rectangular casing.

Figure 2-2

An integrated circuit

Following the incorporation of ICs into digital devices, Gordon Moore, the cofounder of Intel, observed a trend in smaller transistors on ICs. In 1965, Moore made a prediction about the number of transistors that could be integrated into a chip within the next 10 years. This prediction proved to be correct, and his observation later became known as **Moore's Law**. In its current version, the law states: *The number of transistors that can be integrated on an integrated circuit doubles every 18 months.* A close version of this statement remains true today and is considered to hold true for the next several years. However, current technology trends indicate that Moore's Law will reach its limit in the future, so research on alternative technologies is under way. Chapter 4 discusses these technologies.

From vacuum tubes to integrated circuits, digital technology has evolved tremendously over the past few decades. This technology continues to grow as the demand for faster, smaller, cheaper, and more versatile devices continues to rise.

THE DIFFERENCE BETWEEN ANALOG AND DIGITAL REPRESENTATIONS OF INFORMATION

ANALOG INFORMATION

Digital devices process many forms of information in combinations of bits, but most information we encounter in daily life is expressed in analog format, not in terms of ones and zeros. When we speak, for example, we exchange information in analog format. What we hear varies proportionally, or *analogously*, to the sound produced by the person who is speaking. Hence, we call this **analog information**. When we look at a cat, we are receiving analog information. The image captured by our eyes is *analogous* to the actual cat. Basically, analog information is captured and presented in its original form, and does not exist in combinations of ones and zeros. However, there is more to the explanation of analog information. The following example will help clarify this point.

Consider the classic example of a speedometer that measures the speed of a vehicle on a stretch of highway for 10 seconds. Assume that an observer is sitting in the car and visually monitoring the speedometer needle. The information that the person receives by observing the needle is analog information, because the needle's continuous movement is analogous to the car's continuous speed variations. Now assume that the speed of the vehicle is recorded and plotted on a graph, as depicted in Figure 2-3.

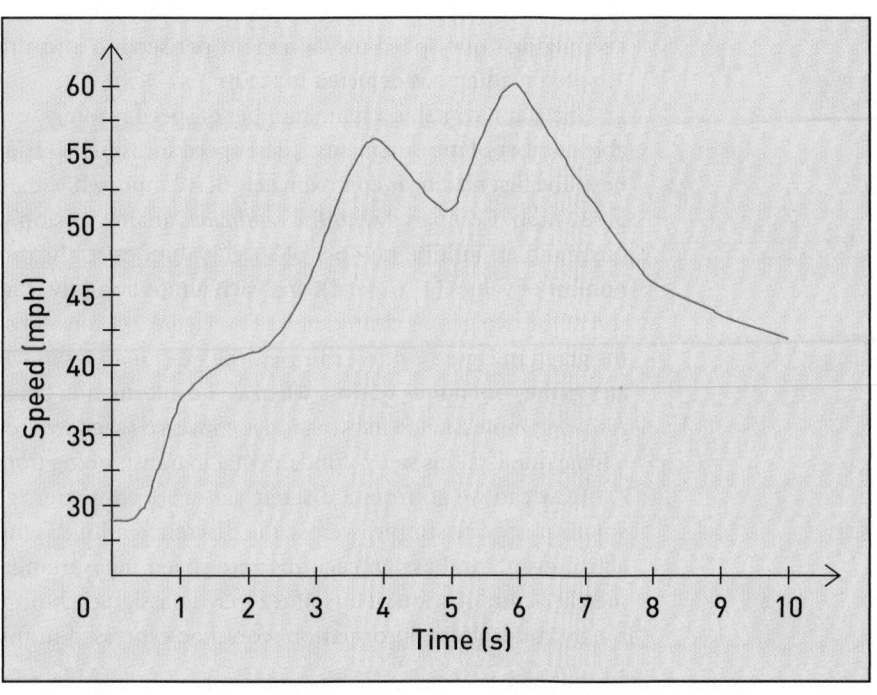

The horizontal axis denotes time in seconds (s), and the vertical axis represents speed in miles per hour (mph). The speedometer begins measuring time at 0 seconds and continues measuring for the next 10 seconds. The speed varies continuously within a range of approximately 30 to 60 mph for the 10-second duration.

As the speed varies *continuously* within this range for 10 seconds, there are an *infinite* number of actual speed measurements along the way. The speed is said to vary continuously because it does not jump abruptly from one value to the next. It changes gradually. Although we cannot read exact values directly from the graph, the actual speed at 1 second could be 37.82 mph. Similarly, at 4.554422 seconds, the speed could be 59.87 mph, indicating the infinite number of speed measurements that exist within the 10-second duration. The information conveyed by the speedometer is analogous to the actual event taking place. The speed variation is continuous, and an infinite number of speed measurements exist within the 10-second time frame. Hence, the information conveyed is analog information. Because the speedometer produces analog output, it is classified as an **analog device**.

Now consider the case of human speech. When a person speaks, the variation of sound is continuous and the number of sound intensities that make up the sound wave is infinite for the duration of the speech. What we hear is analogous to what the speaker is saying. Thus, information is being conveyed to us in analog format. Similarly, while looking at a cat, the image we see is analogous to the actual cat. The infinite number of colored dots whose intensities vary continuously come together and make up the image of the cat.

DIGITAL INFORMATION

Although analog information is all around us, we often choose to convert information from its natural form to a combination of ones and zeros. In other terms, we digitize information using **analog-to-digital converters** and use digital devices for communication, storage, and processing. To accomplish this task, we first capture the information in some form and reduce the infinite number of values that make up the analog information to a finite number. For example, an infinite number of readings captured by a speedometer may be reduced to a finite number of measurements. Similarly, an infinite number of sound levels may be captured by a microphone, and the infinite number of dots that make up an image may be captured by a camera lens and reduced to a finite number during digitization.

To reduce the infinite number of measurements in our speedometer example, we can take *samples* of the speed measurements once every second. This approach would correspond to keeping just one speed measurement per second and throwing away the rest, resulting in 11 speed readings, as depicted in Figure 2-4.

The graph contains a finite number of speed readings, or in other terms, a **discrete** or countable number of measurements. These speed measurements jump abruptly from one value to the next. The first reading is approximately 31.42 mph, while the reading at the second point leaps to 37.82 mph. Compared with the *continuous* analog version, in which the speed varies continuously and an infinite number of speed values exist, the *discrete* version now contains a finite number of values (11 in total), and each jumps abruptly from one value to the next. A comparison of the two graphs clearly shows that Figure 2-3 conveys considerably more information than the graph in Figure 2-4. We can view the exact manner in which the car accelerates and decelerates in the continuous version, whereas the information between each second is lost in the latter. Although more limited information is conveyed when we reduce the infinite number of values to a finite number, this step is fundamental in transitioning from analog to digital.

How can we go from a discrete set of speedometer readings to a sequence of bits? The solution is quite simple: we use the discrete readings, which represent only a portion of an actual event, and assign each discrete speed measurement a special binary code, which is merely a combination of bits. If we were to assign each numeric speed reading an appropriate binary code, all the information would be expressed in the form of binary symbols, and the

Figure 2-4

Graph of the finite
number of speed
measurements

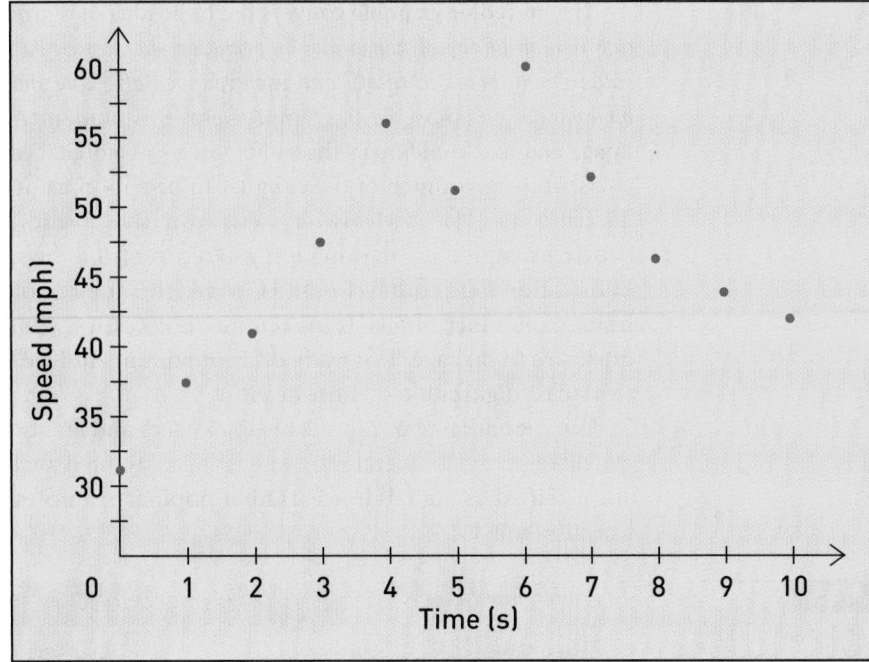

information for the speed variations of the car over 10 seconds would exist as a series of ones and zeros. Figure 2-5 depicts a list of the binary codes assigned to each speedometer reading. The first reading is assigned a binary code of 0000, the second sample is assigned a binary code of 0100, and so on. Although Chapter 3 explains how each reading can be assigned the proper binary code, it is useful to mention here that the actual speed readings are first rounded off to one of a close number of readings, and the rounded-off speed values are actually assigned a binary code. Analog speedometers, which are found in most cars, therefore convey analog information with their smoothly rotating needle. Digital speedometers have analog-to-digital converters that take regular samples of the speed, round them off to a close value, and assign a binary code to each value. These binary codes are then displayed on the dashboard as regular numbers, such as 56.34 mph.

Figure 2-5

Graph of the binary codes
assigned to the finite
number of speed
measurements

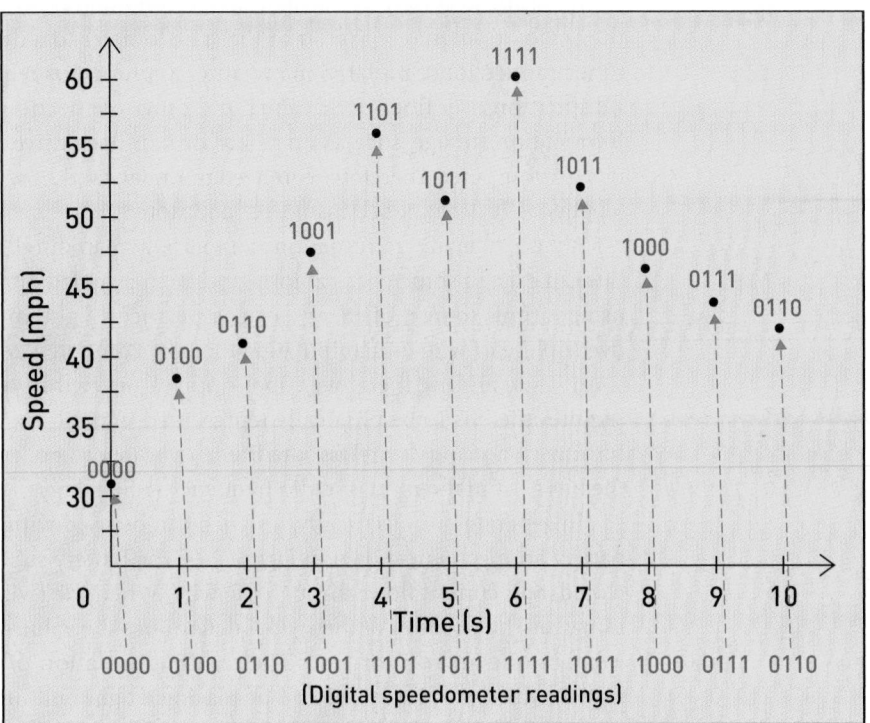

The preceding example conveys the basic idea of digitization: *To convert any form of analog information to digital, the analog information should first be reduced to a finite set of values, each value should be rounded off, and each rounded-off value should then be assigned an appropriate binary code.* An audio digitizer implements this fundamental process in a cell phone, and an image digitizer implements the same process in a digital camera or scanner.

Capturing as much of the original information as possible requires a large number of discrete values. The more discrete values you have to work with, the less information you lose between samples; the information you recover from the series of bits resembles the original information much more closely. However, the number of discrete values directly affects the resulting number of bits. If the number is excessive, your task becomes more difficult and expensive to manage. This trade-off is important, and we will examine it in later chapters as we discuss digitization in more detail.

Traditional cameras capture analog images and are therefore classified as analog devices. Digital cameras take pictures by converting a captured analog image to digital format, so they are classified as digital devices. Other popular examples of analog and digital devices are listed in Table 2-1.

Table 2-1 Examples of analog and digital devices	**Analog devices**	**Digital devices**
	Record player	CD player
	Videocassette	DVD
	Traditional mercury thermometer	Digital thermometer

MANIPULATING BITS

The speedometer example conveyed how logically information could be represented by a string of bits. The logical representation of the speedometer measurement, for example, is a sequence of 44 bits:

00000100011010011101101111111011100001110110

It may be necessary to transmit this information from one point to another, such as from the analog-to-digital converter of the speedometer to a display device over a copper wire or other connection. Similarly, many other applications require transmitting various types of information over fiber-optic cables and other connections. These applications also require information storage, such as on a hard disk or flash drive, and manipulation of information using a computer. Therefore, you need to understand how these logical sequences of bits can be physically generated, transmitted, and stored.

Although analog information can logically map into bits, this ability is useless unless the bits can exist in some physical form and be manipulated by digital devices. Bits are generated using various sources of energy such as electricity, light, or radio frequency. Transmitting bits over a piece of wire or manipulating them by computers requires the bits to exist in electrical form. Transmitting them over fiber-optic cables within telephone and computer networks requires the bits to be physically represented in the form of light. Similarly, sending digital information using a wireless satellite telephone or wireless computer network requires that the bits exist in the form of radio frequency (RF) energy.

To generate bits electrically, a switch can be used with a source of electricity, such as the battery arrangement shown in Figure 2-6. The battery supplies electricity when the switch is closed, and current flows through the wire. When the switch is turned off, the current is terminated. The position of the switch determines the value of the bit; it can be turned on for a value of one and off for zero. The resulting variation of electricity is called the **electrical signal**, which may be monitored by placing a light bulb in the loop. You will see light when the switch is closed, but not when it is open. This simple scenario illustrates the basic idea of

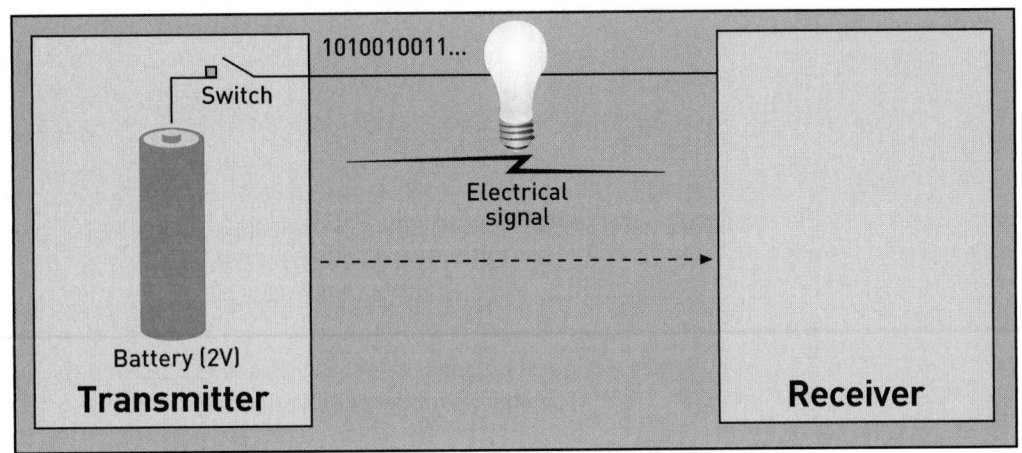

Figure 2-6

A scheme to physically generate and transmit bits

how a logical stream of ones and zeros can exist in some physical form, how it can be transmitted from one point to another, and how it can be used by digital devices.

The preceding example discussed how an electrical signal flowed through copper wires. Universally, a **signal** refers to the variation of a form of energy such as electricity or light. In many instances, signals require physical media called **transmission media** to carry them from one point to another. Electrical signals, for example, are relayed over copper wires. Metallic wires, optical fibers, and air all fall under the category of transmission media. The signal in the preceding example is the pulses of electricity produced as the switch is turned on and off. The transmission medium is copper wire. In the case of fiber-optic communications, the signal is an **optical signal**, and the transmission medium is fiber-optic cable. In the case of cellular communications, the signal is a **radio frequency signal**. Note that radio waves do not require any transmission medium for propagation. They can even travel through a vacuum, so air is not considered a transmission medium for radio waves. However, it is considered a transmission medium for other types of signals, such as sound waves.

Figure 2-7 depicts the electrical signal following the copper wires in the circuit. If the switch is turned on for 1 second, the action is represented by a one; when the switch is turned off for 1 second, a zero is shown. The y-axis denotes **voltage** (a measure of electricity) and the x-axis denotes time in seconds. The switch is first turned on for 1 second, and electricity flows through, creating an electrical pulse. The switch is then turned off, which corresponds to a zero, and then turned on again. The process continues in this manner until the last bit of the logical binary sequence 1010010011 is generated electrically. The electrical signal is classified as a **digital signal** because it varies only between two distinct voltage values: 0 V and 2 V.

The duration of the pulses is also called the **bit period**, and is measured in terms of seconds. In Figure 2-7, the bit period is 1 second (1 s). A pulse of electricity first flows for a period of 1 second to represent a one and is stopped for 1 second to represent a zero. The pulse then flows again for 1 second to represent a one and is stopped for 2 seconds to represent two consecutive zeros. The transmission continues in this manner.

Figure 2-7

A digital electrical signal

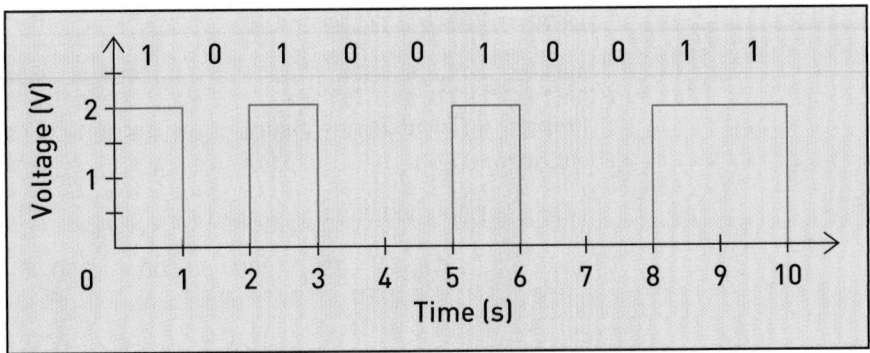

This scheme also generates bits in the form of light because the bulb turns on and off according to the setting of the switch. Light pulses generated by an appropriate light source can be incorporated into a fiber-optic cable that can carry light from the source to a destination, or the pulses can be stored or processed by various optical systems.

Radio frequency energy can carry bits in wireless applications. To carry information, digital wireless devices such as cellular telephones and modems rely on radio frequency energy that travels through air or in a vacuum, such as outer space. An electrical circuit, used with an antenna such as that on a cellular telephone, can radiate radio frequency waves to carry digital information to a receiving station.

Digital information can be carried across communication systems via many more sophisticated methods than pulses of electricity or light. This discussion is covered in a later chapter.

DATA RATE

The rate at which bits are sent over a transmission medium is called the **data rate**, and is expressed in bits per second (bps). Most people have encountered this terminology in the context of Internet connections. The rate at which a cable modem sends digital information over the cable might be 300,000 bps. In the example depicted in Figure 2-7, the data rate is 1 bps because 1 bit is transmitted in 1 second. If 2 bits were transmitted in 1 second, the data rate would be 2 bps. There is a clear inverse relationship between data rate and bit period:

Data rate = 1/bit period

The smaller the bit period, the higher is the data rate. Obviously, transmitting more bits in a smaller amount of time requires pulses that last for a shorter period of time. In the speedometer example, if we transmitted 1 bit in 1 second over a copper wire to another unit in the car, the data rate would be 1 bps and the bit period would be 1 second. Similarly, if we transmitted all 44 bits in 1 second, the data rate would be 44 bps and the bit period would be 1/44 = 0.022 seconds.

Multipliers for Expressing Data Rates

In the IT world, large numbers are usually expressed using shorthand nomenclature. For example, speeds are characterized as 56 Kbps (kilobits per second) instead of 56,000 bps. The *K*, called a multiplier, stands for *kilo*, and it represents a large number—1000 in this case. Common multipliers include the following:

- *Kilo*bits per second (Kbps) 10^3 = 1000 bps (thousand)
- *Mega*bits per second (Mbps) 10^6 = 1,000,000 bps (million)
- *Giga*bits per second (Gbps) 10^9 = 1,000,000,000 bps (billion)
- *Tera*bits per second (Tbps) 10^{12} = 1,000,000,000,000 bps (trillion)

The following examples illustrate the use of multipliers.

Problem—Calculate how many *bits* per second are transmitted over a 384-Kbps cable modem connection.

1 Kbps = 1000 bits per second

Answer: 384 Kbps = 384 × 1000=384,000 bits per second

Problem—Calculate how many *bits* per second are transmitted over a 1.25-Gbps fiber-optic connection.

1 Gbps = 1,000,000,000 bits per second

Answer: 1.25 Gbps = 1.25 × 1,000,000,000 = 1,250,000,000 bits per second

STORING BITS

In addition to generating and transmitting digital information, storing bits on a physical medium is an essential requirement. In the speedometer example, you might want to store the 44 bits in some form. Storing bits on compact discs (CDs) or any other form of **storage media** is essential so that you can retrieve the digital information later. We briefly introduce storage technologies in this section; for a more detailed discussion of the many types of storage media and their operating principles, see Chapter 4. Most forms of storage media traditionally fall into one of the following categories:

- **Mechanical storage**
- **Magnetic storage**
- **Optical storage**
- **Magneto-optical storage**
- **Electronic storage**

The mechanical approach to data storage is no longer popular, but it was the first method for storing digital data in the early days of computing. Punch cards and paper tapes fall under this category. The early storage of digital information was made possible by assigning a small space to each bit on either a card or roll of paper. Each section of the paper was either punched with a small hole or left intact, representing the physical storage of either a logical one or a zero at that location. A reader later retrieved information by detecting the presence or absence of a hole. The capacity of mechanical storage media was very limited, and depended on how small a hole you could make in the material. The more holes you could make, the more bits you could store.

The next advancement in storing bits was magnetic tapes or disks. Each bit is allocated a specific position on the surface of the medium, which is magnetized in a precise way, depending on the logical value of the bit. Magnetic storage media have a relatively large storage capacity; they are commonly used in computer hard disks, floppy disks, and digital audio and videotapes. The area that can be magnetized to hold a single bit of information is much smaller than the area allocated on paper for a hole, which explains why magnetic storage media have a larger storage capacity than mechanical storage.

Optical storage is also an effective and widespread means of storing information, especially in the entertainment industry. CDs and digital versatile disks (DVDs) fall under the category of optical storage media. To store data on the surface of a CD, tiny areas with different light reflectivities are created based on the logical bit pattern to be stored. The laser light that shines on the surface of the disc when it is being read is reflected differently from each area, which is then detected by a sensor and interpreted into bits. The use of lasers and other optics to focus the light on a disc's surface explains why this type of storage is called optical.

Magneto-optical storage media, such as magneto-optical drives, save bits on a magnetic disk using a laser. The disk is also read by a laser later, which explains its name. These media are not as popular as other types of storage media because they are more expensive.

Flash memory, a form of electronic storage, has gained great popularity over the past few years. Devices such as digital cameras store digital images and other data on **flash cards**, which are a type of flash memory. Digital files are easy to save and store using **flash drives** that connect to a computer and contain flash memory. Flash memory uses an array of small electronic elements to store electricity. The level of electricity in each element is detected and interpreted as bits when the contents of the memory are retrieved.

Mathematics of Storage

Digital information for all the storage media described in the preceding sections is stored in groups of bits; the most fundamental grouping is a **byte**. A byte of data is a grouping of eight bits. File sizes and hard-disk capacities are expressed in terms of bytes. A binary stream comprising 32 bits would be 4 bytes in size, for example. In the speedometer example, the digitized speedometer readings would require a storage space of 44/8 = 5.5 bytes. Because information storage usually involves massive amounts of bits, shortcuts are needed to express large numbers of bytes. For example, hard-disk capacity can be expressed in gigabytes, and CD capacity is typically expressed in megabytes.

As with transmission rates, multipliers are used as shortcuts for expressing large numbers. However, there is an important distinction between the kilo value used for transmission and that used for storage. Similarly, there is a difference in the mega value used in transmission and that used in storage. The multipliers used for storage capacity are derived in powers of 2, and those for transmission are derived in powers of 10. In other words, 1 kilobyte is not exactly 1000 bytes; it is 1024 bytes, which is derived from 2^{10} = 1024.

The following multipliers are used for expressing storage capacities and their values.

- *Kilo*byte (KB) 2^{10} = 1,024 bytes
- *Mega*byte (MB) 2^{20} = 1,048,576 bytes
- *Giga*byte (GB) 2^{30} = 1,073,741,824 bytes
- *Tera*byte (TB) 2^{40} = 1,099,511,627,776 bytes

The following examples demonstrate the use of multipliers for storage:

Problem—Calculate the storage capacity (the number of bits) that can be stored on a 700-MB CD.

1 MB = 2^{20} bytes = 1,048,576 bytes
700 MB = 700 × 1,048,576 bytes = 734,003,200 bytes
1 byte = 8 bits

Answer: 734,003,200 bytes = 8 × 734,003,200 = 5,872,025,600 bits

Problem—Calculate the number of bits in a 68-KB digital file stored on your hard disk.

1 KB = 2^{10} bytes = 1024 bytes
68 KB = 68 × 1024 bytes = 69,632 bytes
1 byte = 8 bits

Answer: 69,632 bytes = 8 × 69,632 = 557,056 bits

Problem—Calculate the number of bytes in a 2-MB digital image file you received through e-mail.

1 MB = 2^{20} bytes = 1,048,576 bytes

Answer: 2 MB = 2 × 1,048,576 bytes = 2,097,152 bytes

ADVANTAGES OF DIGITAL TECHNOLOGY

The preceding sections discussed the difference between analog and digital as well as the physical means by which bits can be generated, transmitted, and stored. This section extends the discussion to provide you a deeper understanding of the benefits of digital compared with analog. The most striking aspects of the digital format are summarized in the following list and then described in more detail:

- Ability for noise removal
- Capacity for error control
- High speed
- High level of security
- Amenability to compression
- Reliable storage of information
- Ease of reproduction
- Simplicity in transmission

ABILITY FOR NOISE REMOVAL

In IT, **noise** is defined as an effect that disrupts information—an unwanted occurrence that interferes with a signal and hence the information carried by the signal. In cellular communications, noise can present itself in the form of random fluctuations within the radio wave signal that carries information, resulting in poor audio quality. In storage systems, it can appear in the form of scratches on a DVD or CD. Any effect that interferes with the signal that carries the information can be classified as noise, whether the information is being transmitted or stored.

Noise is highly prevalent in all analog and digital systems used for information communication, storage, and processing, and is very difficult to remove, especially from analog systems. In digital systems, however, noise can be effectively filtered out if its level is not too great. The following discussion demonstrates why this is true.

Reconsider the speedometer example, and assume that an electrical signal proportional to the speed variation is output from the speedometer, as depicted in Figure 2-8.

The electrical signal generated by the speedometer is classified as an **analog signal**, as it is analogous to the continuous speed variation. The analog signal is shown in Figure 2-8, where the y-axis is voltage and the x-axis is time. A rise in the speed produces a corresponding surge in electricity levels, and the converse is true when the vehicle decelerates.

Now assume that the electrical speedometer output signal is transmitted over a copper wire to a secondary unit in the car, and that it experiences interference from a neighboring electrical unit, resulting in a **noisy signal** that might look like the one depicted in Figure 2-9.

Noise in this signal exhibits itself in the form of unwanted fluctuations in the electrical signal levels, which do not correspond to the actual speed variations and have to be removed. These unwanted glitches can be removed to some extent by an electronic filter, but it is very difficult to extract these spikes from the original signal in their entirety, and so the noise usually remains as a permanent part of the signal.

Now assume that the speedometer output is digitized as before and that it exists in the form of 44 bits:

00000100011010011011011111111011100001110110

In order to transmit this digital information to the secondary unit of the car over a copper wire, a switch and a circuit may be used to generate pulses of electricity (see Figure 2-6). The digital electrical signal to be transmitted to the secondary unit might resemble the signal in Figure 2-10.

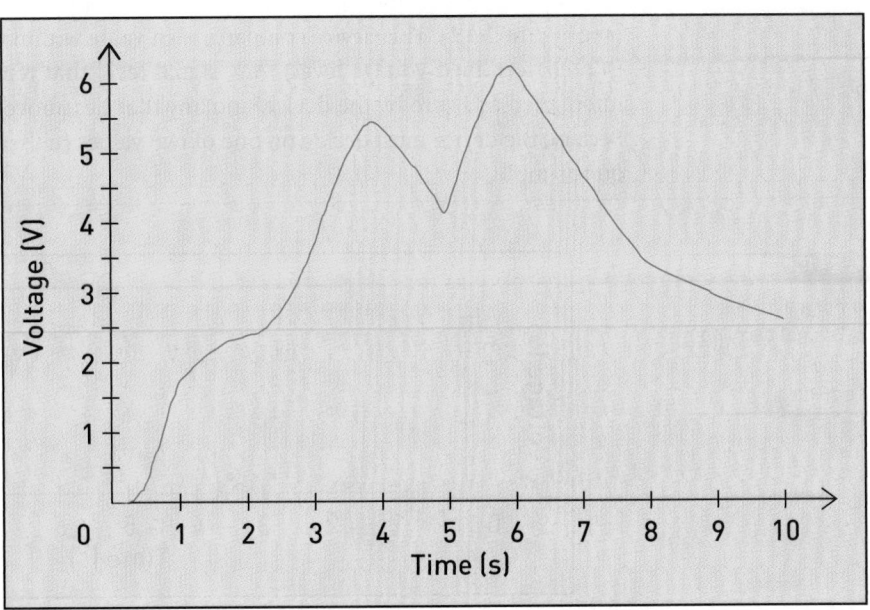

Figure 2-9

The noisy analog
electrical signal
proportional to the speed
variation

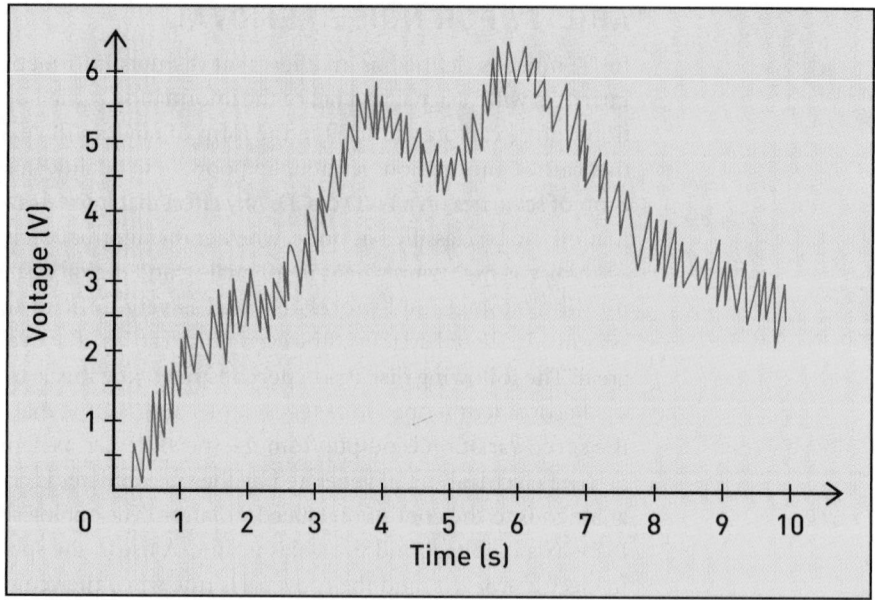

Figure 2-10

The digital electrical
signal carrying digitized
speed measurements

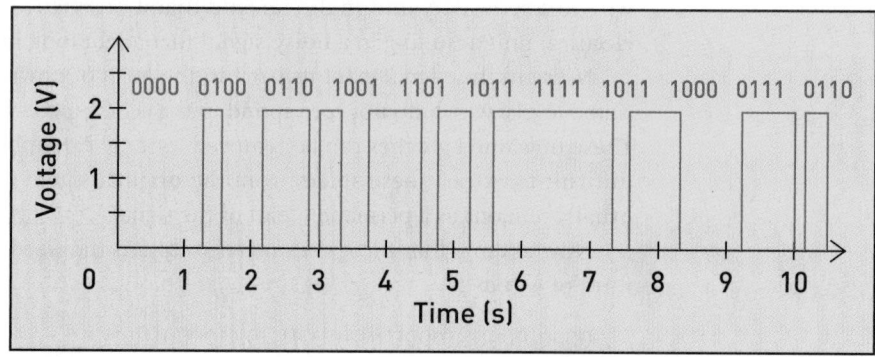

After traveling over a copper wire, the signal that arrives from the speedometer at the secondary unit in the car, along with accumulated noise, might look something like the one in Figure 2-11.

Although this signal still has significant levels of unwanted fluctuations, this noise can be removed by passing the signal through an electronic device that is set at a certain threshold (see Figure 2-12). This device compares each value within the signal to a given threshold level (a predetermined voltage level). Any signal level that remains below a certain threshold is interpreted as a zero by the device, and any that lies above the threshold is detected as a one. Because the device must decide on one of two values (either a one or a zero), its task is actually quite simple.

Figure 2-11

The noisy digital
electrical signal carrying
digitized speed
measurements

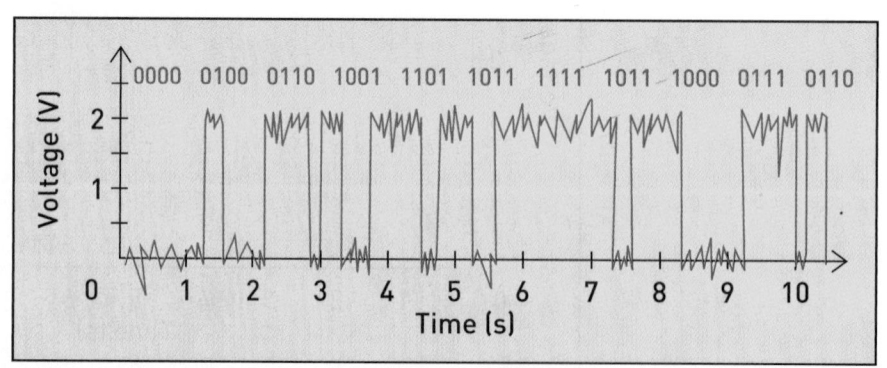

Figure 2-12

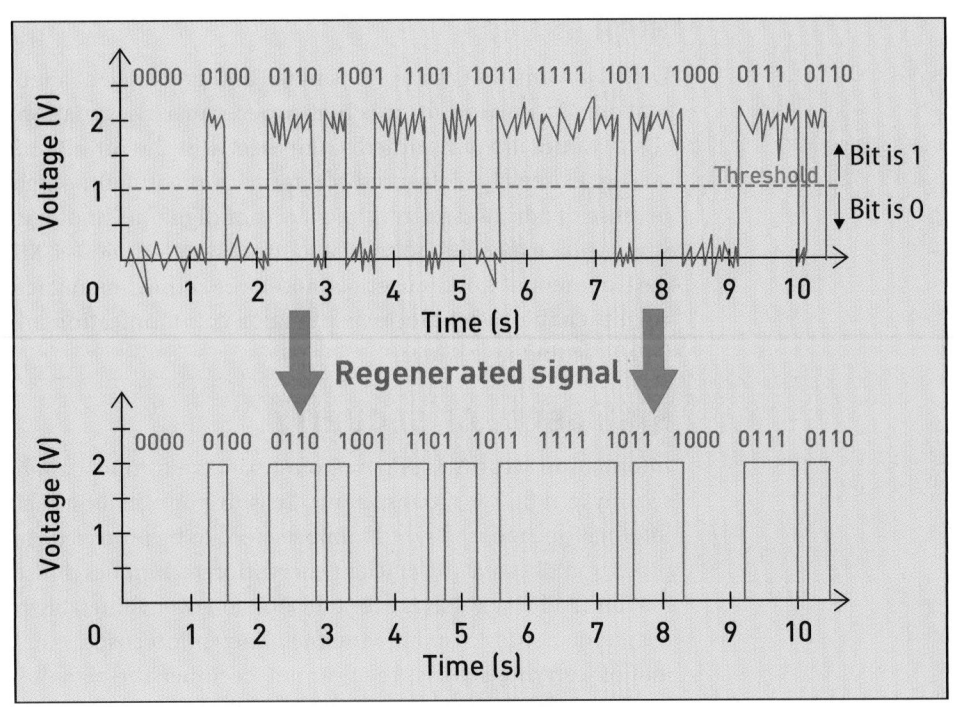

This real-world example clearly demonstrates the digital system's ability to remove noise and the reason that digital signals and systems are much easier to use than their analog counterparts, where noise becomes a significant issue.

CAPACITY FOR ERROR CONTROL

Although digital systems are highly immune to noise, not all levels of noise can be eliminated from a digital signal. If excessive amounts of noise are acting on the transmitted signal for any number of reasons, the threshold device can make an incorrect decision that results in an **error**, meaning that a one may be interpreted as a zero and vice versa. In the speedometer example, a lightning strike can cause sudden unwanted fluctuations in the digital electrical signal as it is transmitted to the secondary unit in the car, resulting in a high level of noise and errors at the receiver. Fortunately, in almost all digital systems, special **error control** schemes are applied to detect and sometimes even correct errors when they occur. Error control schemes rely on the principle of adding extra bits to the digital information prior to transmission, so that errors can be detected if they occur after reception. Detecting errors is imperative because unreliable information is discarded and not even considered; the receiving party can then ask for a retransmission of the information. Correcting an error is better than detecting it, as many schemes can be applied to implement both these functions. For example, computer networks transmit information with extra bits for error control. If the receiving end detects an error, the system has the option of requesting that the information be retransmitted. Similarly, music on a CD is stored with extra bits for error detection and correction. Even if the CD contains slight scratches that cause errors, the CD player can detect and correct some or all of these errors using the extra bits.

Error control is unique to digital systems and is considered a major advantage over analog, where it is difficult to recover information once it is disrupted by noise. A common example is music stored on old-fashioned audiocassettes. As the cassette is played over and over again, the cassette player head scratches the magnetic tape that stores the music as an analog signal. This corresponds to a degradation of audio quality, which is apparent when the tape is replayed. Although digital remastering and other techniques can restore sound quality, these techniques are difficult, and often the unwanted effects cannot be entirely removed. Error control schemes are discussed in greater detail in Chapter 8.

HIGH SPEED

Digital information can be transmitted and processed at a much faster rate than analog information. One reason for this high-speed capacity is that only a finite number of sampled signal values are transmitted, compared with the infinite number of values transmitted in analog. Extremely high-speed communication and ultra-high-speed computing are possible because of advances in multiple facets of digital technology, including improved transmission media materials and special transmission techniques. Miniaturization of transistors, the development of novel materials, and more refined manufacturing techniques for integrated circuits enable faster processing. Analog communication and processing have significantly lagged behind in this area.

HIGH LEVEL OF SECURITY

Information security is of paramount concern for most applications, especially for national security and financial transactions. Sensitive information such as account numbers, personal information, and credit card numbers need to be protected, and digital systems can offer such protection. By applying efficient **encryption** techniques, bit streams that carry sensitive information can be made secure for transmission over vulnerable communication systems, such as telephone networks and the Internet. Encryption ensures that if third parties intercept information-carrying signals, they cannot decipher the signals. For example, computers on a wireless network transmit information that is carried across the wireless channel by radio waves. These radio frequency signals can easily be intercepted by unauthorized parties due to the openness of the channel, so encryption algorithms are frequently applied to scramble the transmitted bits and protect the information as much as possible. One encryption method is to reverse the order of the bits before transmission; for example, a bit stream of 1100 is transmitted as 0011. As long as the transmitting and receiving ends agree on the scheme, the receiver can decipher the information.

AMENABILITY TO COMPRESSION

Because information is generated constantly and demand for it continues to grow, efficient information storage has been a major focus in IT. Information files can be stored in ever-smaller amounts of space, such as on portable devices like MP3 players, and transmitted in smaller amounts of time, such as across Internet connections. Digital information, especially digital images and movies, is highly amenable to **compression** because it contains an abundant amount of repetition. Repeating patterns of bit streams, such as long strings of ones and zeros, are frequently found in audio and image files. Applying an assortment of compression algorithms to binary streams can achieve varying degrees of compression. Efficient compression algorithms can reduce a 20-KB file to 1 KB, which means that the file can occupy a fraction of its original storage space and require a fraction of the time for transmission, compared with its uncompressed form. Analog information is more difficult to compress.

RELIABLE STORAGE OF INFORMATION

Analog storage approaches are highly susceptible to degradation because of aging and environmental factors, so digital storage formats are usually preferable. If music stored on a traditional audiocassette is played over and over again, it becomes degraded, reducing the overall audio quality. Music stored in digital form, such as on a flash drive, is less susceptible to these detrimental effects. Similarly, a letter written on paper can easily degrade over time due to moisture in the air and other factors. If the letter is scanned and stored in digital format on a hard disk, there is less chance for degradation.

EASE OF REPRODUCTION

Analog information is highly susceptible to degradation during reproduction. A tape copied in a double cassette recorder usually does not have the same quality as the original tape. On the other hand, information on digital media such as flash drives can be reproduced with the same quality as the original, and with greater ease. A digital audio file on a flash card can be easily duplicated with the same quality as the original; all the user must do is copy the contents of the card to the computer's hard disk. The device that duplicates analog information, such as the double cassette recorder, has the difficult task of duplicating an infinite number of possible signal values stored on the magnetic tape. On the other hand, digital devices only have to duplicate two values: a one or a zero.

SIMPLICITY IN TRANSMISSION

Digital information is easy to transmit, because the transmitter needs to generate a signal with only one of two values, such as 0 V or 2 V. Consequently, the receiver needs to follow a signal with only two values. Transmitters that have to transmit analog signals must be able to generate a signal with an infinite number of values, and the receiver has to follow this complex signal. This is clearly a more challenging task than that faced by a digital counterpart.

CHAPTER SUMMARY

- Digital devices are technologies that process information in the form of ones and zeros, otherwise known as binary language. Digital information is discrete and comprises a finite number of information values, while analog information is continuous and comprises an infinite number of values.
- The development of the transistor and the integrated circuit—a layer of silicon containing numerous interconnected transistors—were major breakthroughs that led to smaller and faster digital technologies.
- Moore's Law, the observation that the number of transistors contained on an integrated circuit doubles every 18 months, has held true for years.
- Any form of information, including sound, text, or images, can be represented digitally using ones and zeros. The bits can be physically generated using various sources of energy, such as electricity, light, or radio frequency. The rate at which bits are transmitted over any medium, such as air, fiber-optic cable, or copper cable, is called the data rate.
- Bits are usually stored on magnetic media such as hard disks, on optical media such as CDs and DVDs, on magneto-optical media such as magneto-optical drives, and on electronic media such as flash cards.
- Digital technology has many advantages over analog technology, including resistance to noise, higher speeds, and greater reliability. Digital technology also offers more efficient security, error detection and correction, compression, and ease of reproduction.

KEY TERMS

analog device	error control
analog information	flash card
analog signal	flash drive
analog-to-digital converter	flash memory
binary digits	integrated circuit
binary language	magnetic storage
binary symbols	magneto-optical storage
bit period	mechanical storage
bits	Moore's Law
byte	noise
compression	noisy signal
data rate	optical signal
digital device	optical storage
digital information	radio frequency signal
digital signal	signal
discrete	storage media
electrical signal	transistor
electronic storage	transmission media
encryption	vacuum tube
error	voltage

REVIEW QUESTIONS

1. What is a transistor? Explain how the invention of the transistor contributed to modern computing.
2. Sketch an example of an analog signal.
3. Describe one way in which information can be presented in analog format.
4. Calculate how many bits per second are transmitted over a connection with a data rate of 721 Kbps.
5. Calculate how many bits per second are transmitted over a connection with a data rate of 55 Mbps.
6. Sketch the digital electrical signal for transmitting the following bit stream: 10101. Each electrical pulse should have a value of 2 volts and last for 2 seconds.
7. Sketch the noisy version of the digital electrical signal described in the previous question.
8. Explain how noise can be removed from the noisy signal described in the previous question.
9. Calculate the bit period corresponding to a data rate of 50 Kbps.
10. Given that the bit period of a transmission is 0.005 seconds, calculate the data rate of the transmission.
11. List two types of optical storage media.
12. Calculate the number of bits that can be stored on an Apple iPod with a 40-GB hard drive.
13. Calculate the number of bytes in a 4-MB file attachment sent via e-mail.
14. Calculate the number of bits in 2 bytes of data.
15. Write the numerical values of three multipliers commonly used for storage capacity.
16. Briefly explain why you would want to compress digital information.

DISCUSSION QUESTIONS

1. Moore's Law states that the number of transistors on an integrated circuit will double every 18 months. Do you think this law will hold for the foreseeable future? Why or why not? What natural limitations might eventually constrain this law?
2. Discuss the sources of noise that you've noticed can disrupt wireless communications. Have you personally noticed noise while talking on a cell phone or listening to the radio? What do you think were the sources of this noise?
3. What information that you send on the Internet would you like to be encrypted, and why? How can you determine whether your Internet purchases, text messages, or transmitted documents are encrypted when transmitted over a network?
4. Do you use more analog or digital devices on a day-to-day basis? For example, is your watch analog or digital? Is your speedometer analog or digital? Name all the analog devices you use every day. Name all the digital devices you use.
5. How many different types of transmission media do you use on a day-to-day basis? Could you consolidate all these different types of transmission media onto a single medium? What would be the financial, technical, and logistical advantages or disadvantages of this consolidation?

CASE PROJECT

Consider the various technologies you use to store information. Do you use a flash drive, an iPod, CDs, DVDs, a digital camera, and a laptop?

List the four devices you use most often and find out how much storage capacity each one contains. For example, you might pick up a CD and see that it can store 700 MB of information. Once you determine the capacity of your storage devices, do an Internet search to determine the maximum amount of storage these devices provide in today's market. For example, what is the largest hard-disk capacity you can find for a computer? What is the largest storage capacity on a DVD?

If products are available that provide greater storage capacity than the ones you use, calculate the percentage difference between the products' capacities.

Finally, list some information storage devices that you no longer use on a regular basis because they have been replaced by newer technologies.

3

REPRESENTING NUMBERS AND TEXT IN BINARY

In this chapter you will:

- Understand the binary numbering system

- Mathematically convert numbers between decimal and binary

- Understand binary coded decimal (BCD) representation

- Learn about alternative numbering systems such as octal and hexadecimal and explain their significance in information technology

- Provide real-world examples of binary and hexadecimal representation in information technology

- Convert alphanumeric text into binary

INTRODUCTION

Just as people from different countries require translation between languages to understand each other, digital devices require human languages to be converted into binary code. People can understand a decimal number such as 256, but digital devices can't—they only understand ones and zeros. How does a human-readable number such as 256 translate into a binary number? How does a piece of text such as "good morning" translate into binary? This chapter introduces techniques for converting between the decimal and binary numbering systems, an important foundation in IT. The chapter also introduces other numbering systems such as octal and hexadecimal, and explains why they are sometimes used as a shorthand notation in IT. Finally, this chapter explains how the binary numbering system can represent text and alphanumeric characters using standard codes such as ASCII.

Before thinking about the binary numbering system that computers use, it is helpful to consider the **decimal numbering system** that we use in everyday life. Decimal is a counting system that uses the 10 distinct numbers from 0 to 9, which in various combinations can represent any imaginable number. For example, we can count higher than 9 after expending all 10 symbols by adding a number and starting to count from 0 again, resulting in a number with multiple digits. Each digit of the number occupies a "placeholder," and each placeholder has a specific weighting, based on powers of 10, that contributes to the number's overall value. For example, the number 5627 is made up of four digits, each occupying a placeholder with a different weighting. The digit on the right, 7, occupies the so-called "1s place" because its placeholder's weighting is $10^0 = 1$. Similarly, the 2 occupies the "10s place" because its weighting is $10^1 = 10$, the 6 occupies the "100s place" because its weighting is $10^2 = 100$, and the 5 occupies the "1000s place" because its weighting is $10^3 = 1000$. When we multiply each number by its weighting and add them, we get $5000 + 600 + 20 + 7 = 5627$. That is why we read a number such as 5627 as "five thousand six hundred and twenty-seven." The leftmost digit has the highest weighting and contributes the most to the number's overall value. Similarly, the rightmost number has the smallest weighting and contributes the least.

In addition to integer numbers in the decimal system, such as 5627, we often encounter noninteger numbers that are identified by the presence of the **decimal point**. A six-digit number such as 4268.25 is a noninteger number. It contains an integer portion (4268) and a fractional portion (0.25) that are separated by a decimal point.

Similar to the weightings of the placeholders in an integer number, the number 4268.25 is made up of six individual digits, each occupying a placeholder with a different weighting. The location of the decimal point is taken as a reference to determine the weighting of each placeholder. These weightings start from 0.1 at the right of the decimal point and decrease by a factor of 10 from left to right. The weightings of the placeholders to the left of the decimal point start from 1 and increase by a factor of 10 from right to left. Based on this scheme, the rightmost digit, 5, occupies the so-called "0.01s place" because its weighting is $10^{-2} = 0.01$, the 2 occupies the "0.1s place" because its weighting is $10^{-1} = 0.1$, the 8 occupies the "1s place" because its weighting is $10^0 = 1$, and so on, as explained earlier. When we multiply each number by its associated weighting and add them, we get $4000 + 200 + 60 + 8 + 0.2 + 0.05 = 4268.25$.

Because the decimal system has only 10 different digits ranging from 0 to 9, it is referred to as the **base 10** numbering system (see Figure 3–1).

This counting system is intuitive because we learn it from childhood, but nothing is preordained about using 10 symbols to count. Why not use five symbols, or eight or nine, for that matter? We probably count with 10 distinct symbols because we have 10 fingers. Counting in other numbering systems can become just as easy as counting in decimal, as long as we understand that the decimal system is only one of many possible approaches to counting.

Similar to the decimal system's use of 10 numbers, the binary system uses two numbers: 1 and 0. Any binary number can be created by combining the basic numerals 1 and 0, just as any decimal number can be created by combining the numerals 0 through 9.

However, how do we count higher than 1 in binary if we have already expended both symbols? Counting above 1 simply requires adding a binary digit and starting to count from 0 again using the new placeholders. As in the decimal system, each placeholder has its own weighting, but the weightings are based on powers of 2 in the **binary numbering system** instead of powers of 10. For example, the number 1010 is made up of four digits, each occupying a placeholder with a different weighting. The rightmost 0 occupies the "1s place" because its weighting is $2^0 = 1$, the next digit to the left has a value of 1 and occupies the "2s place" because its weighting is $2^1 = 2$, the next digit has a value of 0 and occupies the "4s place" because its weighting is $2^2 = 4$, and the last digit has a value of 1 and occupies the "8s place" because its weighting is $2^3 = 8$.

Figure 3-1

The base 10 numbering system

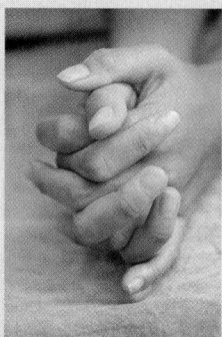

THE DECIMAL SYSTEM
To understand binary, think about
our everyday numbering system.

0, 1, 2, 3, 4, 5, 6, 7, 8, 9

We have ten fingers
and use ten digits!
Coincidence?

Decimal digits are combined to create larger numbers
$4268.25 = (4 \times 10^3) + (2 \times 10^2) + (6 \times 10^1) + (8 \times 10^0) + (2 \times 10^{-1}) + (5 \times 10^{-2})$
10 raised to the power of...
$10^{-2} = 1/(10 \times 10) = 0.01$
$10^{-1} = 1/10 = 0.1$
$10^0 = 1$
$10^1 = 10$
$10^2 = 10 \times 10 = 100$
$10^3 = 10 \times 10 \times 10 = 1000$
$10^4 = 10 \times 10 \times 10 \times 10 = 10,000$
and so on...
Therefore, decimal is also called the base 10 system.

Because the binary system has only two numbers, it is referred to as the **base 2** numbering system. Etymologically, words that start with the prefix *dec*, such as decimal or decathlete, derive from the Latin *decem*, indicating "ten." Words that begin with *bi*, such as binary or bicentennial, derive from the Latin prefix *bi*, indicating "two."

In addition to a binary number such as 1010, we can use binary numbers with a **binary point**, such as 1011.11. As before, the location of the binary point is taken as a reference to determine the weighting of each placeholder. The weightings of the placeholders to the right of the binary point start from $2^{-1} = 0.5$ and decrease by a factor of 2 from left to right. The weightings of the placeholders to the left of the binary point start from $2^0 = 1$ and increase by a factor of 2 from right to left. Based on this scheme, the rightmost 1 in the binary number 1011.11 occupies the "0.25s place," the number 1 to the left occupies the "0.5s place," the next number to the left occupies the "1s place," and so on, as described earlier (see Figure 3-2).

BINARY TO DECIMAL INTEGER CONVERSION

Suppose that a computer produces a binary number made up of a series of five 1s and 0s, such as 11001. This number is called a "5-bit" number because it has 5 **binary digits** (bits) and has a corresponding decimal equivalent. How would we convert the binary number 11001 into a decimal number? Calculating the value of a binary number is identical to calculating the value of a decimal number, except that we use powers of 2 instead of powers of 10, as mentioned previously. The rightmost place is the "1s place." Moving to the left, the next place is the "2s place," the next is the "4s place," then the "8s place," and finally the "16s place." The following example shows the mathematical conversion of the binary number 11001 into its decimal equivalent.

Figure 3-2

The base 2 numbering system

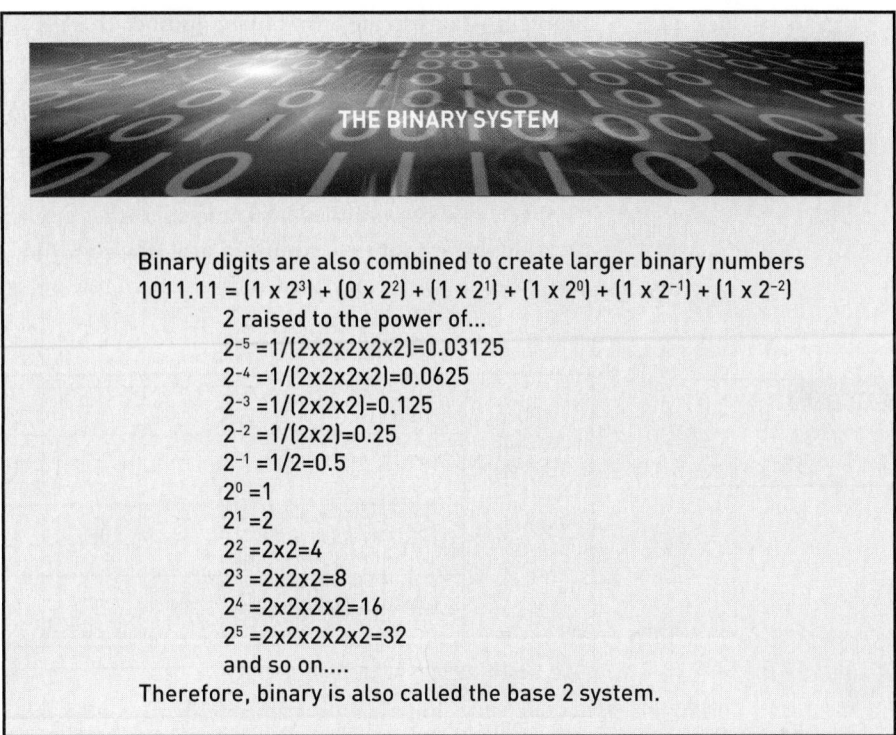

Binary digits are also combined to create larger binary numbers
$1011.11 = (1 \times 2^3) + (0 \times 2^2) + (1 \times 2^1) + (1 \times 2^0) + (1 \times 2^{-1}) + (1 \times 2^{-2})$
2 raised to the power of...
$2^{-5} = 1/(2\times2\times2\times2\times2) = 0.03125$
$2^{-4} = 1/(2\times2\times2\times2) = 0.0625$
$2^{-3} = 1/(2\times2\times2) = 0.125$
$2^{-2} = 1/(2\times2) = 0.25$
$2^{-1} = 1/2 = 0.5$
$2^0 = 1$
$2^1 = 2$
$2^2 = 2\times2 = 4$
$2^3 = 2\times2\times2 = 8$
$2^4 = 2\times2\times2\times2 = 16$
$2^5 = 2\times2\times2\times2\times2 = 32$
and so on....
Therefore, binary is also called the base 2 system.

Binary number: 1 1 0 0 1

First, multiply each binary number by its placeholder weighting:

$$(1 \times 2^4) + (1 \times 2^3) + (0 \times 2^2) + (0 \times 2^1) + (1 \times 2^0) =$$
$$(1 \times 16) + (1 \times 8) + (0 \times 4) + (0 \times 2) + (1 \times 1)$$

Next, add it all together:

$$16 + 8 + 0 + 0 + 1 = 25$$

This process is shown in Figure 3-3. As you can see, the binary number $11001_{(2)}$ equals the decimal number $25_{(10)}$. The subscript after each number indicates the base of the numbering system. Because decimal is a base 10 system, decimal numbers have a subscript of 10. Similarly, binary numbers have a subscript of 2. The subscripts, however, may be dropped occasionally for ease of presentation in this text.

To quickly convert numbers from binary into decimal, it helps to memorize some powers of 2.

Figure 3-3

Binary to decimal conversion

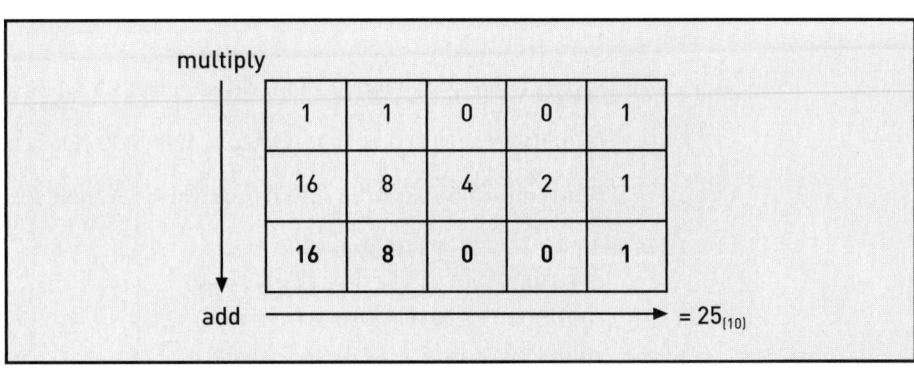

Problem—Convert the 8-bit binary number $11011110_{(2)}$ into its decimal equivalent.

As shown in Figure 3-4, the binary number $11011110_{(2)}$ is equal to the decimal number $222_{(10)}$. The number in the rightmost place, the "1s place," is multiplied by 1. The next number to the left occupies the "2s place" and is multiplied by 2. The next number to the left occupies the "4s place" and is multiplied by 4. The next number to the left occupies the "8s place" and is multiplied by 8, and so forth. Once each of the binary numbers is multiplied by the appropriate multiple of 2, the resulting numbers are added to obtain the answer to the problem, as shown in the final row of Figure 3-4.

Figure 3-4

Converting an 8-bit number into decimal

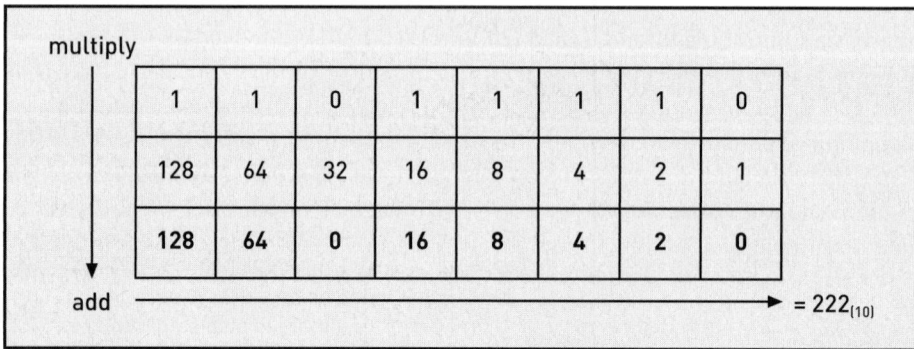

REAL-WORLD EXAMPLE OF BINARY TO DECIMAL CONVERSION

Information technology workers must understand how to convert binary numbers into decimal, and sometimes must make these conversions in real-world situations. For example, every computing device connecting to the Internet uses a unique identifier, known as an IP (Internet Protocol) address, which is analogous to every home having a unique postal address. Some IP addresses are 32-bit numbers, while others are 128-bit numbers (the difference is explained in Chapter 14). A 32-bit number is a combination of 32 1s and 0s, and a 128-bit number is a combination of 128 1s and 0s. This section explains how to convert 32-bit IP addresses into decimal.

The following example shows a 32-bit IP address:

01000111001111001001100010100000

Dotted decimal format condenses the addresses into a more compact and readable format (for example, 71.60.152.160). Converting a 32-bit Internet address to dotted decimal format requires breaking the binary address into four groups of 8 bits, converting each group of 8 bits into its equivalent decimal number, and separating each of the four resulting decimal numbers with dots. This method applies only to converting binary IP addresses, not to the regular conversion of a 32-bit binary number to decimal. The following problem illustrates the dotted decimal method.

Problem—Convert the following binary IP address into dotted decimal format:

01000111001111001001100010100000

1. Separate the IP address into four octets (groups of 8 bits).

   01000111 (Octet 1) 00111100 (Octet 2) 10011000 (Octet 3) 10100000 (Octet 4)

2. Convert each binary octet into its equivalent decimal number.

   01000111 = 0+64+0+0+0+4+2+1 = 71
   00111100 = 0+0+32+16+8+4+0+0 = 60
   10011000 = 128+0+0+16+8+0+0+0 =152
   10100000 = 128+0+32+0+0+0+0+0 = 160

3. Write the decimal values separated by dots.

Human-readable IP address: 71.60.152.160

The dotted decimal result, 71.60.152.160, is much easier to read and manage than its 32-bit equivalent. Network administrators frequently enter these decimal numbers through the keyboard to configure their networks. It is obviously easier to enter a few decimal integers than to type out 32 bits.

DECIMAL INTEGER TO BINARY CONVERSION

How do you convert decimal integer numbers back into binary numbers? This conversion is slightly more difficult than converting from binary to decimal, even though it only requires reversing the mathematical process. Some instances of conversion are intuitively obvious. For example, in Table 3-1, you can convert the decimal integer numbers into their binary equivalents.

Table 3-1
Decimal to binary conversions

Decimal to binary conversion	Explanation of conversion
The decimal number $1_{(10)} = 1_{(2)}$	The rightmost bit is the 1s place, so a 1 in that placeholder equals 1.
The decimal number $2_{(10)} = 10_{(2)}$	The second bit from the right is the 2s place, so a 1 in that spot equals 2.
The decimal number $3_{(10)} = 11_{(2)}$	Combining the preceding examples, a 1 in the 2s place and a 1 in the 1s place add up to 3.
The decimal number $4_{(10)} = 100_{(2)}$	The third bit from the right is the 4s place. Placing a 1 in this position equals 4.
The decimal number $8_{(10)} = 1000_{(2)}$	The fourth bit from the right is the 8s place, so placing a 1 in that spot equals 8.
The decimal number $9_{(10)} = 1001_{(2)}$	To the preceding example, a 1 in the 8s place, add a 1 in the 1s place (8+1 = 9).
The decimal number $16_{(10)} = 10000_{(2)}$	Placing a 1 in the 16s place equals 16.
The decimal number $32_{(10)} = 100000_{(2)}$	Placing a 1 in the 32s place equals 32.
The decimal number $64_{(10)} = 1000000_{(2)}$	Placing a 1 in the 64s place equals 64.
The decimal number $128_{(10)} = 10000000_{(2)}$	In an 8-bit number, the leftmost bit is the 128s place, so placing a 1 in that position equals 128.
The decimal number $129_{(10)} = 10000001_{(2)}$	Placing a 1 in the 128s place plus a 1 in the 1s place equals 128+1 or 129.
The decimal number $130_{(10)} = 10000010_{(2)}$	Placing a 1 in the 128s place plus a 1 in the 2s place equals 128+2 or 130.

These examples are intuitively easy to compute. However, for less obvious examples, what is the mathematical procedure for converting from a decimal number to a binary number? For example, what is the binary equivalent of the decimal integer 30? You can make this conversion using two methods, which are explained in the following sections.

Method 1 for Converting Decimal to Binary

The first method involves repeatedly dividing a number by two, noting the presence or absence of a remainder, and using this information to derive the binary representation of the decimal number. To find the binary equivalent of a decimal number, it is helpful to construct a table with two columns. The first column contains the quotient—in other words, the integer part of the division results. The second column contains a decision bit to indicate the presence or absence of a remainder as a result of the division.

To convert the decimal number 30 to a binary number using Method 1:

1. First, divide the number to be converted by 2. For example, dividing 30 by 2 results in 15. The integer part of the division is 15, so its value is noted in the first column of Figure 3-5.

2. Next, the result of the division is examined to determine whether there is a remainder. The result of the division is 15, with no remainder. Therefore, a decision bit with a value of 0 is entered in the second column to represent the absence of a remainder. If there were a remainder as a result of the division, you would enter a decision bit with a value of 1 in the table.

3. Next, the quotient from the previous row is divided by 2. In other words, 15 is divided by 2, resulting in 7.5. The quotient of this number is 7, so the value of the quotient is entered in the table. Because the result of the division has a remainder, its presence is indicated by entering a 1 as the decision bit in the table.

4. The quotient from the previous row, 7, is then divided by 2. The result is 3.5. The quotient of this number is 3, so a 3 is entered in the table. Because the division results in a remainder, its presence is indicated by noting a 1 in the second column of the table.

5. The quotient from the previous operation, 3, is then divided by 2. The result is 1.5. The quotient of this number is 1, so a 1 is entered in the table. Because the result of the division has a remainder, a decision bit with a value of 1 is entered in the table.

6. The quotient from the previous operation, 1, is then divided by 2. The result is 0.5. The quotient of this number is 0, so a 0 is entered in the last row of the table. Because the result of the division has a remainder, a decision bit with a value of 1 is entered in the last row of the table.

7. The conversion process ends when the division results in a quotient of 0. To find the corresponding binary number, read the decision bits in column 2 of the table from bottom to top, and write them from left to right. The last value in the second column is 1, so the first bit is read as 1. The next row from the bottom is 1, so the next bit is 1. Similarly, all of the bits are written out, resulting in a binary string (series of bits) with the value 11110. Thus, the binary equivalent of $30_{(10)}$ is calculated to be $11110_{(2)}$.

Figure 3-5

Converting the decimal number 30 into binary using Method 1

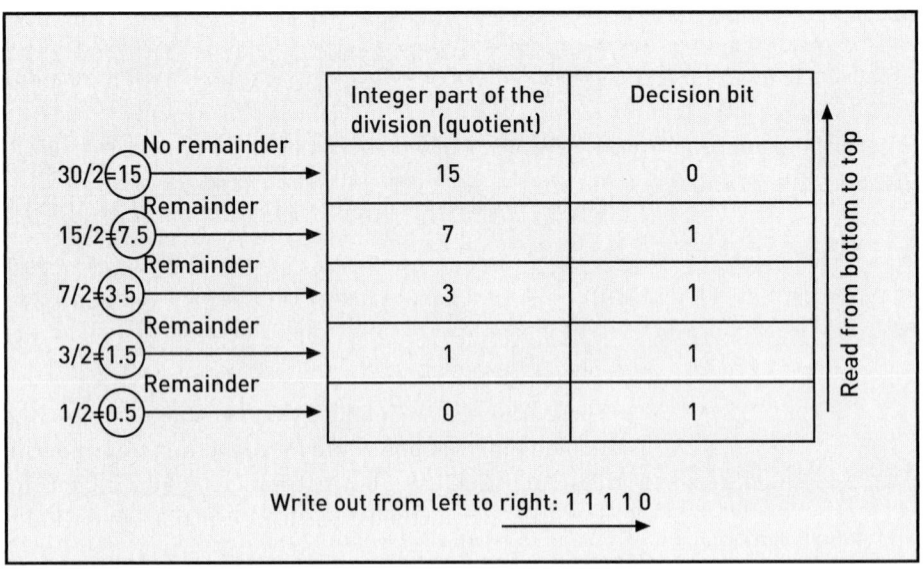

Method 2 for Converting Decimal to Binary

The second method involves determining how many bits are found in the binary equivalent of the decimal number and iteratively comparing and subtracting numbers. A single bit can be used to represent only two values: 1 and 0. Two bits can represent four values: 00, 01, 10, and 11, or the decimal numbers 0, 1, 2, and 3. Three bits can represent 8 values: 000, 001, 010, 011, 100, 101, 110, and 111, or the decimal numbers 0 through 7.

In other words, *x bits can represent 2^x decimal numbers ranging from 0 to 2^x-1*. Therefore, four bits can represent 2^4 or 16 values (decimal values from 0 to 15), which is still not enough to represent the decimal value 30. Five bits can represent 2^5 or 32 values (decimal values from 0 to 31), a sufficient amount to represent the decimal value 30. In other words, converting the decimal number 30 results in a five-bit binary number. Similarly, if we wanted to convert the decimal number 64 into binary, the binary equivalent would have 7 bits, because they can represent decimal values ranging from 0 to 127. It would not be enough to use 6 bits because they would only represent decimal values from 0 to 63. Figure 3-6 illustrates the conversion process.

Figure 3-6

Converting the decimal number 30 into binary using Method 2

Problem—Convert the decimal number 30 into binary.

16s place	8s place	4s place	2s place	1s place

Step 1: Compare 30 to 16. Because 30 is larger than 16, place a 1 in the 16s place. Subtract 30−16 = 14.

1				
16s place	8s place	4s place	2s place	1s place

Step 2: Compare the remainder 14 to 8. Because 14 is larger than 8, place a 1 in the 8s place. Subtract 14−8 = 6.

1	1			
16s place	8s place	4s place	2s place	1s place

Step 3: Compare the remainder 6 to 4. Because 6 is larger than 4, place a 1 in the 4s place. Subtract 6−4 = 2.

1	1	1		
16s place	8s place	4s place	2s place	1s place

Step 4: Compare the remainder 2 to 2. Because 2 is equal to 2, place a 1 in the 2s place. Subtract 2−2=0. There is no remainder, meaning that the entire original value, 30, has been represented. Therefore, place a 0 in the 1s place.

1	1	1	1	0
16s place	8s place	4s place	2s place	1s place

Answer: The decimal number $30_{(10)}$ is equal to the binary number $11110_{(2)}$.

Notice that the procedure involves four steps:

- Iteratively comparing the decimal number to the weighting of each placeholder starting from the left
- Placing a 1 in the placeholder if a number's value is greater than or equal to the weighting of the placeholder, and placing a 0 in the placeholder if a number's value is less than the weighting of the placeholder

- Using subtraction if a number's value is greater than or equal to the weighting of the placeholder to find a remainder
- Comparing the remainder to the weighting of the next placeholder to the right

If the comparison results in a number that is less than the value of a placeholder's weighting, a 0 is placed in the placeholder, and the remainder is compared to the next lowest weighting.

Consider one more example. How would we convert the decimal number 9 into its 4-bit binary equivalent? You can do this conversion intuitively by recalling that placing a bit in the leftmost position equals 8 and then adding a 1 in the rightmost position equals 1, resulting in 8 + 1 = 9. The iterative mathematical procedure is shown in the following example.

Problem—Convert the decimal number 9 into binary using Method 1.

Using Method 1, we divide 9 by 2 to get a quotient of 4 and a remainder. We next divide 4 by 2 to get a quotient of 2 with no remainder. We then divide 2 by 2 to get a quotient of 1 with no remainder, and finally divide 1 by 2 to get a quotient of 0 and a remainder. By noting the presence or absence of the remainder as decision bits in the second column of the table and reading the column values from bottom to top, we see that the binary equivalent of 9 is 1001 (see Figure 3-7).

Figure 3-7

Converting the decimal number 9 into binary using Method 1

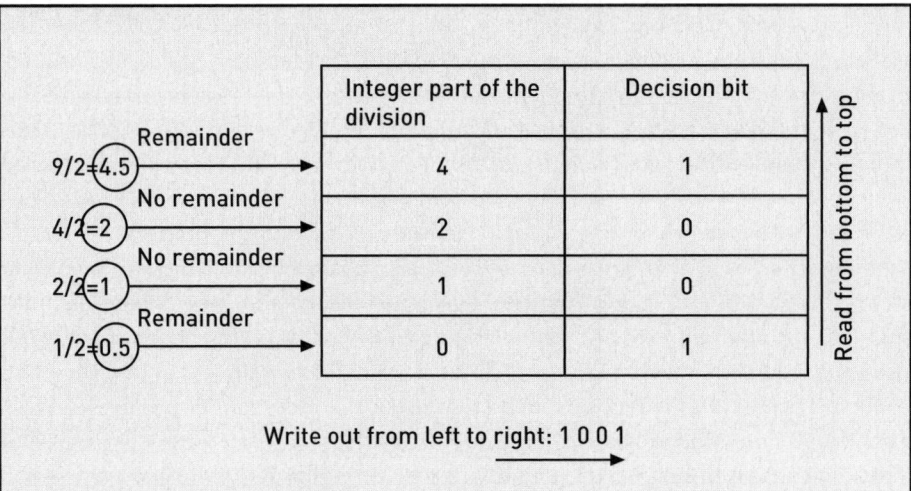

Problem—Convert the decimal number 9 into binary using Method 2.

1. Using Method 2, we know that 4 bits can represent 2^4 = 16 numbers from 0 to 15, so we know that the binary equivalent of 9 will comprise 4 bits.
   It would not be enough to use 3 bits because they only represent 2^3 = 8 different values from 0 to 7. Therefore, we write out four placeholders and their corresponding weightings below the placeholders.
2. Next, we compare 9 to 8. Because 9 is greater than 8, we place a 1 in the "8s" place and subtract 8 from 9, resulting in 1.
3. We then compare the remainder 1 to the weighting of the next placeholder. Because 1 is less than 4, we place a 0 in the "4s place" and continue to the next placeholder.
4. We compare 1 to 2. Because 1 is less than 2, we place a 0 in the "2s place" and continue to the next placeholder.
5. Finally, 1 is equal to 1, so we place a 1 in the "1s place" and subtract 1 from 1, resulting in a 0. This completes the conversion process, which is illustrated in Figure 3-8.

BINARY CODED DECIMAL

Some applications, such as certain electronics systems, use a different approach to representing decimal numbers in binary. In the previous conversion process, an entire decimal number was converted into binary. Another approach, called **binary coded decimal** (BCD), encodes each digit in the decimal number individually rather than converting the entire number.

Figure 3-8

Converting the decimal
number 9 into binary
using Method 2

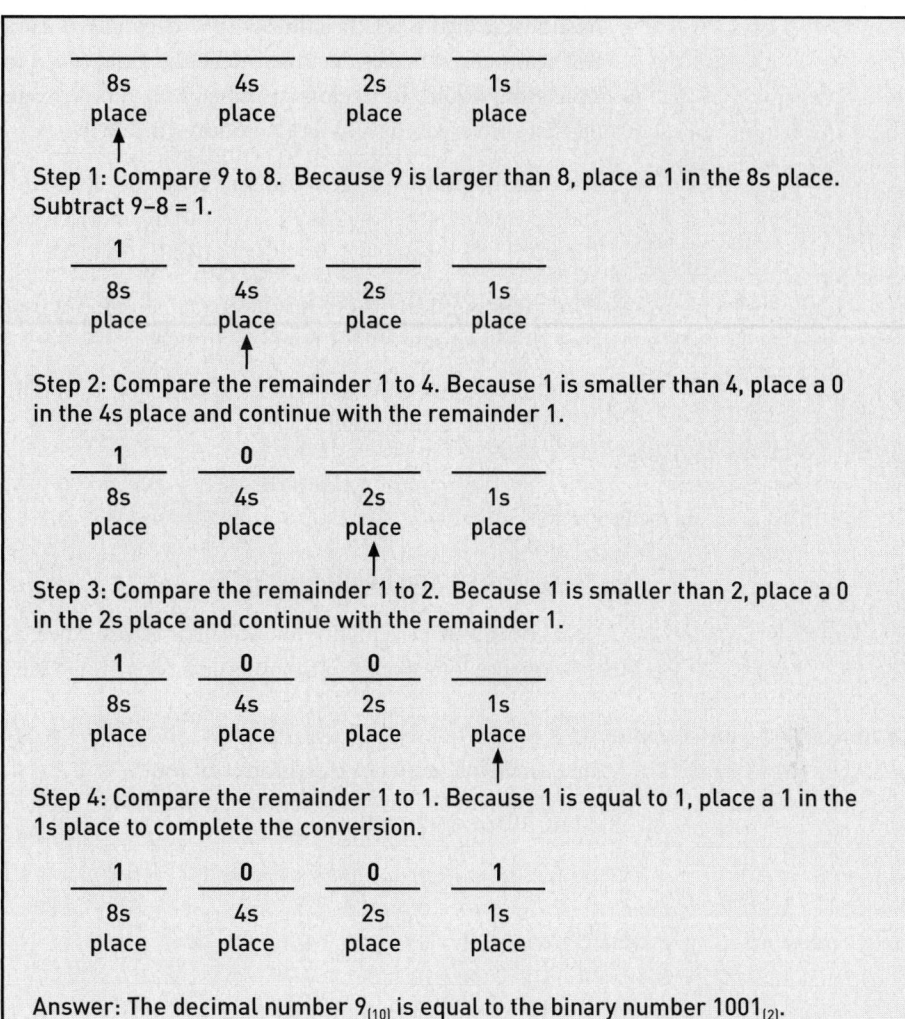

Step 1: Compare 9 to 8. Because 9 is larger than 8, place a 1 in the 8s place. Subtract 9−8 = 1.

Step 2: Compare the remainder 1 to 4. Because 1 is smaller than 4, place a 0 in the 4s place and continue with the remainder 1.

Step 3: Compare the remainder 1 to 2. Because 1 is smaller than 2, place a 0 in the 2s place and continue with the remainder 1.

Step 4: Compare the remainder 1 to 1. Because 1 is equal to 1, place a 1 in the 1s place to complete the conversion.

Answer: The decimal number $9_{(10)}$ is equal to the binary number $1001_{(2)}$.

Therefore, to convert 30 to BCD, the 3 would first be converted into binary, and 0 would then be converted into binary. The two binary strings would be concatenated, or brought together, to represent the number 30 in BCD. Four bits are used to represent each of the 10 numbers in decimal: 0, 1, 2, 3, 4, 5, 6, 7, 8, and 9. Why four bits and not three or five bits? Three bits provides 2^3, or 8 different combinations, which is not enough to represent 10 values. Five bits provides 2^5, or 32 different combinations, which is far more than necessary to represent the 10 decimal numbers. Four bits can represent 2^4, or 16 combinations, which is more than enough to represent the 10 decimal numbers. Figure 3-9 lists the 4-bit code that represents each decimal number in binary coded decimal.

Figure 3-9

Binary coded decimal

Decimal Number	BCD Representation
0	0000
1	0001
2	0010
3	0011
4	0100
5	0101
6	0110
7	0111
8	1000
9	1001

As shown in Figure 3-9, the number 30 in decimal would be represented as 0011 0000 in BCD.

BCD can represent any decimal integer by assigning a four-bit code for each number. For example, how would the decimal number $234_{(10)}$ be encoded using BCD representation? You would first convert each number (2, 3, and 4) to BCD:

The decimal number $2_{(10)}$ is equal to $0010_{(2)}$ in BCD.
The decimal number $3_{(10)}$ is equal to $0011_{(2)}$ in BCD.
The decimal number $4_{(10)}$ is equal to $0100_{(2)}$ in BCD.

When you combine these results into a single binary stream, you see that the decimal number $234_{(10)}$ equals $001000110100_{(2)}$ in BCD.

Problem—Convert the decimal number $7244_{(10)}$ into binary coded decimal.

7 = 0111
2 = 0010
4 = 0100
4 = 0100
$7244_{(10)} = 0111001001000100_{(2)}$ (BCD)

Converting a BCD sequence into decimal is easy. Reversing the process involves breaking a binary sequence into groups of four bits and converting each group into its decimal equivalent.

Problem—Convert the BCD sequence 0001100001111001 into decimal.

First break the sequence into groups of four:

0001   1000   0111   1001

Then convert each group into decimal:

$0001_{(2)} = 1_{(10)}$
$1000_{(2)} = 8_{(10)}$
$0111_{(2)} = 7_{(10)}$
$1001_{(2)} = 9_{(10)}$

The answer is $1879_{(10)}$.

BINARY REPRESENTATION OF POSITIVE NONINTEGERS

So far, this chapter has explained how integers such as 2, 9, or 5627 can be converted to binary, but how can binary represent noninteger numbers such as 30.333333?

Converting a noninteger number into binary is similar to converting an integer into binary. For example, consider the number 42.4375. Finding the binary equivalent of this non-integer number requires the following steps:

1. The integer part of 42.4375, which is the number to the left of the decimal point (42), is first converted into its binary equivalent using the procedures described previously. The binary equivalent of 42 is calculated to be 101010.
2. Next, the fractional part of the decimal number (0.4375) is converted into binary by multiplying 0.4375 by 2 and comparing it to 1.
3. If the result of the multiplication is greater than or equal to 1, a bit with a value of 1 is noted, the fractional part of the multiplication result is extracted and multiplied by 2, and the result is compared to 1. If the result of the multiplication is less than 1, a bit with a value of 0 is noted, and the number is multiplied by 2 again.
4. This procedure repeats until the result of the multiplication is 1.

You might understand the conversion process better by constructing a table with two columns, as shown in Figure 3-10. The left column contains the fractional part of the multi-plication result, and the right column contains a decision bit with a value equal to 1 or 0, indicating whether the result of the multiplication is greater than or equal to one (≥1) or less than one (<1).

Figure 3-10

Converting the decimal number 42.4375 into binary

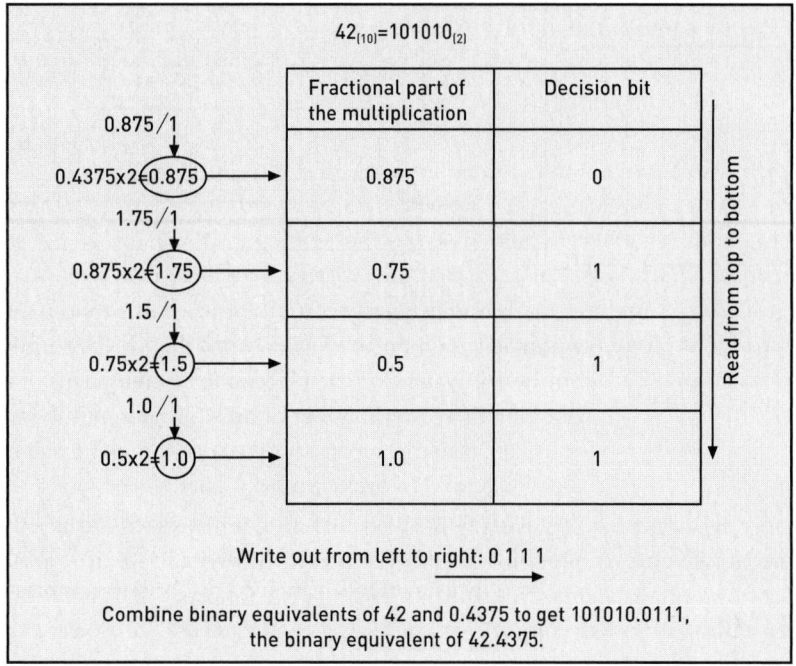

$42_{(10)} = 101010_{(2)}$

Fractional part of the multiplication	Decision bit
0.875	0
0.75	1
0.5	1
1.0	1

0.875/1
0.4375x2=0.875
1.75/1
0.875x2=1.75
1.5/1
0.75x2=1.5
1.0/1
0.5x2=1.0

Read from top to bottom

Write out from left to right: 0 1 1 1

Combine binary equivalents of 42 and 0.4375 to get 101010.0111, the binary equivalent of 42.4375.

The following points summarize Figure 3-10:

- The number to be converted (0.4375) is initially multiplied by 2, resulting in 0.875, which is less than 1. Therefore, a value of 0 is entered in the right column.
- The number 0.875 is then multiplied by 2, resulting in 1.75. Because 1.75 ≥ 1, a 1 is entered in the right column and the fractional part (0.75) is entered in the left column.
- The number 0.75 is then multiplied by 2, resulting in 1.5. Because 1.5 ≥ 1, a 1 is entered in the right column and the fractional part (0.5) is entered in the left column.
- The fractional part (0.5) is multiplied by 2, resulting in a value of 1. Because 1 ≥ 1, a 1 is entered in the right column.
- Whenever the result of the multiplication is equal to 1, the conversion stops. The binary number corresponding to 0.4375 is determined by reading the right column from top to bottom and writing the number from left to right. By reading the right column, you can see a binary string of 0111.
- The equivalent of the number 42.4375 is then written by combining the binary equivalent of 42 with the binary equivalent of 0.4375, separated by a binary point. This results in the binary string 101010.0111.

A similar process is used to convert binary strings into their corresponding noninteger decimal equivalents. Consider converting the binary string 111.110 into its decimal equivalent, as shown in Figure 3-11. Each bit is multiplied by its appropriate weighting.

- From left to right, the first bit, which is equal to 1, has a weighting of 4 (2^2 = 4) and is therefore multiplied by 4.
- The next bit has a weighting of 2 (2^1 = 2) and is multiplied by 2.
- The third bit has a weighting of 1 (2^0 = 1) and is multiplied by 1.
- The fourth bit has a weighting of 0.5 (2^{-1} = 0.5) and is multiplied by 0.5.
- The fifth bit has a weighting of 0.25 (2^{-2} = 0.25) and is multiplied by 0.25.
- The sixth and final bit has a weighting of 0.125 (2^{-3} = 0.125) and is multiplied by 0.125.
- The results of all the multiplications are added together to produce the decimal number 7.75.

Figure 3-11

Converting the binary number 111.110 into decimal

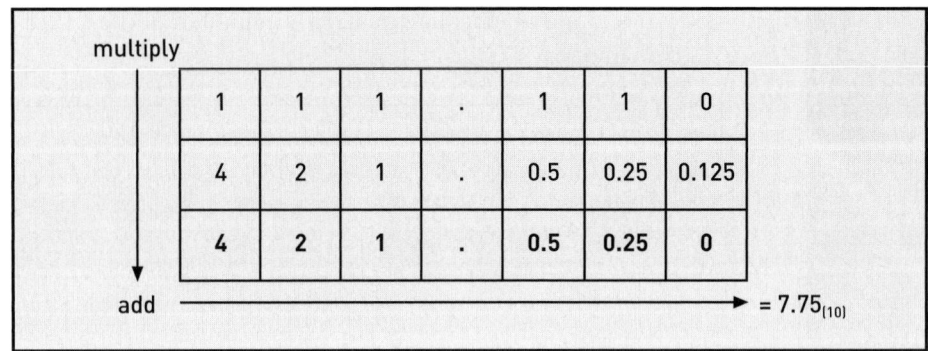

In some instances, you might want to convert a series of decimal nonintegers that vary within a certain range of values into their binary equivalents without having to calculate the binary equivalent of each decimal number. This step saves considerable time and effort, resulting in a lower number of bits to represent the same information.

As an example, reconsider the speedometer discussion of Chapter 2. The discussion noted that converting information from analog to digital required representing the analog information in discrete quantities. The infinite number of analog speed measurements was reduced to 11 discrete measurements by taking a selected number of samples. Consider the following 11 speed measurements, which correspond to samples that vary within a range of 30–60 mph:

Sample 1: 31.5 mph
Sample 2: 37.75 mph
Sample 3: 41.66 mph
Sample 4: 46.96 mph
Sample 5: 56.02 mph
Sample 6: 51.03 mph
Sample 7: 59.80 mph
Sample 8: 51.76 mph
Sample 9: 45.21 mph
Sample 10: 44.93 mph
Sample 11: 42.33 mph

To represent the speedometer readings in binary, you could convert each speedometer reading into its corresponding binary equivalent. For example, the value of the first speedometer reading is 31.5. You could calculate the binary equivalent of 31.5, which is 111111.1, a binary number with 7 bits. Similarly, the binary equivalent of the second reading, 37.75, is 100101.11, a binary number with 8 bits. Although you could convert every reading to its binary equivalent, it is better to convert the series of speed measurements that vary within a range of 30–60 mph using a different approach—one that involves approximation and a smaller number of bits.

The first step to realizing the advantage of this approach involves deciding how many bits to use to represent each speedometer reading. As you have seen, the number of bits for each speed measurement can vary. Seven bits represent the first speed measurement, and 8 bits are used to represent the second speed measurement. Similarly, if we calculated the binary equivalents of the remaining speed measurements, we would find that 20 bits might be used to represent a single measurement.

However, what if we could not afford more than 4 bits per speed measurement? We would need to approximate the speed measurements in an effective way. Suppose we decide to use 4 bits to represent each reading; we know that 4 bits can represent 16 different speed measurements because $2^4 = 16$. To determine the binary codes that correspond to each speed measurement, we can use one of several methods. One method involves calculating a **step size** to find the value to which each actual speed reading will be rounded off or approximated. Assuming that we want to encode each speed reading with a 4-bit pattern, we divide the speed range of 30–60 mph into 2^4 = 16 distinct speed readings:

Step size = Range/$2^{\text{number of bits}}$ = (30–60 mph)/2^4 = 30 mph/2^4 = 1.875 mph

Breaking the speed range into 1.875-mph steps enables us to calculate how to round off each speed measurement. We start with the lowest value of the range and increment it by the step size. Because the speed varies between 30 and 60 mph, we increment the lowest value of the range (30 mph) to 1.875 mph to get 31.875 mph. Next, we increment 31.875 by 1.875 and get 33.75 mph. We repeat this process until we reach the high end of the range—16 increments are needed to get from 30 mph to 60 mph. This process is illustrated in Figure 3-12. The first column contains the 16 speed increments, and the second column contains the midpoints of these increments.

Figure 3-12

Converting speed measurements into binary

Speed Increments (mph)	Midpoint of the increments (mph)	Binary Representation
30.000–31.875	30.9375	0000
31.875–33.750	32.8125	0001
33.750–35.625	34.6875	0010
35.625–37.500	36.5625	0011
37.500–39.375	38.4375	0100
39.375–41.250	40.3125	0101
41.250–43.125	42.1875	0110
43.125–45.000	44.0625	0111
45.000–46.875	45.9375	1000
46.875–48.750	47.8125	1001
48.750–50.625	49.6875	1010
50.625–52.500	51.5625	1011
52.500–54.375	53.4375	1100
54.375–56.250	55.3125	1101
56.250–58.125	57.1875	1110
58.125–60.000	59.0625	1111

For example, the midpoint of the 30–31.875 increment is calculated by (30+31.875)/2 = 30.9375 mph. The midpoint of the second increment is calculated by (31.875+33.75)/2 = 32.8125 mph. After all of the midpoints are found, each midpoint value is assigned a 4-bit binary code, as shown in the third column of the table in Figure 3-12. The assignment of binary codes can follow any order and may be arbitrarily assigned.

The table is then used to determine the binary equivalent of each speed reading. The first column is used to determine which range of values the actual speed measurement falls between. The second column contains the value to which each speed measurement should be rounded off, and the third column contains the binary code assigned to each rounded-off value. For example, the first sample has a value of 31.5 mph. We see from Figure 3-12 that 31.5 mph falls within the range of 30.000 to 31.875 mph, and decide that it should be rounded off to 30.9375 mph and assigned a binary code of 0000. The second reading, which falls within the 37.500–39.375 mph range, should be rounded off to 38.4375 mph and assigned a binary value of 0100. The rest of the measurements are rounded off to their appropriate values and assigned binary codes in this manner. As a result, the 11 speed measurements are represented by a total of $11 \times 4 = 44$ bits:

0000 0100 0110 1001 1101 1011 1111 1011 1000 0111 0110

Figure 2-5 in Chapter 2 originally presented this sequence of binary digits, and now you see how they came about. If we had converted the actual speed measurements directly into their corresponding values, it would have taken a much greater number of bits to represent the same sequence of speed measurements. The trade-off is that the actual speed measurements were rounded off from their actual values. The actual value of the first sample was 31.5 mph, but the value was rounded to 30.9375 mph. Although there is some loss of information by using a smaller number of bits, this method is usually better than converting each sample value to its exact binary equivalent whenever you have a range of values, such as speed measurements or voltages.

To provide a better approximation, we could assign a longer code, such as an 8-bit pattern. An 8-bit code would provide 2^8 unique patterns, or 256 increments. The step size in this case would be 30–60 mph/256 or 0.117 mph, which would provide a much closer approximation than the step size produced by the 4-bit pattern, 1.875 mph.

REPRESENTING NEGATIVE INTEGERS IN BINARY USING "2'S COMPLEMENT NOTATION"

Besides positive decimal numbers, negative decimal numbers are also used in computing. How can negative integers be represented in binary? One method is to incorporate an extra bit in front of the binary equivalent of the integer to specify its sign. For example, the binary equivalent of 5 is 101. If we were to insert a sign bit in front of the binary equivalent, +5 might be represented as $\underline{0}101$ and –5 might be represented as $\underline{1}101$, where the leftmost bit is the sign bit. Any digital device would take this fact into account. This scheme, however, does not work well in real life due to the complications it poses for arithmetic operations. To understand its limitations, you need to understand how addition is performed in binary. The following rules apply to binary addition:

0 + 0 = 0
0 + 1 = 1
1 + 0 = 1
1 + 1 = 0 with a carry of 1
1 + 1 + carry of 1 = 1 with a carry of 1

Based on the preceding rules, the addition of 0101 (+5) and 1101 (–5) is:

```
 0101
+1101
10010
```

Considering that the leftmost bit is the sign bit, the decimal equivalent of 10010 would be –2. We know, however, that the result of adding +5 and –5 should be 0. The use of the sign bit, therefore, is often not always feasible. Instead of using a sign bit, the binary equivalent of a negative integer may be represented by using **2's complement notation**. Taking the complement means changing all 0s to 1s and all 1s to 0s in the binary pattern. You can obtain the binary representation of –5, for example, by computing the 2's complement of 5. Note that the 2's complement process requires you to specify the number of bits, as the following examples illustrate.

Problem—Express –5 in 8-bit 2's complement form. In other words, compute the 8-bit 2's complement of 5.

1. Determine the 8-bit binary representation of 5.

   00000101

   Note that the binary representation of 5 is 101. The 8-bit binary representation of 5 is obtained by inserting five leading 0s.

2. Take the "complement" of the binary pattern determined in Step 1. Taking the complement means changing all the 0s to 1s and changing all the 1s to 0s in the binary pattern.

   11111010

3. Add 1 to the binary pattern determined in Step 2.

   11111010
   +   1
   11111011

   The 8-bit 2's complement of –5 is therefore 11111011.
   Note that the addition of +5 and –5 should be equivalent to 0:

    00000101
   +11111011
   100000000.

   The leftmost bit is called the overflow and can be disregarded. The end result is therefore 00000000, illustrating the usefulness of 2's complement notation.

Problem—Express –7 in 16-bit 2's complement form.

1. Determine the 16-bit binary representation of 7.

   0000000000000111

2. Take the "complement" of the binary pattern determined in Step 1.

   1111111111111000

3. Add 1 to the binary pattern determined in Step 2.

   1111111111111001

   The decimal number –7 in 16-bit 2's complement form is therefore 1111111111111001.

ALTERNATIVE NUMBERING SYSTEMS

People use the decimal numbering system in everyday life and computers use the binary system, so you might wonder why other numbering systems would be necessary. The answer is that most IT activities result in streams of ones and zeros that are extremely long and difficult to manage. To make these long binary streams more manageable and compact, it is sometimes helpful to condense them using other numbering systems, such as octal and hexadecimal. These systems are sometimes called "shorthand notations."

OCTAL

The **octal** numbering system, also called **base 8**, uses eight numbers. Counting in octal is like counting in decimal or binary, but octal uses eight numbers (0 through 7) instead of 10 or two, as in the decimal and binary systems. Because we run out of numbers at 7, the next number after 7 is 10 in octal. The octal numbering system is as follows:

0, 1, 2, 3, 4, 5, 6, 7, 10, 11, 12, 13, 14, 15, 16, 17, 20, 21, 22, 23, 24, 25, 26, 27, 30, and so on

As in the decimal and binary numbering systems, the rightmost place in octal is the 1s place. In decimal, the second place from the right is the 10s place; in binary, the second place from the right is the 2s place. Similarly, in octal, the second place from the right is the 8s place, and the third place from the right is the $8^2 = 8 \times 8$ or 64s place. Because octal uses powers of 8, it is called the base 8 system.

Consider the octal number 167. Starting from the right side of the number, the seven is in the 8^0, or "1s place"; the six is in the 8^1, or "8s place"; and the one is in the 8^2, or "64s place." Just as you convert binary to decimal using the placeholder weightings, the octal number 167 can be converted to decimal as follows:

$$(1 \times 64) + (6 \times 8) + (7 \times 1) = 64 + 48 + 7 = 119_{(10)}$$
The octal number $167_{(8)}$ = the decimal number $119_{(10)}$.

How does octal provide a shorthand notation for binary? Recall that each octal place-holder can have a value from 0 to 7. Because 8 is a power of 2, or 2^3, each octal number can represent three binary digits (bits), as shown in Table 3-2.

Table 3-2

Octal as a shorthand notation for binary

Octal number	Binary pattern
0	000
1	001
2	010
3	011
4	100
5	101
6	110
7	111

Using this association of octal and binary numbers, we can use the octal number 7 rather than writing out the binary pattern 111. Instead of writing out the binary number 011, we can use the octal number 3. A longer binary stream such as 111000 can easily be divided into groups of three (in this case, 111 and 000) and then converted to octal, or 7 and 0. Therefore, the binary stream 111000 is equivalent to the octal number 70. This example illustrates how octal provides a shorthand notation for binary streams.

How does octal provide a shorthand notation for the following long binary stream?

$011111000011100101010010000111110_{(2)}$

To convert this binary number into an octal shorthand number, break the bit stream into groups of three, starting from the right, and convert each bit grouping into its octal equivalent, as follows:

110 = 6
111 = 7
000 = 0
010 = 2

010 = 2
101 = 5
100 = 4
011 = 3
000 = 0
111 = 7
011 = 3

As shown, the binary number is broken into groups of three, and each group is then translated into octal.

The final step is to place the resulting octal numbers in sequential order, as follows:

$37034522076_{(8)}$

As you can see, the binary stream has been significantly shortened by condensing it to its octal equivalent.

Problem—Convert the binary stream 111101000000 into octal.

First, break the number into groups of three digits starting from the right: 000, 000, 101, and 111. Convert each group into its octal equivalent and then place the resulting octal numbers in sequential order.

000 = 0
000 = 0
101 = 5
111 = 7

When you place the resulting octal numbers in sequential order, you see that the answer is $7500_{(8)}$. To convert from octal back into binary, you must use the reverse process: change each octal number into its associated binary pattern and then place the resulting binary groups in sequential order. For example, to convert the octal number 7500 into binary, note that the octal number 7 equals the binary pattern 111, the octal number 5 equals the binary pattern 101, and the octal number 0 equals the binary pattern 000 (twice). When you string the resulting groups together, the resulting binary pattern is $111101000000_{(2)}$.

If the binary sequence does not contain a number of bits that you can evenly multiply by 3, add one or two 0s in front so the final group of bits totals three.

HEXADECIMAL

The **hexadecimal** numbering system, also called Hex or **base 16**, uses 16 characters. Given that there are only 10 unique numbers, 0–9, the hexadecimal system has to introduce alphabetic letters. The 16 characters in hexadecimal are 0, 1, 2, 3, 4, 5, 6, 7, 8, 9, A, B, C, D, E, and F. The letter "A" symbolizes the tenth number, "B" symbolizes the eleventh number, and so forth. To continue counting in hexadecimal once these 16 symbols are used, you add a second placeholder with the numeral 1, just as in decimal. In other words, the number after 9 in decimal is 10, and the number after the hexadecimal number "F" is 10.

Therefore, the hexadecimal numbering system is as follows:

0, 1, 2, 3, 4, 5, 6, 7, 8, 9, A, B, C, D, E, F, 10, 11, 12, 13, 14, 15, 16, 17, 18, 19, 1A, 1B, 1C, 1D, 1E, 1F, 20, 21, 22, 23, 24, 25, 26, 27, 28, 29, 2A, 2B, 2C, 2D, 2E, 2F, 30, 31, and so on

How do you determine the decimal equivalent of a hexadecimal number? Just as in the decimal, octal, and binary numbering systems, the rightmost place is the 1s place. In decimal, the second place from the right is the 10s place; in binary, the second place from the right is the 2s place. Similarly, in hexadecimal, the second place from the right is the 16s place. In hexadecimal, the third place from the right is the 16×16 or 256s place. Consider the hexadecimal number 1A2:

1	A	2
256s place	16s place	1s place

Similar to converting binary or octal to decimal using the placeholder weightings, the preceding hexadecimal number can be converted to decimal. The 1 that occupies the 256s place has a value of 1×256, or 256. The 2 occupying the 1s place has a value of 2×1, or 2. How do we calculate the value of the "A" in the 16s place? Recall that in the hexadecimal system, "A" comes after the numeral 9 and thus is numerically equivalent to 10. Therefore, A $\times$ 16 is really 10×16 or 160. The decimal value of the hexadecimal number 1A2 is therefore:

$$(1 \times 256) + (10 \times 16) + (2 \times 1) = 256 + 160 + 2 = 418_{(10)}$$
The hexadecimal number $1A2_{(16)}$ = the decimal number $418_{(10)}$.

Recall that the primary purpose of using alternative numbering systems such as hexadecimal is to provide a shorthand notation for binary. Each hexadecimal number has one of 16 possible values. Because 16 is a power of 2, or 2^4, each hexadecimal number can represent four binary digits (bits), as shown in Table 3-3.

Table 3-3

Hexadecimal as a shorthand notation for binary

Hexadecimal number	Binary pattern
0	0000
1	0001
2	0010
3	0011
4	0100
5	0101
6	0110
7	0111
8	1000
9	1001
A	1010
B	1011
C	1100
D	1101
E	1110
F	1111

Each hexadecimal character acts as a shorthand notation for 4 bits, resulting in a greatly condensed representation of information. Instead of writing out the binary pattern 1101, we could use the hexadecimal character "D." Similarly, we could use the hexadecimal character 7 to represent the 4-bit pattern 0111. When you combine these two examples, you see that the hex pattern D7 is the equivalent shorthand notation for the binary pattern 11010111.

Problem—Convert the following binary stream into hexadecimal shorthand:

$1111101010110001_{(2)}$

To make this conversion, break up the binary bit stream into groups of four, starting with the rightmost bit (the least significant bit). Next, convert each bit grouping into its hexadecimal equivalent, as follows:

0001 = 1
1011 = B
1010 = A
1111 = F

Next, place the resulting hexadecimal numbers in their original sequential order: FAB1.

The binary stream has been significantly shortened by condensing it to its hex equivalent. If the binary sequence does not contain a number of bits that you can evenly multiply by four, simply add one, two, or three 0s in front so the final group of bits totals four.

Converting back from hexadecimal into binary simply requires the reverse process of changing each hex number into its associated binary pattern and then placing the resulting binary groups in sequential order. For example, to convert the hexadecimal number 2D into binary, note that the hex number 2 equals the binary pattern 0010 and the hex number D equals the binary pattern 1101. Placing the binary quartets in sequential order produces the binary pattern 00101101.

Table 3-4 illustrates the representations of 16 numbers in all the numbering systems discussed in this chapter. Observe that the hexadecimal version provides the shortest representation and the binary version is the longest.

Table 3-4

Comparison of numbering systems

Hexadecimal number	Decimal number	Octal number	Binary number
0	0	0	0000
1	1	1	0001
2	2	2	0010
3	3	3	0011
4	4	4	0100
5	5	5	0101
6	6	6	0110
7	7	7	0111
8	8	10	1000
9	9	11	1001
A	10	12	1010
B	11	13	1011
C	12	14	1100
D	13	15	1101
E	14	16	1110
F	15	17	1111

REAL-WORLD EXAMPLE OF HEXADECIMAL AS SHORTHAND NOTATION

Computing devices process, store, and transmit information in binary, producing long groups of bits that can be cumbersome for people to read or manage, even when expressed in the decimal numbering system. This section describes an example of using hexadecimal as a convenient shorthand in information technology.

Later chapters will discuss the equipment and software that comprise local area networks (LANs), an information network that connects computing devices within a limited geographical vicinity. One component of a LAN is a computer's network interface card (NIC), such as an "Ethernet" card, that acts as a physical interface between the computer and a network. Within a LAN, the network requires a way to determine where to route information. Each NIC has a unique binary address, which is often 48 bits long. (This address is different from the binary IP address discussed previously.) Rather than writing 48 1s and 0s on each physical piece of equipment, administrators usually write the address in shorthand hexadecimal.

For example, suppose the following represents a 48-bit NIC address:

101000011111000001011011001010101100010000000001

We could write this address in shorthand hexadecimal notation by breaking it into groups of four and converting each group into its equivalent hexadecimal character, as follows:

1010 = A   0010 = 2
0001 = 1   1010 = A
1111 = F   1100 = C
0000 = 0   0100 = 4
0101 = 5   0000 = 0
1011 = B   0001 = 1

The hexadecimal shorthand representation of the 48-bit address is A1F05B2AC401, a much more compact description than 48 1s and 0s.

REPRESENTING TEXT AND OTHER CHARACTERS IN BINARY

We have seen how a decimal number can be converted into a binary number. What if we need to represent non-numeric information such as letters and other characters (the asterisk sign, for example) in binary code? This section describes how binary code can represent text and alphanumeric characters, primarily through two standards called ASCII and Unicode.

ASCII

Pronounced *ask-key*, **ASCII** stands for American Standard Code for Information Interchange. You can use ASCII to translate alphanumeric English text to binary in personal computers. Alphanumeric text includes letters, numbers, and other characters such as !, %, ", (, and &. Anything you can type on a keyboard may need to be translated into binary code that a computer can read. In other words, pressing the spacebar or typing the letter "m" must somehow translate into a group of ones and zeros (see Figure 3-13).

ASCII
American Standard Code for Information Interchange

A computer user enters characters into a computer via a keyboard. For example: "Hello!"

The computer translates each typed character into an 8-bit binary code. For example:
0100100001100101
0110110001101100
0110111100100001

Fortunately, there are standards for translating the alphanumeric characters that people understand into binary code. For example, the Extended ASCII standard assigns an 8-bit code for each alphanumeric character. (Note that the simple ASCII standard only assigns a 7-bit code.) Recall that an 8-bit code can represent 2^8, or 256, unique items. This number is sufficient to represent the 26 lowercase characters in the English alphabet, the 26 uppercase characters, the numbers 0 to 9, punctuation marks, and a host of command characters that can be entered via a keyboard. Table 3-5 lists the 8-bit binary codes assigned to numbers, lowercase and uppercase letters, and an assortment of other characters.

Table 3-5 also illustrates how we could represent a word such as "Hello!" in binary. Looking up each character in the Extended ASCII chart produces the following:

H = 01001000
e = 01100101
l = 01101100
l = 01101100
o = 01101111
! = 00100001

We can see that "Hello!" = 010010000110010101101100011011000110111100100001. The hexadecimal shorthand for this binary sequence is 48 65 6C 6C 6F 21.

Note that the binary code for "H" represents an uppercase letter. As shown in Table 3-5, the binary representation of a lowercase "h" is different from "H."

The binary representation of "Hello!" is readable to a computer, which can store, process, or transmit the binary code as appropriate. Because the binary code is a long and cumbersome sequence of ones and zeros, people can use hexadecimal to make the sequence more readable. Table 3-5 also lists the hexadecimal and decimal code associated with each binary octet (groupings of 8 bits).

Table 3-5

Extended ASCII chart

Character Name	Char	Decimal	Binary	Hex
Space		32	00100000	20
Exclamation point	!	33	00100001	21
Double quote	"	34	00100010	22
Pound/number sign	#	35	00100011	23
Dollar sign	$	36	00100100	24
Percent sign	%	37	00100101	25
Ampersand	&	38	00100110	26
Single quote	'	39	00100111	27
Left parenthesis	(	40	00101000	28
Right parenthesis	)	41	00101001	29
Asterisk	*	42	00101010	2A
Plus sign	+	43	00101011	2B
Comma	,	44	00101100	2C
Hyphen/minus sign	-	45	00101101	2D
Period	.	46	00101110	2E
Forward slash	/	47	00101111	2F
Zero	0	48	00110000	30
One	1	49	00110001	31
Two	2	50	00110010	32
Three	3	51	00110011	33
Four	4	52	00110100	34

Table 3-5

Extended ASCII chart
(continued)

Character Name	Char	Decimal	Binary	Hex
Five	5	53	00110101	35
Six	6	54	00110110	36
Seven	7	55	00110111	37
Eight	8	56	00111000	38
Nine	9	57	00111001	39
Colon	:	58	00111010	3A
Semicolon	;	59	00111011	3B
Less-than sign	←	60	00111100	3C
Equals sign	=	61	00111101	3D
Greater-than sign	→	62	00111110	3E
Question mark	?	63	00111111	3F
At sign	@	64	01000000	40
Capital A	A	65	01000001	41
Capital B	B	66	01000010	42
Capital C	C	67	01000011	43
Capital D	D	68	01000100	44
Capital E	E	69	01000101	45
Capital F	F	70	01000110	46
Capital G	G	71	01000111	47
Capital H	H	72	01001000	48
Capital I	I	73	01001001	49
Capital J	J	74	01001010	4A
Capital K	K	75	01001011	4B
Capital L	L	76	01001100	4C
Capital M	M	77	01001101	4D
Capital N	N	78	01001110	4E
Capital O	O	79	01001111	4F
Capital P	P	80	01010000	50
Capital Q	Q	81	01010001	51
Capital R	R	82	01010010	52
Capital S	S	83	01010011	53
Capital T	T	84	01010100	54
Capital U	U	85	01010101	55
Capital V	V	86	01010110	56
Capital W	W	87	01010111	57
Capital X	X	88	01011000	58
Capital Y	Y	89	01011001	59
Capital Z	Z	90	01011010	5A
Left bracket	[	91	01011011	5B
Backward slash	\	92	01011100	5C

Table 3-5

Extended ASCII chart
(continued)

Character Name	Char	Decimal	Binary	Hex
Right bracket	]	93	01011101	5D
Caret	^	94	01011110	5E
Underscore	_	95	01011111	5F
Back quote	`	96	01100000	60
Lowercase A	a	97	01100001	61
Lowercase B	b	98	01100010	62
Lowercase C	c	99	01100011	63
Lowercase D	d	100	01100100	64
Lowercase E	e	101	01100101	65
Lowercase F	f	102	01100110	66
Lowercase G	g	103	01100111	67
Lowercase H	h	104	01101000	68
Lowercase I	I	105	01101001	69
Lowercase J	j	106	01101010	6A
Lowercase K	k	107	01101011	6B
Lowercase L	l	108	01101100	6C
Lowercase M	m	109	01101101	6D
Lowercase N	n	110	01101110	6E
Lowercase O	o	111	01101111	6F
Lowercase P	p	112	01110000	70
Lowercase Q	q	113	01110001	71
Lowercase R	r	114	01110010	72
Lowercase S	s	115	01110011	73
Lowercase T	t	116	01110100	74
Lowercase U	u	117	01110101	75
Lowercase V	v	118	01110110	76
Lowercase W	w	119	01110111	77
Lowercase X	x	120	01111000	78
Lowercase Y	y	121	01111001	79
Lowercase Z	z	122	01111010	7A

UNICODE

Unicode is an important standard that uses 16 bits, allowing for a representation of 2^{16} (more than 65,000) unique characters. Many technology platforms, standards, and applications have adopted Unicode because it provides sufficient characters to encode languages such as English, Arabic, and Chinese that use different character sets. Unicode charts are available at *http://unicode.org*.

EBCDIC

ASCII and Unicode are not the only standard codes used in IT. Extended Binary Coded Decimal Interchange Code (**EBCDIC**) is a standard associated with IBM computers. EBCDIC assigns eight bits per character and is an extension of binary coded decimal.

CHAPTER SUMMARY

- Combinations of binary digits (bits) can encode any type of information, including the decimal numbers that we use in everyday life and alphanumeric text. You can convert decimal numbers into binary, including using binary coded decimal. A real-world example of binary to decimal conversion is a unique Internet address, which is encoded in dotted decimal format to make a long binary string easier for people to read.
- You can also represent negative integers and positive noninteger numbers in binary.
- Other IT numbering systems are octal, which uses 8 numbers, and hexadecimal, which uses 16 numbers. These alternative numbering systems are not used by digital devices, but by people as a shorthand convention to more easily read a binary number. For example, the addressing of devices connected to a LAN is usually represented in hexadecimal to shorten the number considerably.
- Binary code can also represent text and alphanumeric characters through standard conventions such as ASCII, Unicode, and EBCDIC. Anything you can type on a keyboard can be translated into a binary code that computers can read. Later chapters will extend this discussion to describe how images, voice communications, music, and video can be encoded in binary and stored, digitally manipulated, compressed, or exchanged over a communications network.

KEY TERMS

2's complement notation	decimal numbering system
ASCII	decimal point
base 2	dotted decimal format
base 8	EBCDIC
base 10	hexadecimal
base 16	octal
binary coded decimal	step size
binary numbering system	Unicode
binary point	

REVIEW QUESTIONS

1. Convert the 8-bit binary number $01110011_{(2)}$ into its decimal equivalent.
2. Convert the decimal number $100_{(10)}$ into binary representation.
3. Convert the decimal number $130_{(10)}$ into binary representation.
4. Convert the decimal number $843_{(10)}$ into binary coded decimal (BCD) representation.
5. Convert the BCD representation 011101011001 into decimal.
6. Convert the decimal number $68.09375_{(10)}$ into its binary equivalent.
7. Convert the binary number $10110101.1011_{(2)}$ into its decimal equivalent.
8. Express -9 in 8-bit 2's complement notation.
9. Why are the hexadecimal and octal numbering systems used in IT?
10. Count to 30 in the octal numbering system.
11. Count to 30 in the hexadecimal numbering system.
12. Name a real-world example of using hexadecimal in IT.
13. Convert the binary code $100110101011_{(2)}$ into hexadecimal.
14. Convert the binary code $0011110100100010_{(2)}$ into hexadecimal.
15. Convert the binary code $001010011111_{(2)}$ into octal.
16. Convert the octal number $237_{(8)}$ into decimal.
17. Convert the hexadecimal number $9B_{(16)}$ into decimal.
18. Convert the hexadecimal number $A2B_{(16)}$ into binary.
19. Using the Extended ASCII character code chart, convert the alphanumeric text "IT 101" into binary. Remember to take into account the space between IT and 101.
20. What is the binary equivalent of the decimal number $56.5_{(10)}$?
21. Convert the 32-bit binary Internet address 00100110011101011000100010101111 into dotted decimal format.

DISCUSSION QUESTIONS

1. Why do human beings use a numbering system with 10 digits? Can you find any examples of cultures that use a different numbering system?

2. Explain why it is helpful for IT professionals to understand binary representation. Provide examples of IT professions that might not benefit from this information.

3. What is the purpose of octal and hexadecimal?

4. What examples of binary or hexadecimal representation might you encounter in your everyday computer usage?

5. Why is it desirable to use a small step size, given a range of several values that need to be converted to binary? What is the trade-off of using a smaller step size?

CASE PROJECT

Every domain name (for example, *myspace.com*) has an associated numeric IP address. In this project, you determine the IP addresses of a Web site you visit frequently. Do the following:

1. Select a Web site you visit on a regular basis.

2. Look up the IP address of the Web site.
   If you use a PC, you can use the command prompt in Windows. Select Start in the lower-left corner of your desktop, select Run, and then enter "cmd" in the Open box. At the prompt, type "nslookup" followed by the Web site name (for example, nslookup course.com). Your computer will return the Web site's IP address in binary coded decimal.

   If you use a Macintosh, open the Applications folder and then open the Utilities folder. Click the Network Utilities folder and go to the "lookup link." Type the domain name and press the lookup key to return the IP address.

3. Convert the dotted decimal address you found into its corresponding 32-bit binary IP address.

PART
2

FUNDAMENTALS OF COMPUTERS

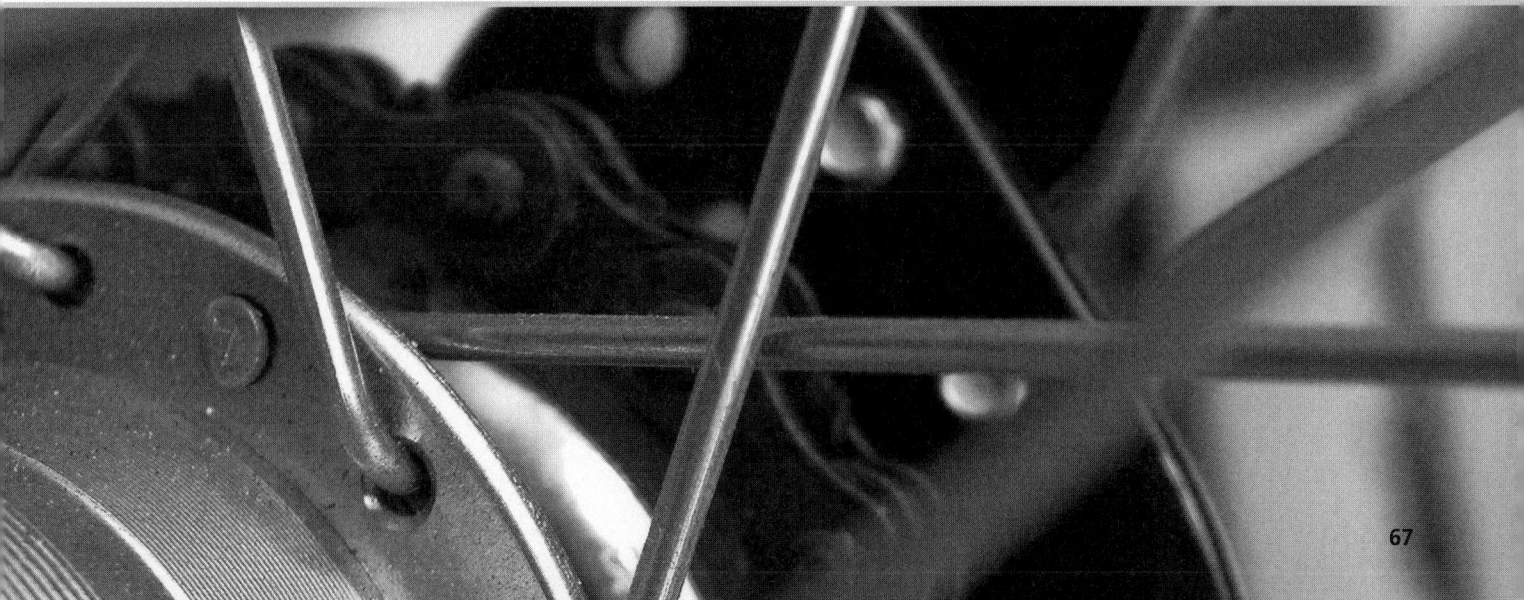

COMPUTER HARDWARE

In this chapter you will:

- Identify some important historical milestones in the development of computers

- Understand logic gates and how computers use them to process information

- Identify the fundamental components of a computer

- Understand how computer performance depends on factors such as processor speed, chip set, bus width, bus speed, number of CPUs, and instruction set

- Examine future trends in computing and societal issues related to these advances

- Understand the physical principles of data storage and the difference between mechanical, magnetic, optical, and electronic storage

INTRODUCTION

Computers permeate every facet of modern life, but they were practically nonexistent little more than half a century ago. Although the notion of using devices for computation spans centuries, a fully electronic digital computer was not demonstrated until the mid-1900s. Until the 1980s, only large institutions such as universities, businesses, and government agencies could afford computers. Most people understand how to use computers but have little understanding of their inner workings. This chapter simplifies the seemingly complex workings of computing devices by explaining how transistors can be combined to create logic gates, the basic building blocks for integrated circuits.

All computers contain a few basic components, such as a central processing unit (CPU), main memory, storage, input/output devices, and an interconnection system. This chapter describes such components and introduces some future computing trends. The final section addresses storage technologies, which enable governments, businesses, and individual users to store and access massive amounts of digital information. It describes the evolution of storage media, including magnetic, optical, and electronic storage; explains how storage technologies work; and describes future possibilities for meeting ever-increasing demands for digital storage.

If you consider a "computer" to be any machine that counts or makes mathematical calculations, then computing devices trace back at least to the abacus, an ancient calculation tool made of stone or wood with beads that slid along wires. As humans began to comprehend the advantages of using mechanical devices for computation, they developed increasingly sophisticated tools. Several mechanical calculation tools were invented in the 1600s. William Oughtred invented the slide rule to implement arithmetic calculations, including multiplication and logarithms. Other mechanical devices included Blaise Pascal's arithmetic machine, which could perform operations such as addition and subtraction, and Gottfried Leibniz's step reckoner, which could take square roots and perform other arithmetic calculations.

In the early 1800s, British mathematician Charles Babbage conceived of a mechanical calculating machine that would automatically perform repetitive computations. He named his conceptual device "the difference engine." He began prototyping the design in 1822 and subsequently developed a new idea called the analytical engine, which was more sophisticated and complex. The analytical engine was a programmable, general-purpose computing machine that relied on the use of punch cards—small cards with holes punched in them to input instructions and data. While Babbage was developing his computers, Augusta Ada Byron, the Countess of Loveless, worked with him in programming the analytical engine. Babbage is often called "the father of modern computing" because his design included many of the elements of modern computing, including memory, a control unit, and input/output devices (all of which are discussed in this chapter). Similarly, Ada Byron is often called "the first computer programmer."

In the late nineteenth century, a rapid influx of immigrants swelled the population of the United States, and the government sought a way to more efficiently tally and manage the enormous amount of census information. A U.S. Census Bureau employee named Herman Hollerith invented an automated computation device, or tabulator, that used punch cards to store census information. His computational device could read the information stored on the cards and process it in a very short amount of time. The punch card concept was first used by Joseph Marie Jacquard in the early 1800s for weaving patterns onto a fabric. Hollerith went on to start his own business, which eventually became International Business Machines (IBM).

In the 1940s, World War II spurred major advances in computing. John Vincent Atanasoff and Clifford Berry, two scientists at Iowa State College, built the Atanasoff-Berry Computer (ABC), regarded as the first automatic digital electronic computer. A German engineer named Konrad Zuse built the Z3, which was considered the first digital, programmable, general-purpose computer. IBM scientist Howard Aiken built the Mark I, a fully automatic electro-mechanical computer that was later programmed by Grace Murray Hopper, a prominent figure in programming history. British scientists invented a digital electronic computer named Colossus to help decode German encryption algorithms. Colossus was so secret that it wasn't included in historical accounts of computing for many years.

In the United States, the military needed to calculate artillery firing tables for ballistic missiles during World War II. In response, John Mauchly and J. Presper Eckert at the University of Pennsylvania developed the now-famous ENIAC (Electronic Numerical Integrator and Computer), an enormous electronic computer that used more than 17,000 vacuum tubes and could be manually programmed by moving dials and cables. The vacuum tubes took up a great deal of space and failed often, requiring constant maintenance and replacement.

The inventions of the transistor and integrated circuits in the mid-1900s were great advancements over vacuum tube technology, and they laid the groundwork for the next generation of digital, electronic computers. (These inventions were described in Chapter 2.) Computers made by IBM and Digital Equipment Corporation (DEC) pervaded business environments, but in the 1970s, computers began to enter the mainstream of society after *Popular Electronics* ran a feature story about a home "do-it-yourself" computer kit called the Altair. Shortly thereafter, Radio Shack began to sell the TRS-80 home computer, and Apple Computer Inc. introduced its own line of home computers. These advances, and the introduction of IBM's personal computer (PC) in 1981, contributed to the explosion of computers in all facets of modern life.

A problem still existed, however: computers made by different manufacturers could not communicate with each other at first. The proliferation of computer networks, based on open communications standards such as the Internet Protocol (IP), enabled interoperability between previously disparate computing platforms. A variety of social and economic factors, along with advances in microprocessors and distributed networks, led to the introduction of mobile, wireless, and handheld computers.

DIGITAL LOGIC

As you should know by now, computers are made up of small switches called transistors. These basic switching elements enable complex operations such as addition, subtraction, comparison, word processing, and image editing. A computer can perform any type of processing if it has enough transistor-based electronic circuits and the transistors are switched in an organized manner. To create these integrated circuits, or chips, numerous transistors are laid out on a common substrate made of silicon, interconnected depending on the preferred functionality of the circuit using thin metallic wires, and packaged within a carrier, such as a ceramic casing with metal pins, to connect the integrated circuit to other components. The transistors in the chips can then be switched on and off in a particular manner to perform the needed processing. A computer has one or more main chips called the **microprocessor**, which performs most of the processing, and several other supporting chips.

Note that the design of integrated circuits does not require an engineer to select individual transistors, lay them out on the substrate, and interconnect them. Instead, the designer employs software and uses basic building blocks called **logic gates**, which contain a few transistors that are already interconnected. These basic building blocks can be combined to create successively larger building blocks, which may subsequently be combined to create complex circuitry and packaged with other components within a carrier to create integrated circuits such as microprocessors. Figure 4-1 illustrates a conceptual diagram of this approach and a packaged chip.

Figure 4-1

Logic gates as building blocks for integrated circuits

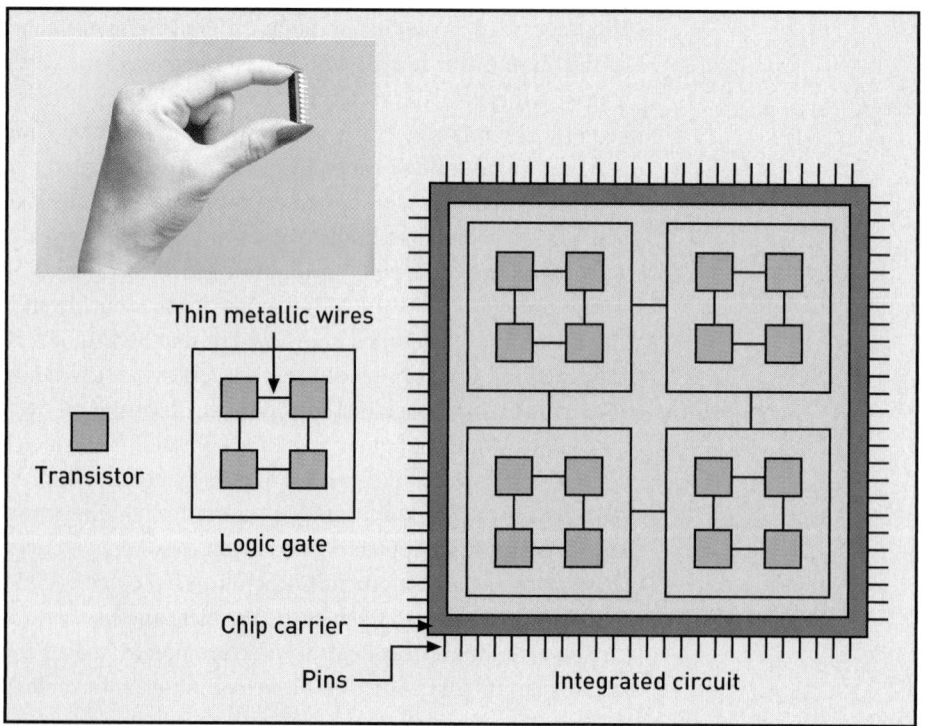

Thin metallic wires

Transistor

Logic gate

Chip carrier

Pins

Integrated circuit

Integrated circuits are built using the following types of logic gates. The difference between these logic gates is essentially the way the transistors within them are connected to each other.

- **NOT gate (inverter)**
- **AND gate**
- **NAND (not AND) gate**
- **OR gate**
- **NOR (not OR) gate**
- **Exclusive OR (XOR) gate**
- **XNOR (exclusive NOR) gate**

With the exception of the NOT gate, each gate is a component with two inputs and one output. The on-and-off switching of transistors within the logic gate produces output that varies depending on the type of gate. For example, if the two inputs to an AND gate are 0s, then the output of the AND gate is a 0. If the two inputs are 1s, then the output of the AND gate is 1. The functionality of each gate is described by a **truth table**, which depicts all of the possible inputs to the gate and all the possible outputs. A truth table is based on **Boolean logic**. For example, according to Boolean logic, the output of an AND gate is always a 1 when the two inputs are 1s; otherwise, the output is 0. Each type of gate has a special associated symbol to distinguish it from the other gates. Figure 4-2 illustrates these symbols, and Tables 4-1 through 4-4 depict the truth table for each type of gate.

Figure 4-2

Symbols representing each type of logic gate

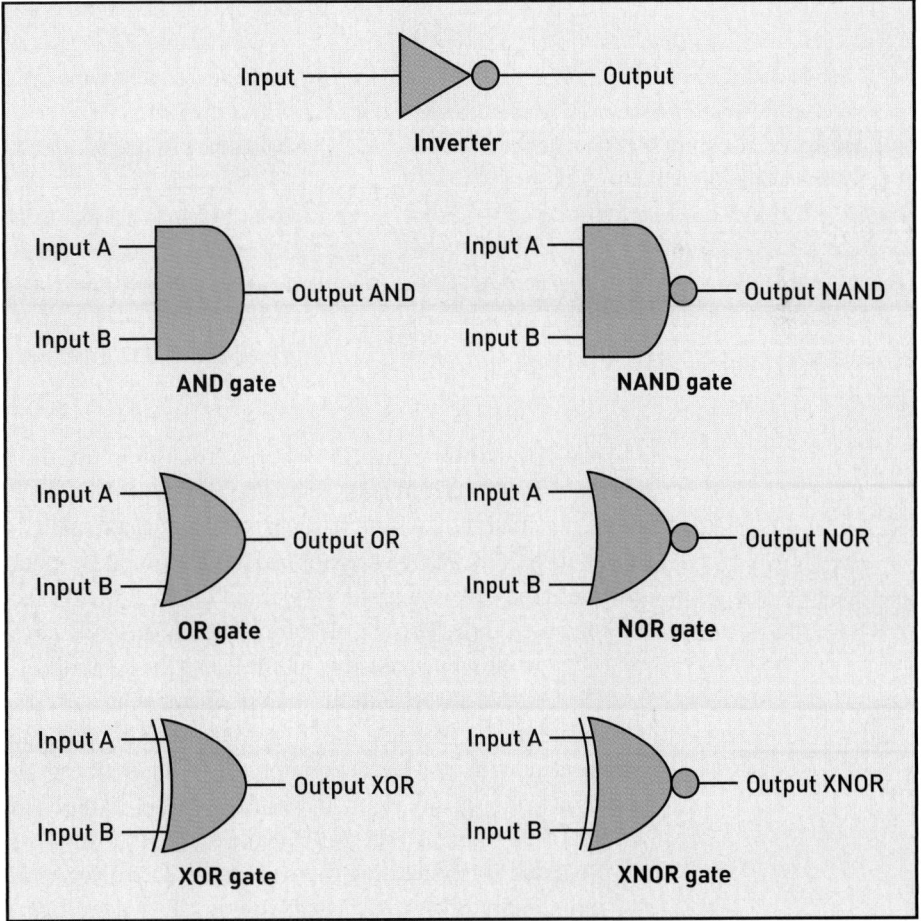

Table 4-1

Truth table for a NOT gate

Type of gate	Input	Output
NOT gate	0	1
	1	0

Table 4-2

Truth table for an AND and NAND gate

Type of gate	Input A	Input B	Output AND	Output NAND
AND/NAND gate	0	0	0	1
	0	1	0	1
	1	0	0	1
	1	1	1	0

Table 4-3

Truth table for an OR and NOR gate

Type of gate	Input A	Input B	Output OR	Output NOR
OR/NOR gate	0	0	0	1
	0	1	1	0
	1	0	1	0
	1	1	1	0

Table 4-4

Truth table for an XOR and XNOR gate

Type of gate	Input A	Input B	Output XOR	Output XNOR
XOR/XNOR gate	0	0	0	1
	0	1	1	0
	1	0	1	0
	1	1	0	1

The truth tables depict the value of the input and output of each gate. A NOT gate, for example, merely outputs the opposite value of the input. In other words, it inverts, or complements, the input, which is why a NOT gate is also called an **inverter**. If a 0 is applied at its input, it outputs a 1; if a 1 is applied at its input, it outputs a value of 0. The rest of the gates require at least two inputs. As explained before, the AND gate always outputs a 0 unless both input values are 1. The NAND gate is merely an AND gate with an inverter at its output, as depicted by the circle (called a bubble) after the symbol.

The OR gate always outputs a 1 unless both values are 0, in which case it outputs a 0. The NOR gate is an OR gate with an inverted output, meaning that it outputs a 0 unless both inputs are a 0, in which case it outputs a 1. The XOR gate only outputs a value of 1 if the two inputs are of opposite value. If input A is 1, for example, and input B is 0, the output of the XOR gate is 1. Similarly, the XNOR gate outputs the opposite of the XOR gate.

Note that implementing these operations involves actual voltage levels. For example, a 1 might be input to the logic gate by physically generating a signal with a voltage level of 5 V, and a 0 may be represented physically by a voltage level of 0 V. The on-and-off switching of the transistors produces the physical output signal, which may be a voltage of 5 V for a 1 and 0 V for a 0.

Problem—Find the output of the logic gates depicted in Figure 4-3.

Figure 4-3

Two types of logic gates

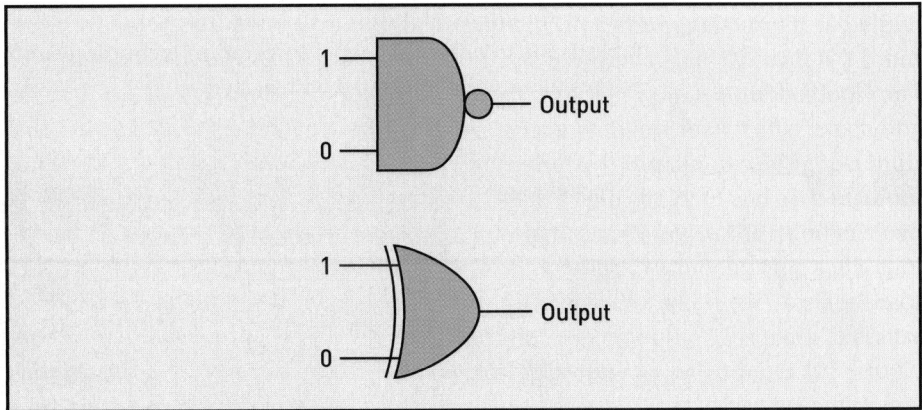

Based on the symbols, you can infer that the upper gate is a NAND gate and that the lower one is an XOR gate. Based on the truth table, the output for the NAND gate is 1, given that one input is 1 and the other is 0. The output of the XOR gate is 1, given that one input is 1 and the other is 0.

As mentioned before, by selectively combining logic gates you can perform any type of operation. The combinational circuit shown in Figure 4-4 implements the addition operation on two 1-bit binary numbers. (The rules for binary addition were discussed in Chapter 3.) Similarly, more complex combinations of logic gates can be used to perform more complex processing operations.

Figure 4-4

Combining logic gates to create a 1-bit binary adder

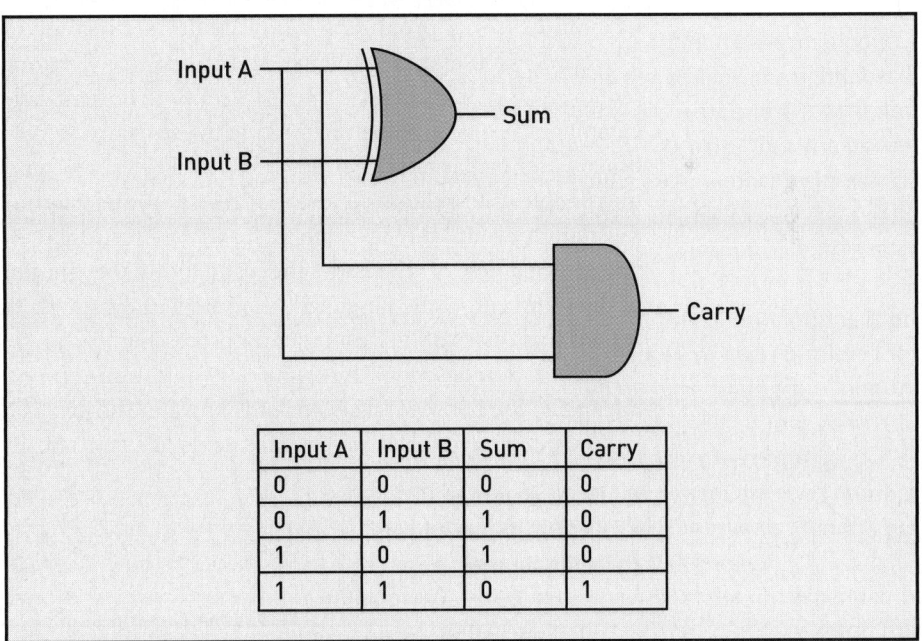

Input A	Input B	Sum	Carry
0	0	0	0
0	1	1	0
1	0	1	0
1	1	0	1

The simple arrangement in Figure 4-4 can only perform addition on two 1-bit numbers. Adding numbers with 32 or 64 bits, which is what modern computers implement, requires more complex circuitry with several more gates—hence the need for millions of transistors and other components. The sophistication and complexity of an integrated circuit is proportional to the number of transistors and other components that can be integrated on the chip. The maximum number of transistors and other components that are manufactured on a single integrated circuit depends on the fabrication technology. When integrated circuits were first put into fabrication, only a few transistors could fit on a single chip. Over the years, this number grew dramatically as chip fabrication techniques and materials became more sophisticated. Hundreds of millions or more transistors can now fit on a single integrated circuit. Table 4-5 lists various levels of integration technology and a rough range of the number of devices that can be integrated with each level.[1]

Table 4-5

Integration technologies

Technology	Number of devices
Small Scale Integration (SSI)	10–100
Medium Scale Integration (MSI)	100–1,000
Large Scale Integration (LSI)	1,000–100,000
Very Large Scale Integration (VLSI)	100,000–1,000,000
Ultra Large Scale Integration (ULSI)	>1,000,000

FUNDAMENTAL COMPONENTS OF A COMPUTER

Computers consist of a few basic components. The components most familiar to users are the keyboard, mouse, and monitor because they provide the interaction between users and computers—in other words, the human-computer interface. Users enter instructions for the computer through an **input device**, often the keyboard. The computer converts these instructions into binary, processes them via transistors on a microprocessor and other integrated circuits, and sends the outcome through an **output device**, which is often a monitor.

Computer **hardware** is any physical device associated with a computer, such as a keyboard or monitor. Other prominent hardware devices include the printer, mouse, motherboard, microprocessor, graphics adapter, sound adapter, network adapter, hard disk, and the CD (compact disc) drive. Besides the instructions that users issue through input hardware devices, computers also receive instructions via the **software** stored in its **memory**. Software is traditionally defined as a series of instructions written by computer programmers in a language that people can understand and that the computer can translate into binary. The computer must interpret and then execute each software instruction, again through the use of its transistors. Regardless of who issues the instructions, they must ultimately be converted to a string of ones and zeros, the language computers understand. The ones and zeros are fed to the various logic gates of the computer for processing.

This section describes the elementary hardware components of a computer and explains the interaction between hardware and software. The next chapter describes software in detail.

The fundamental hardware components of a typical computer include the following:

- Input/output (I/O) devices
- Central processing unit (CPU)
- Main memory
- Storage (secondary memory)
- Interconnection system

A diagram of these components is shown in Figure 4-5.

---

[1] The line between VLSI and ULSI technology is usually a blurry one. Table 4-5 depicts only rough counts of the number of devices that may be integrated using each technology.

Figure 4-5

The fundamental
components of a
computer

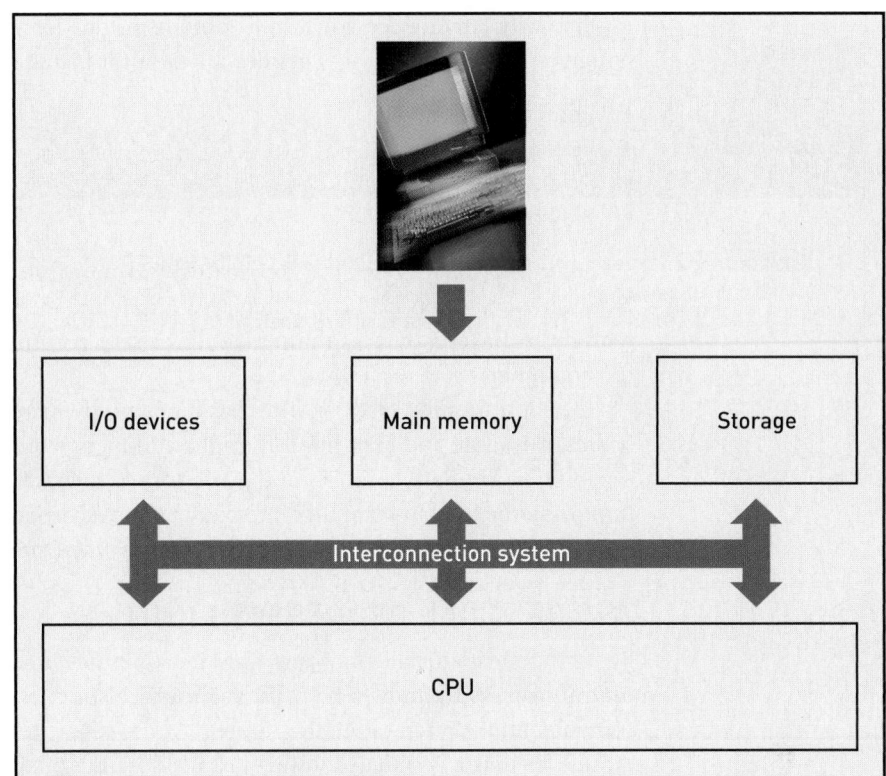

Although I/O devices provide the means for the human-computer interface, the most crucial component of a computer is the CPU, commonly called the microprocessor. The CPU receives instructions from the I/O devices and then interprets and executes them. For example, you enter the instruction to add two numbers by pressing the number keys and the plus sign on the keyboard. These instructions are converted to instructions that the CPU understands. The CPU interprets these instructions, executes them to generate the result of the addition, and sends the result to the monitor for display. The CPU also receives instructions from software that resides in the main memory of the computer. The storage units of the computer, such as the hard disk, contain software and data that a user might need to retrieve later. Hardware components that are not vital to a computer's operation, but are attached to the computer to enhance its capabilities, are called **peripherals**. Peripherals include I/O devices such as a microphone, speaker, or printer. Besides the elementary items described earlier, a typical computer has many other types of associated hardware. In some texts, storage devices are not considered a basic computer component because a computer can operate without storage, but we include them here for the sake of completeness.

INPUT/OUTPUT DEVICES

Input and output devices are hardware components that interface a computer to the outside world. Any device that issues input to the computer is classified as an input device. Other input devices besides the keyboard include the following:

- Mice
- Trackballs
- Microphones
- Web cams
- Styluses
- Touch pads
- Pointing sticks
- Joysticks
- Scanners

Similarly, any device you use to obtain output from the computer is classified as an output device. Output devices besides the monitor include the following:

- Printers
- Plotters
- Speakers
- Projectors

Some I/O devices, such as keyboards and monitors, are integral parts of computers. However, note that not all computers have input and output devices connected to them. For example, some computers can interface to a user over a network and do not need their own I/O equipment.

The sophistication of I/O devices has dramatically increased over the years. Computer keyboards, earpieces, and mice have all improved in terms of their size, ergonomics, and wireless capabilities. Similarly, printers and projectors have better resolution and color capabilities. Improving the human-computer interface enhances corporate efficiency and productivity, and is therefore a sustained goal of companies that manufacture computer components.

THE CENTRAL PROCESSING UNIT

By far the most crucial component of any computer is a chip called the microprocessor or **central processing unit (CPU)**. The microprocessor controls the major functionality of the computer and is often called the "brains" of the computer. The microprocessor's many tasks include fetching a wealth of software instructions, interpreting and executing these instructions, and storing and outputting results. The microprocessor chip is made up of several parts, and all of them help control a critical process. The main units of the microprocessor are the **control unit**, **arithmetic logic unit (ALU)**, **registers**, and **cache memory**, as shown in Figure 4-6.

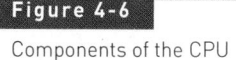

Figure 4-6

Components of the CPU

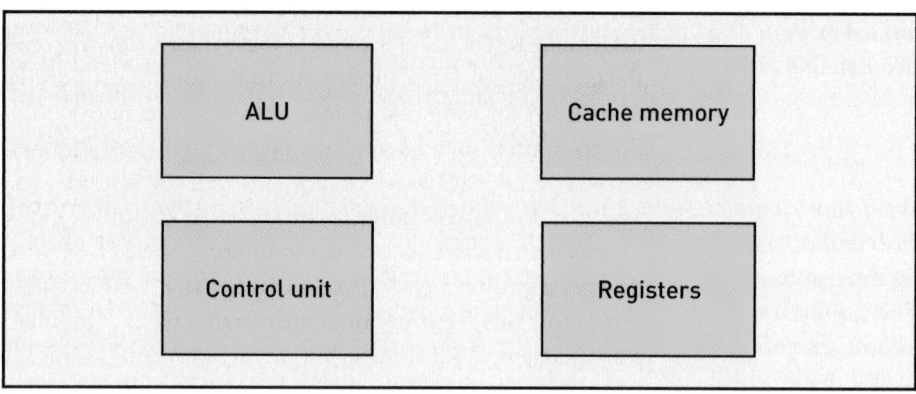

The key component of the CPU is the control unit, where instructions are interpreted and control signals are generated. Based on the result of the interpretation, the control unit supplies crucial control signals to other parts of the computer. For example, if the instruction is to display an image stored on the computer, the control unit issues the necessary control signals to components associated with the display device. If the instruction involves the addition of two numbers, then the control unit issues control signals to a unit called the ALU, which is responsible for arithmetic operations. This unit contains a binary adder circuit like the one in Figure 4-4, except that the circuit is more complex and has several logic gates. A CPU can have more than one ALU. Besides addition, the ALU also performs logical operations such as comparing two numbers and determining which is larger.

The registers within the CPU are small areas for temporarily storing information, such as instructions that are supplied to the control unit and the results of addition and comparison operations implemented by the ALU. The results that the computer needs to output are then retrieved from these registers and sent to the appropriate output device.

Software instructions that the CPU interprets and data that it processes are fetched from a module called main memory that resides outside the CPU. Software programs such as Microsoft Word and Adobe Acrobat contain thousands of instructions that are loaded into main memory from a hard disk or other storage device when a user starts the software.

Unlike main memory, cache memory is internal to the CPU. Cache memory holds frequently used instructions and data. Because some software instructions and data are needed more often than others by the CPU, they can be stored inside the cache memory and retrieved when necessary instead of having to be fetched from the distant main memory every time they are needed. Saving frequently used instructions and data in close proximity saves considerable time and energy. Note also that many computer architectures have cache memory residing outside the CPU; however, they are not discussed in this text.

Numerous manufacturers produce microprocessors, including Intel, Advanced Micro Devices (AMD), Motorola, Via Technologies, and Sun Microsystems. One of the first manufactured microprocessors was the 4004 by Intel in 1971. Microprocessors can differ significantly in many respects. Some microprocessors on the current market include AMD Athlon, Intel Pentium, and Intel Core models.

MAIN MEMORY

As discussed before, main memory is external to the CPU and stores instructions and data, which are ultimately a string of ones and zeros. Memory devices are actually chips made of transistors and other components that can store bits corresponding to instructions and data. Bits are physically stored on a memory chip in the form of electrical charges.

The main memory of a computer consists of two types of memory: **random access memory (RAM)** and **read-only memory (ROM)**. RAM is much more prominent; it retains its contents as long as power is supplied to the computer and loses its contents when the power is switched off. RAM is also called **temporary memory** or **volatile memory**. ROM, on the other hand, retains its contents even after power to the computer is switched off. Therefore, it is called **permanent memory** or **nonvolatile memory**.

Essentially, two types of RAM are used to store information as long as power is supplied to the computer: **static RAM** (SRAM) and **dynamic RAM** (DRAM). Although both types of memory temporarily hold data, the difference between them is that the contents of dynamic RAM must be refreshed several times per second, because the electrical charges representing the bits of DRAM leak out and must be replenished. Alternately, there is no need to refresh the contents of SRAM. The disparity between these two types of RAM is due to the different components used to construct them. DRAMs include transistors and other electronic components called **capacitors** that store electricity, whereas SRAMs are based on transistors. Although capacitors can store electricity, they cannot do so indefinitely. Electricity eventually leaks out of them, which is why it is necessary to refresh the contents of DRAM.

Each type of RAM finds its application within the computer. Memory chips that correspond to the main memory are mostly DRAM, and the cache memory within the CPU is typically SRAM. Due to fundamental differences in how the two types of RAM are constructed, they differ in speed, size, and cost. It is faster to save and retrieve the contents of SRAM, but SRAM is also more expensive and larger than DRAM. Because cache memory has to be fast, it is better to use SRAM to build cache memory. DRAM is used within the main memory because it is the main holding area of the computer and must have a large holding capacity. Main memory also must be cheap to minimize the cost of the computer.

RAM chips retain ones and zeros that correspond to instructions and data within small areas called cells. Each RAM cell can hold one bit of information, regardless of whether it is DRAM or SRAM. Cells within RAM chips are arranged in rows and columns, as shown in Figure 4-7.

Figure 4-7

Organization of bits within RAM

Row/Column Address	000	001	010	011	100	101	110	111
000	0	0	1	1	1	1	0	0
001	1	0	1	1	1	1	1	1
010	0	0	1	1	1	1	0	0
011	0	0	1	1	1	1	0	0
100	0	0	0	1	0	1	0	0
101	0	0	1	1	1	1	0	0
110	0	0	0	0	0	0	0	0
111	0	0	1	1	1	1	0	0

Saving information to RAM is referred to as "writing to RAM," and retrieving RAM contents is called "reading the RAM." Information is written to and read from the RAM by addressing the RAM cells using **addresses**, which again are a combination of bits. Each row and each column within the RAM in Figure 4-7 has a specific address. The first row of the RAM, for example, has an address of 000. Similarly, the second row address is 001 and the first column address is 000. You can find the address of each cell by concatenating the address of the row with its corresponding column. The cell depicted in the bold box in Figure 4-7 stores a bit with a value of 1. This cell resides in row 101 and column 100 of the RAM. Hence, the address of this cell is 101100. The RAM depicted in this example can hold 64 bits of information corresponding to instructions and data.

Note also that each cell is not addressed individually—bits are stored within groups of bytes, and a group of cells is addressed while writing to and reading from RAM. Also, a RAM size of 64 bits is only a simplification used for illustrative purposes. RAM capacities that range from 256 MB to several gigabytes are typical in computers. This capacity is possible by incorporating multiple RAM chips within the computer. The RAM chips actually reside on a small board called a **memory module** (see Figure 4-8), and are connected to the CPU through this module, as we explain later.

Figure 4-8

RAM chips

Although a major part of the instructions that the computer executes are supplied from RAM, the computer also relies on some operations based on instructions stored in ROM chips. ROM chips and RAM chips physically reside at different places within a computer, but together they constitute the main memory of the computer. As mentioned before, ROM is permanent memory; when a computer is switched off, the ROM's contents are retained. When a user starts a software application such as Microsoft PowerPoint, instructions that correspond to the software are loaded into the RAM chips. When the user turns the computer off, the contents of RAM are erased. The next time the user switches the computer on, the user must start the program again to use it.

In some instances, it is necessary and practical to automatically load certain types of software into the RAM for the user. One example is the **operating system**—the software that provides the user with a friendly interface to the computer. The user should not have to load the operating system manually from secondary storage into RAM every time the computer is switched on. However, the computer must be instructed to locate and load the instructions that correspond to the operating system. This is accomplished using special instructions stored permanently on a ROM chip called the basic input/output system (**BIOS**) chip. The instructions stored on a computer's BIOS chip supply the vital steps for the computer to follow as it is switched on. These steps include verifying the proper operation of computer components and locating and loading the operating system, among many other actions.

As with RAM, there are different types of ROM chips. The content of some types can be erased and rewritten, although other types cannot. The major variations of ROMs have historically included the following:

- Programmable ROM (PROM)
- Erasable programmable ROM (EPROM)
- Electrically erasable programmable ROM (EEPROM)

Information on a PROM chip can be written only once; once it is written, the information cannot be erased, and the PROM chip cannot be reused for any other purpose. In contrast, information on an EPROM chip can be erased and written over and over again. However, the user must unplug the chip from the computer and expose it to ultraviolet light to erase its contents. Furthermore, once the contents are actually erased, the user must use a separate writing device such as an EPROM programmer to load contents into the EPROM. These chips are rarely used today, but they were common during the earlier days of computer manufacturing.

The more sophisticated and popular type of ROM is EEPROM. Its contents can be erased electronically without having to unplug the chip from the computer and without having to use ultraviolet light. An electrical signal applied to the EEPROM chip can permanently erase its contents. The contents of the EEPROM chip can be written directly through the computer or another device without having to use a separate writing apparatus.

A device that is similar to EEPROM is **flash memory**. Memory sticks, which are prominently used in digital cameras, are an example of flash memory. Flash memory has many applications in computing, including the BIOS chip within a computer's main memory. Flash memory is also a popular storage device for software and files, because data on a flash drive can be erased and saved quite easily.

STORAGE

Storage and retrieval of digital information is one of the most critical components of an IT system. Without storage, we would have no Web servers, electronic banking, e-mail, or almost any other digital application. See the final section of this chapter for an extensive discussion of storage technologies.

INTERCONNECTION SYSTEM

The physical system that connects I/O devices, main memory, CPU, storage, and other components is called the interconnection system. This system is actually a set of wires; they are often grouped together as a set of parallel wires that transfer signals corresponding to data, instructions, and control information in an arrangement called a **parallel bus**. Some connections between components, however, do not rely on a set of parallel wires. Instead, they use a main connection that transfers bits one after the other, or serially, from one point to another. Such connections are called a **serial bus**. A parallel bus connects main memory such as RAM to the CPU, whereas you can use a serial bus to connect a keyboard to the CPU.

Figure 4-9 illustrates the structures of parallel and serial buses.[2]

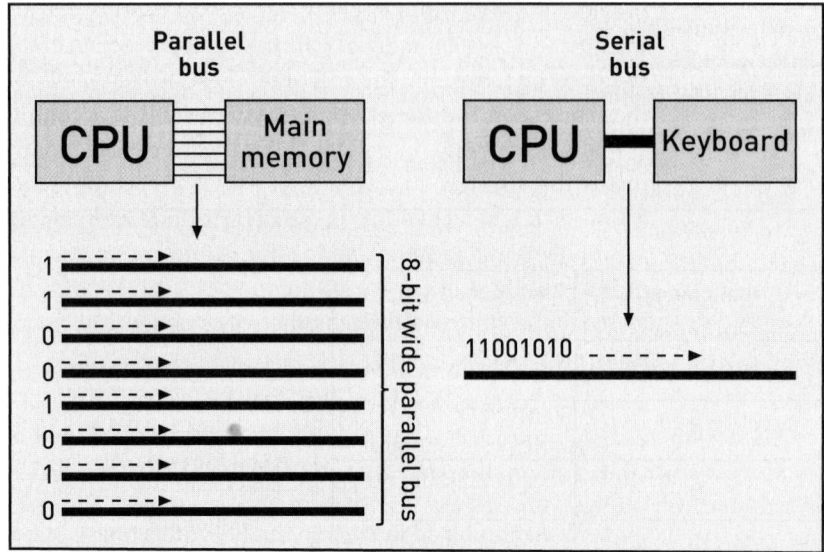

The number of parallel lines within a parallel bus and the transfer speed of bits across each line dictate how many overall bits the bus can carry. The more parallel wires there are, and the faster they can carry bits, the more bits they can carry from one component to another. Buses that can carry 8 bits in parallel are called 8-bit wide buses; those that can carry 64 bits in parallel are called 64-bit wide buses. The larger the **bus width**, the quicker the bits can reach their destination; an example would be the delivery of bits from the main memory to the CPU.

Not all buses carry the same type of information. Some buses carry only instructions and data, and others carry only binary addresses, such as the addresses of bytes of information stored in RAM. Other buses carry control signals between the CPU and the components connected to it.

Types of computer buses include the following:

- **System bus**—A parallel bus that connects the CPU and main memory. The system bus is also called the **front side bus**.
- **Peripheral Component Interconnect (PCI) bus** and **PCI Express (PCI-E) bus**—The PCI bus and its newer and faster version, PCI-E bus, connect **expansion cards** such as network interface cards and sound cards/adapters. Network interface cards are the circuitry that connects a computer to a network. Sound cards, also called sound adapters, are the circuitry that supports audio applications on other computer components.
- **Accelerated Graphics Port (AGP) bus**—This bus connects expansion cards called graphics cards/adapters to the CPU. This circuitry supports computer graphics capabilities.

---

[2] The actual implementation of a serial bus is not a single wire. The explanation is beyond the scope of this chapter.

Expansion cards are connected to the CPU by buses, as discussed before. They are attached to the buses by being plugged into special components called **expansion slots** on the computer's **motherboard**. The motherboard is discussed later in this chapter.

FACTORS THAT AFFECT COMPUTER PERFORMANCE

Although the processing power and speed of a computer are governed by physical aspects, such as the number of transistors and other components that can be integrated within the CPU, several fundamental design parameters play a key role in affecting computing performance for a given integration technology. These parameters affect the speed at which transistors can be switched on and off and the transfer speed of electrical signals between transistors. In other words, even if the CPU contains a large number of transistors that can be switched on and off at a very fast rate, it does not necessarily translate into high-performance computing. Some of the key factors that affect computer performance are shown in the following list and described in the following sections.[3]

- Word length
- Bus width and bus speed
- Memory size and memory access speed
- Processor speed
- Instruction set
- Number of CPUs
- Chip set

WORD LENGTH

Word length is expressed in terms of bits, and corresponds to the maximum number of bits of information that a computer can process at one time. The larger the word length is, the faster the computer. For example, consider adding two numbers in binary: 01 + 01 = 10. If a computer's CPU has been designed with a word length of 2 bits, it can perform the operation without any problems, producing the result in just one step in the ALU. On the other hand, if a 2-bit computer were asked to add two 8-bit numbers such as 10010101 + 00010101 = 10101010, it would need to perform the addition iteratively in the ALU in four steps. The computer would have to break the two 8-bit numbers into groups of two bits each and start by adding the least significant two bits, which would be the two bits on the far right (see Figure 4-10).

Figure 4-10

Addition of two 8-bit numbers with a 2-bit word length

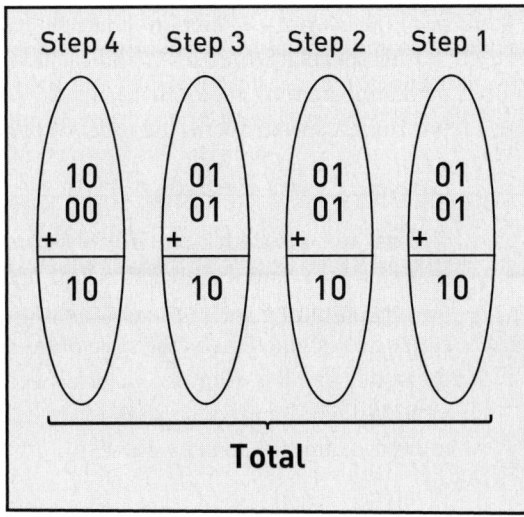

[3] *Understanding and Troubleshooting Your PC.* Shelly, Cashman, Andrews, Jedlicka, 2004. ISBN 0-619-20223-8.

This example clearly illustrates the impact of word length on the speed of the computer. Earlier computers were based on a word length of 8 bits, whereas modern computers have a word length of 32 bits and 64 bits. Engineers are constantly striving to design computers with larger word lengths to achieve high performance.

BUS WIDTH AND BUS SPEED

Another factor that affects performance is the bus size, or parallel bus width. Like word length, bus size is measured in terms of bits. A computer with a large bus width can carry more bits at a time between computer components, such as between main memory and the CPU, making it faster than a computer with a small bus width. Because the buses carry important instructions, data, addresses, and control signals, the speed you can gain by using a larger bus width is important.

The bus size determines how many physical wires are constructed within the bus. The more wires that designers can incorporate within a bus, the larger the bus width becomes, meaning that the bus can carry more bits at one time. Although large bus widths are desirable, they also occupy significant area within the computer, so their size must be carefully considered during the design phase, especially for portable devices.

Bus speed, which is measured in hertz (Hz), also affects computing performance. Even if large bus widths can enable large number of bits to move between computer components in parallel, each bus line must also carry bits as quickly as possible. However, because this speed corresponds to great power consumption in a computer, designers must again weigh the trade-offs. Depending on the type of bus, typical bus speeds range from hundreds of megahertz to gigahertz.

MEMORY SIZE AND MEMORY ACCESS SPEED

Computers need large amounts of memory for high performance and multitasking. As an example, think of the iTunes application stored on a computer's hard disk. The application is loaded into main memory (RAM) when you run it on your computer. Instructions that correspond to the software are supplied by the RAM to the CPU for interpretation. Because every program occupies some amount of space in RAM, a large amount of RAM is essential to operate with multiple programs. Furthermore, RAM size significantly affects computer speed because information is frequently written to and erased from RAM. If there is insufficient RAM space, the computer must frequently resort to using its hard disk, which results in diminished performance. Computers sold today typically include RAM that ranges from hundreds of megabytes to several gigabytes, and RAM can be expanded easily by incorporating extra RAM chips.

Other than RAM, the size of the cache memory (expressed in bytes) is also a consideration. The more cache memory the CPU has, the better the CPU performs.

The speed at which RAM and cache memory contents are written and retrieved helps to determine performance. The higher the read/write speed, the faster the computer is. Typical RAM access speeds are on the order of nanoseconds (ns), or billionths of a second.

PROCESSOR SPEED

The on and off switching of transistors within the integrated circuits of a computer is managed by a central digital electrical signal called a **clock**. The clock signal is akin to a conductor leading an orchestra, and the processing speed depends on the rate of this clock. The conductor essentially sets the pace of the music; if the conductor requests a faster tempo, the orchestra complies. Computer clock speeds are measured in frequency units called hertz. Just as multipliers are used in the IT world to express large numbers of bits, multipliers are used to express large frequency values:

> 1 kHz = 1000 Hz
> 1 MHz = 1,000,000 Hz
> 1 GHz = 1,000,000,000 Hz

The higher the clock frequency, the faster the computer usually is, although there are some exceptions. Typical clock frequencies for modern computers range from hundreds of megahertz to a few gigahertz. Examples of two clock signals are illustrated in Figure 4-11.

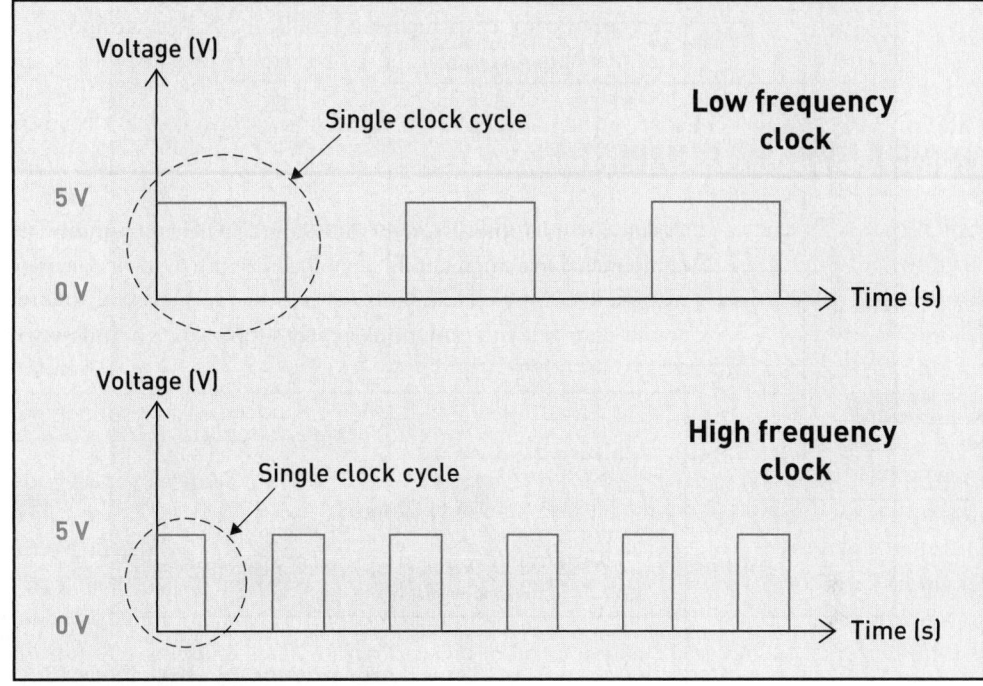

The clock signal is a digital electrical signal that periodically varies between two voltage values, as shown in Figure 4-11. A single clock cycle is defined as the time the signal goes from one value back to the same value—for example, from 5 V to 0 V back to 5 V. The rate of repetition of the clock cycle corresponds to the frequency. The clock signal in the upper portion of Figure 4-11 has a lower frequency than that of the lower signal because it completes a single clock cycle within a longer duration. The cycles of the upper clock do not repeat as often as the lower clock cycles.

INSTRUCTION SET

A computer's CPU is typically based on a **complex instruction set computer (CISC)** architecture or a **reduced instruction set computer (RISC)** architecture. The difference between them is the number of clock cycles it takes to execute a single instruction. Some computer architectures require more cycles to execute the same instruction than others. Engineers who design CPUs have to make complex design choices because each type of architecture has its own advantages and limitations.

NUMBER OF CPUS

Incorporating more than one microprocessor increases a computer's processing power and speed because it can use more transistors and perform multitasking. For example, a CPU with two ALUs can perform additions concurrently, resulting in faster processing. Some computers have dual-core processors and quad-core processors. In dual-core models, the functionality of two microprocessors is essentially integrated within the same microprocessor chip. In quad-core models, four microprocessors are integrated within the same chip. Besides **multicore computers**, **multiprocessor computers** are also available. These two types of computers have only one difference: a multicore computer includes only a single microprocessor chip with multiple integrated processors, and a multiprocessor computer has multiple, distinct microprocessor chips. Computers with multiple cores and multiprocessing capability are at the forefront of current technology.

CHIP SET

The term **chip set** refers to a group of chips that support the microprocessor by controlling the flow of information between it and other components, such as the memory chips, graphics and sound cards, disk drives, and I/O devices. Various manufacturers produce different types of chip sets, which can significantly affect computer performance, due to the limitations that chip sets may impose on memory size, number of processors, and bus speed.

INSIDE A TYPICAL COMPUTER

Besides the microprocessor, memory, and I/O devices, a modern computer contains many other components that augment its capabilities. Some of these components were introduced in earlier discussions. These components include the motherboard, DVD drive, hard drive, graphics card, sound card, network card, ports, power supply, and cooling system (see Figure 4-12).

Figure 4-12

Inside the modern computer

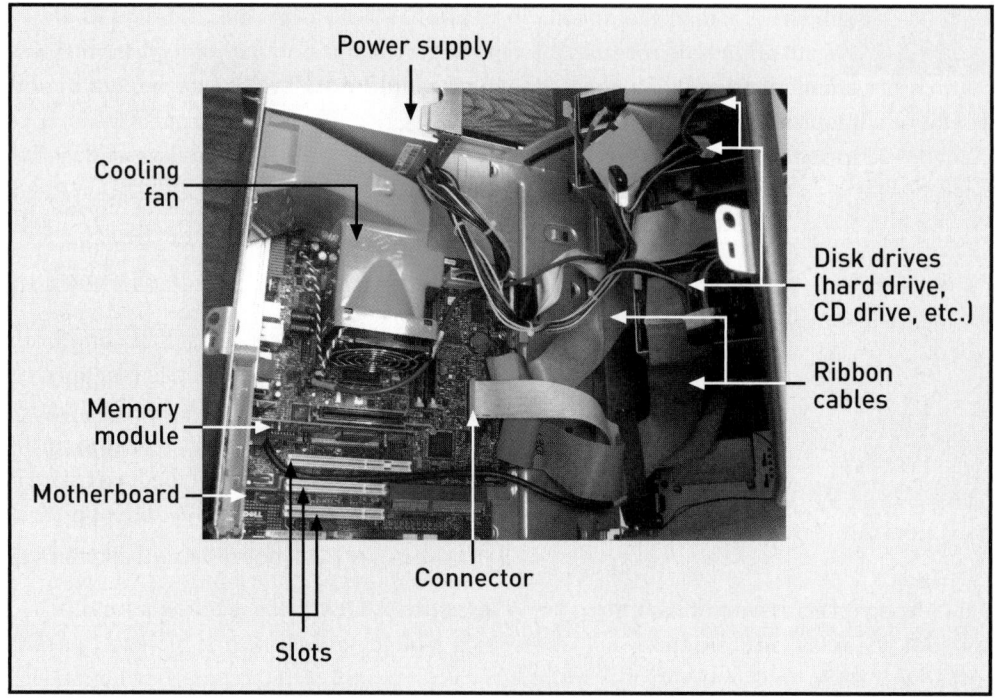

THE MOTHERBOARD

The motherboard is a **printed circuit board (PCB)**—a planar structure made of resin or other materials that contains the microprocessor and many other chips. It also includes the chip set and components that support the operation of a computer, such as capacitors and mechanical switches (see Figure 4-13). A PCB supports and interconnects components through metallic traces called buses printed on the board. You can find several types of buses on the motherboard, including system buses, PCI buses, and AGP buses, as discussed previously.

Figure 4-13

A motherboard

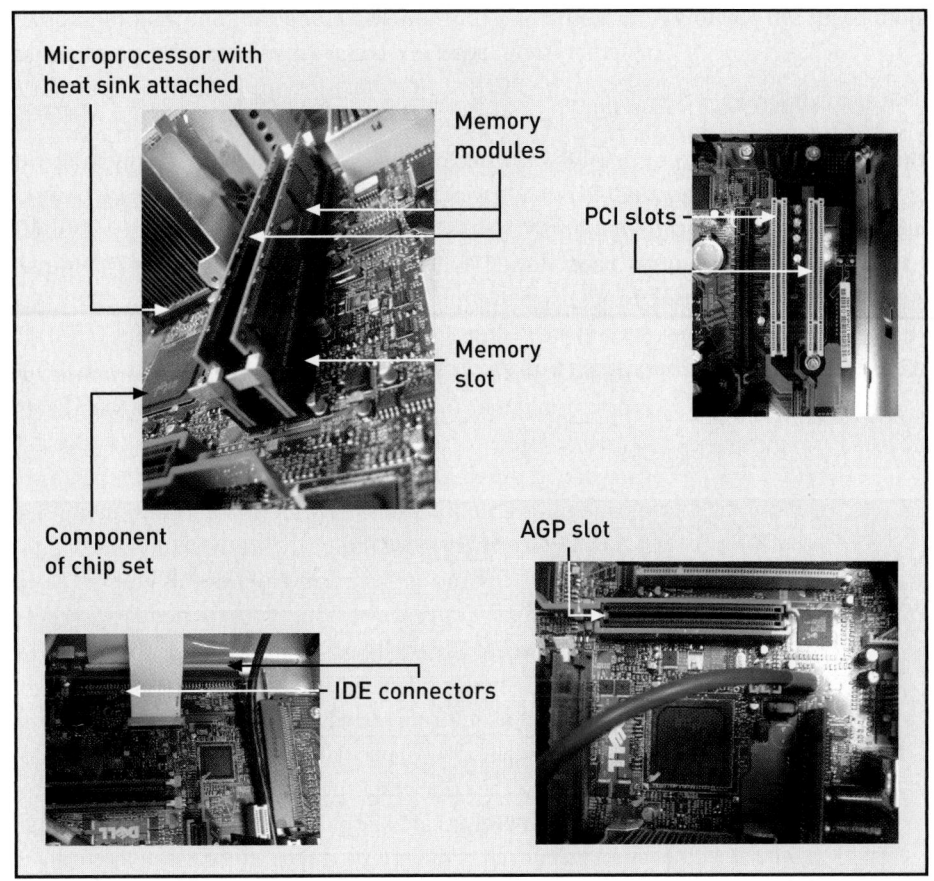

Besides the microprocessor, other components on the motherboard include the BIOS chip; the chip set; slots for connecting RAM modules to the microprocessor; default circuitry for supporting sound and graphics capabilities; various expansion slots for connecting external graphics cards, sound cards, and network cards; **connectors** for attaching hard drives, CD drives, and other disk drives to the motherboard; and **ports** for connecting I/O devices such as the keyboard and mouse.

As explained earlier, the BIOS chip is a flash memory chip that stores permanent instructions required by the CPU during the start-up or **boot-up** phase. The chip set supports the operations of the microprocessor.

RAM chips are also positioned and connected to each other on a printed circuit board that is separate from the motherboard. Boards that carry the RAM chips are called memory modules, and are connected to the microprocessor through **memory slots** on the motherboard. Modern motherboards also come with extra memory slots to allow users to expand the RAM for faster processing. The display of graphical information is possible because of specially designed circuits that frequently come installed on the motherboard, especially with low-end computers.

Circuitry for supporting enhanced graphics may also be incorporated on separate PCBs called graphics cards or graphics adapters, as explained previously. These adapters enable the computer to display sophisticated, high-resolution images and video that might be unavailable using the default graphics circuitry on the motherboard. These types of external graphics adapters connect to the motherboard via a special expansion slot, such as the **AGP slot**. The default graphics circuitry is issued an override if an external graphics adapter is plugged into an expansion slot on the motherboard to provide high-quality support.

Sound adapters enable a computer to output high-quality sound. Like graphics adapters, external sound adapters can be connected to the motherboard via one of its expansion slots, such as the **PCI/PCI-E slots**. This approach may be necessary if certain applications require better audio quality than the motherboard's sound circuitry can provide.

Computer networking is all around us, so motherboards also come with circuitry that supports networking. Alternatively, network adapter cards may be plugged into expansion slots on the motherboard. The motherboard also contains connectors and ports for hard drives, I/O devices such as the keyboard and mouse, and other devices.

DRIVES

Drives are devices that can read and write large amounts of information to and from various types of magnetic, optical, or electrical devices, such as hard disks, CDs, DVDs, and flash memory. These drives fall under the category of storage called secondary memory. They are categorized differently from the main memory found in conventional RAM and ROM chips on the motherboard. You can store information using magnetic energy with hard disks or floppy disks, using optical energy with CDs and DVDs, and using electrical energy with flash memory devices and flash drives. Later in this chapter, you will learn some of the principles by which data is stored on various types of media.

A hard drive essentially reads and writes information from and to a hard disk. Likewise, DVD drives accommodate information on DVDs. If drives are installed in the computer case, they are called **internal drives**. If they are connected to the computer externally, they are called **external drives**. These drives attach to the motherboard using its specifically reserved connectors. One common type of hard drive is the Advanced Technology Attachment (ATA) drive, which is also called a **parallel ATA** drive, an Integrated Device Electronics (**IDE**) drive, or an Enhanced IDE (**EIDE**) drive. These drives include an integrated component called a **controller**, and are attached to the motherboard via special connectors.

Other hard drives include serial ATA (**SATA**) and small computer system interface (**SCSI**) drives, which also attach to the motherboard via special connectors. SATA connections have two advantages over parallel ATA: they can transfer data at a faster rate between the motherboard and the drive, and they are more physically compact. SCSI drives are faster than ATA drives, but because they are more complex, SCSI drives are not a standard feature of most computers. While internal hard drives connect to the motherboard with special connectors, external hard drives attach to the motherboard via Universal Serial Bus (USB) or FireWire ports outside the computer casing.

In some cases, such as with hard drives and flash drives, the drive and storage media are built together as one piece of equipment. With other media, such as DVDs, you must insert the storage media into the drive. Modern computers no longer typically include internal floppy drives, but most have internal DVD drives that can read and write to CDs and DVDs. Like most drives, they attach directly to the motherboard via special connectors.

COMPUTER PORTS

You can connect peripheral devices to a computer through different types of ports that vary depending on the type of computer you use (see Figure 4-14). These ports in turn connect to the motherboard itself. Some examples of ports include **serial ports**, **parallel ports**, **video ports**, **S-video ports**, **USB ports**, **FireWire ports**, **sound ports**, **keyboard ports**, **mouse ports**, **network ports**, and **telephone ports**. You rarely find all types of ports in a single computer.

Figure 4-14

Various computer ports

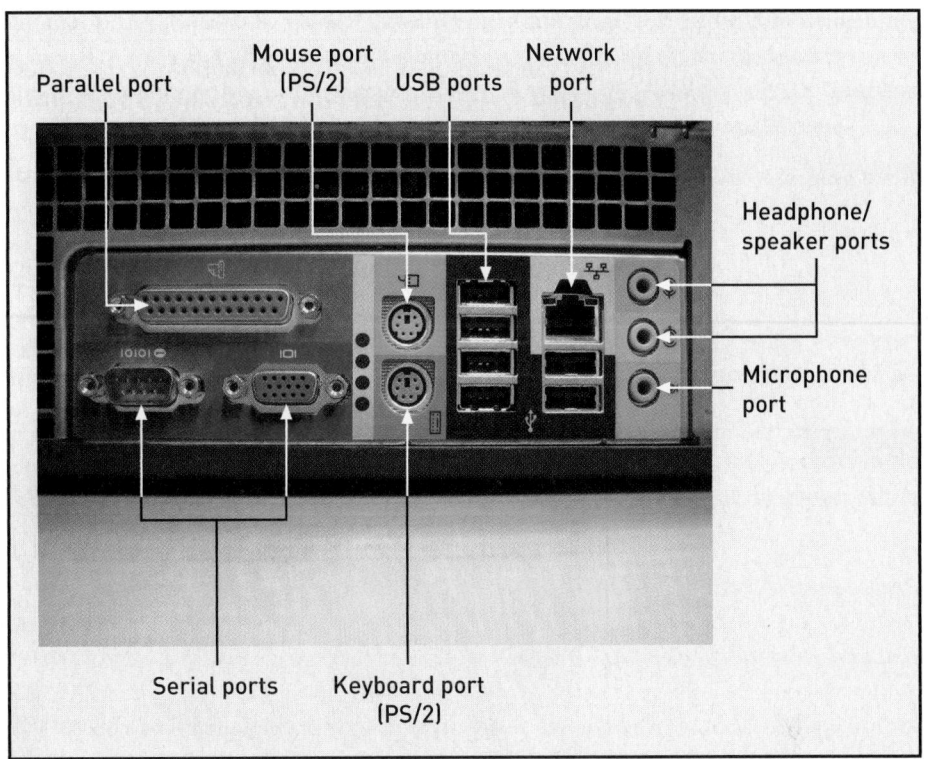

Serial ports are not commonly used anymore, but they are included with some PCs to connect legacy peripheral devices, such as a keyboard or a mouse without a USB interface. Information on these connections is then transferred serially. A parallel port is also used occasionally to connect peripheral devices such as printers; information in such connections is transferred over several wires in a parallel manner.

Video ports and S-video ports are special, fast ports that connect a computer to monitors, projectors, and other display devices. A USB port connects peripheral devices such as digital cameras, USB mice, USB keyboards, and USB printers. USB interfaces are faster than the older serial or parallel interfaces and have essentially replaced them. FireWire is similar to USB—it serially transfers information at a fast rate between a computer and a peripheral device, such as a video camera.

Sound ports connect microphones, headphones, and speakers to the computer, which allows you to listen to music and other audio. Keyboard and mouse ports connect certain types of keyboards and mice that do not have serial connectors or are not USB-compatible. These ports are also called PS/2 ports. The network port connects a computer to a network via the network interface card, and the telephone port connects a telephone line to a computer for dial-up connections via a modem.

POWER SUPPLY

A computer receives its power from electricity via a wall outlet. However, this electricity is supplied by the power company and is not compatible with the type of electricity the computer needs. Therefore, a power supply is included within the computer casing to convert the electricity levels from the wall outlet into levels that the computer can use. (In portable computers, the power supply is outside the casing.) The quality of the power supply is an important consideration, because some computer problems that are difficult to troubleshoot may develop because of faulty power supplies. Besides power supplies, other components such as **surge protection systems** and **uninterruptible power supplies (UPS)** may be employed to safeguard the computer against sudden surges of electricity caused by lightning strikes, power outages, and other phenomena.

COOLING SYSTEM

A significant amount of heat is generated by the many electrical components in a computer, and problems can arise if these components are not sufficiently cooled. The chip that generates the most heat is typically the microprocessor. The microprocessor chip can be cooled by various methods; one popular technique is to install a metallic component called a heat sink. Figure 4-15 shows two types of heat sinks.

Another way to conduct heat away from the chip is to install a fan on the microprocessor or pass **cryogenic materials** or other liquids inside tubes in contact with the microprocessor. Other components within the computer casing generate heat as well, so a fan is often incorporated within the casing to cool the microprocessor chip and drive heat away from the computer.

Figure 4-15

Two types of heat sinks

Heat sinks

TYPES OF COMPUTERS AND THEIR APPLICATIONS

Desktop computers, **laptops** (also called **notebook** computers), and **handhelds** are all popular both for personal and business use. The main differences between desktop computers and laptops are their size and portability, their motherboard, and their human-computer interface. A desktop computer comes with a separate keyboard and monitor, and is larger than a portable laptop computer. Laptops have integrated monitors and keyboards, and often have a touch pad. As for the differences in the motherboard, a desktop computer has expansion slots to plug in external sound and graphics adapters and RAM modules. Upgrading a desktop computer is fairly easy, but updating features on laptops is much more difficult. Handheld computers support many of the features of desktops and laptops, such as word processing and Web browsing, but they have less capacity and processing power.

Mainframes are large, high-capacity computers used by large organizations to process and store colossal amounts of data. **Supercomputers** are extremely fast and expensive computers that can process vast amounts of information within a very short amount of time. Supercomputers have multiple processors and are commonly used by large universities and other organizations to perform research that requires intensive computational power. Such research areas include genetics, chemistry, and physics.

Servers are computers that provide services to other computers called **clients** over a network. **Web servers** and **file servers** essentially supply Web pages and files to computers that need access to them. Similarly, **e-mail servers** supply e-mail requested by other computers over a network such as the Internet. Any user who surfs the Internet accesses a Web server, and any user who logs on to e-mail accesses an e-mail server. Although the user does not directly interact with the Web server or the e-mail server, the user's computer (the client) does. Even though most of the computers discussed previously can take on the role of a server, many organizations use computers that are specifically designed to act as a server. Servers will be discussed in more detail in future chapters that describe computer networking.

Most major computer manufacturers offer specific server computers that include additional functionality not commonly found on computers at the lower end of the spectrum, such as desktops. Servers are not as expensive as other computers, such as mainframes, so they provide a good balance between functionality and cost. Servers usually have more than one processor, considerable memory, and extra hardware, such as multiple network cards and dual power supplies. One popular type of server is called a **blade server**, a computer without a keyboard or a monitor that resembles a big, flat, rectangular pizza box. Blade servers can be controlled in a computer network through a remote access card installed in the server casing. Blade servers have the advantage of saving a significant amount of space. They can be housed in remote locations such as **data centers**, areas that accommodate servers for companies and organizations. These servers generate significant amounts of heat and must be kept properly cooled.

Thin clients are computers that do not have the regular hardware typically found on other computers. They do not have a hard disk or other disk drives, and they have a scaled-down version of an operating system stored on a special memory chip. Thin clients essentially rely on other computers over a network to perform most of their processing functions. Their major advantage is lower cost due to their lack of hardware and low power consumption.

Figure 4-16 shows some of the computers discussed in this section.

Figure 4-16

Various types of computers

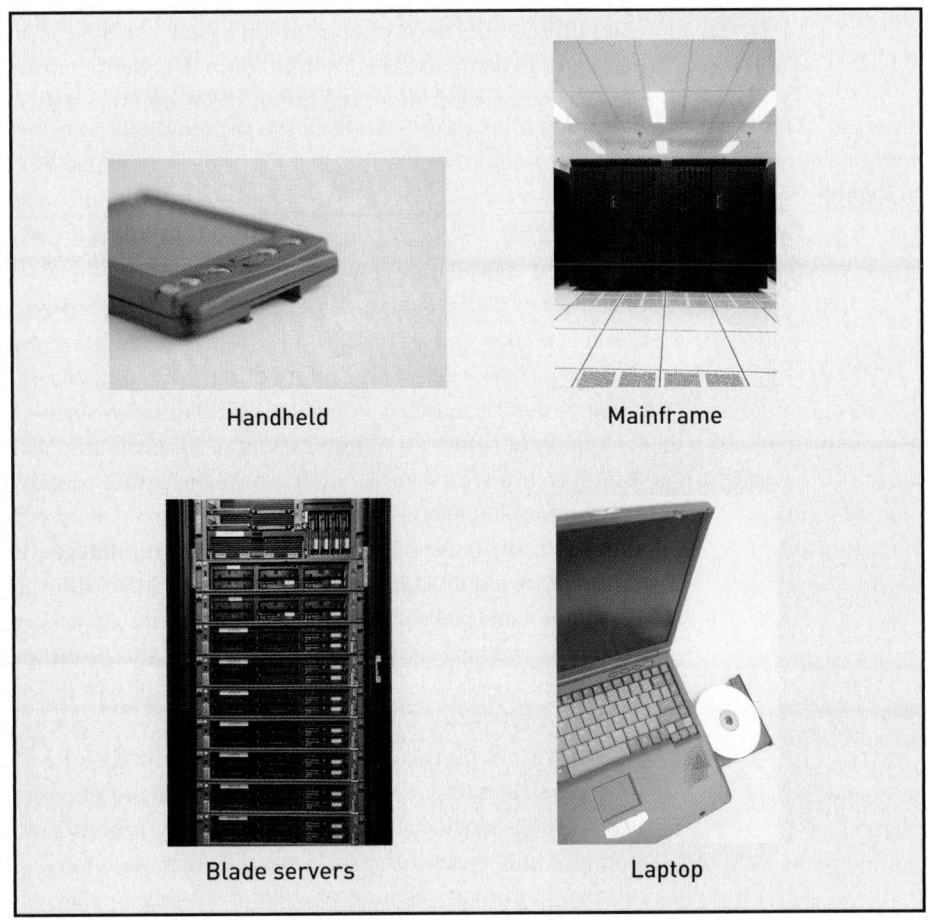

Handheld

Mainframe

Blade servers

Laptop

THE FUTURE OF COMPUTERS

Over the course of a few decades, computers have become faster, smaller, more mobile, and more powerful. Other less visible changes include improved power consumption, reliability, memory size, and graphics and sound capabilities. These advancements can be attributed to four major developments: the shrinking size of transistors, more efficient system design, new manufacturing techniques, and new materials.

Although transistors have continued to shrink over the years in accordance with Moore's Law (explained in Chapter 2), the trend will not hold true indefinitely. Eventually, scientists will reach a physical limit to their ability to shrink a transistor. Similarly, wires that connect transistors to each other continue to diminish in size. This creates a problem as well, because very thin wires can cause a large loss of energy and limit the speed by which signals travel over the wires that constitute computer buses, thereby limiting the computer's speed.

In anticipation of these physical obstacles, researchers have introduced alternative technologies such as **optical computing**. By replacing electricity with light within computers, optical components in place of transistors, and **optical interconnects** in place of wires, researchers hope to overcome the bottleneck that transistors and metallic wires pose.

A significant body of research has developed in the area of **nanotechnology**—technology that permits the manufacture of components on the order of a nanometer, which is one billionth of a meter. Researchers have demonstrated the potential of **carbon nanotubes**, minute interconnections made of carbon molecules with exceptional conducting properties that could replace electrical wires within chips. Another significant area of research is **quantum computing**, which proposes using individual atoms for performing computations. By constructing the computer's basic elements using atoms instead of transistors, groups working on quantum computers hope to change the face of computing. Another alternative, **biological computing**, relies on biological matter such as DNA for processing and storage. DNA computing raises many social questions about the ethical nature of mixing biological and technological material, and about the implications of implanting chips into people and animals.

STORAGE TECHNOLOGIES

Businesses, universities, governments, and individual users produce and consume massive quantities of digital data, images, audio, and video. Storage technologies are critical for businesses to archive and access data, for governments to store important political information, and for libraries to store electronic information. Without inexpensive and high-capacity storage technologies, there would be no digital music players, no video-sharing sites, and no TiVo.

This section presents a short history of the evolution of storage media, a discussion of how storage technologies work, and a glimpse at future possibilities for meeting ever-increasing digital storage demands.

Historically, some of the first data storage technologies used mechanical approaches, such as Joseph Jacquard's "Jacquard loom," and Herman Hollerith's tabulating machine, invented to analyze data from the 1890 U.S. Census. Both Jacquard and Hollerith used punch cards to store data. Jacquard used them to store loom patterns for weaving fabric, and Hollerith stored the collected census information on punch cards, which were fed into a tabulating machine.

Data could be stored on punch cards as long as the card was in good condition. Because the holes were permanent, punch cards were classified as storage media that were not rewritable. The presence or absence of a hole represented the data, which could be read by a punch card reader. In Hollerith's invention, data gathered from the cards was subsequently processed and presented to the Census Bureau for analysis. The combination of punch cards and the tabulating machine enabled the analysis of census information in a few months, compared with the 10 years it took to manually analyze the data from the previous census.

Paper punch cards and paper tapes were the first means of storing computer data, and were used until the 1960s and 1970s, when magnetic media became the dominant approach to data storage. Today, the idea of a mechanical scheme for storing bits seems archaic. However, recent advancements in **microelectromechanical systems** (MEMS)—the technology used to create miniscule machines—have spurred interest in mechanical schemes again.

MAGNETIC STORAGE

Magnetic storage has been a versatile and popular approach for information storage. Magnetic recording was first demonstrated in the late 1800s by Valdemar Poulsen, who showed that sound could be magnetically recorded on a piece of metal wire. Since his discovery, magnetic recording media have made significant advancements, from metal wire to magnetic tape to magnetic disk. Hard disks, floppy disks, and magnetic tapes are all examples of rewritable magnetic storage media. The underlying principle of storing bits is the same for all these devices. To record information, a plastic tape, plastic disk, or ceramic disk is coated with a **ferromagnetic material**—a unique material that can be magnetized. The recording material is positioned close to a writing head that magnetizes small sections of the medium in one of two directions corresponding to ones and zeros, as shown in Figure 4-17. To read information, the head detects the direction of the magnetic field in each spot and recovers data based on the field's orientation in each site.

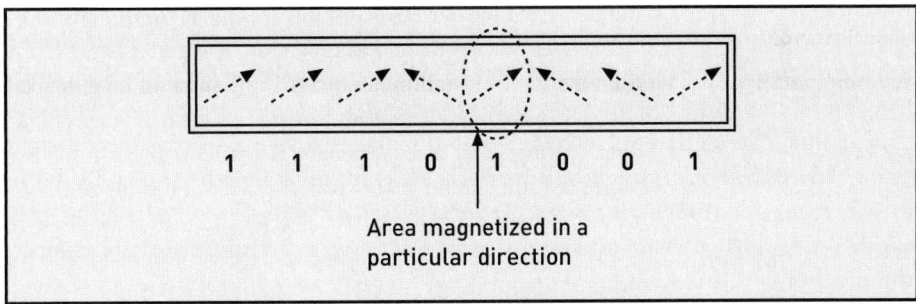

Figure 4-17

Magnetic storage of bits

The underlying mechanism for writing to and reading from magnetic media is essentially the same, but different types of magnetic media have considerably different shapes, sizes, and recording formats. For example, a digital audiotape records audio information on a long plastic tape whose surface is coated with ferromagnetic material. As the tape moves forward in the tape player, the head reads the information on the tape and sends the recovered signal to be processed for playback. In the case of magnetic disks such as hard disks or floppy disks, the recording medium is round, and information is written to it in a circular manner. Data is read and written by spinning the disk around its axis and detecting magnetic field changes in a circular direction.

A hard disk stores information using the same principle as magnetic tape, but there are major differences in their recording formats. A hard disk stores bits on a circular disk (also called a platter) made of highly polished glass or another material. A typical computer hard disk has many of these circular disks stacked on top of each other. Each disk stores information on circular **tracks** and wedge-like **sectors** (see Figure 4-18). Before recording, the hard disk is formatted; the platters are prepared for writing by partitioning their surface into sectors and tracks.

Figure 4-18

Structure of a hard disk

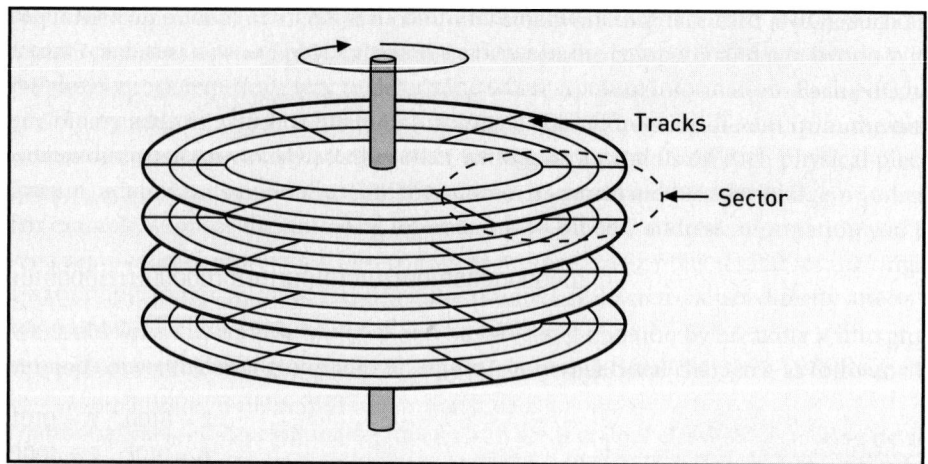

After formatting, bits are written magnetically to the tracks using a writing head. To read data, the appropriate platter is selected, and the magnetic heads detect the string of bits in the relevant sector and track. As the head reads the data, the disks spin at an extremely fast speed—usually thousands of revolutions per minute (RPM). The faster the platters spin, the faster the information can be retrieved from the disks. The spin rate is hence a critical performance factor for hard disks.

Because data is written to magnetic tape and magnetic disks in fundamentally different ways, the way the information is accessed differs as well. In the case of magnetic tape, information is accessed serially (**sequential access**), whereas a magnetic disk can provide **random access** to data. During serial access, the tape is sequentially scanned from one end to the other to find the requested information. This process does not apply to magnetic disks because the appropriate sector and track can be selected directly and read to provide immediate information.

Hard disks have very high capacity, on the order of tens to hundreds of gigabytes. Soon these capacities will increase to the terabyte range. Hard disks are limited mainly by their complex and sensitive mechanics for spinning, writing, and reading data. They are also sometimes affected by external magnetic fields, which can corrupt stored data.

Floppy disks are external storage devices that are not widely used anymore except for file backups in some legacy systems. Floppies are conceptually similar to hard disks, but they have a much lower storage capacity. For example, a 3½-inch floppy has a storage capacity of about 1.44 MB. Floppy disks store information on a flexible, round plastic disk with a ferromagnetic coating. The disk is enclosed in a hard plastic casing that protects against physical damage, and has a metal sliding door for access by the reading head. As the plastic disk rotates, the floppy drive retrieves the information from the opening on the plastic casing; the spin rate is lower than that of a hard disk. Most modern computers don't have a floppy drive, so using these older diskettes often requires an external drive.

OPTICAL STORAGE

Optical data storage has replaced many alternative storage schemes. For example, the CD made the cassette tape obsolete for music recording, and the digital versatile disk (DVD) replaced the video home system (VHS) tape. This section discusses the following types of optical disks:

- Compact disc-read only memory (**CD-ROM**)
- Compact disc-recordable (**CD-R**)
- Compact disc-rewritable (**CD-RW**)
- Digital versatile disk-read only memory (**DVD-ROM**)
- Digital versatile disk-recordable (**DVD-R**)
- Digital versatile disk-rewritable (**DVD-RW**)

As in magnetic storage, where the principle of writing and reading bits is common to all magnetic devices, optical storage media also rely on a common fundamental principle. Bits on optical disks are stored by creating small areas or domains with different reflective properties across a long spiral track. Bits are read by shining laser light on these tiny domains and detecting the light reflected back from them. Each domain reflects light in one of two ways, enabling the reader to distinguish between a one and a zero.

Bits on a CD-ROM are recorded by physically creating tiny protrusions, also called **lands**, on the CD's clear plastic surface during the manufacturing phase. Areas without lands remain shallow; these areas are called **pits**. The next few steps involve depositing several layers of material onto these lands to protect the CD, provide durability, and enable light to be reflected back from the surface of the CD. Figure 4-19 illustrates the structure of a typical CD.

Figure 4-19

Structure of a CD

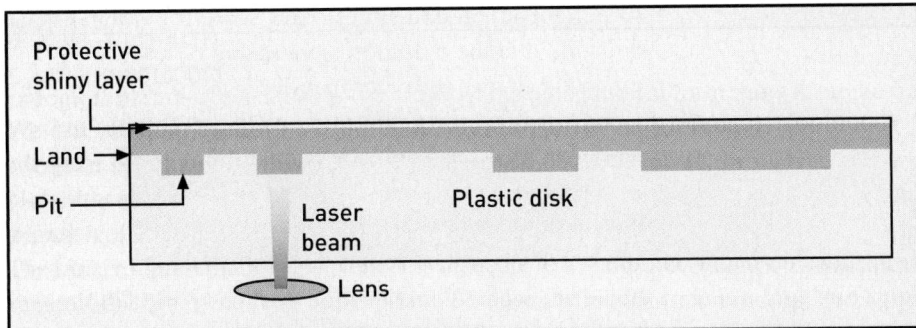

The lands and pits on a CD are created in a long line that begins from its center and spirals outward, as shown in Figure 4-20. The bits are read by focusing a laser beam from a lens system onto the track of lands and pits. As the disk spins, the laser illuminates each domain, which contains either a land or a pit. Light reflected back from the shiny coating in each domain is detected by a sensor. A domain that contains a land reflects light differently from a domain with a pit. The sensor detects the differences in reflectivity and outputs ones and zeros accordingly.

Figure 4-20

Spiral track of a CD

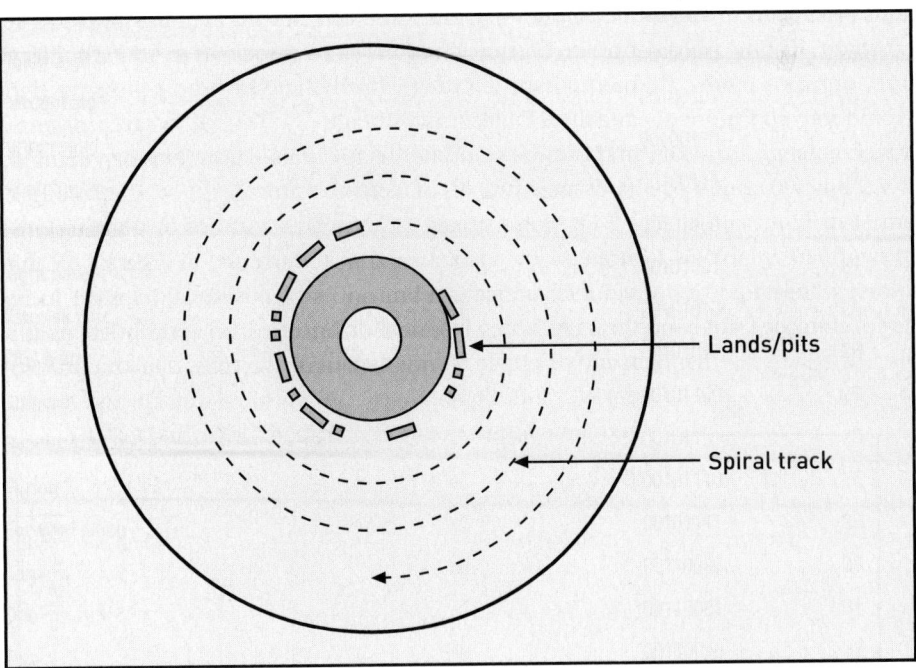

Due to the way CD-ROMs are manufactured, data can be written to the disks only by the manufacturer. Music and software CDs, for example, are mass produced using the physical processes described earlier in this section. Other types of CDs enable users to write and rewrite data. CD-Rs can be written or recorded once by the user and read many times, although they cannot be erased. CD-RWs can be repeatedly written, read, and rewritten, so a single CD-RW disk can be used many times.

Recordable CDs such as CD-Rs and CD-RWs are constructed by coating the disk's surface with a special material, such as a dye. This material can change its reflective properties when an external phenomenon is applied, such as heat from a laser. To record information to these disks, a process called burning, a laser beam in the CD drive is focused onto the surface of the CD. As the disk spins, the laser is turned on to record a 1 and turned off to record a 0. Heat from the laser causes the material to change its property and consequently alter the reflectivity of light at that domain. As data is read from the disk, a laser beam is focused on the CD domains, and the reflection is measured by a sensor. CD-R technology uses the laser to permanently change the color in a dye layer beneath the clear plastic disk surface. In contrast, CD-RW technology temporarily alters another property of the disk surface, which can be repeatedly changed from its original form to a new form and back to its original form.

CDs can record many types of information, including computer data, software, text, images, audio, and video. For a device to properly read and save the information stored on a CD, the device must use standard formats. For example, music on an audio CD is recorded following the **Red Book standard** developed by Philips and Sony.

DVDs rely on the same principles as CD technology. They are used mainly to store movies and audio, but they are also used for storing computer data. DVDs can store much more data than CDs, primarily because the size of lands and pits can be diminished and spaced in closer proximity, and because more than one spiral track may be used to store data due to multiple layers.

Data is directly impressed on a DVD-ROM during manufacture, but users can record information on other types of DVDs. Just as CD-Rs and CD-RWs refer to recordable and rewritable CDs, you can use DVD-Rs and DVD-RWs to record and rewrite information.

ELECTRONIC STORAGE

Electronic storage, also called semiconductor storage, has become an extremely effective technology for quick and efficient information storage. Electronic storage lets users save data quickly without the relatively lengthy burn process required for optical media such as CDs and DVDs. Electronic storage can be used to save images in digital cameras and cell phones and to store computer-related data. Flash memory, memory sticks, flash drives, and memory cards are examples of electronic storage.

Electronic storage technologies store bits by trapping electricity within an array of transistors arranged on a silicon chip, as illustrated in Figure 4-21. When data is recorded, an electrical signal is applied to the metal lines; this signal traps electricity within the corresponding cells of the array. When information is erased, an appropriate electrical signal is applied to the lines to remove trapped electricity within the cells. The absence of moving parts in electronic storage devices, and their ability to be rewritten quickly and repeatedly, makes them highly practical and more robust than magnetic hard drives and CD drives.

Figure 4-21

Structure of electronic
storage media

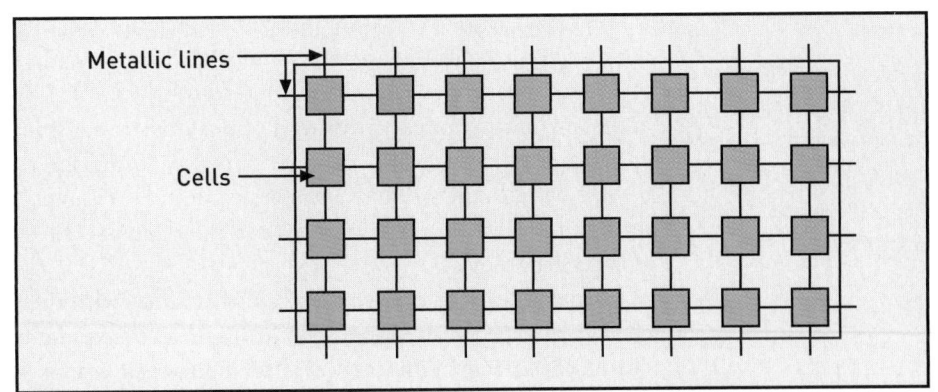

COMPARING VARIOUS TYPES OF STORAGE MEDIA

All the storage technologies discussed in this chapter have found their niche in the IT industry. Hard disks provide very fast recording and access to information and can store large amounts of data. Because they can be written over and over again without encountering the limitations of CDs or DVDs, hard disks are currently the primary storage medium for computers. Optical disks are portable, lightweight, and can store information with greater stability than magnetic media. Optical disks are not affected by magnetic fields, so they are considered more robust than magnetic devices, which are prone to data loss over time. Electronic storage media are compact, fast, and versatile because they do not contain moving parts. Figure 4-22 illustrates various types of storage media.

Figure 4-22

Various types of storage
devices

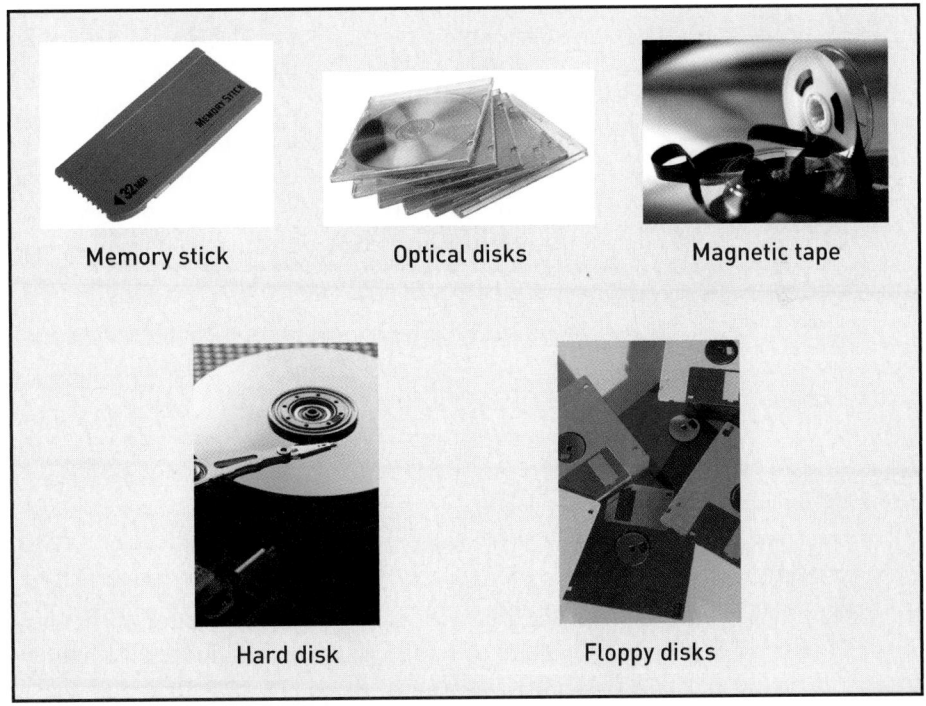

All varieties of storage media have increasingly greater storage capacities. Manufacturers can achieve greater capacity by reducing the physical area occupied by each bit on the media, by incorporating (packing) bits more densely on the same area, or even by storing more than one bit at a time in each area. However, each storage technology has a limit to its packing density. On magnetic media, the **superparamagnetic effect** causes random fluctuations toward the magnetic field if bits are packed too closely to each other. The ability to pack these bits closely on optical media is governed by the wavelength of light and the quality of the optics used to focus the laser beam onto the tracks. Because the beam should be focused to a smaller spot with increasingly large bit packing densities, designing optical systems is a daunting task. Nevertheless, researchers have begun using blue lasers, which can be focused to a very small spot to achieve greater recording capacities. Blue lasers are already used to read and write information on **advanced optical disks** and **Blu-ray disks**.

Integrated circuit fabrication technology limits the density of semiconductor storage devices because the size of transistors is limited. With advancements in chip manufacturing techniques, flash memory devices are now considered to be at the forefront of portable storage technology.

Constant increases in storage capacity, along with the ubiquity of network access, have created an expectation of perpetual progress. Unfortunately, the flip side of this progress is obsolescence. Digital information stored on a disk, computer, or server only 10 years ago may already be inaccessible. Universities have boxes of records saved on punch cards that are no longer readable. Government agencies have vital public information stored on magnetic tape that is not easily readable.

The economic and political consequences of this phenomenon are serious. Businesses that cannot readily access their older electronic records can lose both information and productivity. Democratic governments that are obliged to archive public records and make them available to citizens run into a similar problem. This obsolescence problem might force companies to store paper records even longer than digital records unless information is rearchived every five years to upgrade it to the latest medium, application, and formatting standards.

Some of the technologies mentioned in this chapter will become obsolete as newer technologies emerge. The ultimate limitations of these storage technologies have spurred a large amount of research for alternatives, including **holographic storage** and **molecular storage**.

Holographic storage is a completely different technology from optical disks. Holography systems create an interference pattern between different sources of light and record information in three-dimensional space based on this pattern. Holographic storage offers extremely high volumetric storage densities (storage in three-dimensional space) and the ability to read data very quickly, which could put this technology in the forefront of large-capacity storage.

Another feasible alternative could be molecular storage, in which molecules might be used to retain digital information. This technology still seems futuristic, but the possibility that molecules may be used as a cheap, high-density storage method is appealing.

CHAPTER SUMMARY

- Transistors can be combined to create logic gates, which can be used as the basic building blocks for integrated circuits.
- A computer's fundamental components include input/output (I/O) devices, a central processing unit, main memory, storage (secondary memory), and an interconnection system.
- Factors that affect the performance of computers include word length, bus width and bus speed, memory size and memory access speed, processor rate, instruction set, number of processors, and chip set.
- Components you can expect to find in a modern computer include the motherboard, CD drive, hard drive, graphics adapter, sound adapter, network adapter, ports, power supply, and a cooling system.
- Computers come in various types, including desktop computers, laptops, handhelds, mainframes, supercomputers, servers, and thin clients.
- Future possibilities for computing include nanotechnology, quantum computing, and biological computing.
- Magnetic media store digital data by magnetizing a ferromagnetic medium in one of two directions.
- Optical media store digital data by creating domains with different reflectivities.
- Electronic storage media save data by storing different levels of electrical charge through transistors.
- Magnetic tapes are classified as serial access devices. Hard disks, floppy disks, CDs, DVDs, and electronic memory are classified as random access devices.
- Possible alternatives for future storage media include holographic and molecular storage.

KEY TERMS

address	data center
advanced optical disk	desktop computer
AGP bus	DVD-R
AGP slot	DVD-ROM
ALU	DVD-RW
AND gate	dynamic RAM
biological computing	e-mail server
BIOS	expansion card
blade server	expansion slot
Blu-ray disks	external drive
Boolean logic	ferromagnetic material
boot up	file server
bus width	FireWire port
cache memory	flash memory
capacitor	front side bus
carbon nanotube	handheld
CD-R	hardware
CD-ROM	holographic storage
CD-RW	IDE/EIDE
central processing unit (CPU)	input device
chip set	internal drive
CISC	inverter
client	keyboard port
clock	lands
connector	laptop
control unit	logic gate
controller	mainframe
cryogenic material	memory

memory module

memory slot

microelectromechanical system

microprocessor

molecular storage

motherboard

mouse port

multicore computer

multiprocessor computer

NAND gate

nanotechnology

network port

nonvolatile memory

NOR gate

NOT gate

notebook

operating system

optical computing

optical interconnect

OR gate

output device

parallel ATA

parallel bus

parallel port

PCB

PCI bus

PCI-E bus

PCI/PCI-E slot

peripheral

permanent memory

pit

port

quantum computing

RAM

random access

Red Book standard

registers

RISC

ROM

SATA

SCSI

sector

sequential access

serial bus

serial port

server

software

sound port

static RAM

supercomputer

superparamagnetic effect

surge protection system

S-video port

system bus

telephone port

temporary memory

thin client

track

truth table

UPS

USB port

video port

volatile memory

Web server

word length

XNOR gate

XOR gate

REVIEW QUESTIONS

1. Describe the main components of the CPU.
2. What is the functionality of the chip set?
3. What is the difference between main memory and storage? Give two examples of each.
4. Explain the differences between SRAM and DRAM.
5. Explain the functionality of the ALU. Where does the ALU reside?
6. Explain the purpose of incorporating cache memory within the microprocessor.
7. Describe the various types of nonvolatile memory.
8. Construct a truth table for a NOR gate and an XNOR gate.
9. What would be the output of an OR gate, NAND gate, and an XOR gate, respectively, if one of the inputs to each gate was zero and the other input was one?
10. Explain three factors that affect the performance of a computer.
11. Describe the functionality of the BIOS chip. Where is this chip located?
12. Explain the difference between a SATA and an ATA drive. How are these drives connected to the motherboard?
13. Describe the different types of computer buses.
14. List the different types of computer ports and explain what devices may be connected through each port.
15. Explain the difference between a dual-core computer and a dual-processor computer.
16. What are ferromagnetic materials?
17. Explain how serial access to data differs from random access.
18. Give two examples of serial-access media and two examples of random-access storage media.
19. Explain how optical disks store information.
20. How are lands and pits created on the surface of an optical disk?
21. How do CD-ROMs differ from CD-Rs?
22. How can an optical disk be manufactured so that it is rewritable?
23. Why do DVDs store more information than CDs?
24. Give one example of an electronic storage medium.
25. Name one storage technology that is rewritable.

DISCUSSION QUESTIONS

1. Research optical computing on the Web. Do any commercially available computers employ optical computing?
2. Research Moore's Law and Ray Kurzweil's comments about Moore's Law. Summarize your findings in a couple of paragraphs.
3. Research two different types of memory modules on the market today. What is the capacity of each module? What types of gates are these memory modules based on?
4. What are some valid ethical concerns about the possibility of biological computing?
5. Will electronic, magnetic, and optical storage capacities continue to increase at a rapid rate? What physical constraints limit the bit-packing density of these storage media? Do you think these constraints will be overcome, or will a new storage medium emerge?
6. Do you think flash memory will replace magnetic hard drives? Support your answer with research from the Web.
7. Try to itemize all the storage technologies you use in daily life. Categorize these technologies into mechanical, magnetic, optical, and electronic storage. Do you use all four types of storage?
8. Have you ever been unable to access an electronic file because it is stored on an obsolete medium? Do you think your current files will ever become inaccessible? Why or why not?

CASE PROJECT

Imagine that you are in the market for a new desktop computer. Visit a Web site that sells computers online. List the following specifications for two different types of computers.

- Type of microprocessor and processor speed
- Type of hard drive and its capacity
- Type of RAM, current memory size, and how much the memory can be expanded
- Other types of internal drives
- Graphics adapter specifications

- Sound adapter specifications
- Network adapter specifications
- System bus speed
- Types of ports included with the computer
- Types of expansion slots on the motherboard

What is the price difference between the two alternatives? What types of applications do you need for your new computer, and how will these needs influence your choice of computer?

SOFTWARE

In this chapter you will:

■ Learn what exactly constitutes software

■ Gain familiarity with the various types of programming languages

■ Distinguish between different types of operating systems

■ Understand the difference between system software and application software

■ Learn about the software development process

■ Understand the concept of open source software and related issues

INTRODUCTION

The versatility and widespread use of computers have led to a great diversity of software. Software allows computers to be used for almost any purpose, including image and sound editing, word processing, entertainment, scientific research, data management, Web surfing, and chatting. Surreal creatures in movies, such as *The Lord of the Rings*, are created by software. Associated movie sounds can be artificially created using specialized software.

Early computers did not have this versatility; most were used to perform arithmetic computations. Programming early computers required physically rewiring their electronic circuitry. Long cables had to be unplugged from their sockets and plugged into different sockets to reprogram computers for new computations. Fortunately, such cumbersome programming is no longer necessary. You can program modern computers to perform many tasks simply by loading software from a flash drive or downloading it from the Internet. In other cases, IT professionals develop software to perform more specific or customized business functions. Even IT workers who are not directly involved in software development must understand how to install, use, support, and manage various types of applications.

This chapter begins by introducing the concepts of high-level and low-level programming languages. It then describes different types of software, including system software and application programs, and discusses the software development process. The chapter ends by exploring current development trends such as the open source software movement.

WHAT IS SOFTWARE?

Software is traditionally defined as *a series of data and instructions written by programmers (or sometimes automatically generated) using a programming language to perform a certain task*. Gaming software such as Civilization is just a collection of instructions written in a programming language; data corresponding to graphics and sound are integrated into the instructions. When software is loaded from a CD to a gaming console such as an Xbox 360°, the instructions are processed by the processor within the console, allowing the gamer to play the game; the graphics and sound data embedded in the software provide the associated images and audio. The gaming console is therefore classified as a **programmable device**—it is instructed by the software to enable people to play a game and display a multitude of graphics and sounds. Similarly, word processing software contains a series of instructions and data that allow the user to format and read words on a computer, which is also a programmable device.

In contrast to hardware, which is a tangible device that physically performs particular tasks, software does not physically perform these tasks itself. Instead, software provides a series of instructions to the hardware to perform a task. Thus, software enables the hardware to serve many functions, or in other terms, to be *programmed* without having to be physically reconfigured like early computers.

PROGRAMMING LANGUAGES

Programmers use a multitude of **programming languages** to create software. These languages can be classified in many ways, but they generally fall into two basic categories: **low-level programming languages** and **high-level programming languages**. The software industry often uses another common taxonomy: first-generation languages (1GLs), second-generation languages (2GLs), and so on, up to fifth-generation languages (5GLs).

It is widely accepted that the main difference between low-level and high-level programming languages is their level of abstraction. High-level programming language instructions are more abstract. Instructions written in a low-level language implement short, targeted operations, while instructions in a high-level language are more conceptual and correspond to more elaborate and complex operations. Furthermore, the words used to compose instructions in most high-level languages are closer to sentences used regularly in the English language, and they are easier for people to understand than low-level language instructions.

An analogy can further clarify the disparity in the level of abstraction between the two categories of languages. If someone tells you to play a movie on a DVD player, the person has several options for giving the instructions. An instruction such as "Play the movie" is similar to instructions in a high-level programming language. However, the same function can be implemented by a series of low-level programming language instructions. These less abstract instructions could include "Take the DVD out of its casing," "Switch on the DVD player," "Switch on the TV," "Insert the DVD into the player," and "Press the Play button." High-level programming language instructions are hence more abstract, and may correspond to several low-level programming language instructions.

One example of a low-level programming language is **assembly language**, while an example of a high-level programming language is **Java**. A single line of Java code might correspond to multiple lines of assembly language instructions. Most programmers are familiar with several high-level programming languages, and they can understand that assembly language is not as common for several reasons. First, assembly language and other low-level languages are used for highly specialized programming, such as manipulating the contents of the microprocessor and directly controlling the functions of the arithmetic logic unit (a computer component discussed in the previous chapter). Programmers who write in assembly language, therefore, are familiar with the details of the microprocessor; each type has a different architecture and requires different instructions for decoding and execution.

In contrast, high-level language instructions do not deal with the details of microprocessor architecture. They are written only to implement a high-level operation using a relatively simple line of code. High-level language programmers can be unaware of a microprocessor's architecture—for example, they need not know whether the microprocessor is a CISC or RISC model. Thus, it is less esoteric and easier to program using a high-level programming language, such as Java, rather than using assembly language.

Nevertheless, it is sometimes necessary to write instructions in assembly language, as the efficiency of the resulting program can sometimes surpass that of a Java program. The trade-off between these two general classes of programming languages often boils down to simplicity versus the level of optimization.

Assembly language programming is commonly used to create **device drivers** and to create instructions stored on the BIOS chip on the motherboard. A device driver is software that provides an interface between peripheral components, such as printers and mice, and a computer's other hardware components. As described in Chapter 4, the instructions stored on the computer's BIOS chip supply the vital steps that the computer must follow as it is switched on. These steps include verifying the proper operation of the computer's components, such as the memory chips and microprocessor, and locating and loading the operating system, among many other tasks. The instructions that correspond to the BIOS tell the CPU to locate the operating system and load it into RAM for the user. Because the contents of RAM are lost when the computer is switched off, this process must be implemented each time the user switches off the computer. As discussed in the previous chapter, the BIOS chip can store the instructions permanently, and its contents are not lost when the computer is switched off.

When a user launches a piece of software, such as from the hard disk, the instructions and data that correspond to the software and reside in RAM are not in the form of high-level language. Instead, the instructions and data are in the form of ones and zeros, or **machine language**. When programmers write in any assembly language or high-level language, their programs are eventually converted to machine language so they can be interpreted and executed by the microprocessor. Certain utility programs are used to make these conversions. For example, a program written in assembly language is converted to machine language by another program called an **assembler**. Each line of assembly language instruction corresponds to a single line of machine language instruction. Thus, there is a one-to-one correspondence between the two sets of instructions; assembly language is considered a human-readable form of machine language.

Although programmers could write instructions in machine language, it would not be desirable. Dealing with strings of ones and zeros is confusing and likely to cause programming errors. Instead, because each machine language instruction has a corresponding human-readable version called its **mnemonic**, assembly language programmers use them to write programs. The assembler then converts these programs to machine language instructions. Machine language instructions can also be converted to assembly language using a program called a **disassembler** if the instructions require a more human-readable form.

Similarly, a program written in a high-level language such as C++ may be converted to a low-level machine language using a program called a **compiler**. A **decompiler** can perform the reverse operation. Once the program is converted to machine language, it can be saved on a CD for distribution to users. A user who purchases a software CD can then copy its contents to the hard disk. When the user starts the software, the contents of the hard disk are loaded into RAM and then fetched for the microprocessor to use for decoding and execution.

After a programmer writes a program, it is referred to as the **source code**, which can be converted to a form that is executable by the computer. This form is also called **object code**. Object code basically consists of a string of ones and zeros, which is eventually what the CPU decodes and executes.

Programs written in some types of programming languages are interpreted rather than compiled, using a program called an **interpreter**. For example, when a programmer writes Java code, the interpreter can convert each line of source code into object code, so that each line of object code can be executed *before* going on to the next instruction. This technique is in contrast to the compiler, which converts the entire source code into object code that the microprocessor can then execute. The use of interpreters is common; they allow a programmer to fix any mistakes early in the programming process instead of waiting for the computer to compile the entire program before execution.

Finally, a program written in one type of high-level programming language may be translated to another type of programming language by a **translator**. Translators are essential because so many programming languages exist. Just as translation is sometimes needed between different human languages, such as English, Spanish, French, and Chinese, translators are important in programming languages.

THE EVOLUTION OF PROGRAMMING LANGUAGES

Programming languages have evolved tremendously. Before the inception of high-level languages, devices such as the Jacquard loom were directly programmed using instructions stored on punch cards. Machine language was used to program the early digital computers of the 1950s, before developers began using assembly language for writing programs. The first popular high-level programming language was Formula Translator (FORTRAN). Subsequent early programming languages included Common Business Oriented Language (COBOL), Lisp, BASIC, and Pascal. More modern languages followed, including C, C++, C#, Java, Microsoft Visual Basic, and Delphi.

The timeline in Figure 5-1 includes some of the high-level programming languages discussed in this section.

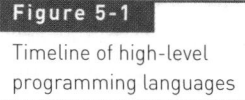

Figure 5-1

Timeline of high-level programming languages

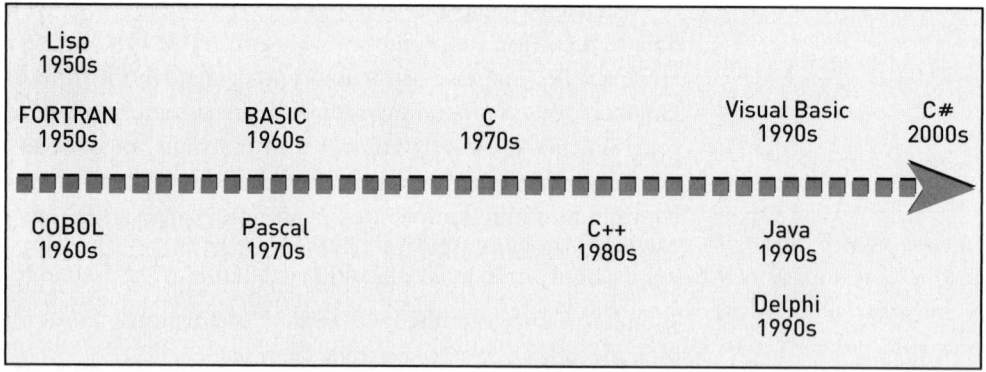

Java, developed by Sun Microsystems, has become an important programming language because Java programs can basically be executed on any operating system or machine. Java programs are first compiled into a form called Java bytecode, which is very close to machine language instructions. After compiling, the Java bytecode is interpreted and executed. This technique is especially attractive for the Internet, which uses data from several types of computers and various types of operating systems. Java also provides the ability to create interactive Web pages through small programs called **Java applets** that are associated with the Web pages.

As mentioned earlier, the software industry sometimes classifies programming languages into a generational taxonomy. The classification of 1GLs almost always refers to machine language. 2GLs refer to types of assembly language. 3GLs, 4GLs, and 5GLs have syntaxes that are progressively easier for people to understand and generally have more sophisticated instructions. These languages include C++, COBOL, FORTRAN, BASIC, Java, and programming languages that are more context specific. Keep in mind that these taxonomies are not entirely precise; there are numerous ways to categorize within a topic as broad as software.

The following example illustrates Java code that displays "Sample Java Code for 'IT in Theory' Book!" on the computer screen when the program is run.

```
class HelloIT
{
    public static void main(String[] args)
    {
        System.out.println("Sample Java Code for \"IT in Theory\" Book!");
    }
}
```

Figure 5-2 shows the output when the program is run.

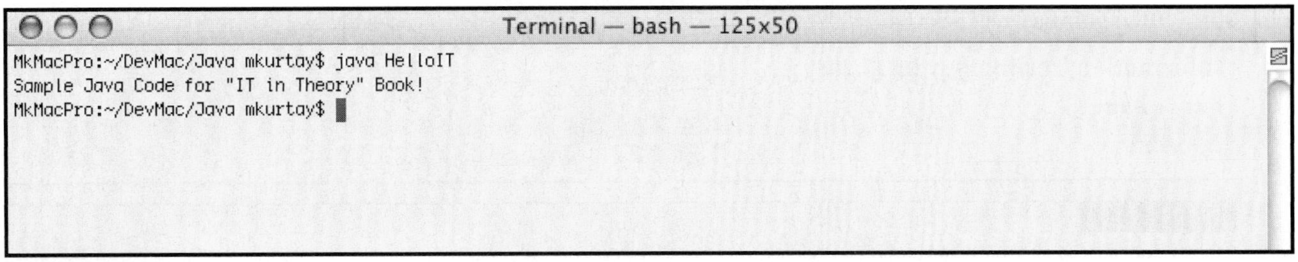

```
● ● ●                    Terminal — bash — 125x50
MkMacPro:~/DevMac/Java mkurtay$ java HelloIT
Sample Java Code for "IT in Theory" Book!
MkMacPro:~/DevMac/Java mkurtay$ █
```

Figure 5-2

Java code output on a computer screen

SCRIPTING LANGUAGES

Some programming languages, including JavaScript, are classified as **scripting languages**. The difference between high-level programming languages and scripting languages is that the resulting script programs can be stored without being compiled into machine language. Programs written in a high-level programming language, such as C++, are first compiled and stored as object code to be executed by the computer processor. **Scripts** are short programs that other applications interpret and execute. Users can summon scripts for some specific purpose, or they might run scripts automatically as a Web page loads, during a log-on process, or at other times.

JavaScript is one example of a language in which you can write scripts to expand the capabilities of Web pages. Another scripting language is VBScript (Visual Basic Scripting Edition).

MARKUP LANGUAGES

You use Hypertext Markup Language (HTML) and Extensible Markup Language (XML) to compose and format Web pages. By using a **markup language** to encapsulate text, images, and audio within special codes called **tags**, you can display the contents of your Web page in a browser using specific formats. For example, you can use special tags to encapsulate a piece of text that should appear in boldface. Other tags allow you to display the text in a specific font.

The following example illustrates HTML and JavaScript code that displays two strings of text: "Information Technology in Theory" and "This is a sample HTML page." The code also displays a "Show Current Time" button generated using JavaScript. When you click the button, the current time appears in a pop-up window. The HTML tags <b> and </b> encapsulate code that you want to appear in boldface on the screen. The output is shown in Figure 5-3.

```
<html>
<head>
<title>Information Technology in Theory: Welcome Page</title>
<script language="javascript">
function OnTimeButtonClicked()
{
alert(Date());
```

```
    }
</script>
</head>
<body>
<h1>Information Technology in Theory</h1>
<p>This is a sample <b>HTML</b> page.</p>
<input type="button" value="Show Current Time" name="button1"
onClick="javascript:OnTimeButtonClicked();" />
</body>
</html>
```

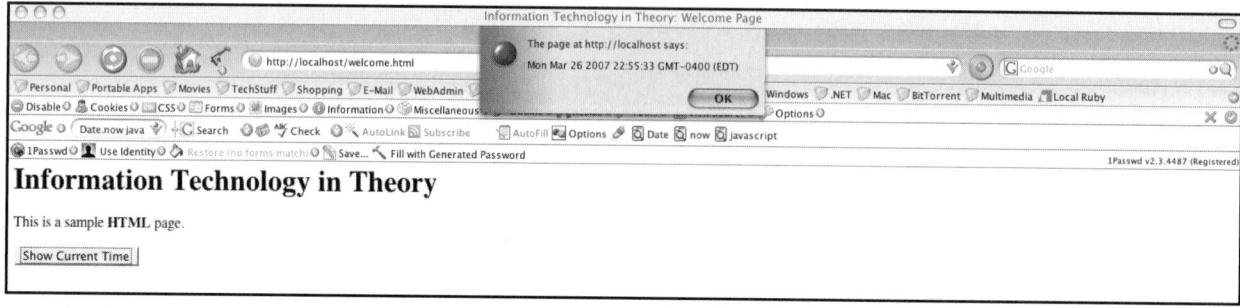

Figure 5-3

HTML and JavaScript output on computer screen

TYPES OF SOFTWARE

In addition to the classification of programming languages described in the previous section, types of software can be essentially categorized into **system software** and **application software**. System software, such as the operating system, supports the various operations of the computer. Compilers, assemblers, and device drivers are also traditionally classified as system software. Application software is used for specific applications, including electronic commerce, gaming, word processing, Web surfing, and image editing.

OPERATING SYSTEMS

Computer users must understand the significance of **operating systems**, which serve many purposes, some of which include:

- Providing a user-friendly interface to the computer
- Managing directories, folders, and files
- Managing memory
- Managing hardware and software
- Providing computer networking functions

The operating system (OS) provides users with an interface to all of a computer's functions. It allows the user to manage folders, access files, and create directories. The operating system manages the hard disk by creating partitions on it to store files and folders. Similarly, the operating system manages how files are stored on CDs and other storage media.

The operating system also manages memory. Software such as application programs occupies a specific amount of area in RAM, and the operating system manages this allocation. Furthermore, it allows the user access to hardware resources and enables the user to configure and troubleshoot hardware.

Users run application programs by interfacing with the OS. They also use the OS to download application programs from a CD or a network to their hard disk, install the programs, and then run them.

When operating systems are used specifically to perform important roles in computer networking, they are sometimes called **network operating systems** (NOS). At one point, operating systems did not support networking, so users had to use a separate NOS to connect their computers to a network. Because all modern operating systems now support networking, the term *NOS* is rarely used anymore.

Though it now seems archaic, the first popular operating system was Microsoft's Disk Operating System, or MS DOS, and many versions were developed over the years. One key feature of DOS was its **command-line interface** (CLI). These operating systems required users to input commands via the keyboard to perform specific tasks. Users who wanted to delete a file had to type a deletion command as well as the file path, which indicated the file's location and full filename. For example, if a file named Figure1.jpg was stored in a folder named Figures on the hard disk, a user would have to type a command such as "DEL Figures/Figure1.jpg" to delete the file.

The first operating system to provide a more user-friendly interface was Apple's Macintosh Mac OS, developed in the mid-1980s. The Mac OS provided a **graphical user interface** (GUI), pronounced "gooey," in which users just needed to click an icon to delete a file or perform similar tasks. Shortly thereafter, Microsoft introduced the popular Windows operating system. Several versions of Windows have been developed over the years, including the recent Vista operating system.

Another significant operating system is **UNIX**. It was originally developed by computer scientists at AT&T Bell Labs, and it was fundamentally different from DOS or Windows in several respects. It could support many users and perform a variety of tasks very efficiently. UNIX is considered a powerful, secure, and stable operating system by today's standards, and it is commonly used on computers that serve a large number of users and mission-critical applications.[1] UNIX also efficiently supports multitasking, the ability to complete multiple tasks at the same time.

Linux, an operating system developed by Linus Torvalds when he was a student, has become very popular. Linux has an **open source code**, which means that the programmers' instructions, or source code, are available for anyone to view and modify. Linux can also be downloaded for free, although not all open source software is free. Linux is frequently used on servers and desktops and on portable devices in its scaled-down version. Another attractive feature of Linux is its security. The basic version of Linux is free, but you can purchase some versions with service contracts and other commercial features through companies such as Red Hat.

The timeline in Figure 5-4 includes some of the operating systems discussed in this section.

Figure 5-4

Operating system timeline

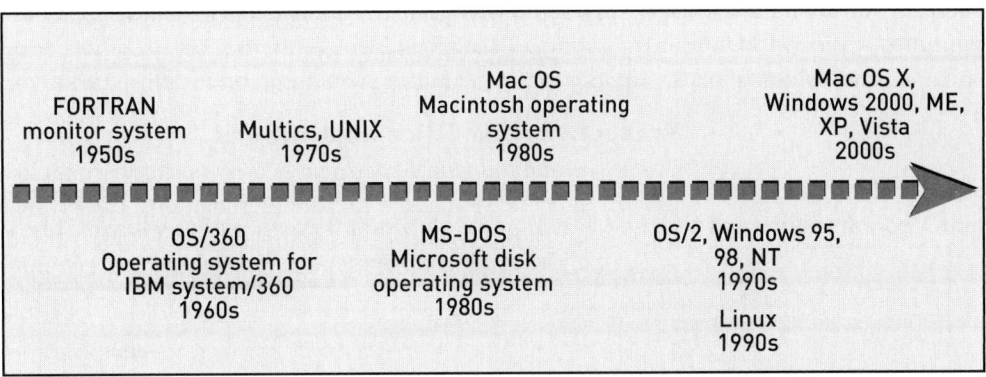

Besides the nature of the interface and features such as security, the selection of an operating system is also governed by the type of computer or device that runs it. Cell phones and handheld computers use operating systems, just like servers and workstations, but they use

[1] http://unix.org

scaled-down versions. Operating systems for smaller devices include Palm OS, Windows Mobile OS, and Symbian OS.

APPLICATION PROGRAMS

Applications, which are also called application programs or application software, are now the most familiar type of software because computer users directly interact with them on a day-to-day basis. The following sections briefly describe some popular applications in general categories.

Enterprise Applications

Enterprise software refers to large applications that assist in business functions. For example, these applications can enable interactions with customers, business partners, and suppliers, and they can help manage internal business processes. Some examples of enterprise software include the following:

- *Salesforce automation*—These applications help salespeople to track their interactions with potential customers. The applications are usually tailored to the needs of specific businesses; they can include product information, contact management systems, and a tracking system for sales leads.
- *Financial management*—This software helps organizations with accounting, financial planning, and business analysis. The software helps to automate the processing of travel expenses, billing, payroll, and asset management.
- *Supply chain management*—Some enterprise software helps companies manage the movement of raw materials, parts, works in process, and finished products. This software helps to ensure that the flow of materials and inventory is optimized at all times.
- *Customer relationship management*—Companies with thousands or even millions of clients rely on enterprise applications that manage and improve customer relationships.

Besides these examples, countless other types of enterprise software are tailored to specific industries: point-of-sale software in retail industries, patient management systems in health care, portfolio management software in financial services, and route management software in transportation.

Many large enterprise applications rely on an underlying **database management system** (DBMS). Enterprise applications usually require enormous stores of data describing inventory items, customers, financial records, and other business transactions. This information is stored in enormous databases—structured collections of records that a computer can access. A DBMS manages the organization of these records and queries them in response to requests from users or other software programs. The records stored in databases are organized in a schema: a specification for how records are arranged in the database and how they are linked in relationship to other records.

Productivity and Office Applications

Businesses and individual users rely heavily on software applications to create and view documents, generate Web pages, prepare presentations, create spreadsheets, and organize data to increase productivity. Such software includes the Microsoft Office Suite (Word, PowerPoint, Excel, Access), OpenOffice, Adobe Acrobat, Adobe Photoshop, Corel DRAW, and Macromedia Dreamweaver.

Communications and Information-Sharing Software

An enormous class of application software is designed to enable communications. These applications facilitate the day-to-day workings of businesses, governments, and individual users. Communication applications include electronic mail software, text messaging, and programs that enable voice communications over the Internet. Other applications, such as Web browsers and FTP access software, enable users to access and share information. Applications that provide these functions include Google Talk, Microsoft Outlook, Yahoo Messenger, MSN Messenger, Internet Explorer, Mozilla Firefox, WS_FTP, and Skype.

Entertainment and Media Development Software

Another major category of application software enables the creation, sharing, dissemination, and accessing of entertainment and media. Media editing software lets users produce digital images, video, music, and graphic designs. Some software is geared more toward media consumption, while other applications enable users to download and listen to digital music and watch video clips. Entertainment software includes iTunes, LimeWire, RealOne, QuickTime, and Windows Media Player.

Other Applications

Software has thousands of applications beyond those described in the previous sections. Other chapters of the book describe software that compresses files, detects viruses, blocks pop-up ads, and remotely accesses other computers.

THE SOFTWARE DEVELOPMENT PROCESS

Software development is often a lengthy and complex process that involves several stages from research to implementation. The level of complexity depends on the type of software. If the software has only a few lines of code, it can be completed with little effort. For a substantial software application that will be marketed to a large audience, the coding effort can be significant. Most people use commercial, off-the-shelf software such as Microsoft Excel, but many software projects require customized development to meet unique needs. Organizations that need such software work either with in-house developers or third-party developers and consultants to define their requirements and constraints. For example, a large hospital might require a customized billing application to manage its billing process.

Regardless of the intended audience or functionality, the creation of most commercial software follows stages that collectively are called the **software development process**. This process ensures that the project is manageable, that it is profitable for the software company, and that the final product meets the customer's needs. For any type of project, a computer programmer does not just start writing code. A great deal of preparation is necessary, and the development involves many people, including business analysts, researchers, system engineers, software engineers, project managers, and marketing personnel.

Developers usually follow a particular model during software development. One common approach is the **waterfall model**, which is shown in Figure 5-5 and explained in this section.[2]

---

[2] Winston Royce, "Managing the Development of Large Software Systems: Concepts and Techniques." In Proceedings of the Western Electronic Show and Convention (WesCon) 1970.

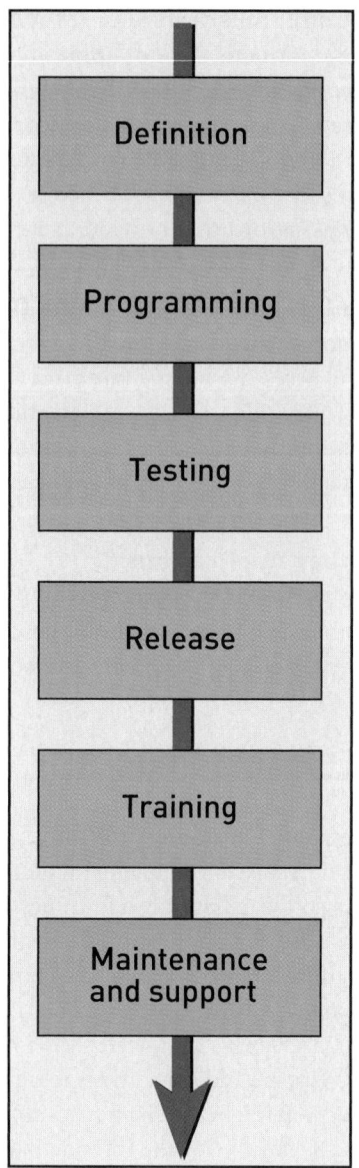

The waterfall model consists of several critical phases, each of which must be fully completed before the next stage can commence. During the definition phase, the software company interacts with the customer to clarify needs and constraints. Before actually writing any code, and in order to design software that fulfills the customer's expectations, the software development team must thoroughly understand the customer's requirements and limitations. Examples of constraints on the customer's side may include economic limitations, time constraints for software implementation, and maintenance costs for managing the product. Requirements may include detailed business-specific functions that the software will support, information about who will use the product, security criteria, and specifications for interacting with existing data stores.

Once these requirements are clearly defined, a design team works to create the software's design architecture, accounting for tools that would be most useful to ensure compatibility with the customer's systems and to minimize problems that the development teams are likely to face. Software programmers and engineers then start to write programs using the selected approach. Many programmers may work in the programming phase, so communication and interaction between them is critical.

Once the bulk of the programming is produced, the software is tested to remove any errors, or **bugs**, and to ensure that it is working correctly. All imperfections are reported and fixed during this phase. Testing may be implemented in two stages called **alpha testing** and **beta testing**. Alpha testing is normally conducted by the software development team, and only a limited version of the product is made available to the customer. Once the bugs are removed from the alpha phase, the software undergoes beta testing, and a **beta version** of the complete software is released to the customer. Customers can then report any bugs to the developer to further refine the software.

After the software is released, the customer must be trained to use the new software. The training phase ensures that the customer has full knowledge of the software's capabilities, can use it effectively, and is satisfied with the end product. The software development team is also responsible for maintaining the software; the customer may pay an annual fee for this maintenance in case problems need to be fixed and updates are necessary.

Although these stages are key components of software development, they need not occur in the order we described. Besides the waterfall model, the software development industry uses a variety of other approaches, including the **iterative model** and the **spiral model**. Each approach has certain advantages and shortcomings. For example, the spiral model is heavily concerned with risk and tries to minimize it during every stage. The development team tries to find alternatives to reduce risk before proceeding with the next phase.

The iterative model relies on breaking down the project into smaller segments and developing each segment in a limited amount of time. The segments may not necessarily be completed in a specific order. The results are combined to fine-tune the end product.

One software development method that is partially based on the iterative model is called **agile software development**. This method promotes flexibility for customers—they may be allowed to change their requirements even if development has commenced or is close to completion. The method also periodically provides the customer with limited versions of the software within a short amount of development time. The customer can then use the product and provide feedback before the project is fully completed. The Agile Alliance principle requires strong interaction between programmers, engineers, business personnel, researchers, analysts, and customers.[3]

OPEN SOURCE SOFTWARE

Some of the most popular types of software are open source, which means that the product's source code is openly available to the public. Anyone can view, copy, or change the software without having to pay licensing fees or royalties. Because of this openness, the development of open source code is usually collaborative, with multiple parties and interests producing the code and improving it for mutual benefit.

This type of decentralized and collaborative development can best be understood in contrast to proprietary commercial software development, in which code is developed within a corporation or by a single person. Once developed, the source code remains proprietary. It is not visible to the public and therefore is not available for copying, changing, and redistribution.

---

[3] *http://agilealliance.org*

Examples of popular open source software include the following:

- *OpenOffice*—A software suite of office applications, including word-processing, spreadsheet, presentation, database, and graphics software. It can be downloaded for free at *http://www.openoffice.org*.
- *Apache HTTP server*—A popular Web server application that is freely available to anyone who wants to establish a Web server.
- *Linux*—A prominent example of an open source operating system. Linux is primarily used for servers, although some people use Linux as their primary operating system.
- *Mozilla Firefox*—A popular open source Web browser that is freely available to download from the Internet. Firefox is probably the most downloaded open source application.
- *MySQL*—A popular open source database used for managing large amounts of data.

As the list indicates, open source initiatives span all types of software, including operating systems, programming languages, and applications. Many examples of open source software are available for free, but others must be purchased. A user with a new laptop could download the Linux operating system, the Firefox Web browser, and the OpenOffice suite and have an impressive collection of free software.

CHAPTER SUMMARY

- Software is the code that systematically instructs hardware to perform certain tasks.
- Software developers use programming languages to create software. These languages can be classified as low-level languages, such as machine code and assembly language, and high-level languages, such as Java and C++, that are syntactically easier for a person to understand.
- The software industry sometimes describes programming languages in historical generations ranging from first-generation languages (1GLs) to higher-level languages, such as 5GLs.
- Programs may be compiled, interpreted, or translated.
- Software is traditionally classified as system software, such as operating systems, or application software, such as Microsoft Office.
- Operating systems are responsible for a variety of functions, including providing an interface to the computer; managing directories, folders, and files; managing memory; managing hardware and software; and providing computer networking functions.
- The software development process involves several stages, including definition, design, programming, testing, releasing, training, and maintenance and support.
- Open source software's source code is openly available to the public. Anyone can view, copy, or change the software.

KEY TERMS

agile software development
alpha testing
application software
assembler
assembly language
beta testing
beta version
bug
command-line interface
compiler
database management system
decompiler
device driver
disassembler
graphical user interface
high-level programming language
interpreter
iterative model
Java
Java applet
Linux

low-level programming language
machine language
markup language
mnemonic
network operating system
object code
open source code
operating system
programmable device
programming language
script
scripting language
software
software development process
source code
spiral model
system software
tags
translator
UNIX
waterfall model

REVIEW QUESTIONS

1. Describe the difference between hardware and software.
2. Explain the difference between high-level programming languages and low-level programming languages, and list one example of each.
3. What are the advantages and disadvantages of programming in assembly language?
4. Explain the difference between source code and object code.
5. What is the difference between a high-level programming language and a scripting language?
6. Explain the purpose of markup languages. Give one example of a markup language.
7. Define the functions of an assembler, interpreter, translator, and compiler.
8. Explain the disadvantages of a CLI-based operating system. Name one CLI-based operating system.
9. What is the significance of a network operating system?
10. What features of Linux make it an attractive operating system?
11. Discuss the difference between system software and application software.
12. What is a Java applet? Which programming language may be used to create a Java applet?
13. What is a tag? Name one language that uses tags.
14. Define open source software. Give an example of a programming language, an operating system, an office application, and a Web browser that were created using open source software.
15. Which language is considered to be the first popular high-level programming language? Is it still widely used today?
16. What is the difference between alpha testing and beta testing?

DISCUSSION QUESTIONS

1. Discuss why Java is an attractive programming language. Besides the features discussed in this chapter, do further research on the Web and identify additional features that make Java attractive, especially for Web programming.
2. Explain the stages of the software development process. Discuss why it is important for developers to go through the various stages. Do these stages have to be implemented in a specific order?
3. Discuss the advantages of the agile approach to software development.
4. Discuss the various functions of an operating system. List three things that you could not accomplish if you did not have an operating system installed on your computer.
5. Do you currently use any open source software? If so, what was your rationale for downloading and using it?

CASE PROJECT

Select an example of open source software, such as the Apache HTTP server. For the software you select, research what organization, if any, is responsible for coordinating the open source development. What is the organization's policy for modifying software, and who is eligible to modify official versions of the software among the decentralized community of developers? If there are membership requirements, what are the criteria for membership, and who creates them?

If you were starting a company that was developing a new software application, would you choose open source development or a more closed approach? Make an economic argument for how your company could be profitable using either approach.

PART
3
CREATING DIGITAL MULTIMEDIA

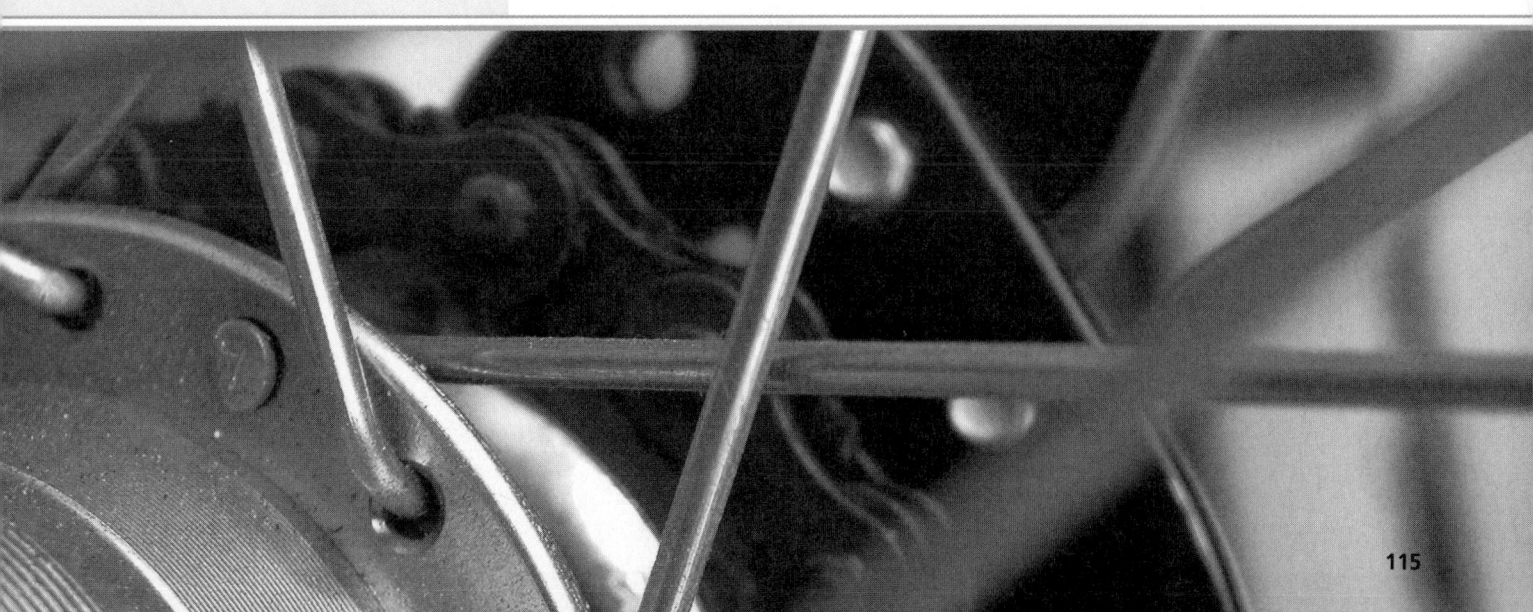

CHAPTER 6

DIGITAL AUDIO TECHNOLOGY

LEARNING OBJECTIVES

In this chapter you will:

- Understand the physical and mathematical basis of sound waves

- Recognize the amplitude, frequency, and phase properties of sound

- Understand the three-step process of audio digitization

- Understand and apply the Nyquist sampling theorem

- Recognize the need for digital audio compression

- Explain how music files can be compressed without degrading sound quality

- Understand the difference between lossy and lossless compression techniques

- Become familiar with popular digital audio formats such as MP3, AAC, WMA, WAV, and AIFF

INTRODUCTION

Satellite radio receivers, computers, mobile phones, iPods, and many other electronic devices share the common denominator of requiring sound digitization—the conversion between analog audio and streams of ones and zeros. The electrical circuitry embedded within some of these devices such as mobile phones converts sound from its natural analog form into digital. To successfully manage systems that embed digital audio, IT professionals must understand the mathematical properties of sound and the process of audio digitization. For example, selecting a sound card for a computer requires understanding such descriptions as "96-kHz sampling rate" and "24-bit sound card."

This chapter explains how sound waves are created, how they propagate through air, and how they can be captured electrically in analog format for digitization by electrical circuits. It explains the underlying physical and mathematical properties of sound waves to establish a solid foundation for understanding the process of audio digitization. The chapter then discusses digitization in detail and associated trade-offs between quality and cost. The chapter also describes techniques for recording audio information in analog and digital forms, explains digital audio compression techniques, and discusses popular digital audio formats.

RECORDING SOUND

The **phonograph**, invented by Thomas Edison in the late 1800s, was the first device created for sound recording and playback. As sound was introduced into the phonograph, a diaphragm vibrated, causing an attached needle to move in proportion to the sound vibrations. The needle inserted grooves along the surface of a tin-coated, rotating cylinder, mechanically cutting the **audio signal** onto the foil of the cylinder. To replay the sound, the cylinder was rotated again with the needle positioned on the grooves. Because the moving needle was connected to a diaphragm, the diaphragm vibrated in proportion to the structure of the grooves and produced sound through the phonograph horn.

The tin-coated cylinders were ultimately replaced by vinyl records (see Figure 6-1). The mechanical vibrations of the record player's stylus along the grooves of the record were converted to electrical variations and subsequently fed to speaker systems.

Another early analog recording technique from the late 1800s relied on magnetism and electricity. The technique, attributed to Danish engineer Valdemar Poulsen, used a magnetizable material such as steel wire to store analog audio signals in the form of varying magnetic fields. As a recording head was moved over the wire, the electrical audio signal sent into the head induced a magnetic field on the metal wire that was proportional to the signal. After the wire was magnetized, the audio signal was stored as a varying magnetic field on the wire. To replay the sound, a reading head was positioned on the wire to detect changes in its magnetic field and convert the changes into a proportional electrical audio signal. The signal was then fed to a speaker, such as that of a telephone. The magnetic recording approach was adopted by the music industry, and magnetic tapes replaced metallic wires as a recording medium. This type of recording technique was later used to store music on audiocassettes, but its popularity diminished in the wake of CDs, flash drives, and alternative storage media.

CREATING SOUND

Distinguishing between early analog sound technologies and today's digital audio devices requires an understanding of a sound wave's physical properties. Sound is caused by physical disturbances of air molecules. Hitting a guitar string, for example, causes it to vibrate, which in turn compresses the surrounding air molecules and creates a variation in air pressure around

Figure 6-1

Phonograph and record player

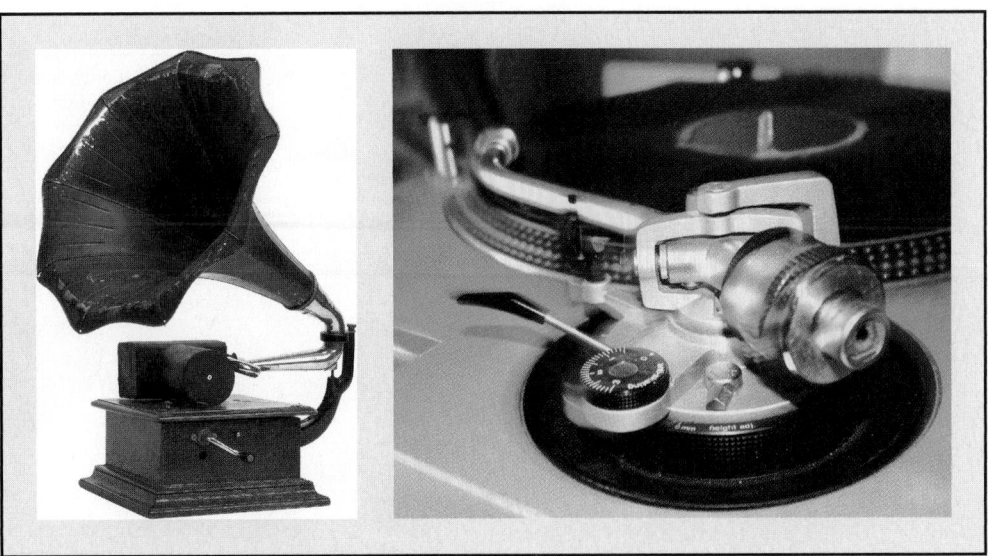

the string. Hitting a drum causes the drum head to vibrate, which compresses the surrounding air and creates an air pressure variation around the drum. The disturbance of air molecules is not confined to the molecules surrounding the string or the drum, but is transferred from molecule to molecule. This transfer of energy among molecules creates a mechanical wave of energy called a **sound wave**, which propagates away from the source of the disturbance.

When sound waves reach the human ear, variations in air pressure of the sound wave cause the eardrum to vibrate in proportion to the pressure variations. The mechanical motion of the eardrum is then converted to a signal that the brain can process. Although this explanation is oversimplified, it helps to explain how sound is physically created and heard.

The creation of sound waves is usually compared to the formation of water waves. Imagine jumping into a swimming pool. As you enter the water, your energy displaces the surrounding water molecules. The disturbed water molecules transfer their energy to neighboring water molecules in turn, and so on. This energy transfer creates a mechanical water wave that radiates out from the point of your contact with the water.

One fundamental difference between sound waves and water waves is the direction of displacement of water molecules and air molecules. Water molecules are displaced in the transverse direction as the water wave radiates away from the disturbance in the longitudinal direction. In contrast, the displacement of air molecules and the radiation of a sound wave occur in the longitudinal direction. Thus, sound waves are classified as longitudinal waves and water waves are a combination of transverse and longitudinal waves. The waves are similar in that both are classified as **mechanical waves** (see Figure 6-2).

CONVERTING BETWEEN SOUND AND ELECTRICITY

To digitize sound, electrical systems must convert *mechanical sound waves* into *electrical sound waves*—in other words, into an *electrical audio signal*. Microphones capture sound waves and convert them into an electrical form, while speakers convert the electrical signal back into

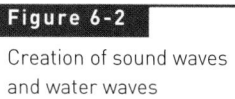

Figure 6-2

Creation of sound waves and water waves

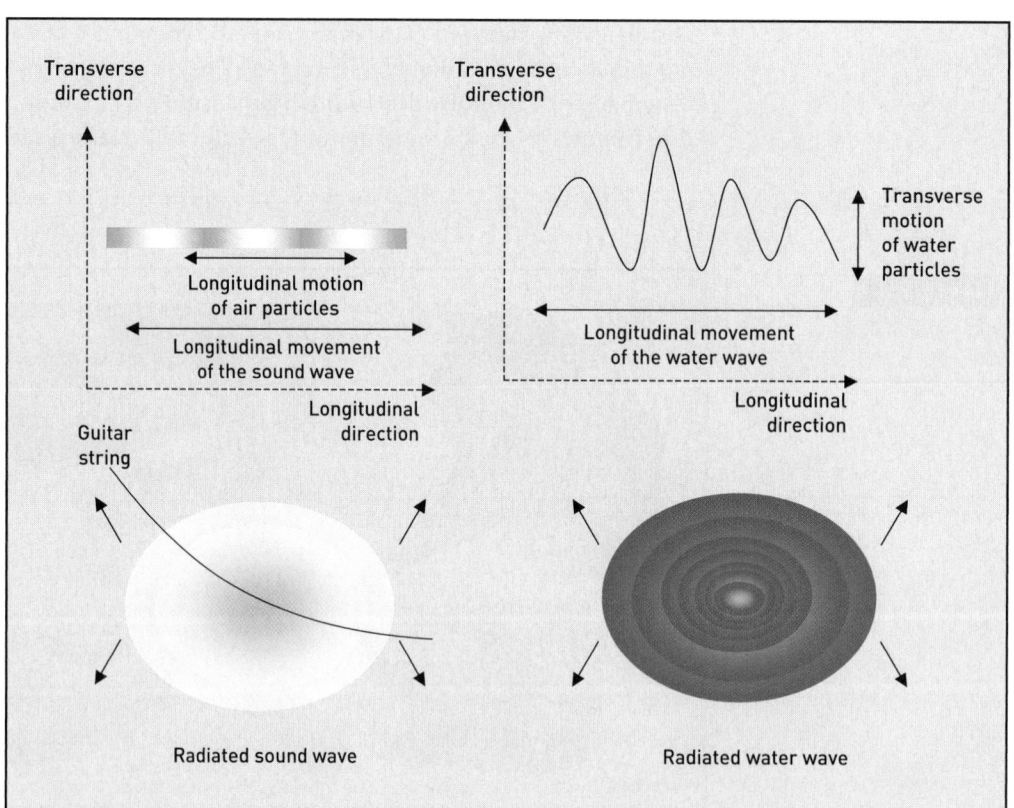

sound waves. When you play a chord on an acoustic guitar, the resulting sound wave may reach a microphone. Part of the microphone, usually called a diaphragm, mechanically vibrates in proportion to the air pressure variations of the sound wave. These mechanical variations are then converted to electrical waves, called an electrical audio signal, which is inherently analog. Similarly, when an electrical audio signal reaches a speaker, it causes the speaker diaphragm to vibrate in proportion to the signal. The mechanical vibrations of the diaphragm disturb nearby molecules and create a sound wave.

Figure 6-3 shows how a microphone converts sound into an electrical audio signal and how a speaker converts an electrical audio signal back into sound. Figure 6-4 depicts a detailed audio signal generated by the microphone when the guitar chord is played.

The audio signal generated at the microphone's output is an analog signal—the electrical variations are analogous to the sound created by the guitar, and the electrical signal varies in a continuous manner, proportional to the sound wave. The electrical fluctuations correspond to the pressure fluctuations of the sound wave. The y-axis represents electricity in volts[1] (V) and the x-axis denotes time in seconds. This electrical signal can then be converted into a stream of bits by passing through an audio digitizer. Such digitizers are found in sound cards embedded within computers and in chips within Cellular phones. To understand the process of audio digitization, you must understand some additional properties of sound.

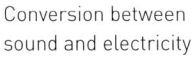

Figure 6-3

Conversion between sound and electricity

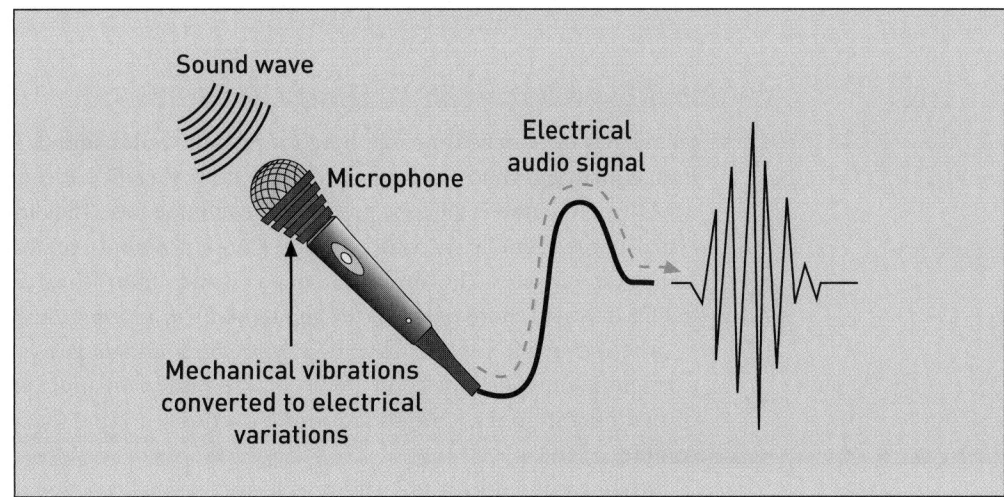

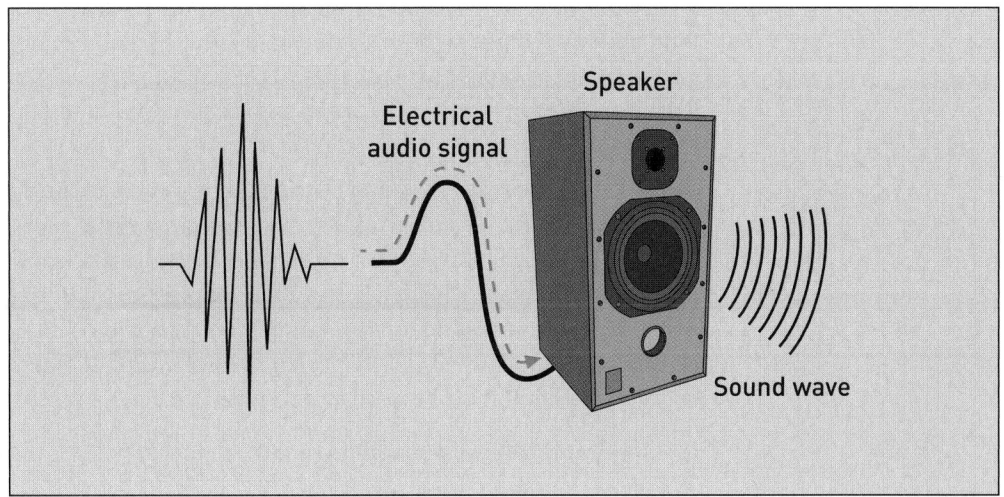

[1] Although variations in current and power can be used to quantify electricity, this discussion concentrates on variations in voltage.

Figure 6-4

Electrical signal
generated by a
microphone capturing
the sound of a guitar

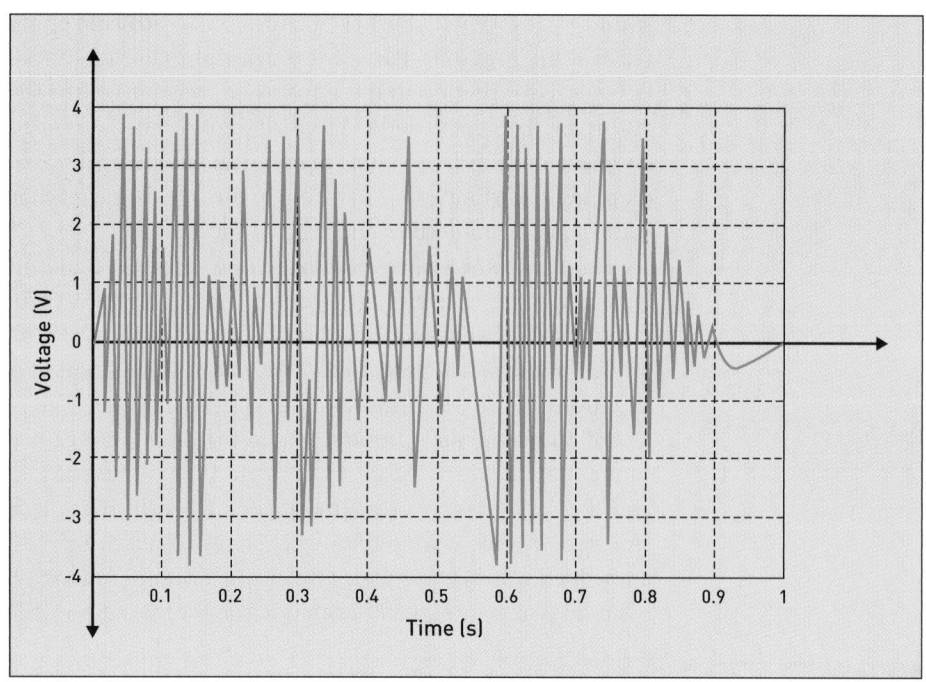

PURE AND COMPLEX SOUNDS

Sound may be classified as either **pure sound** or **complex sound**. Almost all the sounds we hear, from human speech to music to a barking dog, are classified as complex sounds. One example of a nearly pure sound is the tone produced by a tuning fork. The difference between the two types is that a pure sound has a constant pitch. Complex sounds, on the other hand, are perceived as having varying pitch. The tuning fork has a distinct "hum" that corresponds to a constant pitch, but the distorted chord of an electric guitar produces a continuously varying pitch.

The difference between the two types of sound is easier to explain graphically. For example, Figure 6-5 depicts one-tenth of a second of an audio signal produced at the output of two microphones. One microphone captured a pure sound and the other captured a complex sound.

When captured by a microphone, a pure sound is a signal that varies in a **sinusoidal** manner. This term is used because a pure sound wave's electrical variations follow the behavior of the sine function in mathematics. By contrast, a complex sound captured by a microphone fluctuates in a nonsinusoidal and irregular manner. Although most sound is classified as complex, we will first discuss properties of pure sounds and then address complex sound properties. These discussions are useful in understanding the process of audio digitization, which we cover later in this chapter.

Amplitude, Frequency, and Phase Properties of Sound

The audio signal shown in Figure 6-6 corresponds to a pure sound. It begins at a voltage of 0 V, increases to a maximum value of 4 V, starts decreasing, takes on a minimum value of –4 V, and then increases again. This type of variation repeats periodically for pure sounds. The audio signal is analog; in this example, it exhibits a regular and continuous variation between a fixed voltage range of –4 V and 4 V. The voltage value (magnitude) that the signal takes at any point in time is its **amplitude**, a measure of the displacement of the signal from the x-axis.

Amplitude (A) is the magnitude of the signal at a given instant in time (t).

For example, the magnitude of the audio signal shown in Figure 6-6 at t = 0.0125 s is exactly 4 V, so its amplitude at that time is 4 V. Similarly, the amplitude of the signal at t = 0.025 s is 0 V, and the amplitude at 0.0375 s is –4 V. As mentioned earlier, the amplitude starts to increase at t = 0 and reaches a maximum of 4 V at t = 0.0125 s. The amplitude then

Figure 6-5

Audio signals
corresponding to pure
and complex sounds

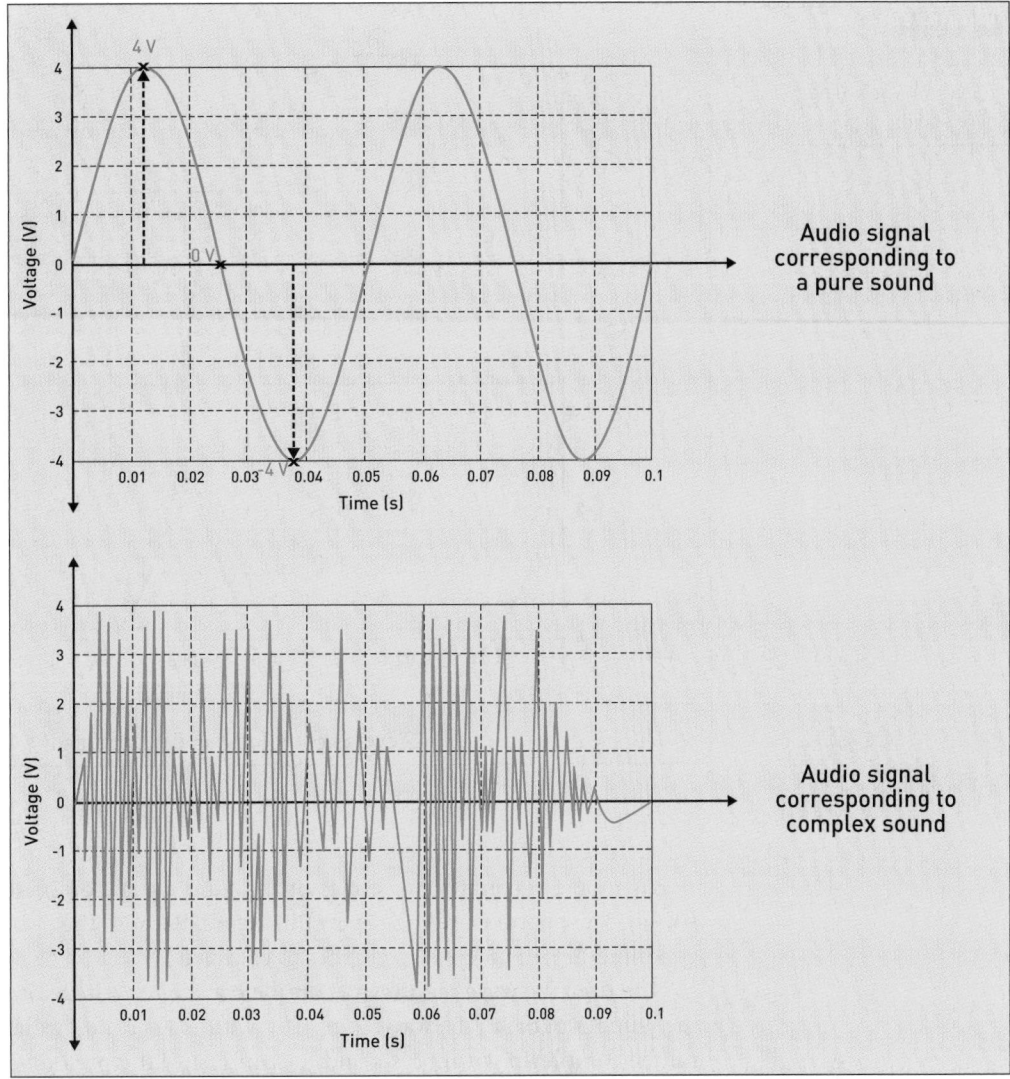

starts to decrease and crosses the x-axis at t = 0.025 s, when the amplitude is 0 V. The amplitude continues to decrease and reaches its minimum value of –4 V at t = 0.0375 s, at which point the amplitude rises and reaches 0 V at t = 0.05 s. A complete cycle of the wave is called a wave cycle, and the time it takes to complete this cycle (0.05 s) is called the **period**. This cycle repeats itself for the duration of the wave.

Period (T) is the time a wave requires to complete a single cycle; it is measured in seconds (s).

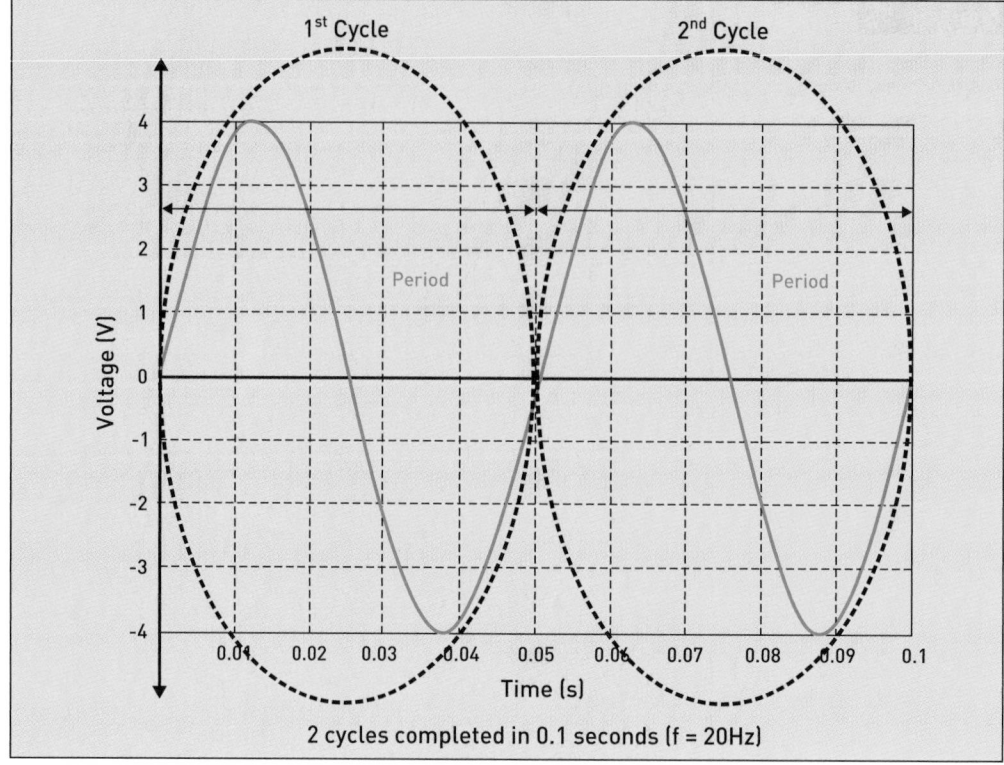

1st Cycle   2nd Cycle

Period   Period

2 cycles completed in 0.1 seconds (f = 20Hz)

Figure 6-6 illustrates two cycles of an audio signal completed in one-tenth of a second (0.1 s). The period is easy to see in this figure. Because the period T = 0.05 s, both the first cycle and the second cycle take 0.05 s to complete.

The time the wave requires to complete a single cycle determines the wave's **frequency**, the number of cycles the wave completes within 1 second. If the period is short, more cycles can be completed within 1 second. In contrast, if the period is long, fewer cycles are completed in 1 second.

Frequency (f), measured in hertz (Hz), is the number of cycles a wave completes in one second (s).

In our example, 20 cycles are completed in 1 second, so the frequency is 20 Hz. Although this point is not immediately obvious from the figure, you can see that if two cycles are completed in 0.1 second, 20 cycles are completed in 1 second.

A high frequency implies that the number of cycles is large. To accommodate the large number of cycles within 1 second, the period has to be correspondingly small. In fact, there is an *inverse relationship* between the period and frequency, such that:

f = 1/T and T = 1/f

Therefore, the frequency of the wave in Figure 6-6 can be calculated as f = 1/0.05 = 20 Hz because the period is known.

Figure 6-7 illustrates two audio signals: one with a frequency of 500 Hz and another with a frequency of 1000 Hz. As the frequency increases, the period decreases and vice versa. Note that 500 cycles are completed within 1 second for the 500-Hz signal, and 1000 cycles are completed per second for the 1000-Hz signal.

Figure 6-7

Audio signals with
different frequencies

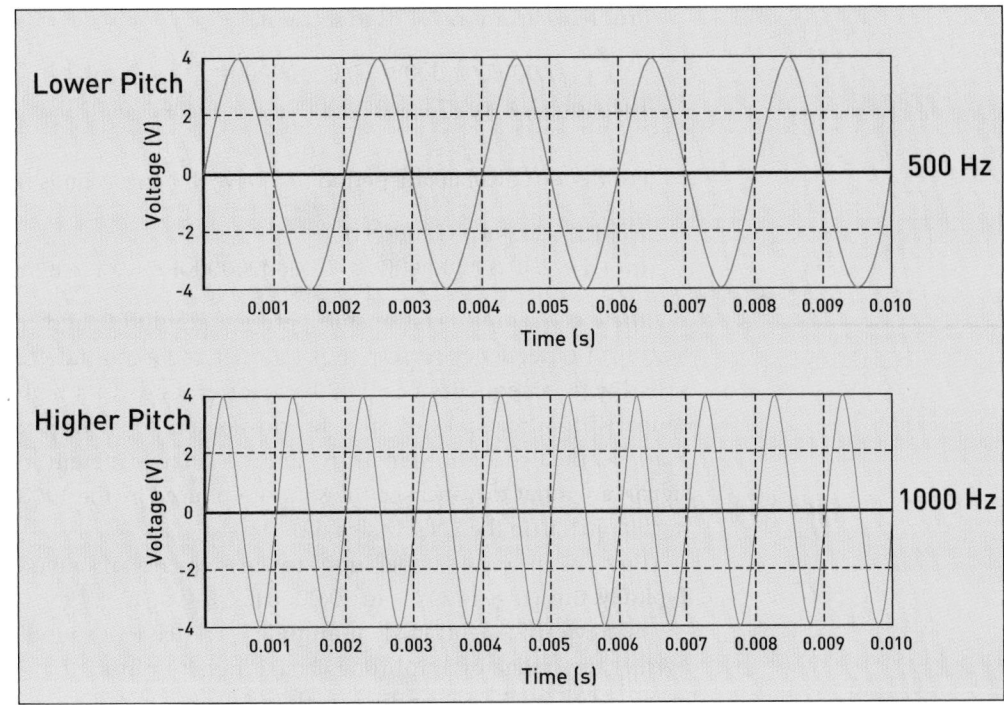

High-frequency signals are commonly encountered in IT, necessitating multipliers as a shortcut to express large numbers. For example, a signal with a frequency of 1000 Hz is commonly called a 1-kHz signal. These multipliers make calculations easier by eliminating large numbers. The following multipliers are commonly used within the context of high-frequency signals:

- *Kilo*hertz (kHz) 10^3 = 1000 Hz (thousand)
- *Mega*hertz (MHz) 10^6 = 1,000,000 Hz (million)
- *Giga*hertz (GHz) 10^9 = 1,000,000,000 Hz (billion)

Problem—Calculate how many cycles of a wave are completed within 1 second if the frequency of the wave is 3 MHz.

1 MHz = 1,000,000 Hz
3 MHz = 3 × 1,000,000 = 3,000,000 Hz
3,000,000 cycles are completed in 1 second.

Problem—Express 5,000,000,000 Hz in terms of GHz.

1 GHz = 1,000,000,000 Hz
Therefore, 5,000,000,000 Hz = 5 GHz

Because frequency and period are inversely related, a very high frequency corresponds to a very small period. Multipliers eliminate the need to use a long string of zeros in front of a small number when referring to the period.

- *milli*seconds (ms) 10^{-3} = 0.001 seconds (1/1000 seconds)
- *micro*seconds (µs) 10^{-6} = 0.000001 seconds (1/1,000,000 seconds)
- *nano*seconds (ns) 10^{-9} = 0.000000001 seconds (1/1,000,000,000 seconds)

Problem—Convert 23 ms to seconds (s).

1 ms = 1/1000 s = 0.001 s
Therefore, 23 ms = 23 × 0.001 = 0.023 s

Problem—Convert 7.4 ns to seconds.

1 ns = 1/1,000,000,000 s = 0.000000001 s
Therefore, 7.4 ns = 7.4 × 0.000000001 = 0.0000000074 s

Problem—Calculate the period of a wave in nanoseconds if the frequency is 10 GHz.

10 GHz = 10,000,000,000 Hz
T = 1/f = 1/10,000,000,000 s = 0.0000000001 s = 0.1×10^{-9} s = 0.1 ns

There is a simple relationship between amplitude and frequency and audible sound. Simply put, frequency corresponds to **pitch** and amplitude corresponds to volume. A pure sound with a frequency of 10,000 Hz corresponds to a high-pitched sound. By contrast, sound with a frequency of 300 Hz corresponds to a low-pitched, rumbling sound. As an example, the two audio signals in Figure 6-7 have different frequencies. When these audio signals are input to a speaker, the sound produced by the 1000-Hz audio signal is heard at a higher pitch than the 500-Hz signal.

There are lower and upper limits to the frequency of sound that the human ear can detect. Typically, this range is 20 Hz to 20,000 Hz.

As previously mentioned, amplitude is related to a sound's volume. A signal that varies over a large range of amplitudes has a higher volume than one that varies over a small range. Consider the two audio signals in Figure 6-8. Although both signals have a frequency of 500 Hz, the upper signal will be louder when applied to a speaker. Because both sounds have the same frequency, they will be heard with the same pitch. Notice that you can directly calculate the frequency of these waves from the graphs. The period is 2 ms = 0.002s, so the frequency f = 1/0.002 = 500 Hz.

Figure 6-8

Audio signals with different volumes

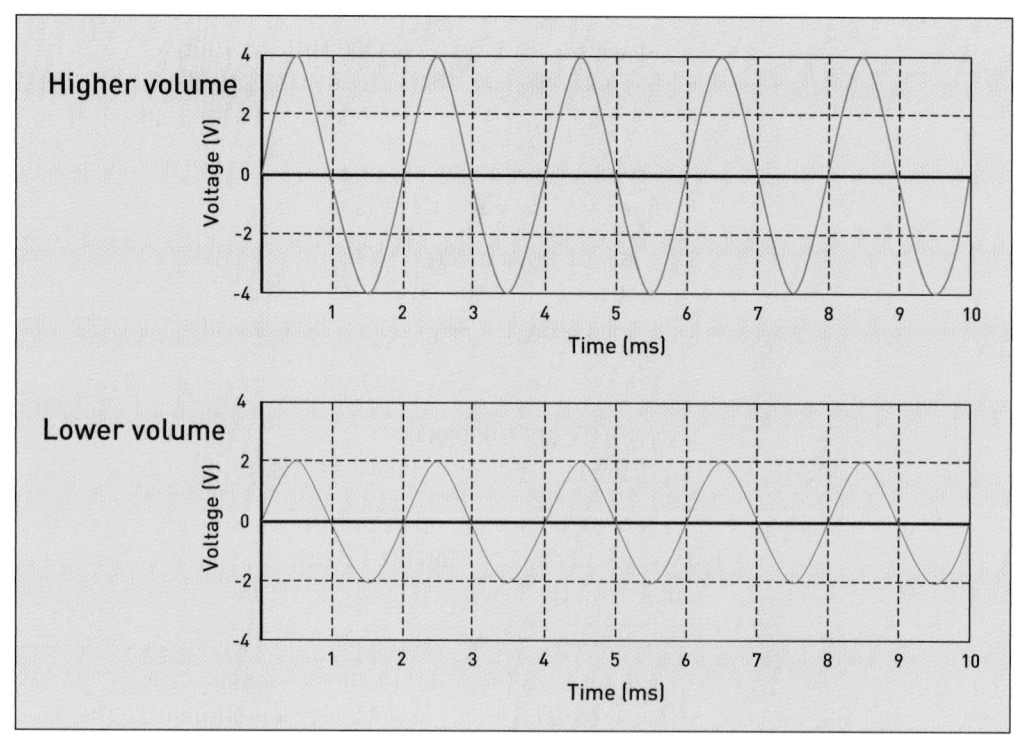

The human ear does not respond to all frequencies with equal sensitivity—it is more responsive to some frequencies than others. For example, the ear would be more responsive to a pure sound with a frequency of 3000 Hz than one with a frequency of 15,000 Hz. Thus, even if the two pure sounds varied between the same amplitude range, such as between –4 V and 4 V, the 3000-Hz pure sound would seem louder than the 15,000-Hz pure sound.

In addition to amplitude and frequency, pure sounds are also described in terms of their **phase difference**. Phase difference is meaningful when comparing two waves, such as two electrical audio waves of the same frequency, because it describes the degree to which the two waves are aligned, or in synch with each other. In other words, it measures the degree to which one wave lags relative to another wave. To understand this concept, consider the arrows A and B in Figure 6-9, which rotate counterclockwise at the same rate starting from their original horizontal position and complete their rotation of 360° within the same amount of time. Also, assume that the two arrows rotate independently of each other. In other words, they do not necessarily start rotating at the same time, even though they complete their full rotation in the same amount of time. If arrow A starts its rotation before arrow B, arrow A will complete its full rotation earlier than arrow B because both arrows rotate at the same rate. Also assume that the start of an arrow's rotation corresponds to the start of the cycles of the sine wave, such that each arrow completes a 360° rotation in the same time it takes to complete one cycle of the sine wave. Therefore, the time the arrows require to complete a full rotation corresponds to the period of the wave.

Figure 6-9

Two rotating arrows 90° apart

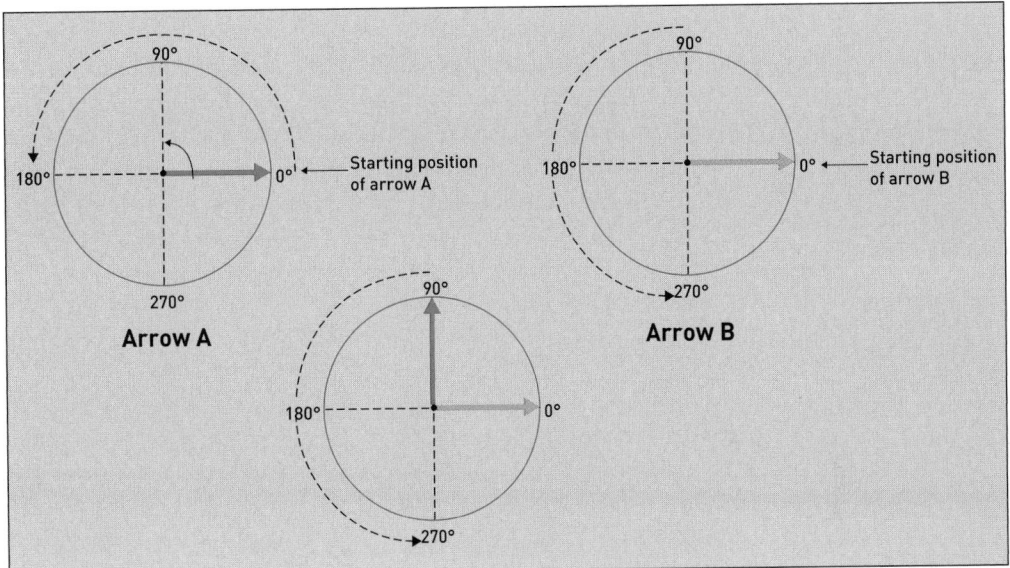

Initially, arrows A and B are at their horizontal positions, as illustrated in the top part of Figure 6-9. Assume that arrow A starts to rotate counterclockwise and that arrow B is stationary. Wave A is generated at the same time the arrow starts rotating, as depicted by the solid line in Figure 6-10. One cycle of wave A is completed within the time the arrow completes a full rotation of 360°, which is within 0.002 s. The first wave cycle thus starts at t = 0 and ends at t = 0.002 s. Next, assume that arrow B starts rotating 0.005 s after arrow A has started. Notice in Figure 6-9 that arrow B begins rotating when arrow A has completed a rotation of 90°. Assume that another sine wave is generated at the start of arrow B's rotation, as depicted by the dashed line in Figure 6-10. Wave A and wave B are said to have a phase difference of 90°, where wave B lags behind wave A. Figure 6-10 illustrates the two waves with a frequency of 500 Hz and a difference in phase of 90°. Note that the phase difference exists even though both waves have the same frequency and amplitude ranges.[2]

> **Phase difference describes the alignment of two waves along the time axis and may be measured in degrees.**

The next section discusses the significance of the phase difference between two waves.

---

[2] Although the concept of phase is not discussed in further detail in this chapter, understanding it here will prove useful in later chapters when we discuss other types of waves.

Figure 6-10

Audio signals with a
phase difference of 90°

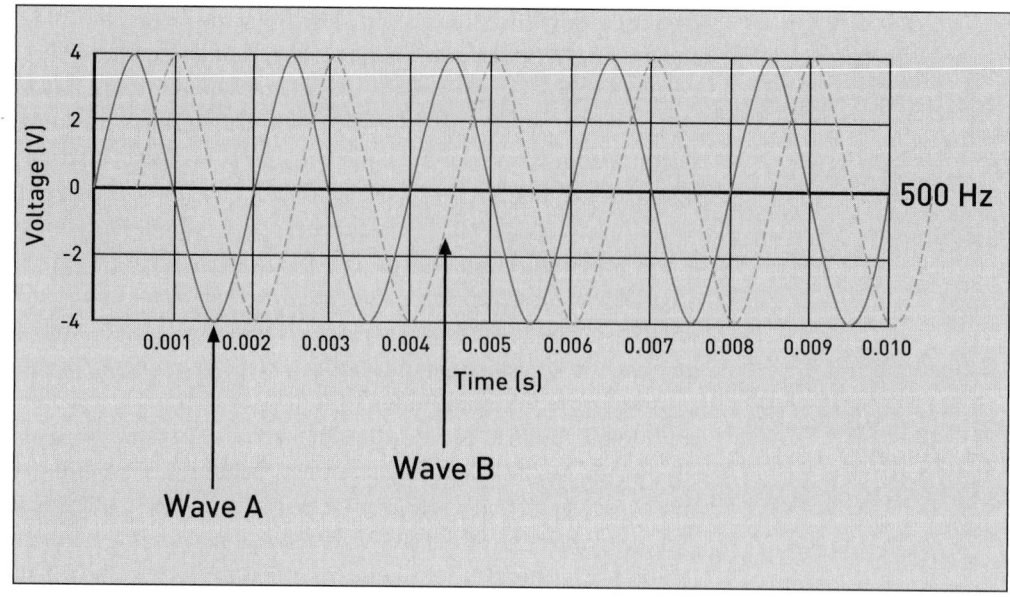

Frequency Composition of Sound

Pure sounds, in simple terms, can be considered the basic components that make up complex sounds. Each pure sound component can differ in terms of frequency, amplitude ranges, and phase differences. Figure 6-11 illustrates how a complex sound is resolved into its pure sound components. Each pure sound component that makes up a complex sound is often called a **frequency component**.

Figure 6-11

Resolving a complex
sound into its frequency
components

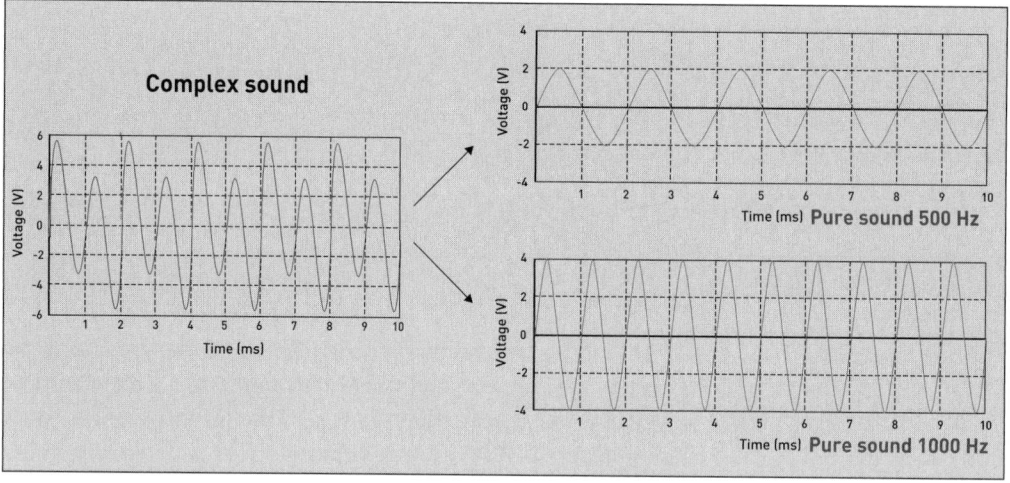

The first component of the complex sound has a frequency of 500 Hz and varies within an amplitude range of –2 V to 2 V. The second component has a frequency of 1000 Hz, varies within an amplitude range of –4 V to 4 V, and has a phase difference of 0° relative to the first component. Any number of components and any combination of frequency, amplitude, and phase differences are possible within the components of complex sounds.

Given that complex sounds can be resolved into several frequency components, you should understand that any type of complex sound can be artificially created by combining two or more frequency components. Some sound synthesis systems rely on this fundamental principle. Music synthesis systems generate electrical audio signals, then combine them and apply the combined signal to speaker systems to create unique complex sounds. Therefore, the resulting complex sound depends on the number of components, the amplitude ranges of each component, and the phase difference between the frequency components. An unlimited number of components can be used to create complex sounds.

Every person's voice has a unique frequency composition. Voice recognition systems use this information to distinguish one person's voice from another's.

In audio digitization, you must have information about the frequency content of the audio signal to be digitized. More specifically, you must know the values of the highest and lowest frequency components. The next section explains the importance of this information.

To determine the highest and lowest frequency values within a complex sound, the audio signal can be applied to a device called a **spectrum analyzer**. A spectrum analyzer separates frequency components and displays this information on a graph called a **frequency spectrum**, with vertical spikes indicating the frequency components. For example, the frequency spectrum of the complex sound shown in Figure 6-11 is illustrated in Figure 6-12. The x-axis corresponds to frequency, and a spike is centered around the frequency of each corresponding component. The frequency spectrum therefore provides information about the number of frequency components and their relative intensities.[3] The phase difference between the frequency components is not shown in the frequency spectra, but you can obtain this information if needed. For the purposes of audio digitization, you only need information about the highest and lowest frequency components.

Note that the frequency composition of the human voice does not resemble the spectrum shown in Figure 6-12. The human voice contains many more frequency components.

Figure 6-12

Frequency spectrum of a complex sound

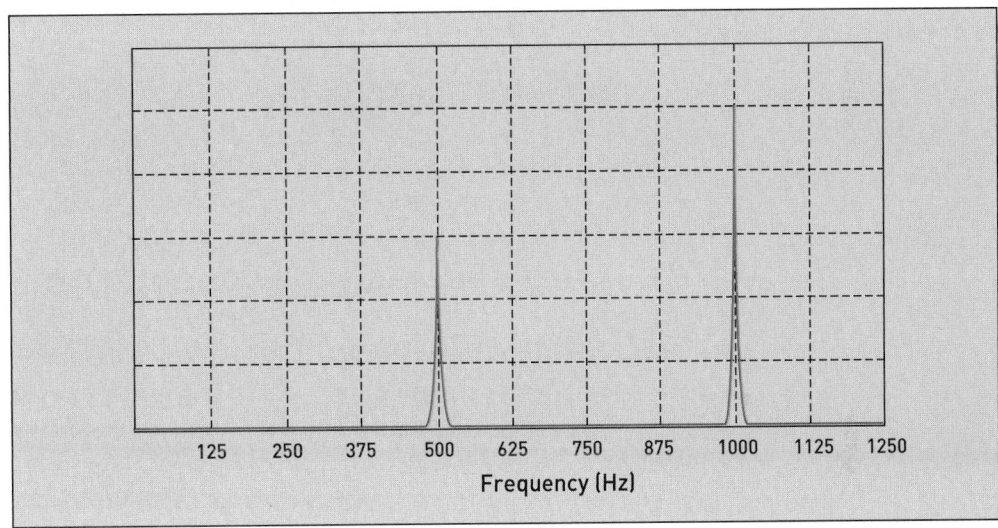

DIGITIZING SOUND

Having been introduced to the basic properties of sound, you are ready to learn how and where audio signals are digitized once they are captured by microphones. The digitization process takes place in circuits called **analog-to-digital converters** (ADCs). For example, digital Cellular phones have ADCs that digitize the audio signal generated by their microphones. The microphone captures sound and converts it into an electrical audio signal. The audio signal is fed to the ADC circuit, where it undergoes digitization. The process produces a stream of bits that are transmitted with extra bits over the wireless channel. The addition of these extra bits serves a specific purpose, as you will learn in later chapters.

Similarly, a computer can digitize audio. A microphone connected to the computer captures the sound and converts it into an electrical audio signal. This signal is then fed to the ADCs within the chips on the motherboard or a sound card. The bits that result from

---

[3] The y-axis of the frequency spectrum typically represents the power of each frequency component, but this concept is not discussed in detail here. However, note that the power of each frequency component is related to the amplitude of each component in the time domain, which is why the 1000-Hz component appears to have a larger spike than the 500-Hz component.

digitization can then be saved on a hard disk or another type of storage medium, possibly after being compressed to reduce the size of the digital audio file.

Sound cards and Cellular phones contain **digital-to-analog converters** (DACs)—circuitry that converts streams of ones and zeros into a corresponding electrical audio signal. This audio signal is then fed to a speaker, where it can be converted to sound before undergoing amplification and additional processing (see Figure 6-13).

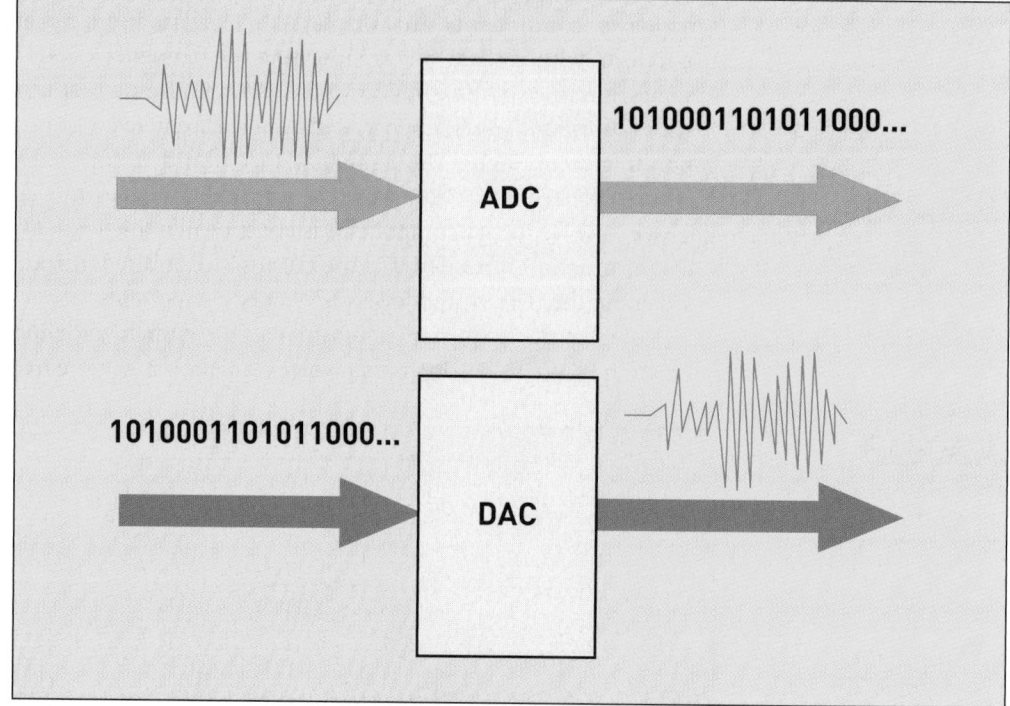

Figure 6-13

Analog-to-digital and digital-to-analog converters

THE THREE-STEP PROCESS OF DIGITIZATION

Theoretically, you can consider audio digitization to be a three-step process. This procedure is often called **pulse code modulation** (PCM), and is commonly associated with digitizing audio in telephony and CD-quality music applications. The three steps of audio digitization are:

- Sampling the audio signal
- Quantizing, or rounding off, the actual voltage value of each sample to the nearest voltage value
- Assigning a binary code to each quantized sample

Sampling

During the first step, the infinite analog audio signal values are reduced to a finite set of voltages so that the audio information is made discrete. Recall that Chapter 3 explained how to digitize speedometer readings by taking samples of them, rounding off each sample to an approximate value, and assigning a binary code to each sample. You can use the same procedure to digitize the electrical audio signal. The process involves sampling the continuous audio signal at certain intervals, retaining only an approximate value of each sample's voltage, and discarding the rest.

Consider the complex audio signal in Figure 6-14 that contains three frequency components, as indicated by the frequency spectrum. The components have frequencies of 83 Hz, 166 Hz, and 332 Hz. During the first step of audio digitization, the infinite number of voltage values that constitute the continuous, analog signal is reduced to a finite number to represent the signal. The technological design requires a compromise between the number of samples and the degree to which these samples represent the original signal. Taking too few samples would lead to an inaccurate representation of the signal. Taking too many samples would drive up costs by increasing the number of bits in the resulting digital file.

Figure 6-14

Audio signal of a complex
sound and its frequency
spectrum

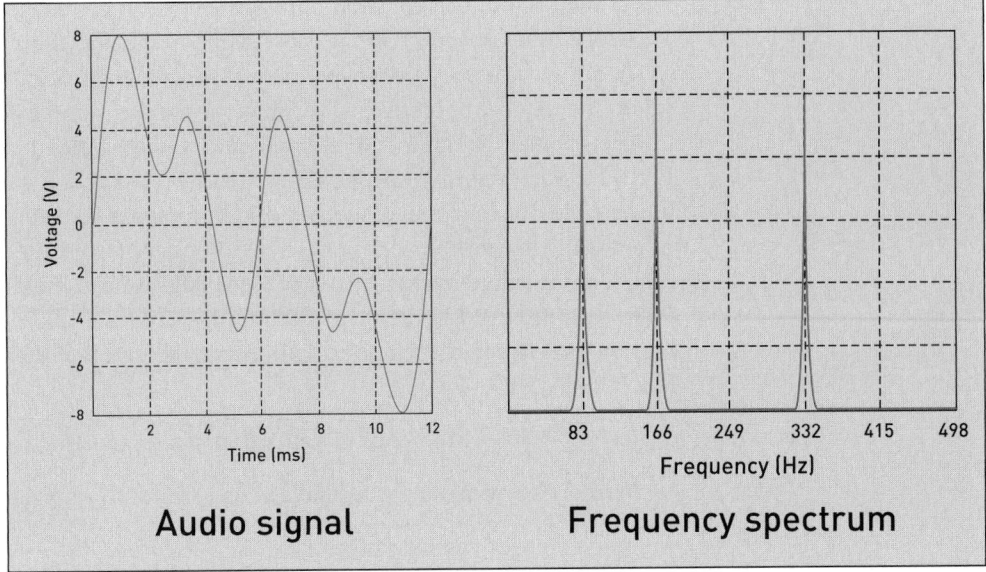

Audio signal   **Frequency spectrum**

The sampling process is implemented by the ADC, which takes samples of the audio signal at regular intervals and records the voltage value of each sample. Figure 6-15 illustrates the result of sampling the audio signal once every millisecond (0.001 second). Sampling starts at t = 0 s, so the first sample at t = 0 s has an amplitude of 0 V. The next sample is taken after one millisecond (t = 0.001 s); the sample's value is 8 V. The process continues in this manner until the entire signal is sampled, resulting in 13 samples represented by the dots in Figure 6-15.

Quantizing

Before assigning binary codes to the samples, each must be quantized, or rounded off, to one of a finite number of voltages. Recall from Chapter 3 that the speedometer measurements were sampled, but each sample could take an infinite number of possible speed values. The measurements had to be rounded off to one of a finite number of values. The same concept applies to digitizing audio signals and rounding them off. Because the amplitudes for each sample can take any value between –8 V and 8 V, an infinite number of different voltages can exist within this 16-V range. We must reduce this complex task into a manageable one, just as we limited the number of samples in the original signal during the first phase of digitization. We must quantize the actual sample values so we can assign a fixed-length binary code to each sample. The number of bits we can assign per sample is limited during the final stage of

Figure 6-15

Samples of the audio
signal

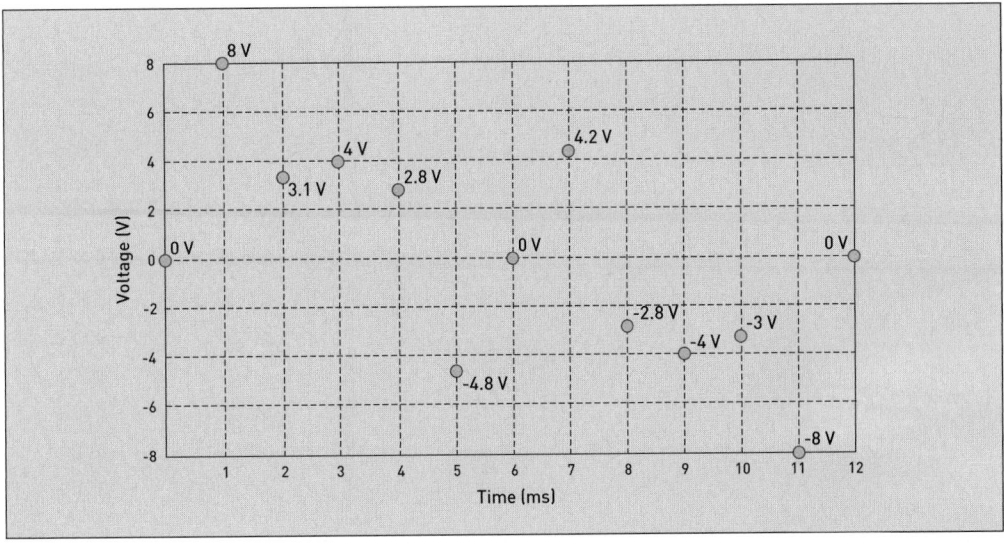

audio digitization. This constraint limits the number of voltage levels that can be represented. Therefore, we must round off the actual sample values to the closest available values.

To achieve this goal, we can represent each actual sample value with an approximate voltage value. The number of bits we are willing to assign per sample during the final step of digitization determines the number of approximate voltage values that are available in the quantization stage. For instance, if we are willing to use two bits per sample, four different voltages would be available. The four voltage levels would be represented by 00, 01, 10, and 11. Similarly, if we were willing to use 4 bits per sample, the number of available voltages for rounding off each sample value would increase to 16. These voltage values would be represented by 0000, 0001, 0010, 0011, 0100, 0101, 0110, 0111, 1000, 1001, 1010, 1011, 1100, 1101, 1110, and 1111. The following equation shows the relationship between the number of bits and the number of voltages that can be represented:

$$2^{\text{Number of bits}} = \textbf{Number of voltages that can be represented}$$

How do we calculate the available voltage values? Two parameters are required:

- Audio **signal dynamic range**
- Number of bits we are willing to use per sample

The first parameter relates to the minimum and maximum voltage values between which the audio signal fluctuates. For example, the audio signal in Figure 6-14 fluctuates within a range of –8 V to +8 V, so its dynamic range is 16 V.

Determining how many bits to assign to each sample depends on the availability of resources. The more bits we use, the better the approximation will be. On the other hand, using a larger number of bits results in a bigger audio file, a longer download time for the digital audio file, and more space required for storage. The trade-off is between performance (digital audio quality) and the cost of transmitting, storing, and processing the digital audio file.

Suppose we are willing to use 4 bits per sample. This means the value of each sample will be rounded off to one of 16 voltage values, because $2^4 = 16$.

Figure 6-16 lists 16 possible voltage values to which each of the actual values may be rounded. Note that these values vary between –7.5 V and 7.5 V, a range that is within the dynamic range of the signal.

Figure 6-16

Possible voltage values for quantization

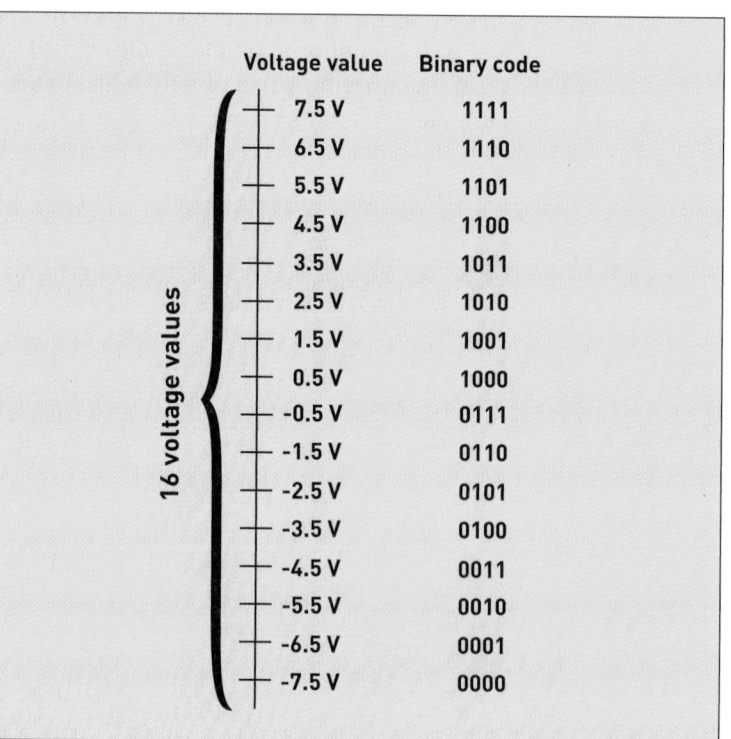

Voltage value	Binary code
7.5 V	1111
6.5 V	1110
5.5 V	1101
4.5 V	1100
3.5 V	1011
2.5 V	1010
1.5 V	1001
0.5 V	1000
-0.5 V	0111
-1.5 V	0110
-2.5 V	0101
-3.5 V	0100
-4.5 V	0011
-5.5 V	0010
-6.5 V	0001
-7.5 V	0000

16 voltage values

After determining the 16 possible voltage values by which each sample value will be approximated,[4] the next task is to develop an assignment scheme so that each rounded-off voltage value has a corresponding binary code. One simple scheme is to start at the bottom of the list and assign a string of four zeros to the last value. In Chapter 2, we learned about binary sequencing: if we start with a string of four ones, all the 4-bit combinations of ones and zeros are exhausted when we reach the top row. Consequently, the first voltage value is assigned a string of four ones, as illustrated in Figure 6-16. The opposite could also be applied—in other words, the all-zeros string could be assigned to the first voltage, and the all-ones string could be assigned to the last voltage value.

Use Figure 6-16 to round off the samples, which corresponds to Step 2 of the audio digitization process. The actual value of the first sample is 0 V. In Figure 6-16, note that 0 V is closest in value to 0.5 V or –0.5 V, so we can round it off to one of the two values. In this example, we round it off to 0.5 V. Thus, the actual value of the sample, 0 V, is rounded off, or quantized to 0.5 V. The actual value of the second sample is 8 V. The closest value is 7.5 V, so we round off this sample to 7.5 V. The rest of the samples are quantized in the same manner so that every sample has a corresponding quantized value.

Encoding

Completing the digitization process requires Step 3, the assignment of a binary code to each quantized value. To assign these codes, look up the binary values in Figure 6-16. For the first sample, the quantized value is 0.5 V. The figure shows that the corresponding code for 0.5 V is 1000, so the first sample is assigned a binary code of 1000. For the second sample, the quantized value is 7.5 V, and the corresponding binary code is 1111. This process is repeated for all the samples, and the corresponding binary codes are entered into Table 6-1.

Table 6-1

Rounded-off (quantized) values

Sample number	Actual value of sample	Quantized value of sample	Binary code
1	0 V	0.5 V	1000
2	8 V	7.5 V	1111
3	3.1 V	3.5 V	1011
4	4 V	3.5 V	1011
5	2.8 V	2.5 V	1010
6	–4.8 V	–4.5 V	0011
7	0 V	0.5 V	1000
8	4.2 V	4.5 V	1100
9	–2.8 V	–2.5 V	0101
10	–4	–3.5 V	0100
11	–3	–2.5 V	0101
12	–8	–7.5 V	0000
13	0 V	0.5 V	1000

Once each sample is converted to a binary string, the digitization process is complete. The digitized audio signal is now represented by $4 \times 13 = 52$ bits of ones and zeros, which correspond to the binary codes assigned to each of the 13 samples:

Sample #:	1	2	3	4	5	6	7	8	9	10	11	12	13
	1000	1111	1011	1011	1010	0011	1000	1100	0101	0100	0101	0000	1000

[4] The list of voltage values to which samples are rounded off can be determined using several techniques. One method is to divide the signal range into 16 equal parts and determine a reasonable voltage value for each. Another approach is to divide the total range into 16 unequal values, which is common in voice digitization. The book does not discuss these techniques in detail.

The entire digitization process takes only a split second in the ADC circuits. As soon as each sample is generated, it is quantized and immediately assigned a binary code. In real examples, the sampling rate and the number of bits per sample are far greater than those in our example. Furthermore, the dynamic range of the audio signal is not always accurately known by the ADC circuit, so an approximate dynamic range is assumed.

Once the ADC process is complete, the resulting bits can either be transmitted, as in a cellular communications scenario, or they can be stored as an audio file, as with a sound card.

Problem—Calculate the number of bits generated from a digital Cellular phone conversation.

The standard sampling rate for digital telephony is 8000 Hz, with 8 bits assigned per sample. Based on these figures, calculate the raw number of bits generated by the ADC circuit in a digital Cellular phone during a call that lasts 4 minutes.

Number of bits generated in 1 second = Number of samples per second × Number of bits per sample = $8000 \times 8 = 64{,}000$ bits per second
64,000 bits are generated in 1 second.
Number of seconds in 1 minute = 60 seconds
Number of seconds in 4 minutes = $4 \times 60 = 240$ seconds
Number of bits in 240 seconds = $64{,}000 \times 240 = 15{,}360{,}000$ bits

Therefore, a digital telephone generates more than 15 million bits from its ADC within 4 minutes.

Problem—Calculate the raw number of bits generated by the ADC of a 16-bit sound card with a 96-kHz sampling rate to digitize an audio signal that lasts for 3 seconds.

Number of bits generated in 1 second = Number of samples per second × Number of bits per sample = $96{,}000 \times 16 = 1{,}536{,}000$ bits per second
Number of bits generated in 3 seconds = $1{,}536{,}000 \times 3 = 4{,}608{,}000$

Therefore, a sound card roughly generates more than 4 million bits from its ADC within 3 seconds.

In IT, systems that use audio digitization rarely store or transmit the raw bit streams of the conversion process from analog to digital. The bit streams are traditionally compressed, and extra bits are usually incorporated for controlling errors and ensuring efficient storage and transmission.

The audio digitization process we have described raises some significant questions:

- How do we know how often to sample?
- How do we decide on the number of bits to use per sample?
- How do we recover the audio signal from the stream of bits?
- What effect does sampling and quantization have on the sound we hear when the digital audio file is played back?

The remaining sections discuss these questions and other related issues.

THE NYQUIST SAMPLING THEOREM

As discussed earlier, the first step of digitization is to take samples. How do you know how often to take samples so that they adequately represent the original signal? The answer lies with the **Nyquist sampling theorem**, or Nyquist-Shannon sampling theorem: *the minimum number of samples per second (the sampling frequency, or f_s) required to perfectly reconstruct the analog signal should equal at least twice the value of the difference between the* *signal's highest frequency component (f_max) and lowest frequency component (f_min).* The theorem is represented by the following equations:

$$f_s \geq 2(f_{max} - f_{min})$$
$$f_s \geq 2B$$

The difference between the highest and lowest frequency components is also known as the **signal bandwidth** (B). Naturally, you must know the frequency content of an audio signal to determine the minimum sampling rate. In Figure 6-14, the complex signal was made up of three frequency components: 83 Hz, 166 Hz, and 332 Hz. By taking the lowest and highest frequency components in the set, we know that this signal should be sampled at a rate of at least 2x(332 – 83) = 498 samples per second. In other words, it should be sampled with a sampling frequency (f_s) of at least 498 Hz in order for the digitized version to accurately represent the original version. Because we took a sample once every millisecond, the corresponding sampling frequency was 1/0.001 = 1000 Hz in our example.

The preceding discussion, in theory, tells us that the sampling rate is more than adequate to accurately capture the original information. It suggests that we could have sampled this signal at a lower rate, corresponding to a sample taken approximately once every 2 ms (1/498 Hz ≈ 0.002s = 2 ms), instead of approximately once every 1 ms. In actuality, however, the more samples we take, the more original information the samples capture. We can therefore consider that the Nyquist theorem only sets a lower limit to the sampling rate, and not an upper limit. The lower limit is also called the **Nyquist rate**. The upper limit is restricted by resources such as storage space and transmission time because as we collect more samples, we need to assign them all binary codes, resulting in more bits that we have to store, transmit, or process.

When digitizing voice information, the Nyquist rate is often taken to be more than twice the value of the highest frequency component, instead of twice the difference between the highest and lowest frequency components, because it is assumed that voice signals have frequency components that are close to 0 Hz.

> **Problem**—Calculate the minimum sampling frequency required to digitize an audio signal generated by a microphone that has captured a pure sound with a frequency of 400 Hz.

If the audio signal were input to a spectrum analyzer, it would show a single spike centered around a frequency of 400 Hz, which corresponds to a single frequency component. The bandwidth of the signal is 400 Hz, because there is only one component. The minimum sampling frequency required to digitize this signal would be 400 Hz × 2 = 800 Hz.

What about digitizing the human voice? The typical voice has several frequency components that span a wide range. Digital communication systems assume that the bandwidth of the human voice is 4000 Hz. Therefore, the standard sampling rate used by ADCs in Cellular phones is 8000 Hz.

Standard Sampling Rates and Number of Bits

Over the public switched telephone network, analog voice signals are also sampled at a rate of 8000 Hz, or 8000 samples per second. Each sample of an analog signal is a measurement of the signal's amplitude at that point in time. The amplitude could have an infinite number of potential values, but it is rounded off to one of 256 possible values. Why 256? Over the telephone network, each sample is encoded with 8 bits, which provides 2^8, or 256, possible values.

Because voice signals are sampled at 8000 samples per second, and because an 8-bit code represents the value of each sample, each voice transmission that uses PCM techniques to convert an analog voice signal into a digital binary stream requires a 64-Kbps transmission channel, as calculated previously.

The standard sampling rate for digital music in the music industry is 44.1 kHz, and each sample is assigned 16 bits, because music contains a larger range of frequency components than voice and requires higher fidelity than telephony applications. The sampling rate and number of bits assigned per sample are correspondingly much higher.

You can see that the more samples we take, the more original information we retain in the digitized version, because the information loss between samples is minimized. To help you choose exactly what sampling frequency to use, answer the following questions:

- With what type of application will the digitized audio be associated? If the application is a high-fidelity digital music system, a high sampling frequency should be used. Sampling rates for most sound cards can be adjusted depending on the desired fidelity.
- What are the associated costs? If the cost of digitizing equipment depends on the sampling frequency, the sampling frequency should be chosen accordingly.
- What are the limitations on resources, such as storage space and transmission rate for the application? If resources are limited, use a sampling frequency closer to the Nyquist rate.

QUANTIZATION ERROR

As we discussed earlier in this chapter, the second step of audio digitization is to round off the samples' actual values so that a finite-length code can be assigned to each sample. However, we did not discuss the effect of quantization on the quality of the resulting digital audio file. After we define the following concept, we can discuss its implications:

Quantization error *is the difference between the actual value of the sample and the value to which the sample is rounded off.*

For example, the first sample of the audio signal in the previous example had an actual voltage value of 0 V and was rounded off to 0.5 V. The difference between the two values, 0.5 V, is the sample's quantization error. The quantization error for the second sample can also be calculated as 0.5 V.

Quantization error measures the amount of information loss that occurs during the second step of audio digitization. During the sampling phase, the portion of the signal between samples is lost. This information loss can be minimized by increasing the sampling frequency. Further information loss occurs during the quantization phase because values are rounded off. Therefore, it is natural to ask how quantization error can be reduced. The answer is to assign more bits per sample. Adding an extra bit to each sample value results in twice as many possible values to which the actual sample values can be rounded. This leads to a smaller quantization error because twice as many voltage values are available for rounding, and the voltage differences decrease between the actual values and the rounded-off values.

If we increased the number of bits from 4 to 5 in our example, a sample value of 8 V would have to be rounded off to 7.75 instead of 7.5 V, resulting in a quantization error of 0.25 V instead of 0.5 V. Clearly, 7.75 V is closer to 8 V than 7.5 V. This example demonstrates the reduced disparity between actual and rounded-off values when more bits are assigned per sample. The trade-off between quality and cost is demonstrated again because more bits per sample means more bits overall. Assigning 4 bits per sample for 13 samples results in 4 × 13 = 52 bits in the digital audio file. Similarly, assigning 5 bits per sample results in 5 × 13 = 65 bits in the digital audio file. To help you choose exactly how many bits to assign per sample, answer the following questions:

- With what type of application will the digitized audio be associated? If the application is high-fidelity digital music systems, a larger number of bits should be used than in low-fidelity applications.
- What are the associated costs? If the cost of digitizing equipment depends on the number of bits assigned per sample, the number of bits should be chosen accordingly.
- What are the limitations on resources, such as storage space and transmission rate for the application? If resources are limited, use the minimum number of bits per sample.

The standard number of bits assigned in telephony systems, for example, is 8 bits per sample. In contrast, the music industry uses a standard of 16 bits per sample. Music needs to have a higher fidelity than human speech.

DIGITAL-TO-ANALOG CONVERSION OF AUDIO

The result of audio digitization is a stream of ones and zeros that can be compressed and stored as an audio file on a CD, transmitted across a network, or digitally altered by a computer. We have not yet addressed how to recover an audio signal from the stream of bits and how we can actually listen to it. The recovered audio signal is often not the same as the original signal, due to the loss in the sampling and quantization phases. If an audio signal has been sampled at a sufficiently high rate and a sufficiently large number of bits are assigned per sample, the audio stream can be reconstructed from the stream of bits to very closely resemble the original audio signal.

The recovery phase takes place in the DAC, which is present in any device associated with digital audio, such as an iPod, CD or DVD player, Play Station Personal (PSP) device, or a Cellular phone. When the bit stream corresponding to the digital audio file is input into the DAC, its circuits generate voltages that correspond to each sequence of bits. For example, a laser in a CD player might read bits on the CD and convey them to the DAC. To implement this process, the DAC needs to know how many bits have been assigned per sample and the sampling rate used to create the digital audio file in the first place.

Because standard bit rates and sampling rates are used in the telephony and music recording industries, the DAC circuits in devices such as CD players and Cellular phones can effectively reconstruct audio signals by segmenting the bits in the digital audio file. A voltage based on the value of the bit group is then regenerated by the DAC circuit. Each voltage value would be regenerated at intervals of 1/8000 Hz = 125 ms for telephony and 1/44,100 Hz = 27 ms for music because of the standard sampling rates that were used during the digitization phase. Figure 6-17 illustrates the reconstructed audio signal of Figure 6-14.

Figure 6-17

Reconstructed audio signal

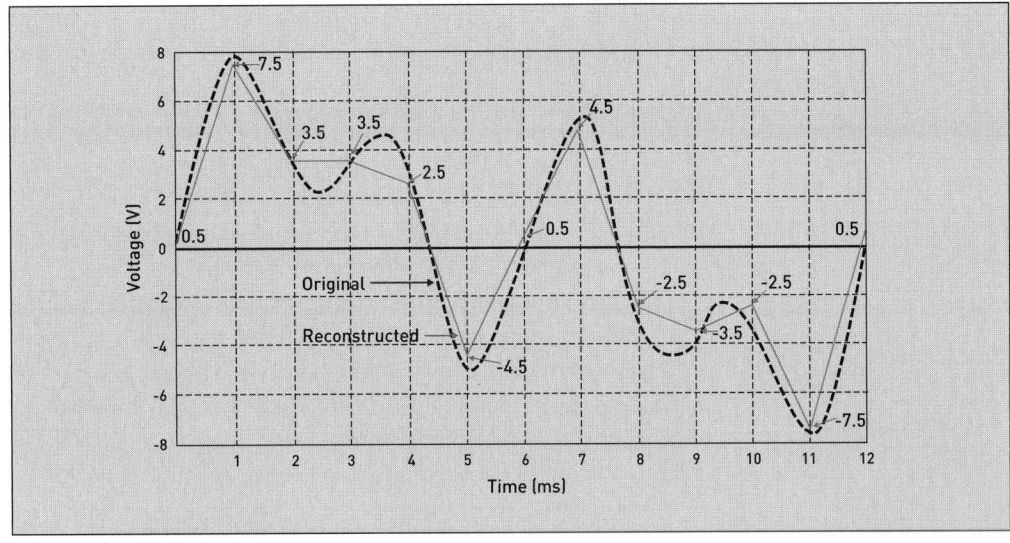

A glance at the figure reveals that the reconstructed signal appears to be "choppy." By postprocessing the signal—for example, smoothing its rough edges by passing it through an electronic filter—the reconstructed sound can be made more like the original. The smoothed-out signal can then be amplified and sent to a loudspeaker to convert the electrical signal into sound.

If a signal is sampled at a rate that is lower than the Nyquist rate during digitization, the reconstructed signal is said to undergo **aliasing** into a new form. Aliasing corresponds to false frequency components found in the reconstructed waveform. To avoid aliasing, audio signals should be sampled with a frequency that is at least equivalent to the Nyquist rate.

Although simple combinations of ones and zeros can encode audio information, these lengthy strings of numbers require enormous amounts of storage space, and require considerable bandwidth to transmit. Compression significantly reduces the size of digital audio files using techniques that are transparent to users. Without compression, users would quickly run out of space on their hard disks and flash drives and be frustrated by time-consuming file downloads. This section demonstrates how compression techniques are a critical component of audio technologies, describes factors that enable compression, and explains how MP3 and other digital audio compression standards work.

AUDIO COMPRESSION REQUIREMENTS

To understand the space and time requirements of digital audio information, consider the following storage and bandwidth calculation.

> **Problem**—Assume that a 4-minute song is sampled 44,100 times per second, that each sample is represented by 2 bytes (or 16 bits), and that the song is recorded to 2 channels.
>
> How large is this digitized song?
> First, determine the number of bits required to represent the song.
> The song is 240 seconds long (4 minutes × 60 seconds/minute).
> Number of samples: 44,100 samples/second × 240 seconds = 10,584,000 samples
> Number of bits: 10,584,000 samples × 16 bits/sample × 2 channels = 338,688,000 bits
> How long would it take to download this song over a 56-Kbps wireless connection?
> 338,688,000 /56,000 = 6048 seconds = 100.8 minutes

This example clearly illustrates the importance of compression. Without it, a song would take almost an hour to download over a slow wireless connection. Fortunately, most songs do not require such download times because they are compressed into fewer bits. Downloaded songs are often referred to as MP3 files. MP3 is a digital compression and formatting standard that is commonly used to reduce the size of music files.

ENABLERS OF COMPRESSION

We routinely use compression techniques in everyday life, whether crushing a soda can to fit better in the recycling bin, pressing clothes into a suitcase so it will close, or folding up a chair for easier storage. But how can digital information be compressed into fewer ones and zeros without losing some of the information content?

The limitations of human abilities, such as hearing, enable the compression of audio information. For example, certain frequencies are more perceivable by human ears than others. Additionally, compression is possible because information often contains intrinsic redundancy. In other words, information sometimes repeats itself numerous times. Rather than re-encoding and retransmitting redundant information, compression techniques can refer to previously encoded information using a shorter code.

Compression techniques fall into two high-level categories: **lossy compression** and **lossless compression**. As its name indicates, a lossless technique compresses information without losing any of the original content. Information is compressed, transmitted, and decompressed on the receiving end in the identical format as the original information prior to compression. Lossless compression is ideal for transmitting or storing information that must retain all the originally encoded data. Examples of information that requires lossless compression include financial data, textual information, and medical research data. Eliminating bits would be problematic for critical information or any data that is numeric or textual. In contrast, lossy techniques compress information by discarding some information detail. For example, information such as digital sound can tolerate information loss without significantly altering human perception. Lossy compression techniques eliminate some frequencies and volumes that human beings cannot readily detect.

Digital audio formats compress audio by actually eliminating some information, but not enough to be readily detectable by human hearing. This diminishment of the original data into fewer bits is called a loss of **fidelity** (from the Latin word *fidelis*, meaning *faithful*). Interestingly, audio formats also employ lossless compression. The process of compressing information in consideration of the limitations of human senses is called **perceptual coding**. Most people who listen to a compressed audio file cannot perceive much difference in quality between it and a CD, but music stored on a CD contains a much more faithful representation of a song. However, this detail contains some frequencies that human beings have difficulty hearing, includes notes or sounds that are so similar tonally or temporally that people cannot process both individually, and may contain softer sounds that are effectively drowned out by louder sounds. Human frequency perceptions and volume tolerances are well known, and are employed during digital audio compression to eliminate bits, resulting in lossy compression.

On average, people cannot hear any frequency lower than 20 Hz (cycles per second) or above 20 kHz (20,000 cycles per second). To compress information, any frequencies outside of this range can be discarded. Also, people can only process so much information simultaneously. If copious amounts of information, such as sound waves, are simultaneously presented to people, their minds perceive some of the prominent information but discard the rest. This perceptual phenomenon of filtering out less dominant information is called **simultaneous masking** (see Figure 6-18).

Compression techniques discard the frequencies that are imperceptible to people and the less prominent sounds that are drowned out perceptually. In addition to this lossy compression approach, digital audio formats also apply lossless compression algorithms that exploit redundancies in information to further compress information.

Figure 6-18

Simultaneous masking

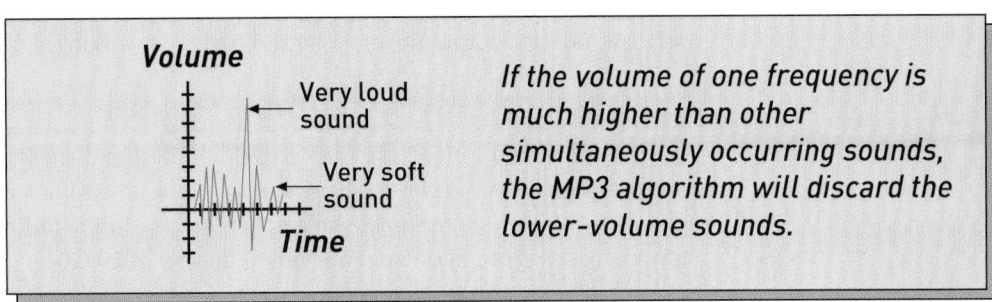

POPULAR DIGITAL AUDIO FORMATS

Audio file formats are standards that dictate the specifications for how audio is formatted and compressed for storage. Some of the most common digital audio formats are summarized in the following list:

- The advent of the Moving Picture Experts Group Audio Layer-3 (**MP3**) format has led to an enormous boost in digital music access and exchange. The MP3 format compresses large digital audio files to very small sizes by applying a combination of lossy and lossless techniques. Lossy compression techniques achieve high degrees of compression by eliminating some characteristics of the audio file at the cost of lower fidelity. Because MP3 can achieve high compression ratios without a significant degradation in quality, audio files can be quickly downloaded from the Internet and stored on a hard drive or in the memory of portable MP3 players within a small space. Audio files in MP3 format are denoted by an .mp3 extension.

- Advanced Audio Coding (**AAC**) is another lossy audio compression format. It is commonly used for audio files and is considered to have better audio quality and higher compression than MP3, making it an attractive alternative. The popularity of AAC is due in part to its adoption by the popular iTunes music store. AAC is based on the MPEG standards for audio and video, where **M4A** is the audio-only variation of the standard. AAC audio files typically have an .m4a or .aac extension.
- Windows Media Audio (**WMA**) is an audio file format designed for the Microsoft Windows Media Player, software that is used to play multimedia on the Windows operating system. WMA essentially achieves lossy compression. Its files have a .wma extension.
- The Waveform Audio Format (**WAV** or WAVE) is another popular format used to play music on computers. WAV files are uncompressed, so they occupy more space on disks than MP3 or WMA files. Therefore, the WAV format is not popular for portable applications. On the other hand, because music formatted in WAV is uncompressed, it has a higher fidelity than music formatted in MP3. Files in WAV format have a .wav extension.
- The Audio Interchange File Format (**AIFF**) is another uncompressed format. Its lossless nature makes it popular for playing high-quality music on Apple computers. Files in AIFF format have an .aif or .aiff extension.

The massive scope of the Internet and the advent of efficient audio formats make it easy to share and access a broad spectrum of music. This technology has inevitably led to a significant reduction in music CD sales and the bankruptcy of some large music stores. Instead of purchasing entire albums, many consumers make purchases on a song-by-song basis and download music instantaneously over the Internet.

This ease of downloading, reproducing, and sharing files has led to what the recording industry calls music piracy, the illegal downloading of music. Parties on both sides of this debate use innovative technical approaches to try to gain an advantage: music downloaders illegally obtain music through peer-to-peer networks, while music copyright holders have instituted digital rights management (DRM) techniques to try to thwart illegal file sharing. This phenomenon is a good example of the complex intersection of technological innovation, law, politics, economic interests, and social networking.

EXAMPLE OF DIGITAL AUDIO STORAGE

Audio compression formats are so effective that thousands of songs can be stored on an iPod or other portable music player. The efficiency of these compression techniques improves each year, allowing even more storage on smaller devices. In the following problem, you calculate how many songs can fit on an iPod with a capacity of 80 GB.

Problem—Confirm that 20,000 compressed songs can be stored on an iPod that has a hard-disk capacity of 80 GB. Assume an average of 4 minutes per song and that the iPod encodes the songs using 128-Kbps AAC formatting.

$80 \text{ GB} = 80 \times 8 \times 2^{30} = 687{,}194{,}767{,}360$ bits

The hard disk can store 687,194,767,360 bits. Assuming that the entire hard disk is dedicated to music storage, how much music can it store?

687,194,767,360 bits/128,000 bits per second = 5,368,709.12 seconds
5,368,709.12 seconds/60 seconds per minute = 89,478.4853 minutes of digital music
89,478.4853 minutes/4 minutes per song = 22,369.621 songs
This number confirms the iPod's advertised rate of accommodating 20,000 songs.

CHAPTER SUMMARY

- Sound waves are created by the mechanical disturbance of air molecules.
- Sound waves can be captured electrically by a microphone as an analog audio signal, which can be digitized using an analog-to-digital converter.
- Audio digitization is a three-step process of sampling, quantization, and encoding. There is an important trade-off between the accuracy of this process and its associated cost.
- Sound is categorized as complex or pure. Complex sounds comprise frequency components that vary in terms of their amplitude ranges, frequencies, and phase differences.
- The minimum sampling frequency for digitizing audio signals is at least twice the value of the highest frequency components. Aliasing occurs with a sampling rate that is less than the Nyquist rate.
- Digital audio files consume considerable storage space and bandwidth, but they can be compressed by exploiting the limitations of human hearing and the built-in redundancy within information.
- The two general approaches to compressing information are lossless compression and lossy compression. In lossless compression, none of the original information is lost. In lossy compression, some information is discarded, such as frequencies. Some compression techniques use a combination of these two approaches.
- Some common digital audio file formats include MP3, AAC, WMA, WAV, and AIFF.

KEY TERMS

AAC	Nyquist sampling theorem
AIFF	perceptual coding
aliasing	period
amplitude	phase difference
analog-to-digital converter	phonograph
audio signal	pitch
complex sound	pulse code modulation
digital-to-analog converter	pure sound
fidelity	quantization error
frequency	signal bandwidth
frequency component	signal dynamic range
frequency spectrum	simultaneous masking
lossless compression	sinusoidal
lossy compression	sound wave
M4A	spectrum analyzer
mechanical waves	WAV
MP3	WMA
Nyquist rate	

REVIEW QUESTIONS

1. Describe how sound waves are physically produced.
2. What is a complex sound?
3. What is the minimum number of frequency components that comprise a complex sound?
4. Draw one cycle of a pure audio signal with a frequency of 100 Hz that varies within a voltage range of −10 V to 10 V. What is the amplitude of the signal at 0.01s?
5. If the frequency of a pure sound wave is 100 Hz, calculate its period in ms.
6. If the frequency of a pure sound wave is 2.2 kHz, calculate its period in μs.
7. If the period of a pure sound wave is 2 ms, calculate the frequency in kHz.
8. Briefly describe the three steps in digitizing audio signals.
9. State the Nyquist sampling theorem.

10. If a complex sound has five components with frequencies of 100 Hz, 200 Hz, 300 Hz, 400 Hz, and 500 Hz, calculate the minimum sampling frequency required to digitize it.

11. Calculate the minimum sampling frequency required to digitize a pure sound wave with a frequency of 2350 Hz.

12. What is quantization error and how can it be reduced?

13. What is the relationship between sampling frequency and reconstructed audio quality?

14. What is the relationship between the number of bits assigned per sample and reconstructed audio quality?

15. Is there any relationship between quantization error and sampling frequency?

16. Calculate the quantization error if the actual value of an audio sample is 5 V but is rounded off to 5.3 V.

17. Briefly explain the function of the ADC and the DAC circuits.

18. Describe the difference between lossy compression and lossless compression.

19. Assume that a 3-minute song is sampled 44,100 times per second, that each sample is represented by 2 bytes, and that the song is recorded to two channels. How large is this digitized song in bits? How long would it take to download the song over a 500-Kbps network connection?

DISCUSSION QUESTIONS

1. Discuss the issue of illegal music downloading from social, legal, law enforcement, and economic perspectives. What are the ethical issues that underlie file sharing? Will the music industry be able to contain illegal music and video downloads? Should it? Discuss technological innovations that might protect copyrighted material, and consider some innovations that might overcome these protections.

2. Why do so many audio formatting standards exist? Make an argument for consolidation into a single audio format and an economic argument for retaining multiple formats.

3. Discuss the relationship between sampling rate and digital audio quality. Why is the standard sampling rate different for digitizing voice and music? Why isn't the standard sampling rate 44,100 Hz for telephony?

4. Can quantization error be reduced by taking more samples per second? Why or why not?

5. Discuss why audio files can be compressed to such a great extent. Can you perceive a difference in sound quality between a compressed MP3 file and a song played directly from a CD?

CASE PROJECT

Search the Internet for two types of computer sound cards with different specifications. Write down their specifications for sampling frequency and the number of bits. Identify which sound card would produce the lower quantization error. What other characteristics are important when choosing a sound card? Are bit specifications on all sound cards equivalent to the actual number of bits assigned per sample during the ADC process? Can sampling rates be selected by the user? How is the sound card connected to the computer?

CHAPTER 7

DIGITAL IMAGES AND VIDEO

LEARNING OBJECTIVES

In this chapter you will:

- Learn how images are digitized

- Understand the technical features and mathematical calculations that determine the bit size of a digital image

- Understand the process by which digital video is made

- Identify the factors that affect the quality of digital images

- Learn about popular digital image and video file formats

- Understand the technical attributes of different types of display technologies

INTRODUCTION

The dazzling sun sparkles across the river and brightens the emerald trees behind the pagoda. The river appears calm and a few clouds flutter in the sky. These two sentences, although highly descriptive, are inadequate to convey all the details in the photograph shown in Figure 7-1. Images communicate enormous amounts of information that are not captured by written descriptions. Images play a fundamental role in disseminating information in society, whether through an advertisement on an electronic billboard, a JPEG file received via e-mail, or a medical diagnostic image. Images must be converted into digital format to be transmitted, displayed, processed, or stored by the numerous versatile digital systems we use in daily life. IT professionals must understand image and video technologies because routine business functions involve imaging applications such as videoconferencing and displays of pictures for marketing and online electronic commerce. Other applications include storage and exchange of industry-specific applications such as manufacturing plans, scanned electronic records of financial documents, or medical diagnostic images.

This chapter discusses the wide range of modern digital imaging technologies that play such an important role in modern life. It contains a general overview of digital imagery, including how still images can be directly captured by a digital camera or indirectly captured by a scanner. The chapter also explains the underlying technical process of producing digital video, the limitations of human vision, and how these limitations are exploited by image and video digitization. Finally, the chapter describes state-of-the-art display technologies and the mathematical properties that allow us to calculate the bit size of digital images and video.

Figure 7-1

Photograph of a pagoda

IMAGING TECHNOLOGIES

The traditional camera, invented more than a century ago, captures images with a few simple components: a lens, a shutter system, and a photographic medium such as film. Light from the object is focused by the lens, the quick shutter action exposes the film to light, and the image is permanently captured on film. The film is then developed and printed on photographic paper. The quality of the photograph depends on the camera type, lens quality, film, and printing paper.

People are sometimes content to insert a few photographs in an album and glance at them occasionally. On other occasions, we want to store hundreds of photos, make multiple copies of them, or distribute them to friends and family. We may want to enhance a photo's color or remove imperfections from it. Commercial photography and imaging require the ability to modify images, store significant quantities of these images, and disseminate them. All of these tasks were possible with analog imaging techniques, but digital imaging that uses devices such as digital cameras and scanners makes them far simpler, more efficient, and cost effective.

Due to the popularity of digital imaging applications, digital cameras have become so ubiquitous that they are integrated into handheld devices such as cell phones, and even come in disposable form. Capturing images in digital format offers significant advantages over traditional imaging techniques. On a personal level, people no longer have to make multiple prints of a photo and send them to friends via "snail mail." Digital images can quickly be attached to an e-mail or posted on a Web site for friends and family to access. Instead of storing many photographs in numerous bulky albums, they can be saved on small flash drives that consume minimal space. Furthermore, instead of using an array of complex procedures and expensive chemicals, photo editing software can be used to alter any aspect of an image. More importantly, the efficiency and ease of capturing and transmitting digital images has dramatically improved applications used for medical diagnostic imaging, military purposes, law enforcement surveillance, and a multitude of business applications. For example, doctors no longer have to wait for a diagnostic film such as a traditional X-ray film to be developed; they can view a medical image in real time and immediately make a diagnosis. The digital diagnostic image can also be quickly transmitted to a distant location to be viewed by other doctors, thereby expediting and confirming the diagnosis. Digital medical imaging is also considered to provide enhanced views of internal organs and systems compared to its analog predecessors.

DIGITIZING IMAGES AND VIDEO

PIXELIZED IMAGES

A digital image is an electronic file made up of a string of ones and zeros when stored on a medium such as on a computer's hard disk or flash drive. When the image file is accessed and shown on a display screen or monitor, it is called a pixelized image because it is composed of many elementary units called **pixels** (short for picture elements), each containing a single unit of color information. A pixelized image is similar to a mosaic, created by piecing together small and generally irregular portions of ceramic or glass in various colors. When viewed from a distance, the small pieces of ceramic look like a full, continuous image (see Figure 7-2).

Figure 7-2

A mosaic image

Pixels in a pixelized image are like the small pieces of ceramic, except that they are all the same size, are regular in shape, usually square, and arranged in multiple rows and columns. A pixelized image appears continuous when viewed from a distance, just like a mosaic.

Each pixel has an associated color, like the color of each piece of ceramic. Pixels in black and white images, also called **grayscale** images, take on a specific shade of gray, and pixels in color images take on an individual color. Because binary code can also represent colors, every pixel has an associated binary number that characterizes its color. For example, a white pixel can be represented by 000000000 in binary code, and a red pixel by 111000000. Figure 7-3 shows an example of a pixelized image, the close-up view of a portion of the image, and the binary code associated with a single pixel.

Figure 7-3

A pixelized image

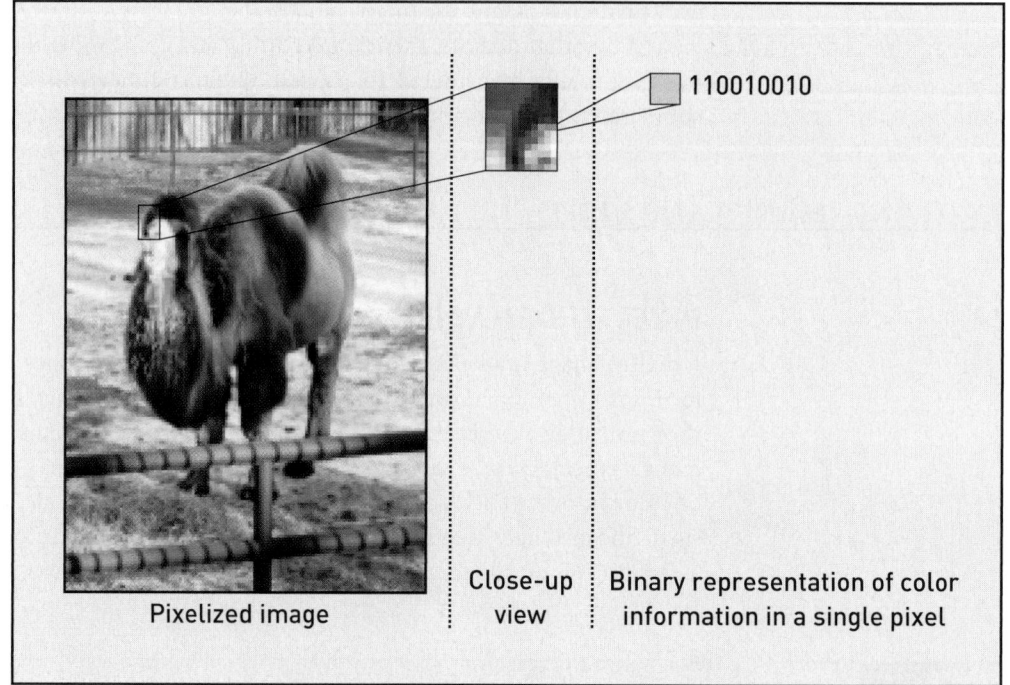

Pixelized image   Close-up view   Binary representation of color information in a single pixel

110010010

In addition to color information for each pixel, the digital image file may contain other types of information, such as the number of pixels in the image, in the form of a row-column multiplication (e.g., 1024 × 768). This information enables the image to be properly displayed on a device such as a monitor. This type of image representation is called a **bitmap image** because each binary code corresponds to a pixel, which can be mapped onto a specific part of the image. In computers, graphics adapters (discussed in Chapter 4) convert the binary numbers corresponding to each pixel color into analog color information and forward them to the monitor for display.

In summary:

- A raw bitmap image file consists of a string of ones and zeros that represent the pixelized image.
- A pixelized image is composed of rows and columns of pixels.
- Each pixel has a unique color associated with it.
- A binary code can express each pixel color.

Other than the bitmap image format explained earlier, images can also be represented in the form of mathematical expressions called vectors. Images represented in this format are called **vector graphics**. Vector graphics generally take up less storage space than bitmap images and can be scaled easily without degradation in quality and resolution. To understand this concept more clearly, consider the bitmap image in the top part of Figure 7-4.

Figure 7-4

A bitmap image

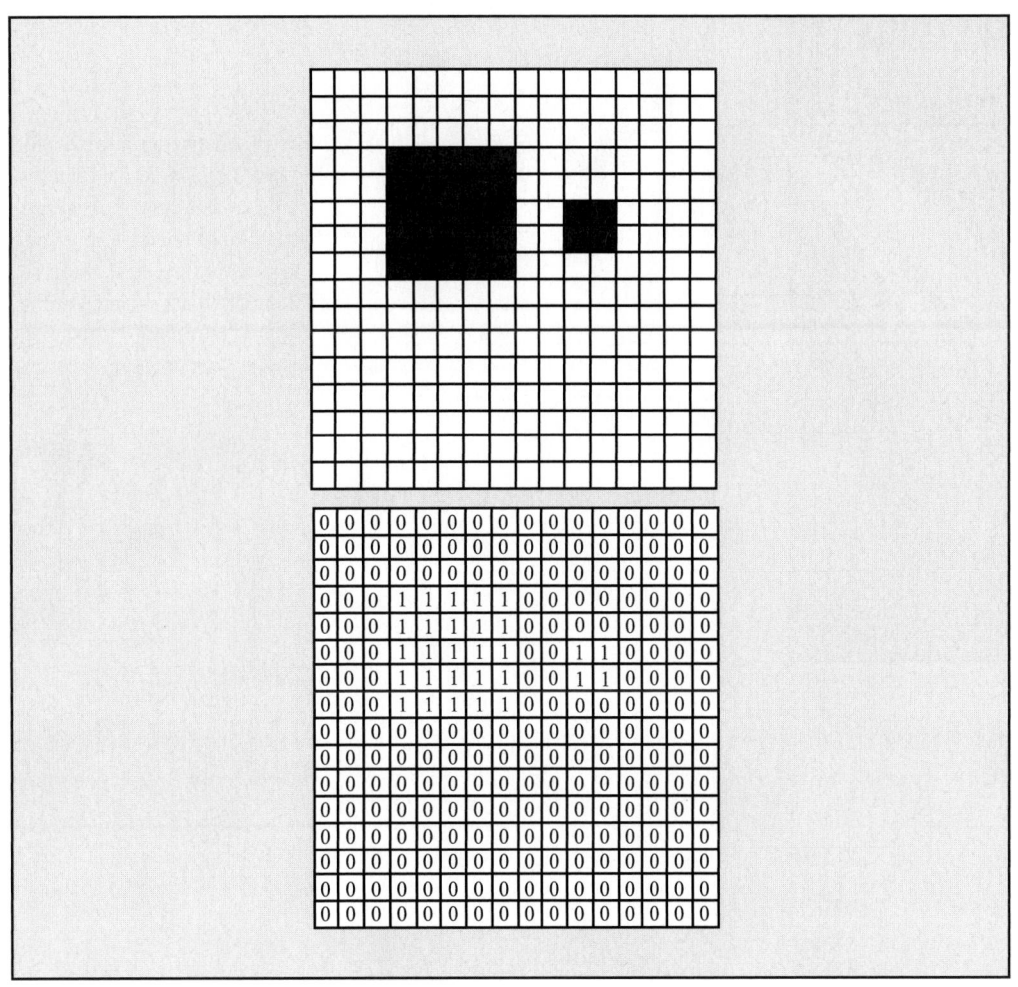

Figure 7-4 is composed of only white and black pixels. The bitmap of this image would consist of a long string of ones and zeros, as shown in the lower part of Figure 7-4, if white pixels were represented by a zero and black pixels by a one. The total number of bits that would need to be stored for this bitmap is $16 \times 16 = 256$ because there are 16 rows and 16 columns of pixels, where each pixel is represented by a single bit. In vector graphics, the same image, which is comprised of two squares in this example, can be described by mathematical expressions with parameters such as orientation, length, width, and color of each square. This description in binary would clearly take less space compared with its bitmap counterpart. Vector graphics are especially useful in representing images that consist of geometric shapes, such as block diagrams, but are not very effective for representing photographs due to the difficulty of expressing the content of a photograph with a mathematical expression. This chapter primarily focuses on bitmap images.

CAPTURING COLOR IMAGES USING DIGITAL CAMERAS

Consider the scene in Figure 7-5, and imagine that we would like to photograph the scene in color and display it on the Web.

Directly capturing the image in digital format would require a digital camera that can take color pictures. Although the capabilities of digital cameras vary from model to model, most of them typically have the following common components:

- Lens system
- Sensor array
- Filter system

Figure 7-5

An image captured by a
digital camera

- Memory
- Processor
- Software
- Viewing screen

These parts, when coupled with other hardware, work together to capture images. When the camera is aimed toward the scenery, it uses its lens system to focus the image (usually automatically) onto a small sensor array called a **CCD** (charge-coupled device) **array**. See Figure 7-6.

The CCD array is comprised of rows and columns of minute sensors that detect light and produce electricity proportional to the intensity of the light impinging on them. For the sake of simplicity, we can consider that each sensor is responsible for forming a single pixel of the pixelized image. A 6-**megapixel** camera, for example, can have sensors arranged in 2816 rows and 2112 columns, and it produces a raw, pixelized image with approximately 6 million pixels ($2816 \times 2112 = 5{,}947{,}392$, or roughly 6 megapixels).[1]

How can a digital camera capture color images if its sensors can only detect light intensities? Each CCD sensor detects the intensity level of the three primary colors—red, green, and blue—within the section of the image on which it is focused. Each sensor produces electrical currents corresponding to the levels of red, green, and blue light falling on each sensor site. These three primary colors form the basis of what is called the **RGB additive color model**. This model specifies that any color of light can be created by combining various proportions of red, green, and blue light (see Figure 7-7). For example, yellow can be created by combining red and green. Similarly, magenta may be created by combining red and blue.

---

[1] Note that in practice, the number of sensors does not correspond to the actual number of pixels that may be produced by the digital camera.

Figure 7-6

A CCD array

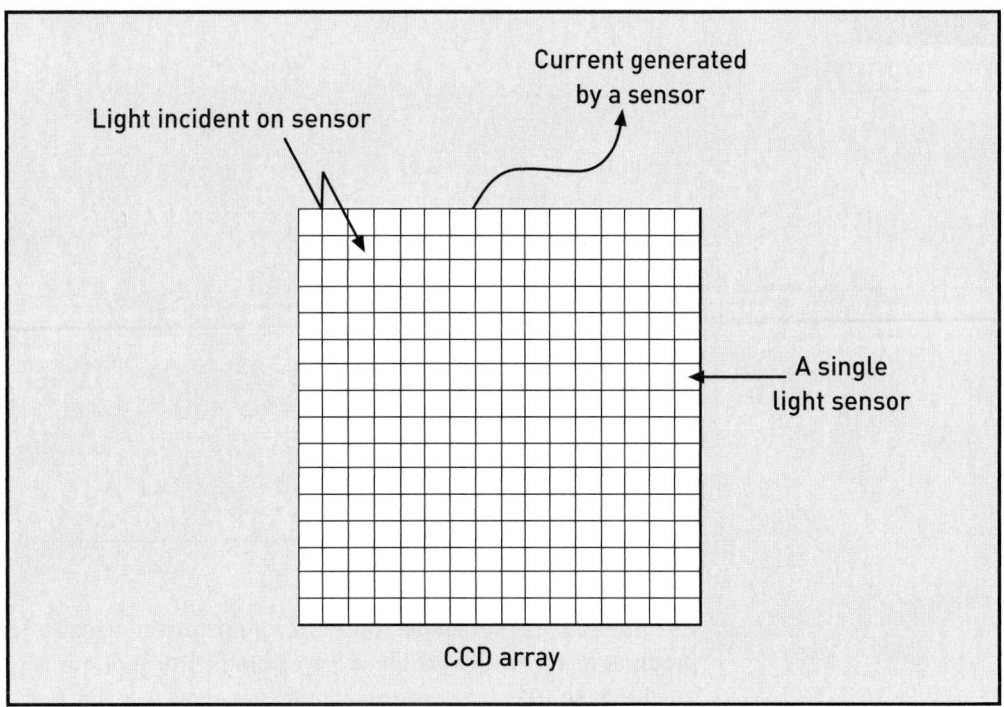

To understand the additive color model more completely, consider three bulbs in a dark room that can emit red, green, and blue light respectively (see Figure 7-8).

Also imagine that the bulbs are connected to a dimmer switch that can adjust the brightness level of each bulb. When the bulbs are lit, all three light sources combine to illuminate part of the ceiling with a unique color that can be altered by adjusting the dimmer switches. For example, if the red bulb is switched to its maximum intensity (a level of 7), and the green and blue bulbs emit at a much lower intensity (such as 1 and 2 respectively), the resulting light on the ceiling will appear reddish. If the blue and red bulbs were adjusted to emit light at much higher intensities than the green one, the resulting ceiling color would appear purplish.

Figure 7-7

The additive color model

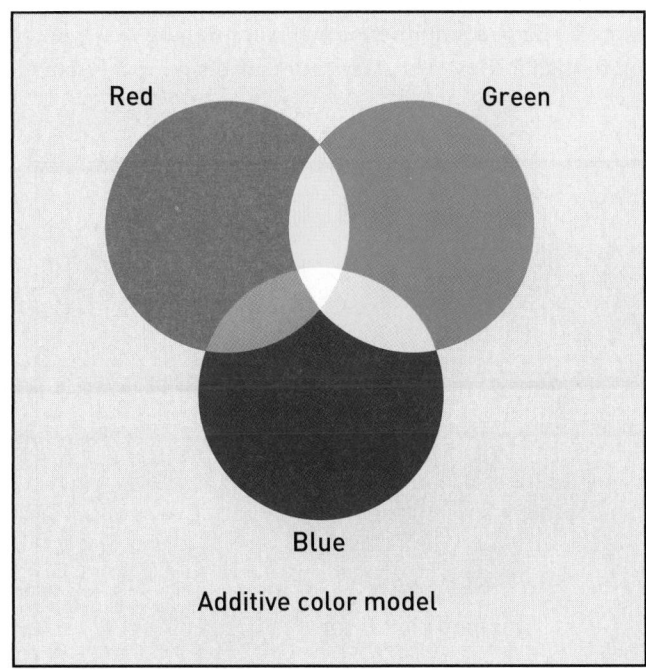

Figure 7-8

Three light sources

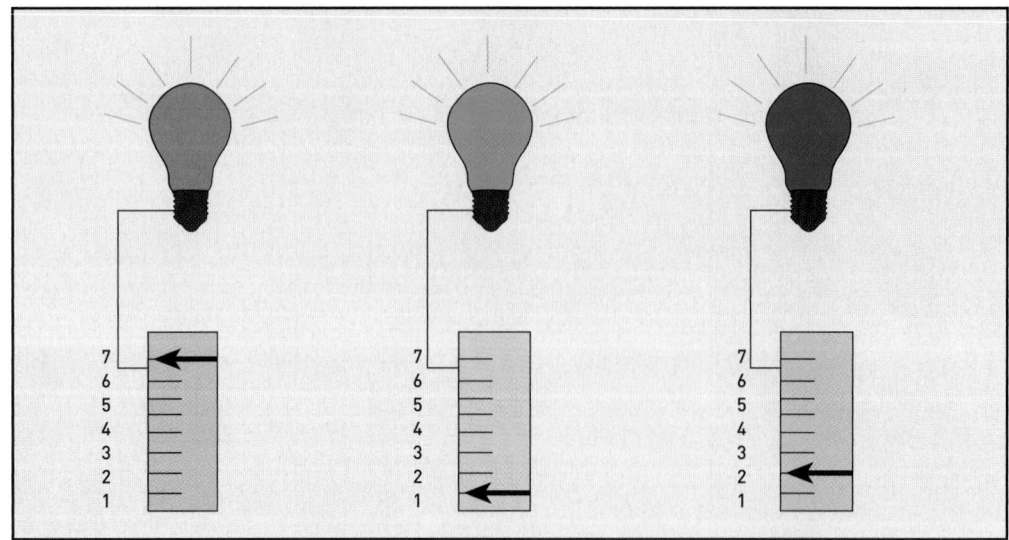

The additive color model forms the foundation of digital color image capture. Each sensor produces an electrical signal level based on the intensity of each color in the light focused on the sensor site. The sensors actually measure light intensity and do not distinguish color; the color information may be obtained by using a circular rotating filter called a **spinning disk filter** positioned on the CCD array.

This filter is made of red, green, and blue sections to filter light (see Figure 7-9). Each section of the filter passes only the corresponding color component and blocks out the other components. For example, the red section of the filter only passes red light and blocks the blue and green components within the light that is applied to the filter. To capture the scene, each filter section is positioned on the CCD array once. When a blue section of filter is positioned on a sensor, for example, only blue light is passed through the filter, and the red and green colors are blocked. The intensity of the blue light is detected at each sensor site and a corresponding current is produced by each sensor. Similarly, as the other filter segments take their turn, corresponding red and green readings are produced by each sensor. This process is invisible to the user because it occurs at a very high speed. The electrical value for each color reading is sent to an amplifier, which amplifies the value and then sends it to the processor's analog-to-digital converter (ADC) for conversion to a binary number. Figure 7-10 illustrates

Figure 7-9

The spinning disk filter

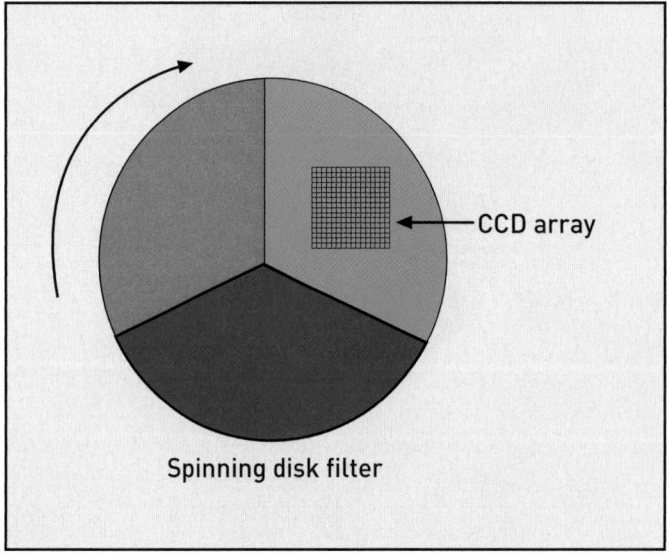

Spinning disk filter

the three images positioned on the CCD array as the disk filter spins. As we explained previously, the sensors of the CCD array output an electrical signal for each color reading.

As discussed in Chapters 3 and 6, the process of digitizing any type of analog information involves taking samples, quantizing (or rounding off) the values of each sample, and assigning a binary code to each rounded-off value. In Chapter 6, an analog audio signal was sampled to reduce its infinite number of values to a finite number, each sample value was quantized to the closest value, and each quantized value was finally converted to a string of ones and zeros. The same process applies to digitizing images. The analog, continuous image, which contains an infinite number of small "dots," is made discrete, or reduced to a finite number of small dots that actually correspond to the pixels. The CCD sensor array therefore performs the sampling process whereby the camera captures only a finite number of samples of the actual image. The electrical levels produced by the sensors are then quantized to the closest available value, and each quantized value is assigned a binary code by the camera's ADC.

The number of quantization levels used by the quantizer in the ADC is related to the camera's **color gamut**. The length of the binary code, or bit depth, that the camera generates for each pixel determines this color gamut. Recall that x bits are needed to represent 2^x different things—in this case, colors. The color gamut of a digital camera is the spectrum of colors it can capture, similar to an artist's palette. For example, if the color gamut of a digital camera has only eight colors, ranging from white to black, any image it captures will contain only eight colors. Any color within the image focused on the CCD will be rounded off to the nearest color in this eight-color palette (see Figure 7-11).

Figure 7-10

An image resolved into its basic color components

Figure 7-11

An 8-color palette

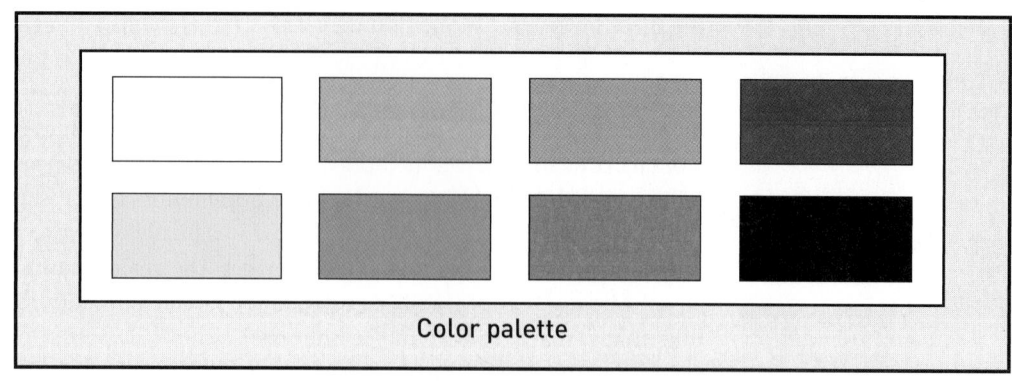

Color palette

As mentioned, the color gamut is directly related to the number of bits assigned per pixel. If 3 bits are assigned per pixel, then the gamut has only 8 colors because $2^3 = 8$, with 1 bit assigned to each color component. The number of bits allocated to each pixel is called the **bit depth**, and directly plays a role in the quality of a camera. Commercial cameras usually have a bit depth of 24, meaning that 2^{24}, or roughly 16.7 million colors can be captured by the camera. This number of bits is necessary because our eyes are sensitive enough to detect approximately 10 million colors. A camera that provides a color gamut of 10 million colors requires 24 bits. Twenty-three bits would represent only $2^{23} = 8,388,608$ colors, so a camera needs a minimum bit depth of 24 to achieve true color representation. A bit depth of 24 is usually referred to as **true color**.

Because each color is represented by 24 bits, each color component (red, green, and blue) is represented by 8 bits, corresponding to $2^8 = 256$ different intensity levels for each color component. Recall that in the three-bulb example, each dimmer switch had eight settings. This would correspond to a 9-bit color depth, with each color component represented by 3 bits. The number of possible color combinations would be 512, because $2^9 = 512$.

Rounding off voltage values produced by the sensors may be thought of as corresponding to rounding off the color of each pixel to the nearest color within the camera's palette. After quantization, the remaining task is to assign a unique binary code (with a length equal to the bit depth) to each color within the palette and use these codes to represent the color value of each pixel. For example, suppose we assign a code of 111 to the first color in the palette, a code of 110 to the second color, and so on. The resulting color-code assignment might resemble the one in Figure 7-12.

If a pixel in the image is white, then the binary code for that pixel is 111. If another pixel is greenish, its color will be rounded off to the green color in the color palette of the camera and assigned a code of 010. This process continues until all the pixels have an assigned binary code, resulting in a long string of bits.

The camera's memory, which is typically some form of flash memory, stores the resulting string of binary numbers in a popular format, such as the JPEG digital image file format. This format, which we discuss later in this chapter, reduces the raw file size of the image. The storage of raw image data is impractical due to the large amount of space it consumes, so each image is compressed by its processor before storage to reduce the amount of space it occupies in the camera's memory. You can view the image later as a pixelized image on the camera viewing screen.

The spinning disk filter approach (see Figure 7-9) illustrates theoretically how a digital camera can capture a color image, but in practice, most commercial cameras use a different type of filter called a **color filter array**. In this array, either a small section of red, green, or blue filter is positioned directly on top of each sensor in a special pattern called a **Bayer pattern**. The details of this method are not explained in this textbook. However, the use of a color filter array eliminates the need for a spinning filter, thereby simplifying the image

Figure 7-12

The color palette and
corresponding binary
codes

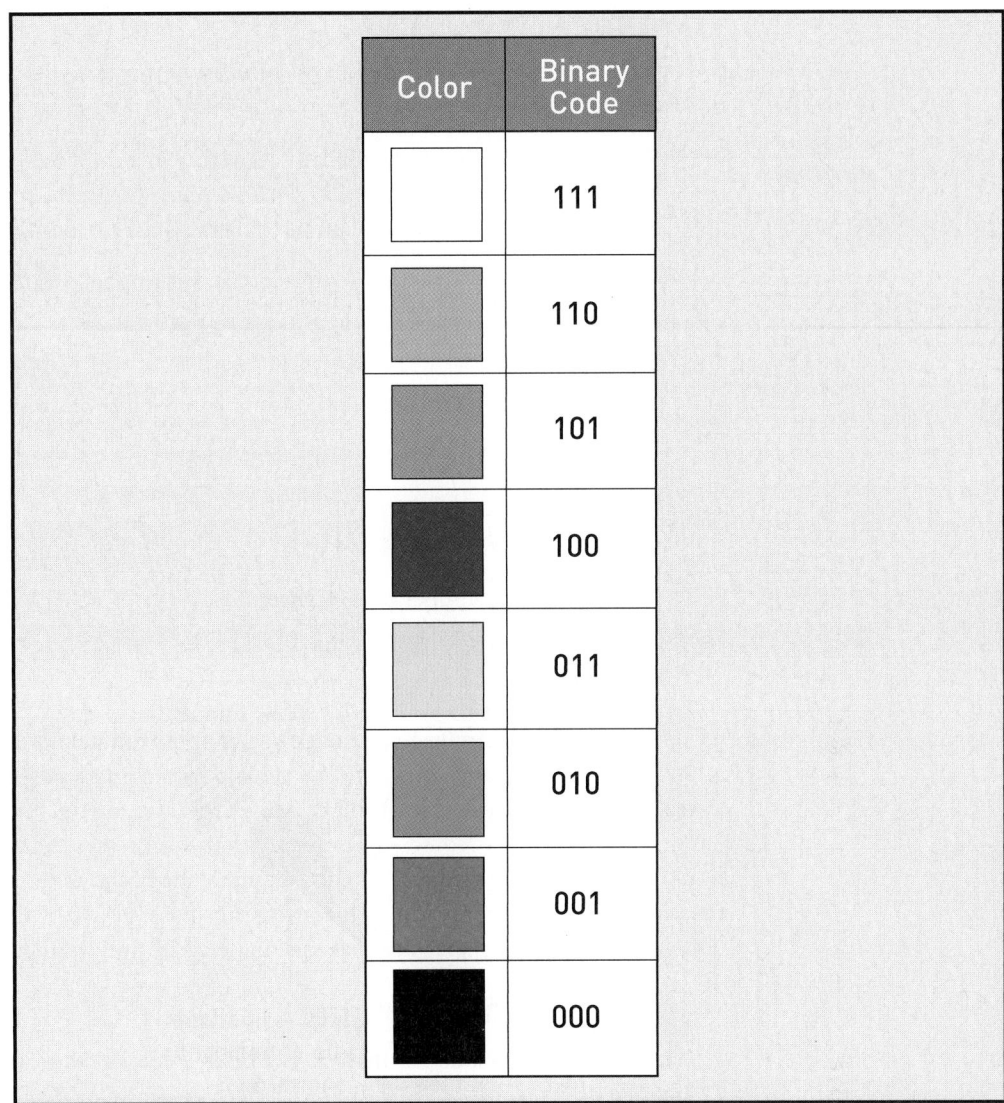

Color	Binary Code
	111
	110
	101
	100
	011
	010
	001
	000

capture process. When you use a color filter array, each pixel in the digital image is created using the output of multiple sensors, instead of the one sensor used with a spinning disk filter. This concept is illustrated in Figure 7-13.

As a final note, some digital cameras use a complementary metal oxide semiconductor (CMOS) sensor array instead of a CCD. CMOS-based cameras and CCDs have similar principles of operation, but CMOS-based cameras are more inexpensive and of lower quality.

Figure 7-13

The spinning disk filter
and Bayer filter approach
to capturing color

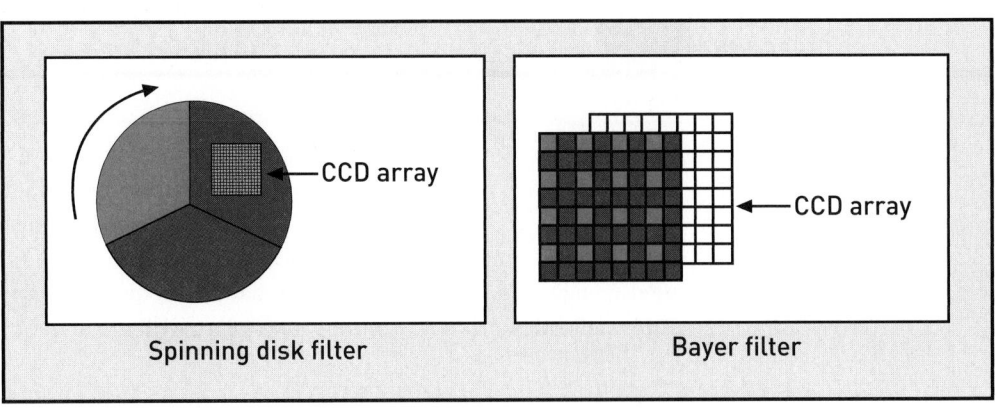

CALCULATING DIGITAL CAMERA IMAGE SIZES

Having been introduced to the image digitization process, you can learn how the raw bit size of a digital image is calculated.

> **Problem**—Calculate the number of megabytes in a color image captured by a digital camera that can produce images with 2272×1704 pixels with a bit depth of 24. Also, calculate how many images can be recorded on the camera if its flash memory capacity is 32 MB.

The raw size of a single image captured by this camera is calculated by the following:

Number of pixels = 2272×1704 = 3,871,488
Number of bits per pixel = 24
Number of bits per image = $3,871,488 \times 24$ = 92,915,712 bits

To calculate the number of bytes:

1 byte = 8 bits
92,915,712 bits/8 = 11,614,464 bytes

To calculate the number of megabytes:

1 MB = 2^{20} bytes = 1,048,576 bytes
11,614,464 bytes/1,048,576 = 11.07 MB

The calculation results in an image size of approximately 11 MB. Because the camera has a 32-MB storage capacity, only two images can be recorded on the camera simultaneously. In reality, the above calculation of 11 MB is the raw, or uncompressed, size of the image. Fortunately, cameras come with software that compresses the digital image considerably. If the image is compressed to 2 MB, for example, the flash memory can store approximately 16 images. Image compression formats commonly associated with digital cameras, as well as digital imaging in general, are discussed in detail later in this chapter.

SCANNING COLOR IMAGES

The process by which digital images may be created using a scanner is similar in principle to that of a digital camera. One obvious difference between the two is that a digital camera is pointed toward an object and immediately captures the image of the object. The object may be located at any distance from the camera and may be any size and shape. In contrast, a scanner captures the digital image of an object (usually a photograph or a document) that is inserted under the scanner and has to be of a certain size to fit in the scanner. To understand the process of scanning, you will learn the composition of a traditional flatbed scanner, although other types of scanners are also prevalent.

A flatbed scanner has several components, some of which include:

- Lens system
- Sensor array
- Filter system
- Processor
- Mirrors
- High-intensity lamp

A scanner usually connects to a computer with software that assists in image capture and storage. When an object such as a photograph is placed on the glass plate under the scanner cover, the scanning process begins by illuminating the photograph with light from a high-intensity light source. A scanning head that carries the CCD array is positioned on the photograph and slides over it as scanning takes place with the help of the scanning belt and motor.

The arrangement of sensors in the CCD array used in scanners is slightly different from the one used in digital cameras. The CCD in a digital camera has a large number of sensors arranged in multiple rows and columns, like a small rectangular box. The CCD array for scanners has a large number of sensors arranged in multiple columns and a much smaller number of rows (see Figure 7-14).

Figure 7-14

The CCD array of a scanner

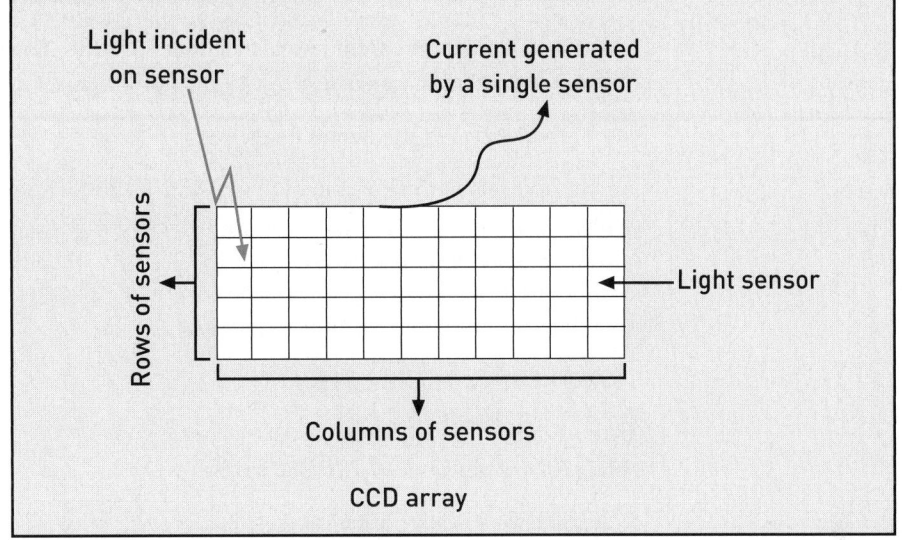

The resulting array resembles a slim ruler; its length approximately equals one of the dimensions of the scanner's viewing glass window. Within the context of scanners, the pixel count in the resulting digital image is related to the **dots per inch** (dpi) specification of the scanner and the physical dimensions of the scanned image. A common specification is 1200 × 2400 dpi[2]; it is also sometimes measured in **pixels per inch** (ppi) or **samples per inch** (spi). The dpi specification essentially corresponds to the number of pixels per inch produced by the scanner in capturing the image. Simply put, if the scanner has a specification of 1200 × 2400 dpi, it will essentially section each inch of the image into 1200 × 2400 pixels. In contrast to the specification of a digital camera, which is expressed as a total number of pixels, scanner specifications are expressed as dpi, ppi, or spi.

As the scanning head slides under the viewing glass with the help of the belt and motor, the image of the illuminated photograph is filtered through red, green, and blue filters and positioned and focused on the CCD array by way of the mirrors and lenses. The limited number of rows of sensors produces electrical signals whose intensities are based on the brightness levels of red, green, or blue light passing through the filters. Each brightness level is then amplified, quantized, and converted to a binary number (see Figure 7-15). Bits generated from the scanning process are then transferred to a computer and stored as a compressed electronic file.

Figure 7-15

CCD array positioned on an image

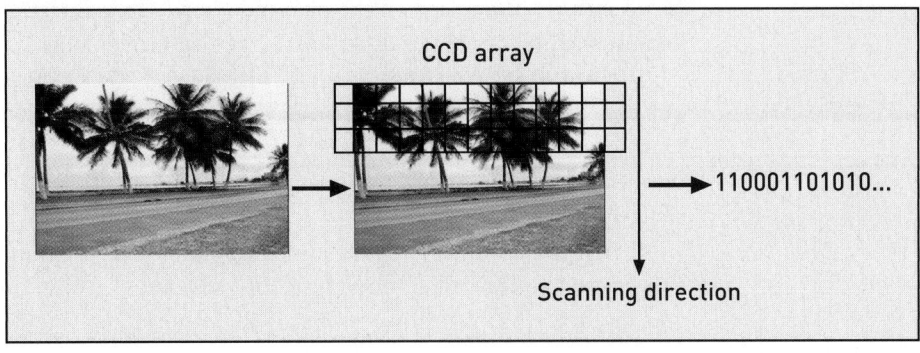

[2] Although dpi is a term more commonly associated with printing, this discussion uses dpi within the context of scanning, as commercial Web sites that sell scanners indicate scanner resolutions in terms of dpi.

The number of colors that the scanner can detect again depends on the scanner's bit depth. Color scanners may have bit depths of 24, 36, 48, or some other value. The larger values are associated with high-quality photograph scanners.

CALCULATING SCANNED IMAGE SIZES

What is the raw bit size of a digital image produced by a scanner?

Problem—Calculate the number of megabytes produced at the output of a 1200×2400 dpi scanner that has scanned an 8.5×11-inch photograph in true color.

The number of bits in the digital image is:

Number of pixels in the horizontal dimension = 1200×8.5 = 10,200
Number of pixels in the vertical dimension = 2400×11 = 26,400
Number of pixels in image = $10,200 \times 26,400$ = 269,280,000 pixels
Number of bits in image = $269,280,000 \times 24$ = 6,462,720,000 bits
(Note that true color corresponds to 24 bits.)

To calculate the number of bytes:

1 byte = 8 bits
6,462,720,000 bits/8 = 807,840,000 bytes

To calculate the number of megabytes:

1 MB = 2^{20} bytes =1,048,576 bytes
807,840,000 bytes/1,048,576 bytes = 770.4 MB

As before, the raw file is large, but is reduced with compression software that works with the scanner.[3]

SOME FACTS ON RESOLUTION

Two characteristics determine the quality of digital images:

- Spatial resolution
- Brightness resolution

Spatial Resolution

Spatial resolution is an indicator of the number of pixels per unit length (such as inches or centimeters) in an image. The more pixels there are per unit length, the higher the spatial resolution of the image. A 6-megapixel image contains a higher spatial resolution than a 4-megapixel image of the same size because the number of pixels per unit length is much greater. Spatial resolution is important because it directly affects the viewing quality of a pixelized image. For example, Figure 7-16 shows two images with the same physical dimensions (6×4 inches). The image on the right has a lower spatial resolution than the one on the left because the number of pixels per unit length is much smaller.

The first image contains 900×600 pixels (with 150 pixels per inch), whereas the second one has 120×80 pixels (with 20 pixels per inch). Obviously, the lower-resolution image loses many of the details. You can discern the contours of the pixels in the image with the lower resolution, but the higher-resolution pixels are not as easy to see because they are much smaller.

Spatial resolution is often meaningful if defined by the number of pixels per unit length. Consider two images that do not have the same physical size; one image has 6 million pixels and the other has 4 million pixels. If the 4-megapixel image is 8.5×11 inches and the

---

[3] Note that this calculation is an oversimplified example. In reality, the raw file size may not actually be calculated following this procedure. Such examples are nonetheless helpful in demonstrating the main concept to the reader.

Figure 7-16

Images with different
spatial resolutions

900 x 600 pixels   120 x 80 pixels

6-megapixel image is 17 × 22 inches, then the 4-megapixel image will look better when viewed from the same distance even though it has fewer pixels, because it has the smaller size and hence the larger number of pixels per inch. Thus, the number of pixels per unit length has an important role in the viewing quality of an image.

Although the two images in Figure 7-16 had noticeable differences in quality when viewed from equal distances, the disparity in quality would not have been as obvious from afar. From a close distance, the difference in quality is apparent, but these images would appear similar if viewed from further away (e.g., 10 feet). Thus, the minimum viewing distance plays an extremely important role in viewing quality. Because our eyes are limited in terms of the level of detail they can resolve, many digital imaging devices on the market have a spatial resolution that is adequate for typical viewing distances. Although a 10-megapixel digital camera has a higher resolution than a 6-megapixel model, the difference in quality is not drastic when the image is printed on standard photographic paper and viewed from a distance of about 3 feet. Printing the photograph on larger paper or viewing it at a closer distance, however, makes the difference in quality apparent.

Finally, the spatial resolution of a digital master image depends on the spatial resolution of the device used to create the image. For example, a 10-megapixel digital camera can produce an image with a maximum theoretical resolution of 10 megapixels. However, after a digital image is obtained, the spatial resolution can be altered using photo-editing software. For example, the resolution of a 200 × 200 ppi image can be reduced to 100 × 100 ppi or even less for more efficient storage.

What effect does varying the spatial resolution of a color image have on file size?

Problem—Calculate the number of kilobytes in a 300 × 300 pixelized color image with a bit depth of 24 bits. What is the total number of KB if the number of pixels is reduced to 200 × 200?

Number of pixels = 300 × 300 = 90,000
Number of bits per pixel = 24
Number of bits in image = 90,000 × 24 = 2,160,000 bits

To calculate the number of bytes:

2,160,000 bits/8 = 270,000 bytes

To calculate the number of kilobytes:

1 KB = 2^{10} bytes = 1024 bytes
270,000 bytes/1024 = 263.67 KB

If the number of pixels is reduced to 200 × 200:

Number of pixels = 200 × 200 = 40,000
Number of bits per pixel = 24
Number of bits in image = 40,000 × 24 = 960,000 bits

To calculate the number of bytes:

960,000 bits/8= 120,000 bytes

To calculate the number of kilobytes:

1 KB = 2^{10} bytes = 1024 bytes
120,000 bytes/1024= 117.18 KB

Reducing the number of pixels from 90,000 to 40,000 produces roughly a 55% decrease in file size, although the quality of the resulting image is much lower.

Brightness Resolution

Brightness resolution is quite distinct from spatial resolution. Brightness resolution is related to bit depth and consequently the number of colors that are inherently present in the image. A black-and-white, or monochrome, image has a low brightness resolution because each pixel can take on only one of two colors. If the scenery in Figure 7-5 was converted into a monochrome image, it would look like the one in Figure 7-17.

Figure 7-17

Monochrome image

Although the spatial resolution has remained unchanged, the fact that each pixel is black or white has resulted in a significant loss of fidelity. The bit depth of a monochrome image is just 1 because each pixel is either black or white. As another example, Figure 7-18 illustrates the effect of changing the image's brightness resolution from 24 bits (true color) to 8 bits while the spatial resolution is held constant.

It is evident that the image on the left contains a greater color variety than the image on the right, due to its greater brightness resolution.

Image brightness resolution also depends on the quality of the device used to capture the image. A scanner with a bit depth of 24 produces a much higher-quality image (in terms of brightness resolution) than one with an 8-bit depth. Again, the bit depth of a master image can be reduced by software, as shown in Figure 7-18.

Figure 7-18

Images with different
brightness resolutions

Because the human eye is limited in its ability to discern color, the bit depth of digital imaging devices should be set accordingly. A bit depth of 1000 for a digital color imaging device would be a waste of resources because our eyes can only discern about 24 bits worth of color information (about 16 million colors). Anything beyond this bit depth would provide a certain level of improvement in quality, but would come at the cost of storing the resulting large number of bits in the image file.

How does varying the brightness resolution of an image affect file size?

Problem—Calculate the raw file size in terms of kilobytes of a 50 × 50 pixelized color image with a bit depth of 24. Calculate the file size if the number of bits per pixel were reduced to 12.

Number of pixels = 50 × 50 = 2500
Number of bits per pixel = 24
Number of bits in image = 2500 × 24 = 60,000 bits

To calculate the number of bytes:

60,000 bits/8 = 7500 bytes

To calculate the number of kilobytes:

1 KB = 2^{10} bytes = 1024 bytes
7500 bytes/1024 = 7.32 KB

If the number of bits per pixel were reduced to 12, then the total number of bits in the new image would be:

Number of bits in image = 2500 × 12 = 30,000 bits

To calculate the number of bytes:

30,000 bits/8 = 3750 bytes

To calculate the number of kilobytes:

1 KB = 2^{10} bytes = 1024 bytes
3750 bytes/1024 = 3.66 KB
The size of the second image has been halved, from 7.32 KB to 3.66 KB.

Storing a high-resolution image is of little use if it cannot be displayed on a correspondingly high-resolution device or printed with a high-resolution printer. Upcoming sections provide more information about the various types of display devices as well as device resolution.

DIGITAL VIDEO

Elementary school acquainted us with a simple method for emulating motion. A stick figure drawn in different positions on a series of pages can be brought to life by flipping the pages in rapid succession (see Figure 7-19).

Figure 7-19

Stick figures for emulating continuous motion

Creating video is based on this simple principle of motion emulation. A traditional, analog video camera takes multiple snapshots per second and records them on film. After the film is developed, the images (called frames) are shown in rapid succession to the audience. Coupled with a synchronized soundtrack, the result is a motion picture. The rate at which these frames are displayed on the screen is critical, and is called the **frame rate**. How often these images should be captured and shown to achieve continuous motion is related to our **visual persistence**. This property relates to the number of distinct images we can detect before they begin to blur with each other. Persistence of vision is what causes us to see an image for a split second even after shutting our eyes. Because human eyes can typically process approximately 20 images per second before losing the ability to distinguish the transition from one image to the next, the motion picture industry uses a standard of at least 24 frames per second (fps). This rate creates a perfect continuous-motion effect.

The videos created by digital video recorders are based on the same principle of traditional movie making, except that each frame is stored as a digital image rather than as an analog picture on film. A CCD array, which is used both in scanners and digital cameras, captures each frame, and sound is captured by a microphone and digitized. A digital video is comprised of a succession of digital images and sound, with each image very slightly different from the next. The result is a lot of bits! Digital recorders, which have become quite inexpensive, differ mostly in the size of their CCD array and the way they store video and audio. The most common storage methods for video on video cameras are flash memory, DVDs, magnetic tape, or hard drives.

How is the bit size of a digital video file calculated?

Problem—Calculate the number of gigabytes in a 2-hour digital movie if each frame in the movie is made up of 512 × 512 pixels, each pixel is assigned a 9-bit binary code, and each second of the movie requires 24 frames.

Number of pixels = 512 × 512 = 262,144 pixels
Number of bits in one frame of the movie = 262,144 × 9 = 2,359,296 bits
Number of bits in one second of the movie = 2,359,296 × 24 = 56,623,104 bits
Number of bits in one minute of the movie = 56,623,104 × 60 = 3,397,386,240 bits
Number of bits in one hour of the movie = 3,397,386,240 × 60 = 203,843,174,400 bits

Number of bits in two hours of the movie = 203,843,174,400 × 2 = 407,686,348,800 bits
Number of bytes in two hours of the movie = 407,686,348,800/8 = 50,960,793,600 bytes
Number of gigabytes in two hours of the movie = 50,960,793,600/2^{30} = 47.46 GB

This is clearly a very large number. In reality, far fewer bits are used to store a digital movie because of efficient compression software that reduces the file size of a video stored on a digital video recorder. (These compression techniques are discussed in the next section.) Furthermore, the motion picture industry uses a greater number of pixels and number of bits per pixel than those in the example. The preceding example considered only the file size of a digital movie, excluding its soundtrack.

Digital video offers considerable advantages over traditional video, such as higher quality, ease of use, reproduction, and low distribution costs. Traditionally, movie theaters have shown motion pictures by projecting a tape onto the silver screen. Repeated replays tend to scratch the tape and degrade the visual quality of the movie. Filming digital motion pictures and storing them on versatile digital storage media eliminate the unfavorable effects of traditional movie display and preserve the film's original quality. Distributors and filmmakers also favor digital filmmaking because it eliminates the costs of duplicating and shipping large rolls of tape. Using digital technology, theaters can directly download movies and present them to an audience without waiting for a tape to arrive by snail mail.

One benefit of digital video technology is the ability of users to make recordings themselves (**videoblogging**) and disseminate it over the Internet on sites like Google video, YouTube, blip.tv, or EyeSpot. People can also submit such recordings to a news Web site like CNN.com. The evolution of YouTube is an example of how quickly IT innovations can progress. YouTube, founded in 2005, was named *TIME* magazine's "Invention of the year" in 2006 and was purchased by Google, Inc. for $1.65 billion the same year.

The phenomena of amateur video dissemination, inexpensive digital video equipment, and online video-sharing sites have had many legal, social, political, and economic consequences. One legal complication is the sharing of copyrighted video material, which occurs when users post their favorite movie clips online. Another legal complexity is the question of what constitutes objectionable material when video is posted online. From a social and economic standpoint, some people, companies, and bands have gained celebrity status from their online video postings. Politically, inexpensive digital video has provided front-line accounts from war zones and provided footage of politicians making controversial statements that may have contributed to lost elections.

DIGITAL IMAGE AND VIDEO FORMATS

Digital images and videos could not be captured, stored, and disseminated without the existence of standard formats for compression. These formats provide the sets of rules, or protocols, required to compress and view images on display devices such as televisions, computer screens, cellular phones, and PDAs. This section introduces many of the most common file formats.

IMAGE FORMATS

JPEG (Joint Photographic Experts Group), which is actually a family of compression standards, is one of the most predominant image file formats. Digital cameras often use the JPEG format to store images on their memory cards, and many of us e-mail pictures in this format to our friends. The JPEG format enables the compression of images so that they occupy a smaller amount of space with some loss of fidelity. Many image editor programs allow you to adjust the level of compression before saving an image according to the tolerated amount of information loss. Other versions of this standard provide for lossless compression, but they do not enable a high degree of compression compared with their lossy counterparts.

The JPEG format supports a large number of colors (24-bit true color) and is ideally suited for photographic-quality images. Most established Web browsers support JPEG images. Digital image files of this format have a .jpg or .jpeg extension.

GIF (Graphics Interchange Format) is an image file format that is mostly encountered on the Web. It is typically used both for still images and animation, and is supported by most browsers. The GIF format provides a relatively high degree of lossless compression, and file sizes tend to be small because the format only supports 8 bits of color information (256 colors) per pixel, making it popular for the Web. The small bit depth makes GIF a suitable format for simple images such as grayscale photographs, simple logos, and cartoons. However, this format is not adequate for photographic-grade images because it cannot support true color. Digital image files of this format have a .gif extension.

PNG (Portable Network Graphics) is a lossless, compressed image file format that stores images in the form of a bitmap. It supports a larger bit depth than GIF but does not support animation. Although this format is not recognized by all browsers, it is still popular due to some of the shortcomings of the GIF standard and its ability to compress files to a greater extent than GIF. Digital image files of this format have a .png extension.

BMP (Bitmap Picture) is one of Microsoft's image file formats, and it stores images in the form of a bitmap. It can support true color, but it is not an attractive standard for storing photographic and print-quality images because it does not provide high compression. Because BMP files are usually large, they are not commonly used on Web pages and are not typically supported by Web browsers. This format is primarily used to view images in Microsoft applications. Digital image files of this format have a .bmp extension.

TIFF (Tagged Image File Format) is frequently used in the printing and publishing industry for many reasons, one of which is that TIFF image files can include multiple images. The TIFF format can support a range of bit depths, including true color, and achieves compression without degradation in quality. It is also highly versatile and flexible because of its use of tags, which are small pieces of information that are stored together with the image file and include information on the characteristics of the image. Although popular for the printing industry, the TIFF format is not a popular format on the Web. Digital image files of this format have a .tif or .tiff extension.

VIDEO FORMATS

MPEG (Moving Pictures Experts Group) is a family of lossy compression standards for moving images and associated audio. The MPEG format provides very high compression and drastically reduces the size of movies. The concept behind this standard is that each frame is not stored as a distinct digital image, instead only differences between two frames are kept. Some key frames are periodically inserted within the file to be used as a reference when reconstructing missing frames. The rest of the information is deemed redundant and is discarded. DVDs and digital broadcast services such as digital cable television use this format to store and broadcast digital moving pictures. The MPEG format is also used extensively for movie files uploaded on the Web.

AVI (Audio Video Interleaved) is Microsoft's format for multimedia applications in the Windows operating environment. It is used to store both audio and video information, but it does not provide high resolution and therefore is not suitable for applications that require high viewing quality. Most AVI files are viewed in a small window on the computer screen due to their low resolution.

WMV (Windows Media Video) is one of Microsoft's video formats, and is prevalent because WMV files can be played on Windows Media Player. The WMA audio file format was discussed in Chapter 6. The WMV format can support both audio and video. Files of this type have a .wmv extension.

MOV is a format developed by Apple Computer. It is primarily associated with Apple Computer's Quicktime media player.

Many video applications use the term **streaming video** or **streaming media**. These terms refer to the ability to view video in near real time as it is transmitted over a network, in contrast to being able to view the content only after downloading the entire video. A media server transmits the video to a player application on a client device, which temporarily and briefly stores (buffers) data until enough video is available and properly assembled to appear as a continuous video to the viewer. Streaming video approaches provide "viewing only" capability rather than depositing a copy of the video on a receiving device. Microsoft's file format, called **Advanced Systems Format** (ASF), is an example of a streaming media format.

DISPLAY TECHNOLOGIES

Display technologies play an integral part in human-computer interaction. High-quality displays can increase productivity and are especially important in the areas of IT that deal with medical imaging, national security applications, Web page design, and multimedia and computer graphics. IT workers in many industries must be familiar with the technical characteristics of displays to be able to make informed procurement and design decisions.

The most common devices for displaying still and time-varying digital images are the **cathode ray tube** (CRT), **liquid crystal display** (LCD), **light emitting diode** (LED), and **plasma display**. The flat displays that accompany desktop computers, laptops, camcorders, cellular phones, and PDAs are commonly based on LCD technology. The bulkier monitors associated with older desktop systems and older television sets use CRTs. Advertisement billboards and electronic traffic signs often use LED technology. Plasma displays are generally used in television sets.

CRT AND LCD DISPLAYS

The image formation methods of display technologies differ considerably. CRT computer monitors, as well as traditional television sets, contain a cathode ray tube that emits three electron beams, whose strength and direction of travel is controlled electromagnetically by a beam deflection mechanism to follow a **raster scan** format. A raster scan procedure forms an image on the screen, one small area at a time, by scanning the screen from side to side and top to bottom. A CRT display has a screen whose surface is coated with three types of phosphor, in a special dotted pattern, that glow in three colors of red, green, and blue. When these three beams strike the three dots of phosphor on the screen surface, the dots glow individually to produce a new color as a combination of red, green, and blue light. Each group of phosphor dots therefore corresponds to a single pixel of the screen. A unique color for each pixel can be created by adjusting the strength of each electron beam (see Figure 7-20). As a high-intensity electron beam strikes a phosphor dot, the beam correspondingly makes the dot glow with high intensity.

As the electron beams scan the screen from left to right and top to bottom at an extremely high speed, a frame is "painted" on the screen. The phosphor glow persists for some time, and the pixel where the electron beams strike remains illuminated until the entire screen is scanned by the beams. Although CRT monitors are still used today, the presence of its large tube makes the displays quite cumbersome, a factor contributing to their rapid obsolescence.

LCDs, such as those on laptop computers, digital camcorders, gaming devices, cellular phones, and PDAs, do not rely on a cathode ray tube for displaying images. Instead, LCD devices use liquid crystal materials, whose properties can be varied by electrical signals for displaying an image. An LCD system contains two plates of glass that sandwich a liquid with tiny crystals filled in small cells, two polarizing filters on each side of the glass plates (one classified as a vertical polarizer and the other as a horizontal polarizer), a color filter, and backlighting for illumination from one face of the plates. The lamp emits light that is passed through the vertical polarizing filter. The polarizing filter passes light based on a property called the **polarization** of the light. The light that is passed through the vertical polarizing filter is said to have a vertical polarization. This light reaches the layer of liquid cells whose crystals are aligned in a twisted pattern and control the polarization of the light incident on the second polarizing filter, based on the electricity level applied to these cells. When no

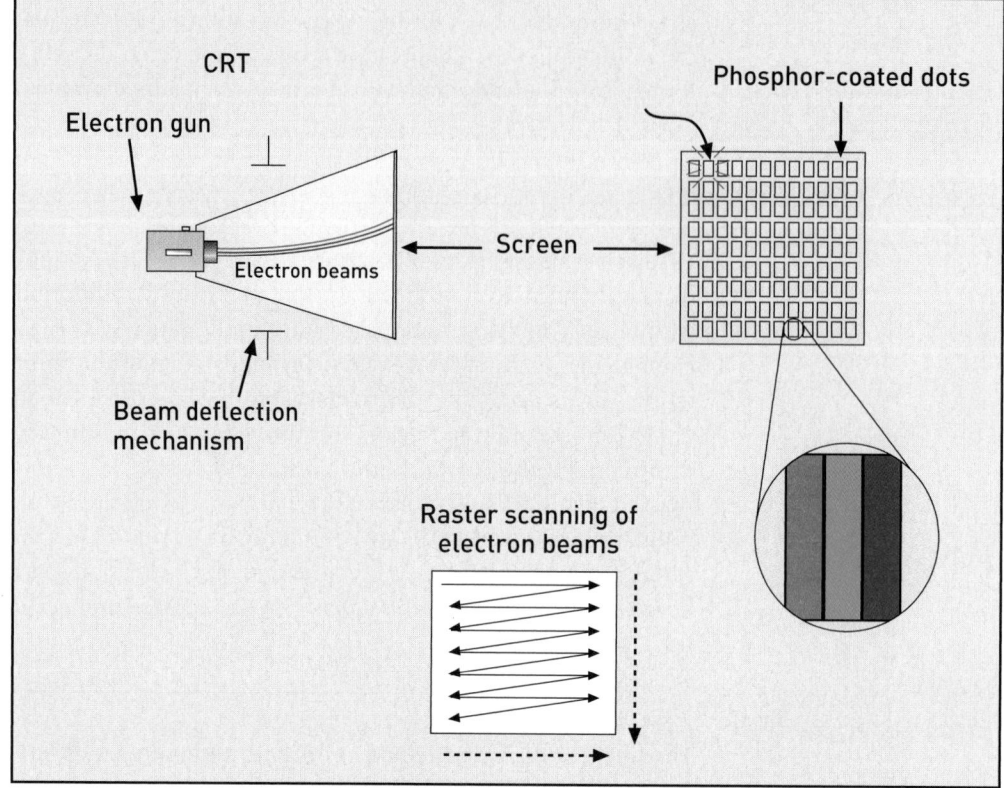

electricity is applied to the cells, the crystals remain twisted and rotate the light from the vertical to the horizontal direction, as depicted in Figure 7-21. The rotated light then reaches the second polarization filter, the horizontal polarizing filter. The filter then passes the light through, as the light that impinges on it is horizontally polarized. As electrical signals are individually applied to each cell, the crystals change alignment to control the level of polarization of the light. Based on the electrical level, the cell will not rotate the vertically polarized light into a perfectly horizontal alignment; hence, the horizontal polarizing filter will not pass all the light through. Depending on the degree by which the polarization of the light has changed when passing through the liquid crystal cells, the light at the other end of the second polarizing filter varies. The light passing through the liquid crystals passes through a color filter before being filtered out by the horizontal polarizing filter. Sections of red, green, and blue filters reside in alignment on top of each cell, and three cells are used to compose the color of each pixel. As the electricity level for each cell varies, the liquid crystals within each cell change alignment, in turn varying the light polarization that reaches each section of the filter (see Figure 7-21).

LED AND PLASMA DISPLAYS

LED displays present both still and time-varying digital images. LED displays are everywhere: Times Square, large billboards, concert hall marquees, and sports events. An LED display is just an array of tiny light emitting diodes, similar to small bulbs, which emit light of a certain color. Their intensity is based on the electricity applied to them. A simple display that can only display red light is based on the use of LEDs that emit red light. In contrast, LED displays that can display a wide array of colors use a combination of red, green, and blue LEDs.

Figure 7-21

Components of an LCD
display

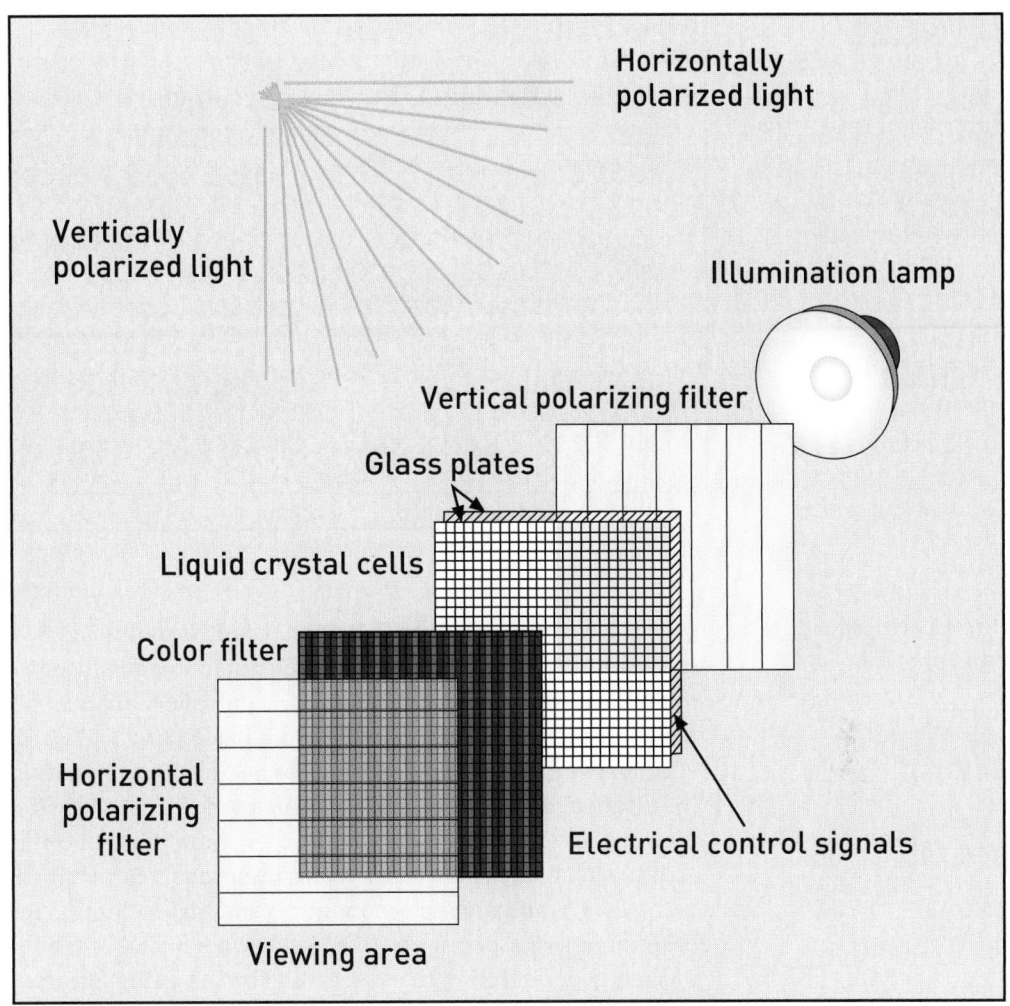

Various assortments of color can be obtained by varying the intensity of each LED. Computer monitors do not use LED displays; LED applications are used for large-scale displays or simple applications such as traffic signs (see Figure 7-22).

Plasma display technology is popular because of the demand for high-quality, thin-panel displays in televisions. Plasma displays use an ionized gas that is trapped inside cells between two glass plates. Each cell acts like a miniature fluorescent bulb. The intensity of light emitted by each cell is controlled with electrical signals, and various phosphors that glow as red, green, and blue painted on one of the glass plates as light from the cells impinges on them. The combination of red, green, and blue phosphors forms the color of each pixel in the display. The major advantages of plasma displays are their wider viewing angles, higher contrast, and better brightness than other display technologies. They are, however, generally more expensive than other display technologies.

TECHNICAL CHARACTERISTICS OF DISPLAY DEVICES

The following design features mainly determine the quality of a display device:

- Spatial resolution
- Screen size
- Pixel/dot pitch
- Brightness resolution (bit depth)

Spatial resolution, or pixel count, is directly indicative of a display's fidelity. A high-resolution display presents images more clearly and with greater detail. Display resolutions are

Figure 7-22

Components of an LED display

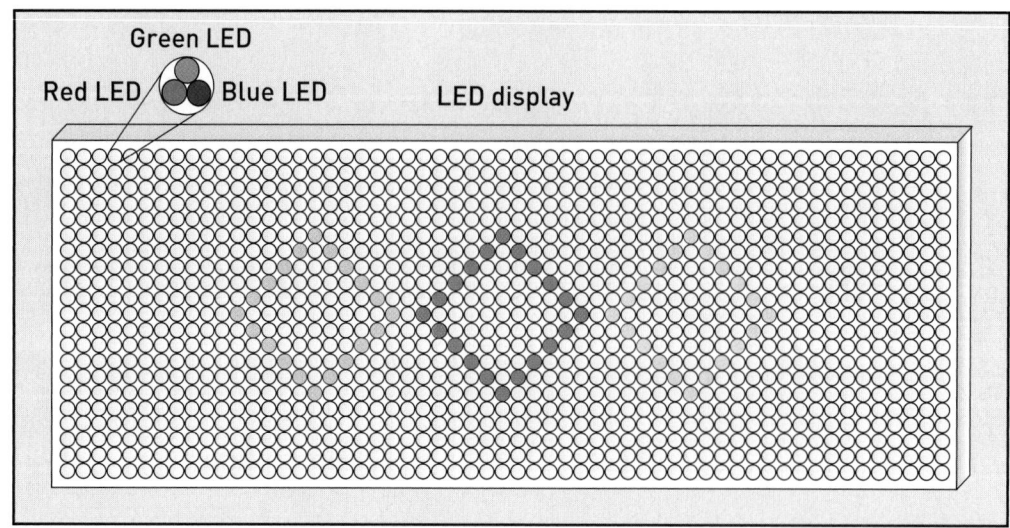

Green LED

Red LED   Blue LED   LED display

usually given as the number of columns of pixels multiplied by the number of rows of pixels, such as 1024 × 768 or 1280 × 1024. The resolution quality of a display device is also relative to the screen size. A design that considers only the number of pixels on a screen and assumes that more pixels mean better resolution is inadequate because screen size also factors into the equation. The screen size for televisions and monitors is traditionally given in inches, and is the diagonal distance between opposite corners of the viewing area (see Figure 7-23). The ratio of width to height is called the **aspect ratio**; it may be 4:3, or 16:9 in the case of **wide-screen** displays, or some other value. Traditional television screens usually have an aspect ratio of 4:3, while wide-screen monitors and high-definition television (HDTV) sets may be based on a 16:9 aspect ratio. **HDTV** refers to a digital broadcasting approach that provides much higher resolution than prevailing broadcasting formats. HDTV does not refer exclusively to the television itself, but to the high-resolution specification for formatting and transmitting the signal. Televisions that are capable of displaying HDTV broadcasts are classified as HDTV sets.

Figure 7-23

Diagonal distance corresponding to screen size

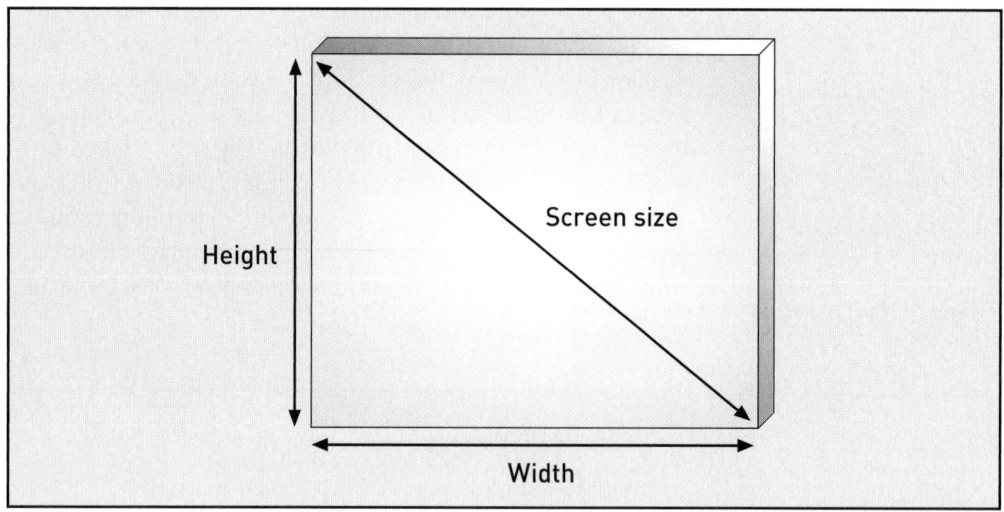

Height

Screen size

Width

Compare two displays with different specifications, as shown in Table 7-1.

Table 7-1		Display 1	Display 2
Display comparison	Spatial resolution	1024 × 768	1600 × 1200
	Screen size	15 inches	21.3 inches

The diagonal screen size or spatial resolution values alone are often not enough to estimate the quality of a monitor. It is also useful to calculate the number of pixels/dots per inch based on these figures, if this value is not already included within the specifications of the display device. This specification is similar to the samples-per-inch specification for a scanner. The vertical (height) and horizontal (width) dimensions of each monitor are calculated to provide the number of pixels per inch (ppi) on the display. The horizontal dimension is divided by the number of pixels in the horizontal dimension, and the vertical dimension is divided by the number of pixels in the vertical dimension (both calculations give the same number, though). The greater the number, the better the quality of the display device.

Display 1:

Because the display has a width/height ratio of 4:3, the first calculation finds the unknown value x in the following equation, which is based on the Pythagorean theorem:

$$(4x)^2 + (3x)^2 = 15^2$$
$$25x^2 = 225$$
$$x = 3$$

Width = $4x = 4 \times 3 = 12$ inches
Height = $3x = 3 \times 3 = 9$ inches
Pixels per inch in horizontal dimension = 1024/12 = 85.33 pixels per inch
Pixels per inch in vertical dimension = 768/9 = 85.33 pixels per inch
Note that both turn out to be the same value.

Display 2:

$$(4x)^2 + (3x)^2 = 21.3^2$$
$$25x^2 = 453.69$$
$$x = 4.26$$

Width = $4x = 4 \times 4.26 = 17.04$ inches
Height = $3x = 3 \times 4.26 = 12.78$ inches
Pixels per inch in horizontal dimension = 1600/17.04 = 93.89 pixels per inch
Pixels per inch in vertical dimension = 1200/12.78 = 93.89 pixels per inch

Because display 2 has a higher pixel density, it will provide a higher-quality picture than display 1.

Pixel pitch, or dot pitch, is another indicator of display quality, and relates to the space between pixels (see Figure 7-24). Displays with a small pixel pitch produce higher-resolution images.

Figure 7-24

Pixel/dot pitch

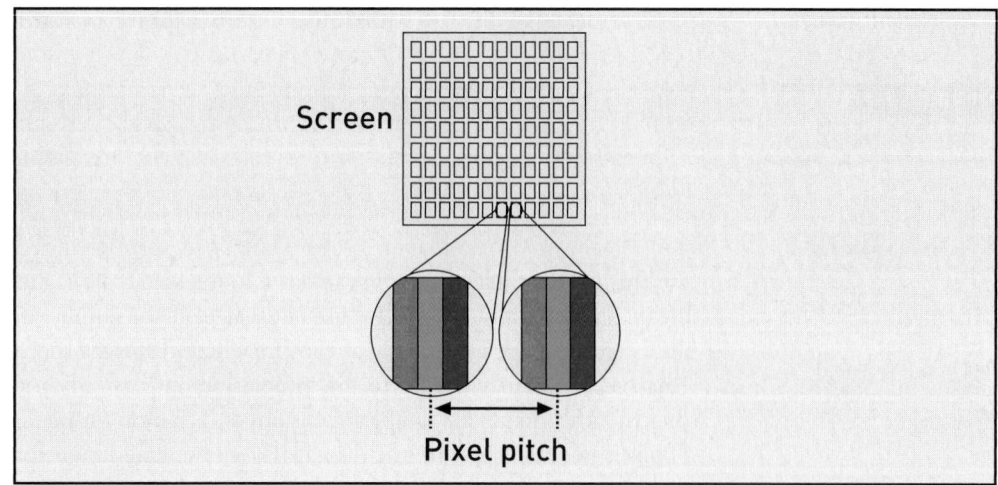

Bit depth signifies the range of colors displayed on the monitor. Technical specifications usually do not include bit depth, as most computer monitors and digital television sets have a bit depth of 24. Older computer monitors could only display 256 different colors, corresponding to a color depth of 8 bits. Most modern monitors and displays can display near-photographic color images by using 24 bits per pixel.

PROJECTION SYSTEMS

For some applications, projection systems provide an alternative to direct-display technologies such as LCDs or CRTs. A projection system relies on sophisticated lens systems and illumination equipment to project still and time-varying images onto a flat surface such as a screen or wall. Because they can display such large images, projection systems are ideally suited to large venues and audiences. These systems are commonly used in classrooms, business centers, conference rooms, movie theaters, and home theaters. Categories of projectors include the following:

- Film-based projectors
- LCD-based projectors
- **Digital light processing** (DLP) based projectors

Traditional film-based projectors are still in widespread use, especially in movie theaters, but they are being replaced by their digital counterparts, as discussed previously. A film or movie projector uses a reel on which the film is wound; each film contains millions of frames that correspond to the images of the movie. Each frame on the reel is illuminated, focused, and projected onto the screen as the film is advanced. The audio signal is synchronized with the frames and a complete motion picture is displayed.

LCD projectors use a strong light source that illuminates an LCD panel from behind, similar to the LCD illumination discussed earlier. Coupled with a lens system, the projector projects an image onto a large screen.

A popular projection technology nowadays is DLP, an approach developed by Texas Instruments in the late 1980s. These projectors are based on **micro-electro mechanical systems** (MEMS) technology, which involves the manufacture of micrometer-sized components such as gears, mirrors, and other parts. A typical DLP projector uses a chip called the **digital micromirror device** (DMD), an array of very tiny mirrors (see Figure 7-25). Each mirror in the DMD corresponds to a single pixel. Electrical signals adjust the position of each mirror to either reflect the light from a lamp or other source onto the lens system, or in another direction that does not reach the screen. To integrate color into the picture, a circular color filter such as a spinning disk filter is positioned between the DMD and the light source and is spun at high speed. The filter is similar to the spinning disk filter shown in Figure 7-9; it consists of red, green, and blue segments. The color information of each pixel in a frame that is to be projected onto the screen is fed to a processor that generates the electrical signals to control the mirrors. Each mirror is controlled three times to generate the exact color of each pixel. As each filter segment passes in front of the light source, each mirror is physically turned from its original position for some duration of time, depending on the brightness of the color at each pixel location. The longer the mirror is displaced from its position, the brighter the light that reaches the filter. The three colors reach the screen sequentially, but the persistence of human vision allows us to see the resulting combined color. A more advanced version of this technology employs three DMDs, one for each color component. These types of projectors, naturally, are more expensive.

Figure 7-25

Components of a DLP projector

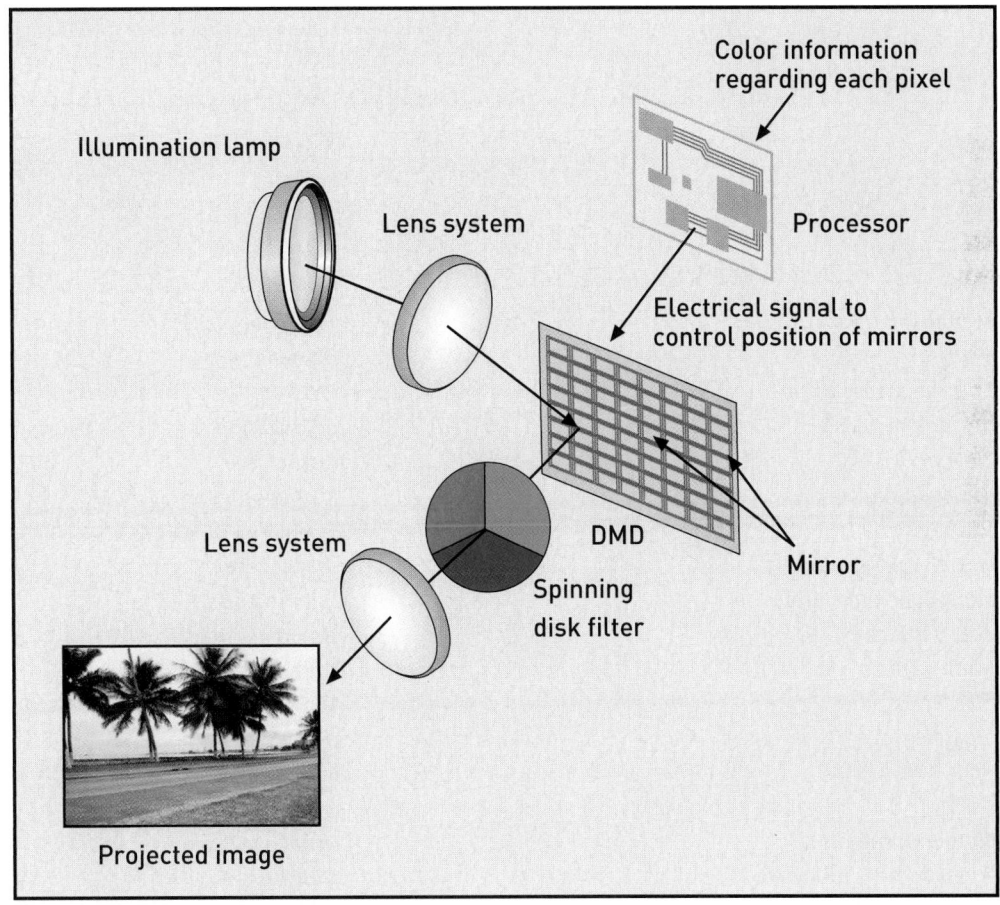

CHAPTER SUMMARY

- A digital image file is made up of a string of ones and zeros that represent the pixelized image. A pixelized image is made up of rows and columns of pixels, each with a unique color associated with it. A binary code expresses each pixel color.
- Two characteristics that determine the quality of images are spatial resolution and brightness resolution. Spatial resolution indicates the number of pixels per unit length (such as inches or centimeters) contained in an image. Brightness resolution describes the bit depth, and is the number of different colors inherently present in an image.
- Digital video is based on the principle of motion emulation, whereby individual digital images are shown in rapid succession. Because the human eye can distinguish at most 20 images per second, the frame rate standard used in the motion picture industry is at least 24 frames per second (fps), which creates the illusion of a perfect continuous-motion effect.
- Formats such as JPEG, GIF, BMP, PNG, TIFF, MPEG, AVI, WMV, and MOV provide the rules for structuring and compressing files that allow us to share and display images and video.
- The most common devices for displaying digital images are plasma displays, LCDs, CRT displays, and LED displays.
- The design features that contribute to the quality of display devices are spatial resolution, screen size, pixel pitch, and brightness resolution (also called bit depth).

KEY TERMS

ASF
aspect ratio
AVI
Bayer pattern
bit depth
bitmap image
BMP
brightness resolution
cathode ray tube
CCD array
color filter array
color gamut
digital light processing
digital micromirror device
dots per inch
frame rate
GIF
grayscale
HDTV
JPEG
light emitting diode
liquid crystal display
megapixel

MEMS
MOV
MPEG
pixel
pixel pitch
pixels per inch
plasma display
PNG
polarization
raster scan
RGB additive color model
samples per inch
spatial resolution
spinning disk filter
streaming media
streaming video
TIFF
true color
vector graphics
videoblogging
visual persistence
wide-screen
WMV

REVIEW QUESTIONS

1. What is a pixel?
2. What is the difference between a true color image and a monochrome image?
3. Roughly sketch an example of a pixelized image.
4. Explain the functionality of a CCD array.
5. What are the three primary colors that form the basis of the additive color model?
6. Calculate the number of megabytes in a color image with 1024×1365 pixels and 9 bits of color information per pixel.
7. Calculate the number of megabytes in a color image with 1024×1365 pixels and 12 bits of color information per pixel.
8. What can reduce the file size of the images described in the previous two questions?
9. How many colors can our eyes detect? Explain how this is related to a bit depth of 24.
10. Calculate the file size, in megabytes, of a 30-minute documentary movie with 256×256 pixels per frame, 9 bits per pixel, and 24 frames per second.
11. What effect does varying the spatial resolution have on an image?
12. How does reducing the brightness resolution affect an image? How can brightness resolution be decreased?
13. What is the color gamut of a camera or scanner, and how is it related to bit depth?
14. Briefly explain how the persistence of human vision enables moviemaking. Name two devices discussed in this chapter that rely on the persistence of human vision.
15. What is the maximum number of images per second that the human eye can discern before images begin to blur together?
16. Name three important image file formats.
17. Explain the difference between bitmapped images and vector graphics.
18. Why is JPEG a better format than GIF for photographic-quality images?
19. Why might GIF be a better format than JPEG for posting images on the Web?
20. What form of display technology is used in laptop computers?
21. Briefly explain how a CRT display can display images.
22. Calculate the width and height of a monitor with a 42-inch screen and an aspect ratio of 16:9.
23. What is an LED? What type of display device employs LEDs?
24. List two important factors in designing or buying a display device such as a computer monitor.
25. Calculate the screen size (diagonal length) of a display that has a width of 4 inches and a length of 3 inches.
26. Discuss how DLP-based systems can display images.

DISCUSSION QUESTIONS

1. Determine how many devices you own or encounter on a daily basis that incorporate display technology. Describe these devices and try to determine which use CRT, LCD, LED, or plasma display technologies. Which type of display technologies do you prefer for various applications and why?
2. What do you consider the best video-sharing Web sites and why? Which of them allow actual sharing of videos, meaning that a user can access and download a video posted by someone else, versus sites that only allow viewing (but not downloading) of videos? What economic, political, and social factors might shape the services that these sites provide or do not provide in the future?
3. If you save images on a flash drive, CD, or hard drive, do you think these storage media will still be viable and available in 10 years? What video or image files do you have that are no longer accessible because they are stored on obsolete media or require software that is obsolete? Will the same thing happen to the video and images you are currently storing?
4. Research some brands of HDTV on the Internet. What specifications are listed by the manufacturer? Compare the screen size and spatial resolution of two HDTVs and decide which is better based on these specifications. Determine whether these HDTVs are based on LCD, plasma display, DLP, or other type of technology. Find out how many HDTV channels are available from your TV service provider and what additional equipment you might need to receive these broadcasts.

CASE PROJECT

In this case project, you create and share digital video.

Write a script for a two-minute video newscast that discusses a technical, legal, economic, or social issue related to sharing video on a Web site for user-generated content. For example, who decides what can be posted, and why? Should video clips be subject to copyright laws? How can a user-generated content site become a profitable business? What technical format for video should become the industry standard and why? How are video-sharing sites changing the way we live together and interact?

Record the video, borrowing your school's equipment or a friend's if necessary, and post it on a user-generated video content site. Did you select a site that allows only posting and viewing videos, or a site that also allows downloading of videos? If you have not posted a video before, how easy was it to figure out how to record and post the video clip? Once the video is posted, e-mail its location to your classmates and ask them to submit comments about the clip to the site. What format did you use to save the video file? Does this format support streaming video? Could you convert the format you selected into another video format?

PART

4

TRANSMISSION OF INFORMATION

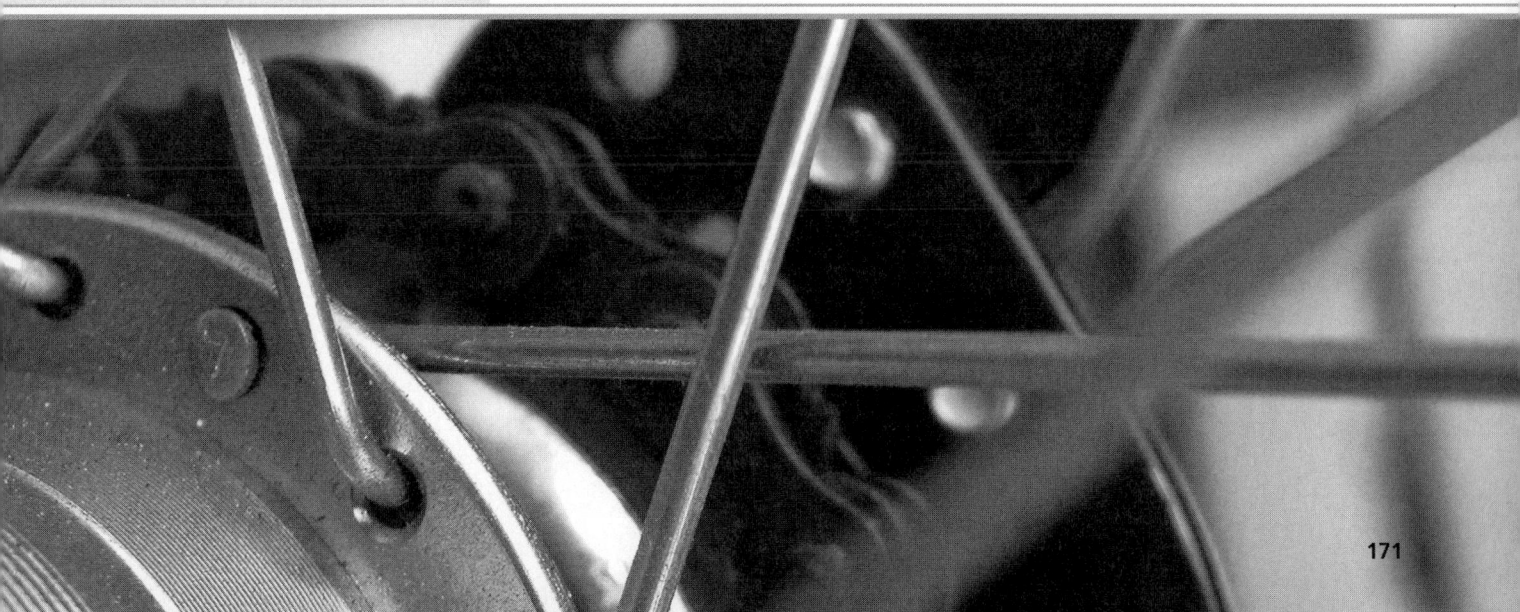

FUNDAMENTALS OF COMMUNICATIONS

LEARNING OBJECTIVES

In this chapter you will:

- Understand how binary streams are physically generated

- Learn how carriers are modulated to carry the binary streams

- Understand important transmission concepts, including attenuation, bandwidth, channel capacity, and multiplexing

- Learn the properties of different types of transmission media

- Identify sources of transmission errors and learn about error detection and correction techniques for digital transmission systems

INTRODUCTION

Although a sequence of binary digits can logically map any type of information using methods explained in earlier chapters, a physical entity such as electrical energy, optical energy, or electromagnetic energy is essential for carrying these bits over transmission channels. This chapter discusses the basics of electrical, electromagnetic, and optical signaling for transmitting analog and digital forms of information over copper wires, optical fibers, and free space, with an emphasis on digital electrical signaling. This chapter also introduces important concepts in transmission such as modulation, multiplexing, attenuation, and bandwidth. You will learn about different types and properties of copper transmission media, such as twisted pair and coaxial cables. Finally, because errors occur when digital signals travel over any transmission channel, the chapter concludes by discussing techniques to detect and correct errors.

For almost two centuries, electricity has supported countless applications, including communication technologies. Electrical signals transmitted through conducting materials, such as metal wires, effectively transmit both analog and digital information. How does electricity flow through a wire and how does analog and digital electrical signaling enable communication?

Metallic conductors, such as copper wires, comprise atoms with loosely attached electrons, or negatively charged particles, around their nuclei. When a voltage/potential difference from an electrical source, such as a battery, is introduced between the two ends of a conductor, the electrons are stimulated to move within the metal from one atom to another (see Figure 8-1). You can think of the applied voltage as an external force pushing the electrons forward through the conductor with a "strength" measured in **volts**. Because materials such as glass or plastic have tightly attached electrons within their atoms, no electron flow occurs, even if a voltage difference is applied to them, which classifies these materials as **electrical insulators**. In conductors, the flow of electrons within the material results in an electric current, with the magnitude of charge flow (size of electrical current) proportional to the value of the potential difference.

Aside from the applied voltage difference, the amount of charge flowing through a conductor also depends on a fundamental property called the **resistance** of the material, which is measured in **ohms**. This resistance property opposes the flow of current through the material. The greater a material's resistance is, the smaller the value of the current flowing through it. The relationship between voltage (V), current (I), and resistance (R) is defined by **Ohm's Law**:

$$V = IR$$

Here, voltage is measured in volts (V), current in **amperes** (A), and resistance in ohms (Ω). Engineers rely on Ohm's Law, as well as many other technical considerations, when designing electrical communication systems.

ANALOG AND DIGITAL SIGNALING

When analog signals are transmitted across a conductor, a continuous voltage difference proportional to the amplitude of the analog signal is applied at the input of the communications circuit. The current flowing through the circuit is proportional to the applied voltage according to Ohm's law, as there is a resistance associated with the circuit. In Figure 8-2, a continuous voltage variation applied at the input produces a proportional variation in current. A light bulb placed in the circuit monitors the current flow. As the applied voltage intensifies, the flow of current and the intensity of light emitted from the bulb increases. As the voltage diminishes to 0 V, the current approaches a value of 0 A, which makes the light bulb stop emitting light. The output end detects the variation in voltage and, hence, receives the

Figure 8-1

Conduction of electricity over a copper wire

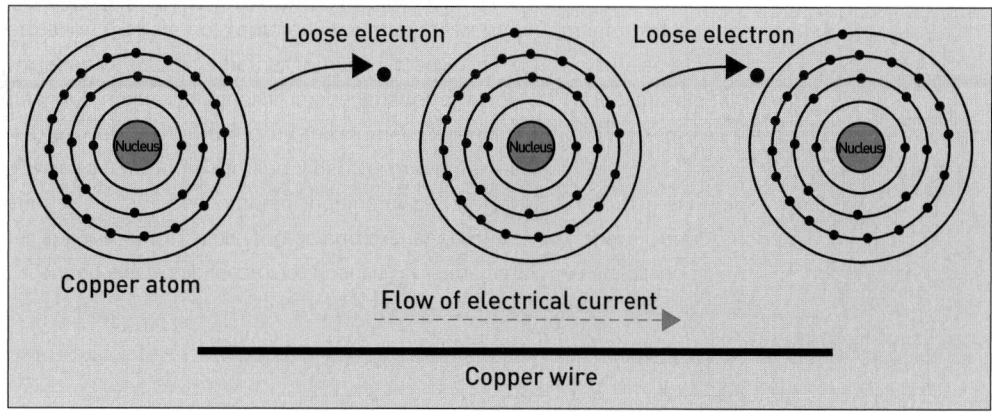

Figure 8-2

Transmission of analog
electrical signals over a
copper wire

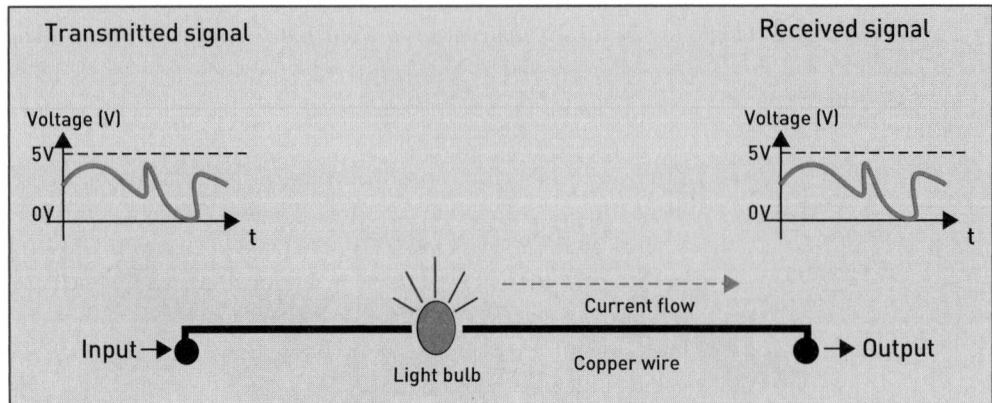

information carried by the signal.[1] The concept of analog signal transmission was briefly introduced in Chapter 2 within the context of the speedometer example.

The transportation of digital information relies on the same fundamental principle, except that the voltage applied to the input of the communications circuit takes on discrete values instead of a continuous range of values. One method for sending digital information across a communication system is called **binary signaling**, which corresponds to sending the data one bit at a time and switching between two discrete voltage values. By applying a voltage of 5 V to represent a 1 and 0 V to represent a 0, a pulsed current may be generated over the wires using binary signaling. The voltage variation is then detected at the receiving end and the binary information is extracted at the output of the system. A battery that supplies a constant voltage (for example, a steady 5 V) connected to a switch can generate the pulsed voltage variation at the input. By positioning the switch on or off, the voltage applied to the circuit can be varied between 5 V and 0 V, creating a pulsed current. The pulsed voltage is then detected at the receiving end for extracting the digital information. In this case, the bulb lights up and turns off as the switch is turned on and off (see Figure 8-3); the rate depends on how fast the switch is turned on and off. The concept of binary signaling was introduced in Chapter 2, when we discussed the transmission of digitized speedometer values.

Figure 8-3

A binary transmission
system

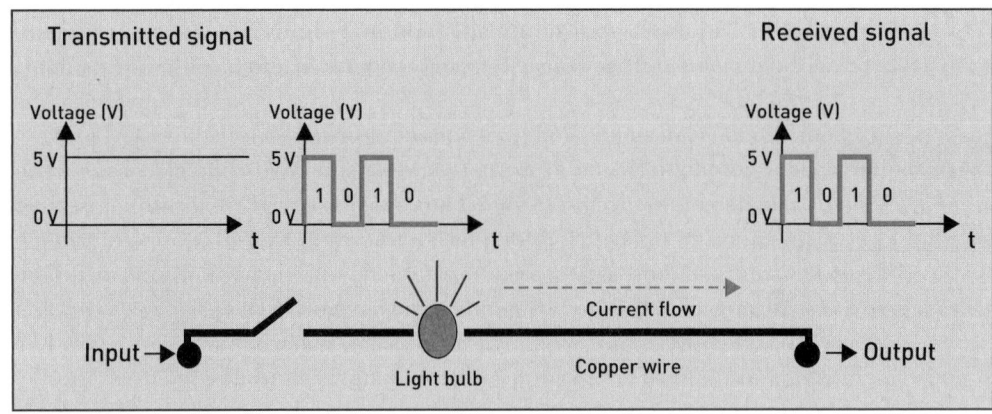

[1] Note that in actual communication systems, the received signal is never exactly the same as the transmitted signal due to many effects, including noise acting upon the signal as it travels down the transmission line. These effects are discussed later in this chapter.

An alternative method to sending digital information across a communication system is 4-ary signaling. The previous example discussed 2-ary signaling, more commonly referred to as binary signaling, where a single electrical pulse was used to transmit a single bit. In 4-ary signaling, four different signal levels are sent over the communication channel, and two bits may be sent over the channel using a single pulse of electricity. Figure 8-4 illustrates this concept.

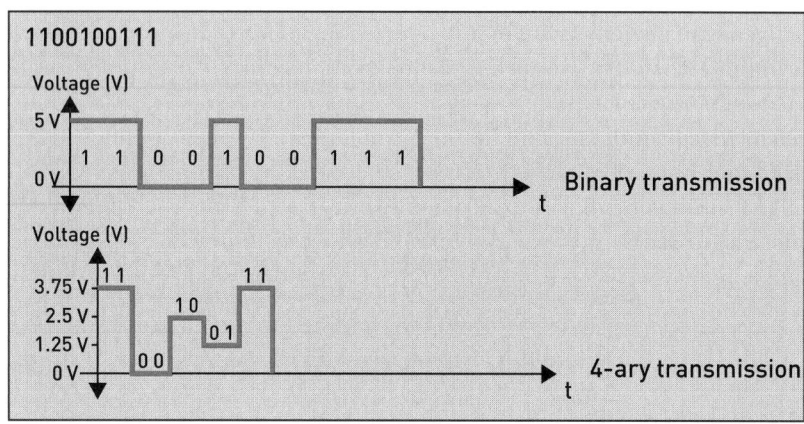

In 4-ary transmission, the transmitter separates the digital information into groups of two bits each and assigns a unique voltage level to each 2-bit group. In Figure 8-4, 00 is assigned 0 V, 01 is assigned 1.25 V, 10 is assigned 2.5 V, and 11 is assigned 3.75 V. Each group of two bits is sent using the corresponding voltage value. The receiver then must distinguish the four discrete voltage levels from each other and recover the digital bit stream. When comparing binary transmission with 4-ary transmission, we see that we can transmit two bits using a single pulse of electricity in 4-ary signaling versus transmitting one bit at a time using binary signaling. The trade-off is that the transmitter and receiver are more complex; the transmitter has to generate four different voltage levels rather than two, and the receiver distinguishes between four voltage values, compared with only two voltage values for binary signaling. Keep in mind that the voltage values in Figure 8-4 are hypothetical—in real communication systems, these voltages can be much different. Furthermore, the voltage may take on negative values.

The 4-ary signaling may be generalized to **M-ary signaling**, whereby M different voltage levels may be used to transmit $\log_2 M$ groups of bits with one signal. The data rate (D) in bits per second for M-ary transmission can be calculated by the following:

$$D = R \log_2 M$$

where R is the signaling transmission rate expressed in terms of signals per second. For binary signaling, the data rate D is equal to R. For 4-ary transmission, the data rate is equal to 2R, and so on.

Some readers of this textbook may not recall logarithms. In consideration of these readers, the "$\log_2 M$" expression corresponds to the logarithm of M according to a base of 2. This means that we need to find a number, x, such that $2^x = M$. For example, $\log_2 2 = 1$ where $x = 1$ because $2^x = 2^1 = 2$. Similarly, $\log_2 4 = 2$ where $x = 2$ because $2^x = 2^1 = 2$.

Problem—Calculate the data rate for a communication system that employs 8-ary signaling if the signal transmission rate is 1000 signals per second.

R = 1000 signals per second
M = 8
According to the equation:
$D = R \log_2 M = 1000 \log_2 8 = 1000 \times 3 = 3000$ bps = 3 Kbps

The data rate is therefore 3 Kbps.

THE ELECTROMAGNETIC SPECTRUM

Besides electrical energy transmitted over conductors, **electromagnetic** (EM) **energy** transmitted over air or a vacuum is also commonly used for analog and digital communications. Systems that use these forms of transmission channels are classified as wireless systems, such as cellular telephony, wireless Internet access, and radio broadcasting. Wireless technologies typically employ EM energy for carrying information across both short and long distances. EM energy travels in the form of **EM waves**, such as **radio waves**, **light waves** (infrared, visible light, ultraviolet), **x-rays**, and **gamma rays**. These waves vary in a sinusoidal manner, taking on a range of frequencies (see Figure 8-5). Sinusoidal waves were first discussed within the context of sound waves in Chapter 6, where you learned that pure sound waves vary in a manner that follows the sine function in mathematics, hence the name sinusoidal. The sinusoidal variation property is not limited to sound waves, but may be found in any wave that varies according to the sine function. Sound waves and EM waves are both waves, but are based on different energy forms. Sound waves are waves of acoustic energy, whereas EM waves involve EM energy. Nonetheless, both types of waves have properties of frequency, amplitude, and phase. Although x-rays, gamma rays, and others are all classified as EM waves, only radio waves and light waves are used in wireless communication systems, due to the dangers associated with x-rays and gamma rays. The EM waves are used as *carriers* to wirelessly carry both analog and digital information by altering certain properties of the wave in proportion to the information signal. This process, called **modulation**, is explained in the next section.

Figure 8-5

A sinusoidally varying
electromagnetic wave

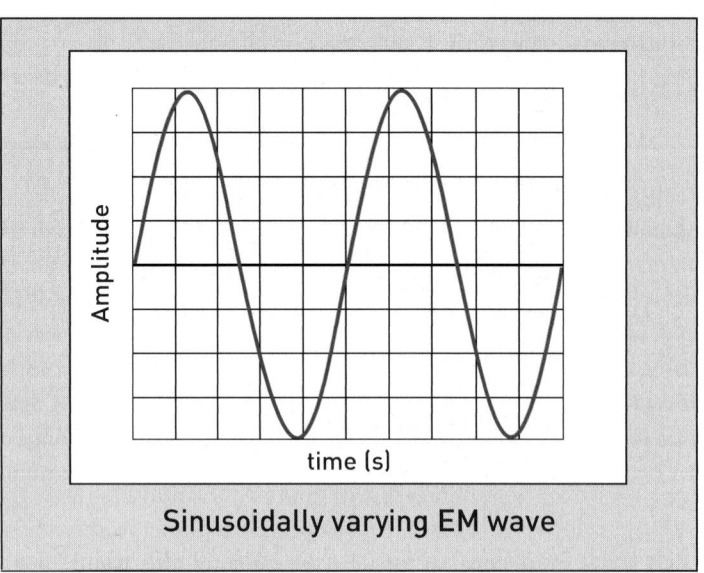

EM waves may be formed by passing a fluctuating current through a conductor (such as a piece of wire), thereby creating electrical and magnetic fields around the conductor. These fields couple to create a wave classified as an electromagnetic wave, which radiates away from the conductor. The various types of EM waves, such as radio waves, light waves, x-rays, and gamma rays, vary within a range of frequencies called the **electromagnetic spectrum**, which extends from approximately 3 Hz to more than 1×10^{20} Hz (100,000,000,000,000,000,000 Hz). Figure 8-6 illustrates the EM spectrum.

Figure 8-6

The electromagnetic
spectrum

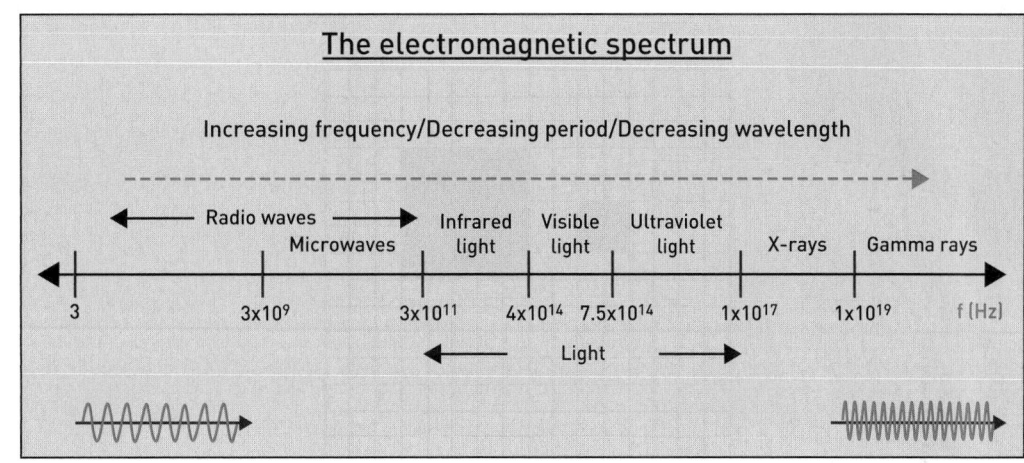

A small part of the EM spectrum classified as "radio waves" is comprised of microwaves. Some textbooks treat radio waves and microwaves separately within the EM spectrum. However, this text classifies microwaves under the category of radio waves. Note also that although Figure 8-6 illustrates clear boundaries that define the various frequency bands, this is not the case in reality. For example, there is an overlap between x-rays and gamma rays, and an overlap between ultraviolet rays and x-rays.

The frequency (f) property of sinusoidal waves was explained in detail in Chapter 6. Frequency was defined to be the rate of repetition of a wave cycle, measured in hertz (Hz). Frequency was defined in Chapter 6 in the context of sound waves, but it is a property of EM waves as well. Besides frequency, EM waves can also be characterized in terms of their **wavelength** (λ). The wavelength is simply the physical length between two peaks or two crests of the EM wave, and is measured in meters (see Figure 8-7). Wavelength is related to the frequency and speed at which EM waves can travel; do not confuse it with the period (T) measured in seconds, which is the time a wave requires to complete a single cycle.

Wavelength can be calculated using the following expression:

$\lambda = c/f$

where c is the speed of EM waves ("the speed of light") in a vacuum measured in meters per second (m/s), and f is the frequency of the EM wave in hertz (Hz). EM waves travel at a maximum speed of approximately 3×10^8 m/s (300,000,000 meters per second) through a

Figure 8-7

The concept of
wavelength

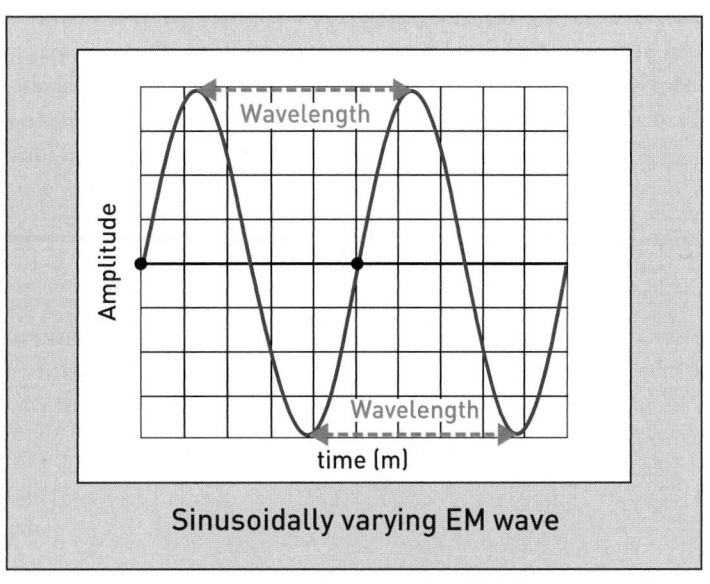

Sinusoidally varying EM wave

vacuum, such as outer space. They slow down as they travel through any other media, such as air, glass, or water. The spectrum, as shown on Figure 8-6, lists EM waves in order of increasing frequency, decreasing period, and decreasing wavelength from left to right. Radio waves have the longest wavelengths and gamma rays have the shortest ones.

Problem—Calculate the wavelength of a radio wave with a frequency of 2.5 GHz traveling through a vacuum.

$\lambda = c/f$
$c = 3 \times 10^8$ m/s = 300,000,000 m/s
$f = 2.5$ GHz = 2,500,000,000 Hz
$\lambda = 300,000,000/2,500,000,000 = 0.12$ m, or 12 cm

The wavelength of the radio wave is 12 cm.

EM energy can occur naturally or be generated artificially. A significant source of natural EM radiation is the sun, which exposes us to ultraviolet rays that can burn our skin, and to infrared radiation that heats the earth's surface. We are also exposed to naturally occurring gamma radiation from natural phenomena in outer space. Besides these natural forms of radiation, we are constantly enveloped by artificial radiation sources, including radio and television broadcasting stations, cellular antennas, satellites, wireless networks, and even microwave ovens. Just as a musical instrument can generate sound waves, radio waves can be generated artificially by transmitting an oscillating electrical current through a conductor such as an antenna. The oscillating current within the conductor creates an EM wave that radiates away from the conductor. Italian scientist Guglielmo Marconi was the first person to successfully demonstrate this concept and apply it to long-distance communication. Marconi's demonstration of the possibility of long-distance radio-wave transmissions gave birth to the field of radio communications.

The Radio Spectrum

The frequency range of the EM spectrum between approximately 3 Hz and 300 GHz is called the **radio spectrum**, and waves within the radio spectrum are called radio waves. The radio spectrum is further divided into regions called **frequency bands**. Radio frequency (RF) systems use EM waves within the radio spectrum for communication and operate on these bands, such as very high frequency (VHF) and ultra high frequency (UHF). Figure 8-8 illustrates the radio spectrum.

Various wireless systems use a specific frequency band or multiple frequency bands. Table 8-1 lists various applications and the frequency bands they employ.[2] For example, AM and FM radio stations operate on different frequency bands, with AM stations broadcasting within the medium frequency (MF) band and FM within the VHF band. Television stations use the VHF/UHF frequency bands for broadcasting television signals. Wireless local area networks operate on the super high frequency (SHF) band.

Figure 8-8

The radio spectrum

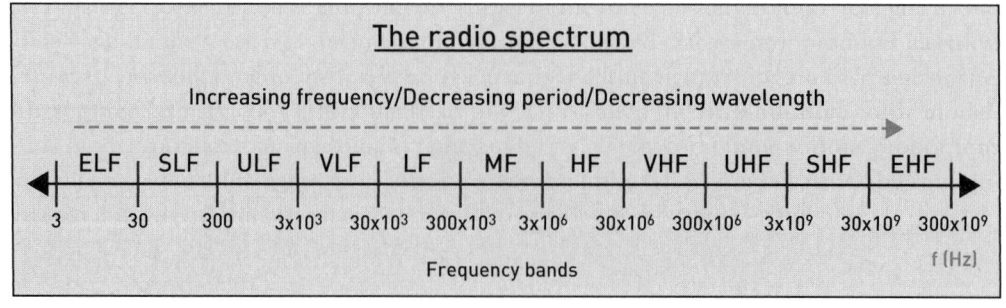

[2] http://www.ntia.doc.gov.

Table 8-1

Radio frequency bands
and their applications

Acronym	Expanded name	Application
ELF	Extremely low frequency	Submarine communication
SLF	Super low frequency	Submarine communication
ULF	Ultra low frequency	Submarine communication, navigation
VLF	Very low frequency	Submarine communication, navigation
LF	Low frequency	Navigation
MF	Medium frequency	AM, amateur radio
HF	High frequency	Citizens band radio, radio control, astronomy
VHF	Very high frequency	FM, TV
UHF	Ultra high frequency	TV, cellular telephony, cordless phones, global positioning system (GPS), aircraft navigation, satellite communication, walkie-talkies
SHF	Super high frequency	Satellite communications, wireless computer networks, microwave ovens
EHF	Extremely high frequency	Space exploration, astronomy, satellite communications

Systems that use radio waves must conform to certain rules, regulations, and standards of operation. For example, music stations are not allowed to broadcast within any frequency band, using any channel (the frequency within the band); the radio spectrum is shared among many users for multiple applications. All parties that use radio-wave communications must operate on a band and channel designated by a central authority. The Federal Communications Commission (**FCC**) allocates radio frequencies in the United States, and most countries have a corresponding organization that assigns frequencies. When a radio station wants to broadcast to the public in a specific area in the United States, for example, the station obtains permission from the FCC. The radio station is then assigned a frequency, if available, within the VHF band. For example, a radio station such as XYZ FM might broadcast at a frequency of 101.1 MHz in Los Angeles and at a frequency of 99.5 MHz in Washington, D.C., because these might be available channels for the radio station in each city. Cordless telephones, walkie-talkies, and wireless network adapters are examples of other systems that have designated sets of operating frequencies but do not require permission for channel use.

MODULATION/DEMODULATION

Sinusoidal radio waves convey both analog and digital information in wireless systems. By varying some of the properties, such as the amplitude, frequency, and phase of these EM waves, both analog and digital information can be represented and sent across the wireless channel. The following discussion describes some common approaches used in radio-wave communications. These methods also apply to communication systems in general.

Modulation techniques superimpose information signals onto a carrier wave, such as a radio wave or a light wave. Conversely, **demodulation** is the process of extracting information-carrying signals from modulated waves. This chapter discusses the following modulation techniques:

Table 8-2

Analog and digital
modulation techniques

Analog modulation techniques	Digital modulation techniques
Amplitude modulation (AM)	Amplitude shift keying (ASK)
Frequency modulation (FM)	Frequency shift keying (FSK)
Phase modulation (PM)	Phase shift keying (PSK)

Other digital modulation techniques exist, including quadrature amplitude modulation (QAM) and minimum shift keying (MSK), but they are beyond the scope of this book. To thoroughly understand modulation/demodulation, revisit the detailed explanations of amplitude, frequency, and phase of a wave in Chapter 6. To superimpose information signals onto carrier EM waves, one of the three properties of the EM wave is varied in proportion to the information signal. For example, analog modulation techniques rely on variation of the amplitude, frequency, or phase of an EM wave in proportion to the continuously varying amplitude of the information-carrying analog signal. Digital modulation techniques rely on the variation of one of the wave's three properties in proportion to the discretely varying amplitude of the information-carrying digital signal. The following sections describe these processes in detail.

Analog Modulation Techniques

Figure 8-9 demonstrates the concept of **amplitude modulation** (AM). The first signal denotes the original information signal, and the wave below it illustrates an EM wave of a specific frequency, called the carrier wave. The amplitude of the carrier wave is varied in proportion to the amplitude of the original analog signal, resulting in the modulated waveform/signal. The rate at which the carrier wave fluctuates must be considerably higher than the fluctuation of the information signal to achieve effective modulation. The modulated signal is then broadcast over the wireless channel by a transmitting antenna to a receiving antenna. To extract the information signal from the modulated waveform, the receiver circuitry (such as that in a car's AM radio receiver) tunes in to the carrier frequency—1300 KHz, for example—and tracks the amplitude changes within the modulated wave.

Similarly, **frequency modulation** varies the frequency of the EM wave in proportion to the amplitude of the analog information signal. Figure 8-10 shows the analog signal from Figure 8-9, now modulated using frequency modulation.

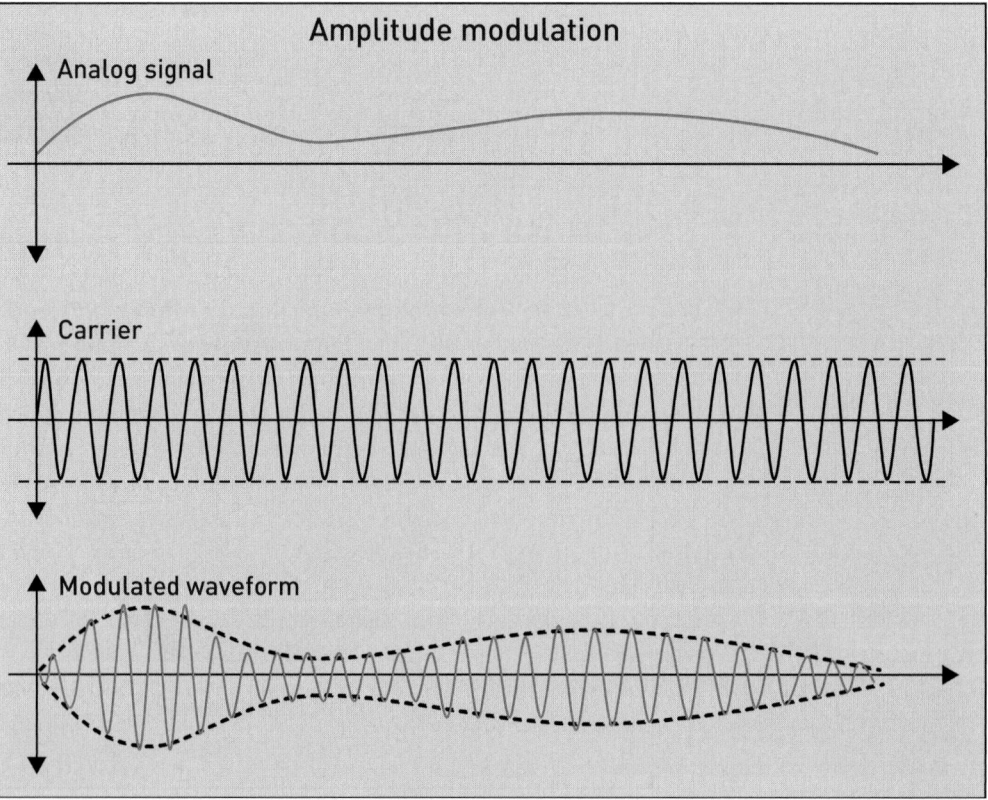

Figure 8-9

Amplitude modulation of a carrier wave

Figure 8-10

Frequency modulation of
a carrier wave

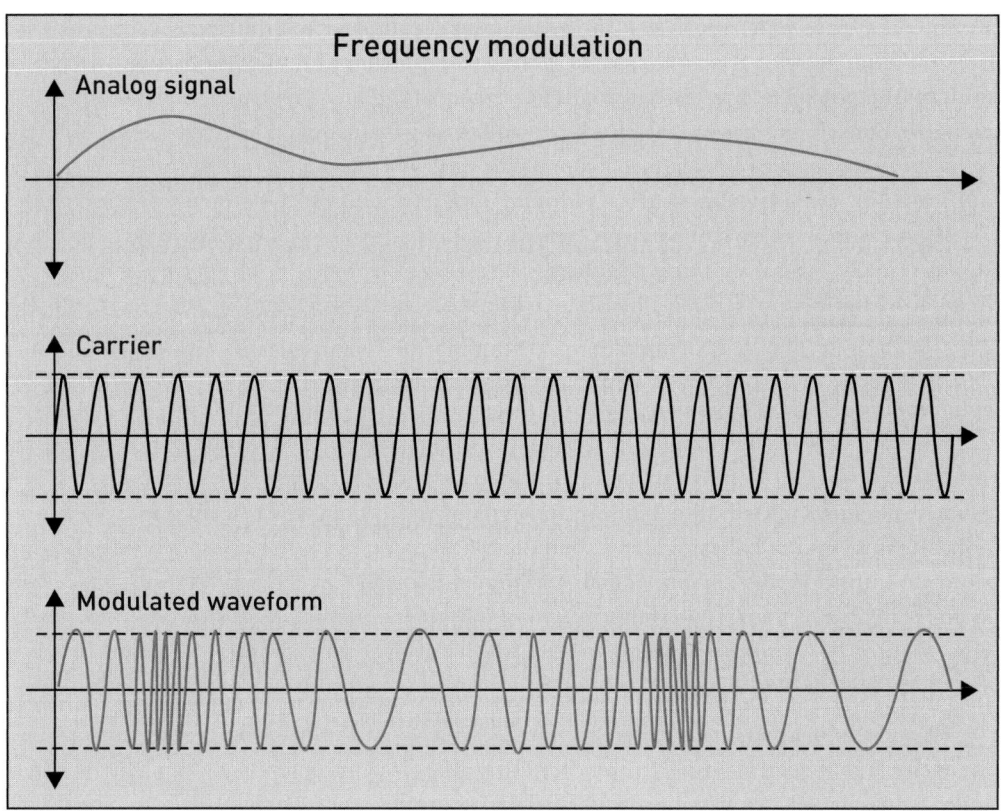

Although the initial frequency of the carrier, called the central frequency, is fixed, modulating the carrier wave causes a deviation from the central frequency by an amount proportional to the varying signal amplitude. The modulated signal is then transmitted over the wireless channel; the receiver extracts the amplitude variation of the information signal by tuning in to the central frequency and tracking frequency variations in the arriving modulated waveform. A radio's antenna receives the signal, and the listener hears the FM broadcast by tuning in to the central frequency. In Figure 8-10, the frequency of the carrier wave increases for increasing values of the amplitude and diminishes as the amplitude diminishes. In contrast to AM, the amplitude of the frequency-modulated signal remains steady, and only the frequency varies.

Phase modulation (PM) is the final analog modulation technique that we will discuss in this chapter. Phase modulation is similar to FM, but instead of varying the carrier frequency in proportion to the original signal's amplitude, phase modulation varies the carrier phase in relation to the signal's amplitude. In other words, the phase of the EM wave varies in proportion to the amplitude of the analog information signal.

A phase-modulated signal looks like a frequency-modulated signal, except that the information is carried in the phase variation of the EM carrier wave. To extract the original analog information signal from the phase-modulated signal, a demodulator that can track phase changes is used at the receiver.

A wide variety of communications systems, most obviously AM and FM radio, use analog modulation techniques. Each implementation varies in the fidelity it provides and the range it covers. FM and PM have higher fidelity than AM, so music channels usually broadcast in FM and talk-radio channels usually use AM. FM and PM have higher fidelity because external noise factors that create undesired amplitude fluctuations in the modulated signal have little effect on the quality of the broadcasted signal, which transmits information through frequency and phase variations. AM is more vulnerable to noise, but systems that employ AM typically consume less power and have a wider coverage area. AM stations use MF waves spread across large areas, compared with the VHF waves that FM broadcasts use. FM and PM are also more expensive to implement because they require a slightly more complex demodulator.

Digital Modulation Techniques

Digital wireless systems, such as digital cellular networks, use digital modulation techniques such as ASK, FSK, and PSK to superimpose digital information. The basic techniques of these three modulation schemes resemble their analog counterparts. The only difference is that the properties of the carrier signal are modulated according to a few discrete values. With binary modulation techniques, the carrier wave is modulated according to only two values. With 4-ary modulation, the carrier wave is modulated according to 4 different values. Similarly, with M-ary modulation, the carrier wave is modulated according to M different values.

Binary **amplitude shift keying** (ASK) varies the amplitude of the carrier wave according to the binary amplitude of the digital information signal. The carrier amplitude may be null to convey a binary 0 and at the maximum level to convey a 1 (see Figure 8-11). The receiver keeps track of the amplitude variation of the received modulated signal to extract the bit stream sent by the transmitting end.

Figure 8-11

Binary amplitude shift keying

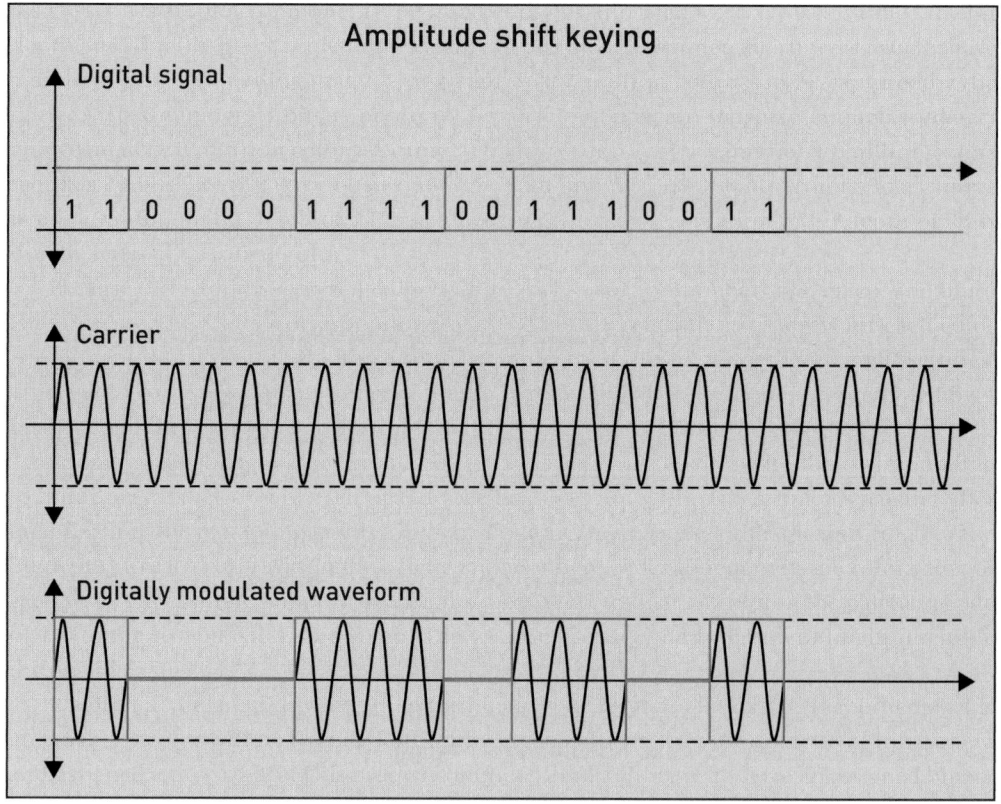

Binary **frequency shift keying** (FSK) varies the frequency of the carrier wave according to the binary amplitude of the information signal. The frequency of the carrier wave is adjusted to one of two values (f1 or f2) corresponding to the two amplitude levels of the information signal. Figure 8-12 illustrates the use of two carriers with two different frequencies. Carrier 1 is used for 1s and carrier 2 is used for 0s. The receiver keeps track of the frequency variation within the received modulated signal to extract the bits.

Binary **phase shift keying** (PSK) varies the phase of the carrier wave according to the binary amplitude of the digital signal. As the amplitude of the information signal changes, the phase of the carrier wave shifts accordingly. Figure 8-13 illustrates a phase shift of 180° with each bit change.

Just as the quality and performance characteristics of the analog modulation techniques differ, so do those of their digital counterparts. The quality of FSK and PSK exceeds that of

Figure 8-12

Binary frequency shift
keying

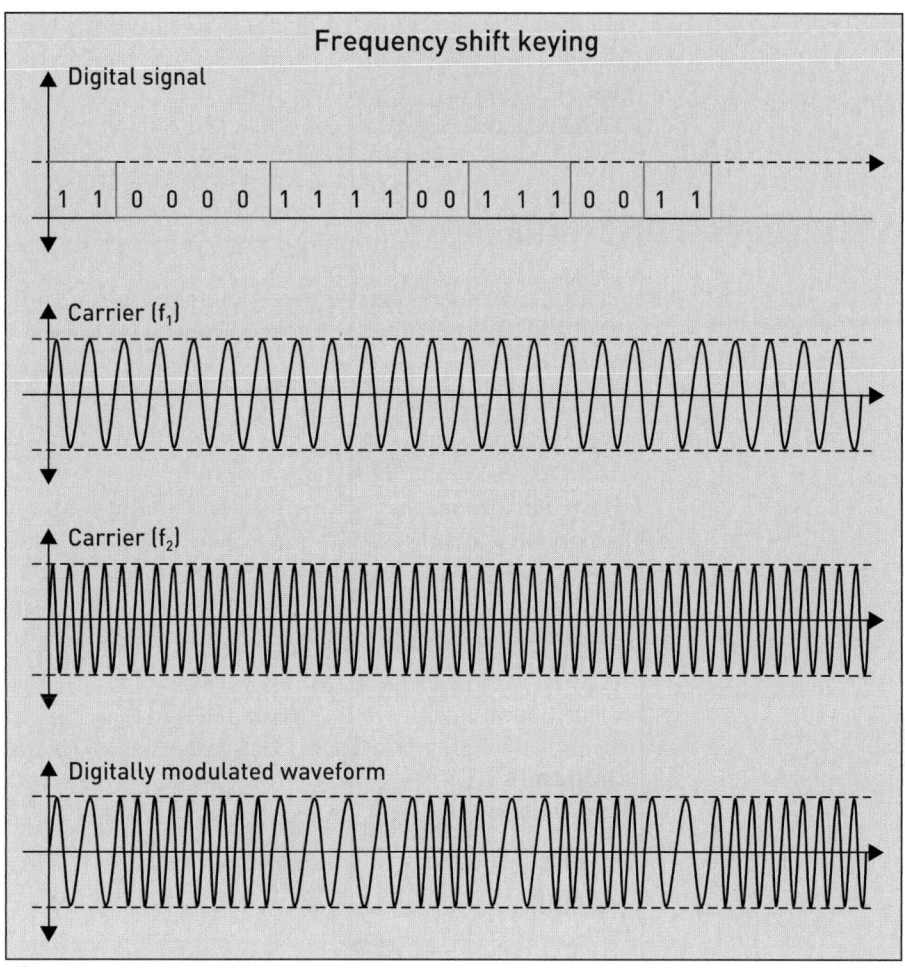

Figure 8-13

Binary phase shift keying

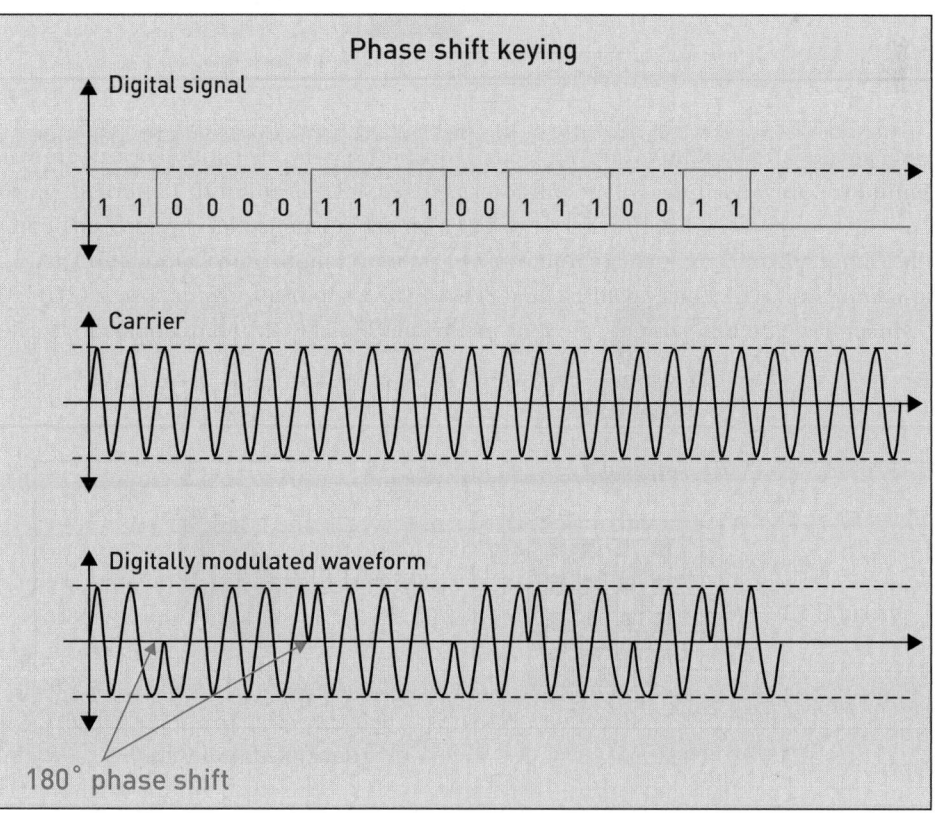

ASK, although ASK consumes less energy. ASK is commonly used in fiber-optic communication systems, and PSK is commonly used in satellite communications, space exploration, modems, and computer networking. FSK is regularly used in facsimile machines to transmit digital information across telephone lines.

LIGHT-WAVE COMMUNICATIONS

Although Chapter 9 addresses optical communications in detail, this chapter introduces some of the background concepts. The previous two sections focused on information transmission using electrical signals over copper wires and radio waves over free space, but many systems, such as telephone and computer networks, use light waves within **fiber-optic cables** to convey information. For example, transatlantic telecommunications channels are fiber-optic cables, as are most Internet backbone links.

Light is an EM wave that resides in a higher-frequency portion of the EM spectrum than radio waves, but a lower frequency range than x-rays, as shown in Figure 8-6. Because light is classified as an EM wave, it exhibits sinusoidal behavior with properties of amplitude, frequency, and phase. Light intensity is related to the wave's amplitude and the type of light is related to frequency. The frequency of blue light, for example, is approximately 7.5×10^{14} Hz, and the frequency of red light is approximately 4×10^{14} Hz. The disparity in frequency is what makes light appear in different colors. In addition to visible light, infrared (IR) and ultraviolet (UV) are forms of light with different frequencies. Infrared occupies a lower frequency than visible light (red, orange, yellow, green, blue, indigo, violet) and UV occupies a higher frequency. The light spectrum, which actually constitutes a small part of the EM spectrum, is illustrated in Figure 8-14.

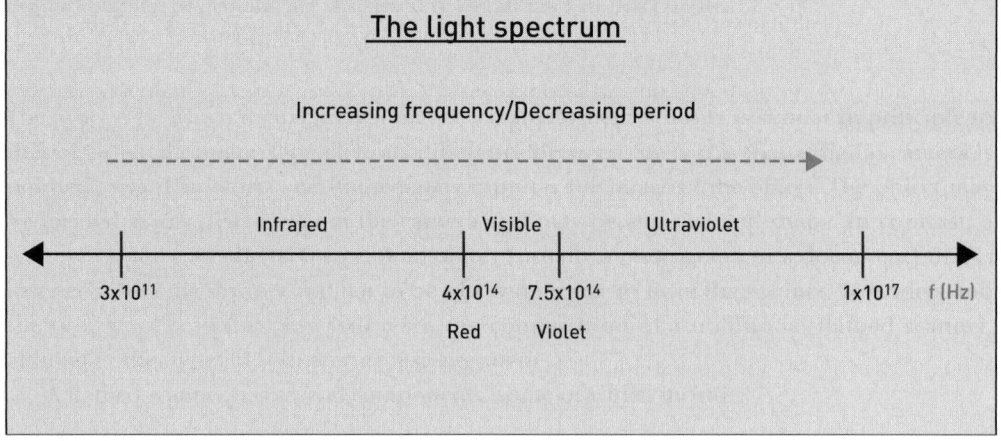

Figure 8-14

The light spectrum

Light-wave communication systems frequently use infrared, as in fiber-optic communication, and infrared/visible light, as in free-space optical communications, for carrying information. For example, cellular phones and TV remote controls commonly use infrared beams to exchange information. In optical communications, one of the three properties of light is modulated for information transmittal; the modulation techniques discussed in the previous section can be applied to light waves. Optical binary ASK (see Figure 8-15) corresponds to switching on a light source, such as a laser or light emitting diode (LED). Similarly, binary FSK corresponds to adjusting a source such as a laser to emit at two different frequencies. PSK, on the other hand, occurs by controlling the phase of light. A receiver detects intensity variations in the case of ASK, frequency variations in FSK, and phase variations in PSK. Analog modulation techniques are used for some light-wave communications as well.

Figure 8-15

Binary ASK-based free-
space optical
communications

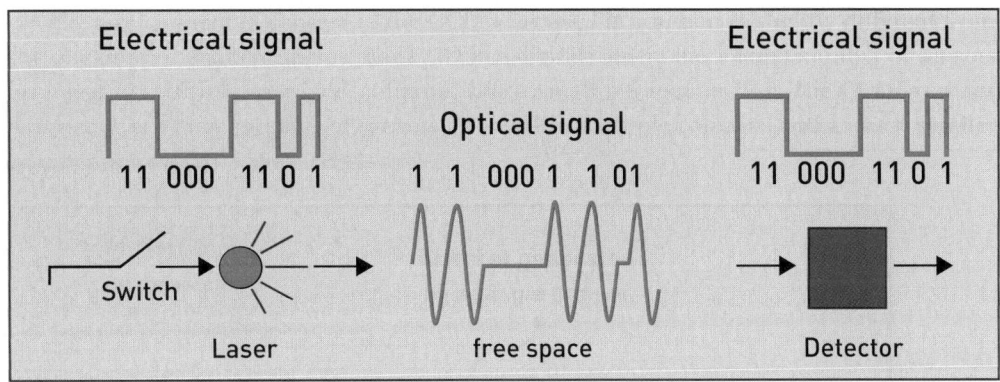

ATTENUATION

Another important transmission concept is attenuation. When signals travel through any transmission medium, including fiber-optic cable, copper wire, or free space, they lose energy (see Figure 8-16). The loss of energy, called **attenuation**, is a significant factor that affects the quality and distance of communications. On transmission channels with high attenuation, information-carrying signals such as digitally modulated EM waves tend to arrive at receivers with very low power, making the receivers highly prone to incorrect decisions (called errors).

Figure 8-16

Attenuation of a digital
signal traveling along a
transmission medium

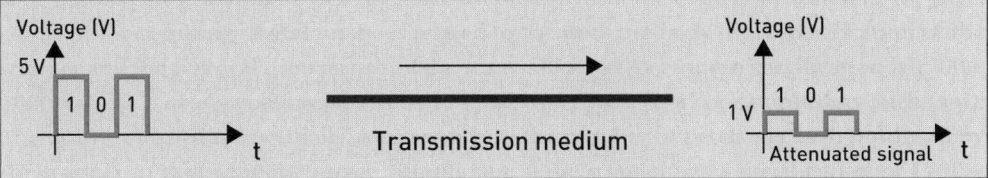

Various types of transmission media experience attenuation to different degrees. Fiber-optic cables have very low attenuation and can therefore carry signals across great distances. Copper wires experience significant amounts of signal attenuation, primarily because of their resistance property. Free space also attenuates signals because of water particles in the air and absorption due to atmospheric gases. As a result, attenuation plays a significant role in the choice of transmission media for any application. Long-distance communication requires materials with low attenuation characteristics, but the trade-off is that lower-attenuation media are usually more expensive.

Attenuation is measured in decibels (dB), and each transmission medium has its own attenuation figure, which is measured in dB per unit length. Some fiber-optic cables, for example, have an attenuation figure/coefficient of 0.2 dB per kilometer (0.2 dB/km), which is considered extremely low. Other types of transmission media such as copper-based media might have attenuation coefficients that are on the order of tens of decibels per kilometer.

Although attenuation is unavoidable in any communication system, there are ways to compensate for the reduction in signal intensity. One approach is to use receiver amplifiers to amplify weak signals. In long-haul communication systems, electronic devices called **repeaters** serve as amplifiers, and are placed at certain intervals to amplify weak signals and relay them along the transmission line (see Figure 8-17). Transatlantic cables are obviously very long and require repeaters at regular intervals. The number of repeaters needed in a communications link depends on the attenuation of the particular transmission medium and the length of the connection. Naturally, the lower the attenuation a transmission medium imposes on a signal, the lower the number of repeaters.

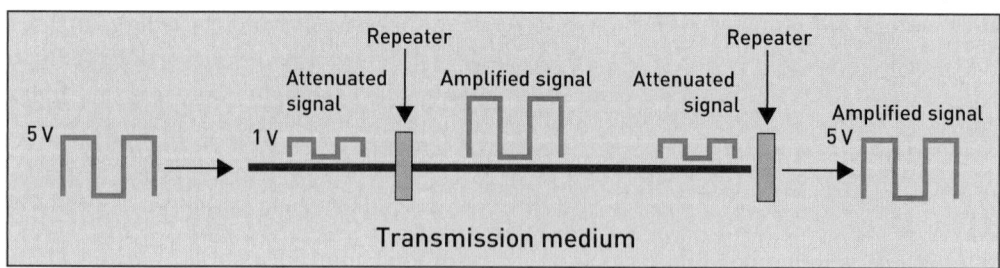

BANDWIDTH

The primary factor that generally governs the choice of transmission media is its bandwidth (expressed in Hz). Bandwidth directly affects the capacity of a transmission medium, which is measured in bits per second. Not all types of transmission media can carry information at the same rate. For example, dial-up modem connections over copper wires usually have a maximum transmission capacity on the order of tens of kilobits per second. A cable modem connection can carry multiple megabits of data per second across a coaxial cable and a fiber-optic communication system can carry data at rates of gigabits per second. High-bandwidth channels can carry larger amounts of information than smaller-bandwidth channels because they have a higher channel capacity. The concept of bandwidth is traditionally explained by comparing it to the diameter of a water pipe. The larger the diameter, the greater the volume of water the pipe can carry within a given time. Bandwidth is often equated with data rate and is used to denote the maximum data rate that can be carried across a channel. For example, a fiber-optic link can be said to have a "1-Gbps bandwidth." However, bandwidth and channel capacity are technically different.

In everyday usage, the terms **broadband** and **narrowband** reflect the information-carrying capacity of a transmission medium. A dial-up Internet connection historically has been classified as narrowband, while connections over cable modem, DSL, or fiber-optic cable are conventionally classified as broadband.

In digital systems, the maximum number of bits per second (channel capacity, or C) that can reliably be carried over a channel depends on the bandwidth B (expressed in Hz) of the channel and a unitless ratio called the **signal-to-noise ratio** (SNR). The formula is:

$$C = B \log_2(1+SNR)$$

where SNR is the ratio of signal power to noise power in the channel. The preceding expression is called the **Shannon-Hartley capacity theorem**, and it illuminates the trade-offs in designing any communications system. In the theorem, maximizing the data rate necessitates choosing a transmission medium with large bandwidth (B). Furthermore, for a given transmission medium with bandwidth B, achieving the largest data rate possible requires maximizing SNR, which, in turn, involves transmitting at higher power and reducing the amount of noise in the communication system. However, noise can be lowered only to a certain extent, and power can be increased only to a certain maximum level.

Noise was introduced briefly in Chapter 2 within the context of the speedometer example. Recall that a digital signal transmitted from one point to another encountered noise, which was graphically observed as unwanted fluctuations in the signal's amplitude. Sources of noise in communication systems are numerous. For example, neighboring electrical devices and other transmission lines can cause noise on a signal as it travels down a transmission medium. Devices and cables in close proximity to the transmission line can induce noise on the signal. Noise can also come from other sources such as radio broadcasting towers, cellular antenna towers, or even lightning strikes. Another source of noise is called **thermal noise**; it arises from random agitation of electrons of the conductor material due to heat. These random fluctuations exhibit themselves as unwanted fluctuations in the signal's amplitude.

In any case, noise is highly prevalent in communications systems, and a signal must be protected as much as possible from the various sources of noise using a combination of techniques, such as protecting cables with insulators, shields, and other means. Besides reducing noise, these techniques are designed to increase SNR as much as possible.

Most communication systems can only deliver limited power within their transmissions. For example, cellular telephones have to transmit with the least amount of power possible both to elongate battery life and because of health considerations. Similarly, satellites transmit at low power to conserve the electrical energy they generate through their solar panels. To minimize transmitted power while maintaining a given level of reliability, sophisticated forward error-correcting schemes can be applied to a signal before transmission. Designing an optimal communications system is a highly complicated task, involving intricate calculations and cost/performance trade-offs.

MULTIPLEXING

A single line can simultaneously transmit multiple information-carrying signals using a technique called **multiplexing**. This technique is extremely useful in telecommunications because it eliminates the need to use a separate transmission line for each information-carrying signal. For example, a single copper cable might be able to carry 24 separate voice conversations, and a single strand of fiber-optic cable might be able to carry more than 120,000 voice channels. Instead of using 24 pairs of wires, a single higher-bandwidth line is adequate to carry all 24 voice conversations. A device called a multiplexer (MUX) combines multiple signals, and a device called a demultiplexer (DEMUX) separates the signals (see Figure 8-18).

Figure 8-18

The concept of multiplexing

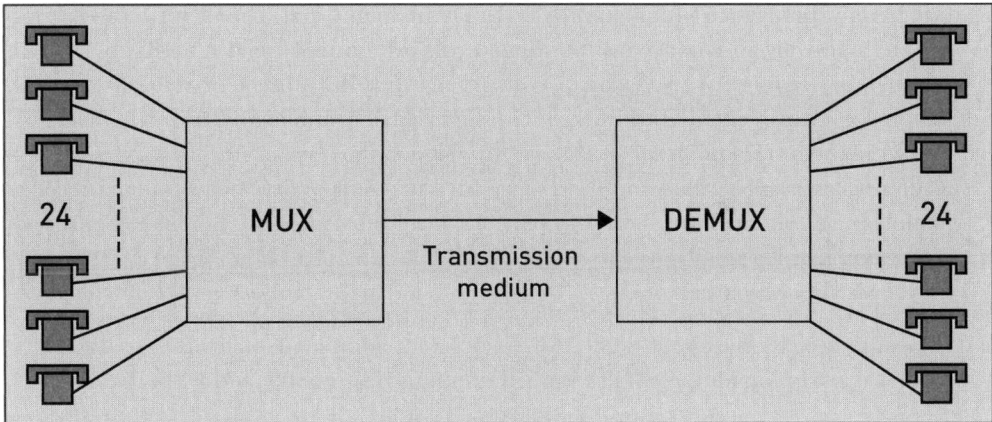

Multiplexing signals over a single transmission line uses one of several possible techniques:

- Time division multiplexing (TDM)
- Frequency division multiplexing (FDM)
- Statistical multiplexing
- Wavelength division multiplexing (WDM)

TDM combines signals from multiple channels onto a single channel by allocating a specific time slot, during which each channel transmits. If two channels are multiplexed over a single transmission line, channel A is first allotted a specific time slot to transmit its bits, such as two bits at a time; channel B is then permitted to transmit at its allotted time slot. Channel A transmits again, then B, and so forth. This way, a MUX allows both channels to alternately transmit, separating each channel in time. A DEMUX at the receiving end synchronizes with the multiplexed bit stream and directs the associated bits to channel A, the other bits to Channel B, and so on. The DEMUX thus acts as an agent that descrambles the transmission and directs each bit to the correct channel (see Figure 8-19).

Figure 8-19

Time division
multiplexing

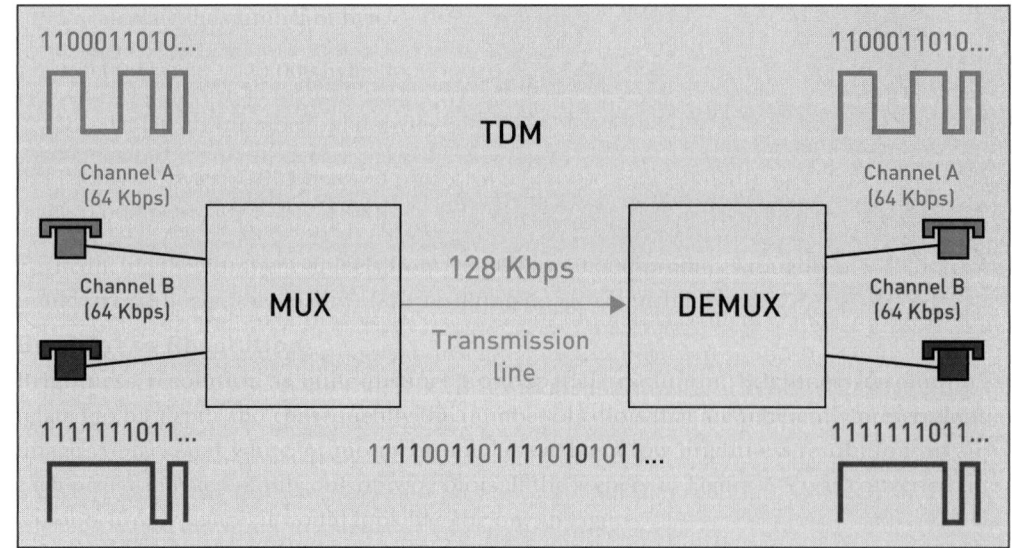

FDM was the first multiplexing technique used in the telephone network. Rather than assigning each channel a specific time slot, each channel is assigned a specific carrier frequency for transmission. The channels are then combined and sent across the transmission medium. Each channel is therefore separated in frequency. FDM is used in TV and radio broadcasting, in which multiple channels are transmitted across a large geographical area. Chapter 17 describes TDM and FDM in more detail and introduces statistical and wavelength division multiplexing. Multiplexing is not described further in this chapter.

COPPER TRANSMISSION MEDIA

Copper is an excellent conductor of electricity. This quality, along with its relatively low cost and great abundance, has made copper the usual choice for electrical wiring. The following sections discuss the three types of copper cables: unshielded twisted pair, shielded twisted pair, and coaxial cable. Each type of cable differs in its attenuation characteristics, bandwidth, and other design considerations.

TWISTED PAIR

Twisted pair cabling is often used to connect computing devices within local area networks (LANs) and for telecommunications services, such as home phone service and digital subscriber line (DSL). This type of cable is categorized into unshielded twisted pair (UTP) and shielded twisted pair (STP). Both types of twisted pair cables enable the transmission of signals across longer distances than parallel wires, with lower attenuation, higher data rates, and lower electromagnetic interference. The main reason for the better performance is the twisting of the two parallel wires (see Figure 8-20), which reduces the amount of **crosstalk** acting on the signals as they travel down the wires. Crosstalk is the effect that a signal traveling on one wire has on a neighboring wire. Because there is more than one wire within the cable, multiple wires induce crosstalk on each other's signals. Twisting the wires around each other helps to reduce the amount of crosstalk. The more the wires are twisted together, the less susceptible they become to crosstalk, although attenuation increases. Cable manufacturers make design decisions based on trade-offs between attenuation, cost, and level of interference. A popular plug for connecting UTP cables to devices is the RJ-45 connector, which supports four twisted pairs and is most commonly used to connect a computer to a LAN.

Figure 8-20

Structure of an
unshielded twisted
pair cable

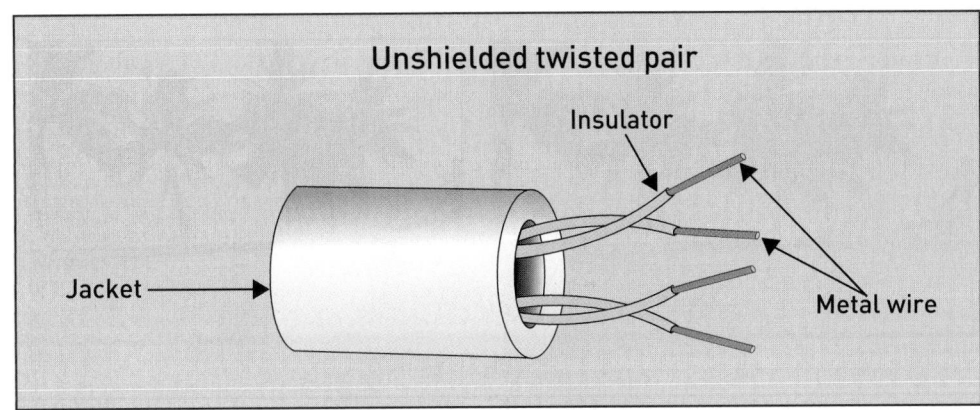

Shielded twisted pair cabling uses the same fundamental wire configuration as UTP, but adds a shield surrounding the twisted pairs. The shield may be a metal foil or thin, braided metal strands (see Figure 8-21). The shielding further reduces the amount of EMI on the signal, among other things, but results in a heavier, more expensive, and less flexible cable than its unshielded counterpart. STP cables are necessary in high-EMI environments such as airports, and some LAN standards use STP cabling.

Figure 8-21

Structure of a shielded
twisted pair cable

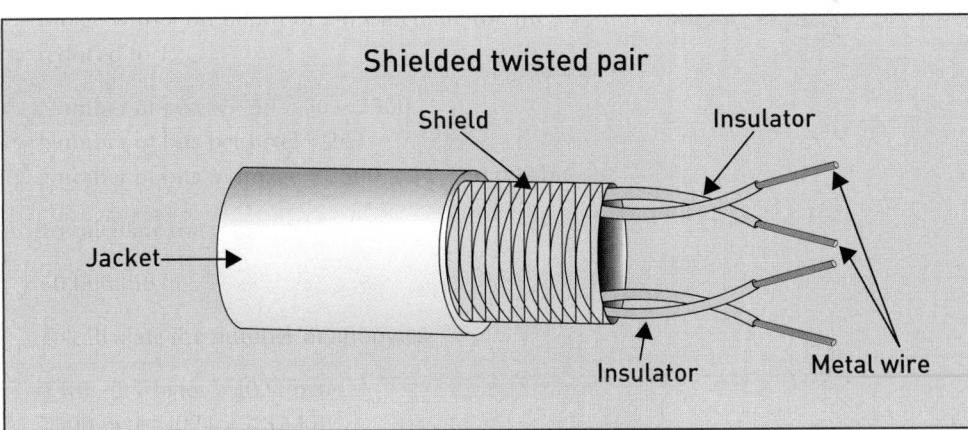

Twisted pair cables are categorized according to their grade. High-grade cables carry higher data rates, but are more expensive than lower-grade cable. The Telecommunication Industry Association (TIA)/Electronic Industries Alliance (EIA) has defined more than six general categories of twisted pair cables, ranging from CAT1 to CAT6, with data rates that range from a few Mbps to about 10 Gbps.

CAT5 cables are the most prevalent type of UTP cable. A single cable is made of four pairs of wires, in which each pair is twisted around the other. This type of cable is shown in Figure 8-23.

COAXIAL CABLE

The structure of a **coaxial cable** is fundamentally different from that of twisted pair cables. It consists of a central, usually solid, cylindrical core conductor made of copper and an insulating material surrounding the core called a dielectric. The structure is further encapsulated in braided metal strands that act as a shield. A tough cable jacket encloses the complete cable structure (see Figure 8-22).

Figure 8-22

Structure of a
coaxial cable

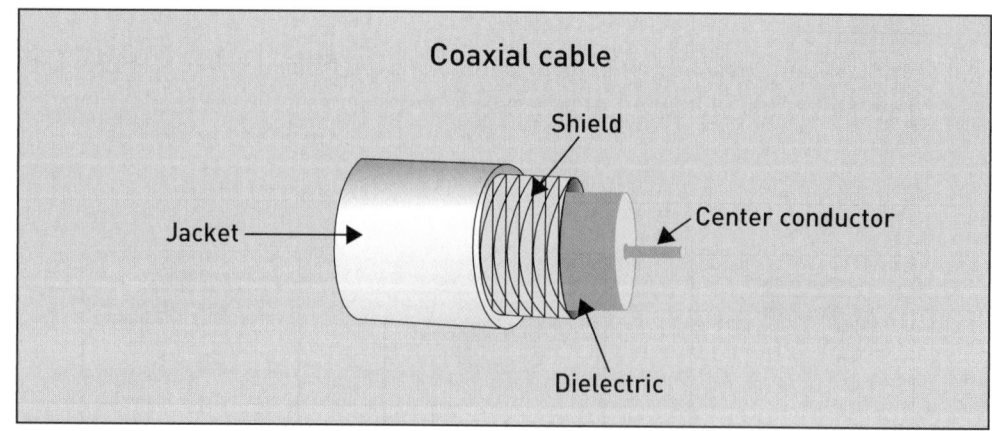

Coaxial cables, or coax, are commonly used for cable television and Internet connectivity via cable modems. Coax provides high bandwidth and is useful for carrying information at high data rates across long distances. Coax cable was used in earlier LAN applications, but it has been replaced by twisted pair cabling.

Figure 8-23 illustrates the various types of transmission media discussed in this section.

Figure 8-23

Photographs of various
cables

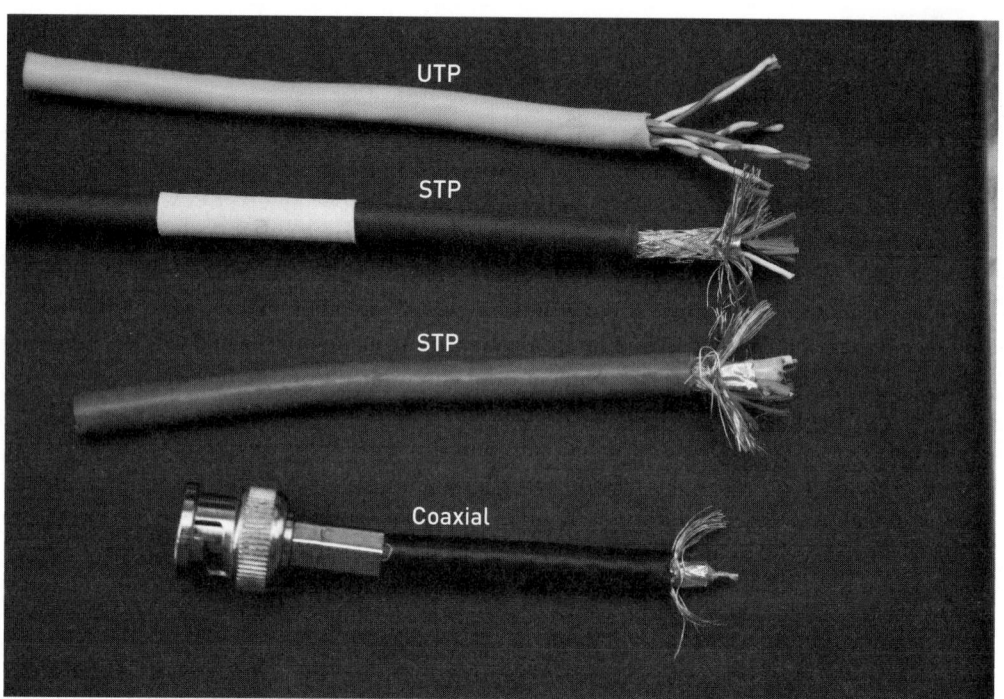

MANAGING ERRORS IN DIGITAL COMMUNICATION SYSTEMS

Unfortunately, there is no such thing as an ideal transmission medium. Whenever digital information is sent across any communication channel, be it twisted pair, coax, air, or optical fiber, there is always a possibility that some bits will arrive at their destination with errors. A 1 may be detected as a 0 and vice versa. The incorrect detection of a binary digit is called an error (see Figure 8-24).

Depending on the type of communication channel you use, a combination of factors contributes to transmission errors. These factors include attenuation, as outlined in the previous sections; noise, such as thermal noise and EMI, distortion, and atmospheric conditions (rain, fog, snow, lightning); and system failures. Fortunately, you can use special coding techniques to detect and sometimes even correct errors. These coding techniques are based on the addition of extra bits, called redundant bits, to the transmission stream. Redundant bits are not part of the original information that is to be transmitted, but are added specifically for

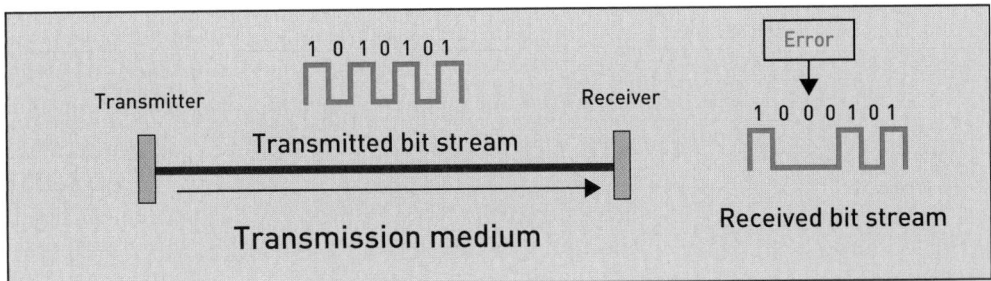

error checking. By encoding a bit stream prior to transmission using an assortment of techniques called **error-control coding** (ECC), the receiver can detect and sometimes even correct errors that may occur at the receiver. Note that errors can occur during storage as well and are not confined to transmission systems. For example, a CD that stores binary information may have a dust particle on it, resulting in errors as the CD player reads that portion of the disk. Therefore, error-control techniques are also used in music and computer storage systems by storing redundant bits together with the actual information.

The extent to which errors can be detected and corrected depends on the sophistication of the coding technique. The complexity of the code is generally proportional to the amount of redundancy added. Adding large amounts of redundancy increases the chances of the receiver detecting and correcting a larger number of errors. Less elaborate coding techniques are used in applications in which errors are tolerable—for example, encoding audio or images. A few errors during the transmission or storage of audio are unlikely to cause a great difference in quality when the digital signal is converted to analog and fed to a speaker.

Similarly, errors can often be tolerated in imaging applications because a few incorrect pixels will not cause a great amount of distortion in the image. More elaborate techniques are essential in applications that would be significantly affected by errors. For example, when transferring software between computers, an error in even a single bit could lead to significant problems. An error in the transmission of financial information would also be unacceptable. The following paragraphs discuss techniques that work at different levels to provide error management in communication systems.

Almost all digital systems employ some form of ECC technique. Cellular telephones digitize voice conversations and apply ECC to the resulting bit stream before transmitting to a base station antenna. Computers employ ECC techniques for transmitting bit streams along system buses that carry data between internal and external units. ECC techniques are also used in transmitting packets (segments of binary data) across computer networks for transferring information. The following sections discuss two types of ECC techniques:

- Block codes
- Convolutional codes

BLOCK CODES

Block codes are a class of error-control techniques, based on adding redundancy to fixed lengths or blocks of data. Some types of block coding techniques include the following:

- Single parity checking
- Rectangular coding
- Cyclic redundancy checking (CRC)

Single Parity Checking

Among the many alternative ECC schemes, probably the easiest to implement is the **single parity checking** scheme. As we will show, this scheme only provides for the detection of errors, not error correction. Nonetheless, error detection is extremely useful, because erroneous messages can be discarded at the receiver and the transmitter may be instructed to retransmit.

Single parity checking is implemented simply by adding a single redundant bit called the **parity bit** to the end of each data block. If you use even parity checking, the parity bit added to the end of the block makes the total number of 1s in the block equivalent to an even number. If you use odd parity checking, the parity bit makes the total number of 1s an odd number.

For example, consider that a 7-bit block of data to be transmitted is 1000111. This block of data contains four 1s, which is an even number. Suppose we wanted to add a parity bit to this data to assign even parity. The original bit block already contains an even number of 1s, so the additional parity bit would be 0 and the resulting 8-bit block to be transmitted would be 1000111<u>0</u>. The last bit would correspond to the parity bit.

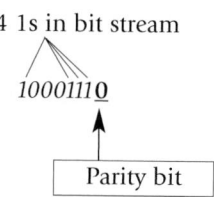

If you wanted to use odd parity for encoding, the added parity bit at the end would have to be 1, bringing the total number of 1s to 5, an odd number.

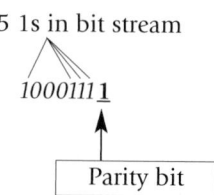

The receiving end can detect an error by counting the number of 1s in the received 8-bit block of data. If the transmitter and receiver agree on odd parity checking and 10000111 arrives in a message, the receiver can detect that an error has occurred because the total number of 1s in the received message equals 4, an even number, and contradicts the agreement of odd parity.

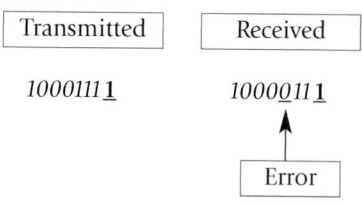

Parity checking has obvious limitations. Although the receiver can detect that an error has occurred, it cannot determine exactly which bit is in error. In other words, parity checking provides error detection but not correction.

Another limitation of parity checking (odd or even) is that it only works if an odd number of errors occur during transmission. Parity checking cannot detect an even number of errors. If two errors occur, they essentially cancel out each other. For example, the following scenario applies odd parity to a block of bits prior to transmission. Two errors occur in transmission, but the resulting count of ones is still odd, erroneously signifying no errors to the receiver.

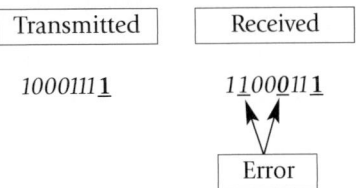

Although this coding scheme has limitations, it does provide for a certain degree of useful error detection.

Problem—Detect errors using parity checking.

If a received bit stream, 11110101, has been encoded with even parity checking, detect whether any errors are present.

The number of 1s in the received binary stream is 6. Six is an even number, indicating to a certain degree that the bit stream has been received correctly.

Rectangular Coding

Rectangular codes are more complex than single parity checking, but they rely on the parity checking method in an interleaved fashion. These codes can detect as well as correct some errors. Consider the following two blocks of 7-bit data that need to be transmitted:

1100000   1000101

If rectangular coding is applied, then the data could be encoded as follows:

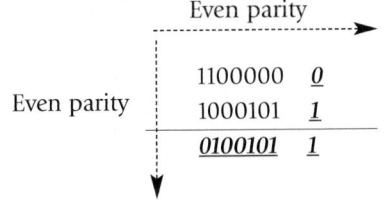

Each block of data is assigned even parity by appending a parity bit to the right end of each row. Each column (the 7 original columns of data corresponding to the 7 bits of original information per block) is also given even parity by appending an even parity bit to the bottom of the column, resulting in 7 extra parity bits. The last row, which contains the appended parity bits, is also assigned an even parity bit at the right end. After applying these additional bits for error management, the resulting bit stream is transmitted:

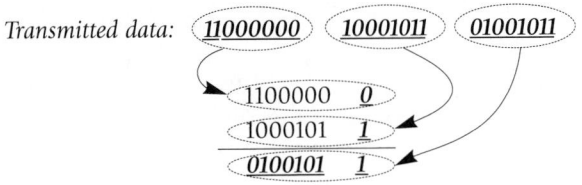

The single parity bits are added to the end of each data block, and the derived third row of 8 parity bits is appended at the end.

Now assume that after transmission, the following bit stream was actually received:

Received data: 10000000   10001011   01001011

The receiver's task is to perform a check on the received bit stream (3 groups of 8 bits each), knowing that the last 8 bits in the stream, as well as the bits at the end of each group of 8, are just parity bits added for error control. The receiver knows that each group of 8 bits should have even parity, so it checks and detects that the first 8-bit group has odd parity, the second has even parity, and the third group has even parity. The receiver knows that each 8-bit group should have even parity, so the error must be in the first group of 8 bits. The exact location of the erroneous bit within the first group requires further checking.

The next task is to check whether the first bits of each group have even parity; this corresponds to checking the first column of bits for even parity. The first bits of each group are 110, which means it contains two 1s and therefore has even parity, as it should. The receiver then checks the second bits of each group corresponding to the second column of bits. The second bits are 001. Because this group has odd parity, the receiver deduces that the error is in the second bit of the first group. The checking can continue, but it is easy to see that the rest of the 3-bit groups have even parity, and only the second bit of the first group is in error.

This scheme provides for both the detection and the correction of errors. The receiver can correct the erroneous bit by changing the 0 to a 1 and discard the redundant bits. This process produces the following stream of bits, exactly corresponding to the original information given to the encoder prior to transmission:

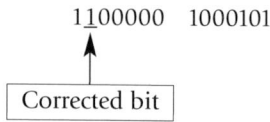

Although this example is based only on a single, simple version of the rectangular coding approach, the method actually has many variations. For example, each row can be assigned odd parity. Keep in mind that much greater error detection and correction sophistication is possible, depending on the ratio of redundancy to the original information.

Problem—Encode bit streams using rectangular coding.

Encode the following blocks of data using rectangular coding based on even parity:

1100011 1010101

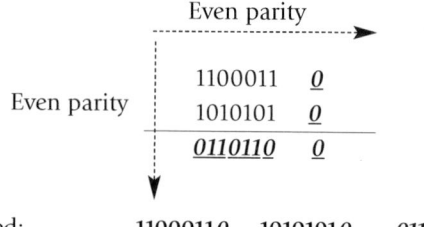

Bit stream to be transmitted: 11000110   10101010   01101100

Problem—Detect and correct errors using rectangular coding.

If a received bit stream, 11000110 10100010 01101100, has been encoded with rectangular coding as described above, detect whether any error has occurred. If it has, make the appropriate correction.

First, express the received bit stream in rectangular format as follows:

```
------1100011------0---➤ Even parity
      1010001      0
      0110110      0
```

The first row is checked for even parity. Because the first row has four 1s, it is unlikely that the first block of data contains an error. Next, the second row is checked for even parity.

```
      1100011      0
------1010001------0---➤ Odd parity
      0110110      0
```

The second row has three 1s, giving it odd parity when it should have even parity. Therefore, the second block of data contains an error. Which of the bits in the data block is in error? To narrow down the error to the individual bit, check each column for even parity. If one of the columns has odd parity, it contains the erroneous individual bit.

```
1100011 | 0
1010001 | 0
0110110 | 0
▼▼▼▼▼▼▼   ▼
EEEEOEE   E
```

Note that the fifth column has odd parity. Thus, you can conclude that the error is in the fifth bit of the second block of data. The bit in this position must be flipped to make the correction. In the received sequence, the fifth bit of the second block is a 0. To correct it, the bit is changed to a 1. After making this correction and stripping off the redundancies, the bit sequence before encoding should be 1100011 1010101.

Cyclic Redundancy Checking

Cyclic redundancy checking is another popular technique for detecting errors. It is slightly more complex than parity checking or rectangular coding and is based on appending a stream of bits to the end of a data block. The appended bits are generated by performing a simple mathematical operation on the original bit stream. You can think of these extra bits as a key that the receiver uses for performing an operation on the received bit stream to detect errors. If an error occurs, the receiver may request a retransmission of the data. This book does not describe cyclic redundancy checking in any detail, but note that the technique is used in satellite communication systems and for sending packets across computer networks, such as Ethernet LANs.

CRC is also used for verifying data integrity within the area of computer forensics. When data is transmitted across a public network, such as the Internet, it is vulnerable to interception and modification by third parties. CRC is used to verify that the data received at the receiver is the data that was actually sent and that a third party did not intercept and modify the data.

CONVOLUTIONAL CODES

Convolutional coding techniques provide powerful error detection and correction. Instead of grouping data into blocks and applying redundancy to each block, convolutional encoders take in a continuous stream of bits at their input and produce a continuous stream of encoded information at their output. The bit stream in the top part of Figure 8-25 enters a convolutional encoder, and an encoded sequence is serially output from it.

Figure 8-25

Convolutional encoding
and decoding of a
bit stream

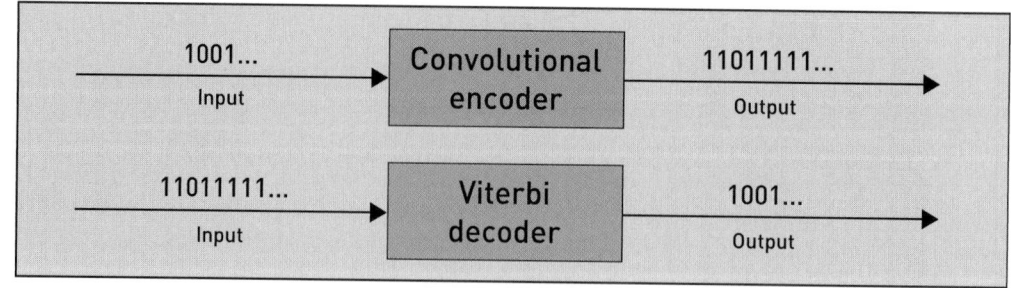

Although the output bit stream looks completely different from the input, the original information is retained in the encoded bit stream because of the nature of the encoding system. The receiver that receives the encoded bit stream uses various algorithms, such as the **Viterbi decoding algorithm**, to extract the original information from the encoder. Cellular telephone systems rely heavily on convolutional codes and Viterbi decoding during voice conversations to encode the digital audio signal and convey it to the transmitter. Convolutional codes are also used extensively in satellite communications and space exploration.

A novel class of codes has emerged based on the idea of convolutional encoding. These special codes, called **Turbo Codes**, encode digital information so it can be retrieved with a very small amount of errors, even under the highest attenuation and noisy environments. Turbo Codes are currently used for mobile telephony applications, especially for sending text messages or images, and their application areas continue to expand.

A DIGITAL COMMUNICATIONS SCENARIO

The previous sections covered some basics of communication systems: the properties of transmission channels, how to physically generate binary data streams, how to modulate carriers to carry these data streams, how to multiplex data streams, and how to apply error-control coding. Figure 8-26 is a block diagram of a typical digital communications system that illustrates how these components are integrated.

Figure 8-26

Components of a digital
communications system

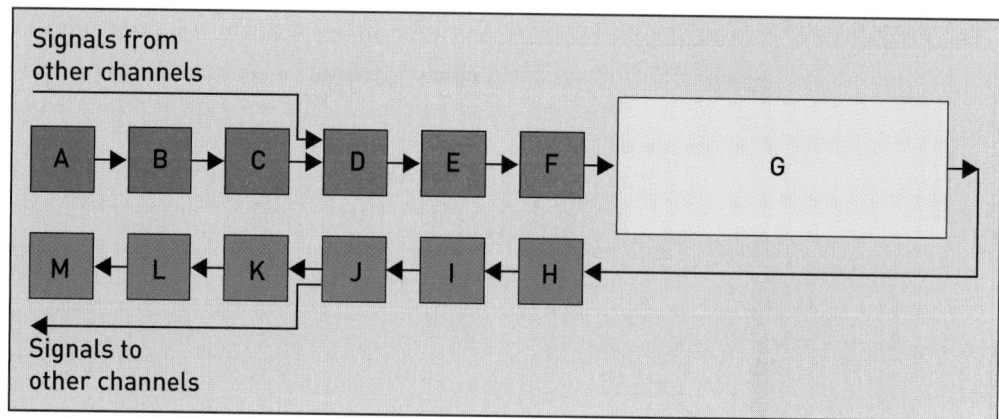

The blocks in the figure correspond to the following components:

- A—The input source is the device that generates the physical information signal. This component could be a microphone that generates the electrical audio signal or a camera that captures images.
- B—The analog/digital (A/D) converter is the device that converts the analog signal into digital format by sampling and quantizing it.

- C—The channel encoder applies redundancy to the bit stream from the A/D converter for error control, such as parity checking or convolutional coding to manage errors at the receiving end.
- D—The multiplexer combines bit streams from multiple channels so that they can be sent across a single transmission medium.
- E—The modulator modulates the carrier wave in proportion to the multiplexer's output to efficiently incorporate the information onto the carrier.
- F—The transmitter receives the modulated waveform from the modulator and generates an electrical, optical, or electromagnetic signal to be carried across the appropriate transmission channel.
- G—The transmission channel carries the physical signal to its destination. During its travel, the signal undergoes attenuation, and noise is added to it.
- H—The receiver detects the signal from the channel, filters the noise and amplifies the signal, and sends it to the demodulator.
- I—The demodulator extracts the digital bit stream from the modulated signal and sends it to the demultiplexer.
- J—The demultiplexer separates all the information-bearing channels and directs the associated bit streams to each channel.
- K—The channel decoder applies an algorithm for detecting and correcting errors that might have occurred at the receiver.
- L—The digital/analog (D/A) converter changes the digital signal into an analog signal.
- M—The output device conveys the analog information to the appropriate party.

Depending on the type of communication system, some of the blocks in the preceding list might be absent. For example, a D/A converter or A/D converter might not be required if the communication system carries computer data, such as software, which is already in digital form. The multiplexer is optional only if a single signal is to be carried over the channel.

CHAPTER SUMMARY

- Binary signaling schemes communicate digital information over a transmission system by corresponding a 1 or a 0 to one of two discrete signal values.
- Signaling schemes may be extended to M-ary signaling, whereby $\log_2 M$ groups of bits may be sent at one time using M different signaling levels.
- The capacity of a channel, expressed in bits per second, depends on the bandwidth and the signal-to-noise ratio of the channel.
- Modulation techniques superimpose information signals onto a carrier wave, such as a radio wave or a light wave, by varying some of its properties, such as the wave's amplitude, frequency, and phase.
- Attenuation (measured in decibels) is the loss of energy occurring over a transmission line. It is a significant factor that affects the quality and distance of communications.
- Multiplexing techniques enable a single transmission line to simultaneously transmit multiple information-carrying signals.
- Sources of transmission errors include electromagnetic interference, distortion, system failures, and atmospheric conditions such as lightning and rain.
- By encoding bit streams prior to transmission using error-control coding (ECC) techniques, a receiver can detect and sometimes even correct errors that may occur at the receiver of a communication system.

KEY TERMS

amperes	metallic conductors
amplitude modulation	modulation
amplitude shift keying	multiplexing
attenuation	narrowband
binary signaling	Ohm's Law
block codes	ohms
broadband	parity bit
coaxial cable	phase modulation
convolutional coding	phase shift keying
crosstalk	radio spectrum
cyclic redundancy checking	radio waves
demodulation	repeaters
electrical insulators	resistance
electromagnetic energy	Shannon-Hartley capacity theorem
electromagnetic spectrum	shielded twisted pair
EM waves	signal-to-noise ratio
error-control coding	single parity checking
FCC	TDM
FDM	thermal noise
fiber-optic cable	Turbo Codes
frequency band	twisted pair
frequency modulation	Viterbi decoding algorithm
frequency shift keying	volts
gamma rays	wavelength
light waves	x-rays
M-ary signaling	

REVIEW QUESTIONS

1. Add a parity bit to the end of each block of data to assign odd parity:
   - 001110_
   - 010101001_
   - 11001101_

2. If a received bit stream, 1100000100, has been encoded with even parity checking, detect whether any errors have occurred during transmission. Can any errors be corrected?

3. If a received bit stream, 10001100, has been encoded with odd parity checking, detect whether any errors have occurred during transmission. Can any errors be corrected?

4. Encode the following blocks of data using rectangular coding based on even parity:
   1000100 0000000

5. Encode the following blocks of data using rectangular coding based on even parity:
   110 001

6. If a received bit stream, 0100101**0** 0000101**0** **00000000**, has been encoded using rectangular coding, detect whether any error has occurred. If it has, make the appropriate correction.

7. If a received bit stream, 111**1** 001**1** **1100**, has been encoded using rectangular coding, detect whether any error has occurred. If it has, make the appropriate correction.

8. Give one example of a type of electromagnetic wave.

9. Briefly explain how the resistance of a conductor affects the amount of current flowing through it.

10. List two frequency bands within the radio spectrum.

11. What is the purpose of modulation?

12. Describe the differences among the three digital modulation techniques covered in this chapter.

13. Briefly compare AM and FM.

14. What is attenuation, and why does it occur? What is it measured in?

15. What can compensate for the attenuation that a signal undergoes while traveling through a communication channel?

16. In your own words, explain the concept of bandwidth.

17. Calculate the data rate for a communication system that employs 4-ary signaling, if the signal transmission rate is 5000 signals per second.

18. What is multiplexing? Describe two multiplexing techniques.

19. List the different types of copper-based transmission media.

20. How does the bandwidth of fiber-optic cable compare to the bandwidth of UTP?

21. What is the difference between UTP and STP?

22. What is the purpose of twisting the pair of wires in STP and UTP?

23. Briefly describe the physical structure of a coaxial cable and draw a simple diagram.

24. Name a common application of UTP.

25. What is the main difference between the way block codes and convolutional codes encode information?

26. How does the amount of redundancy added to a bit stream affect the error-detection capability of a code?

27. List two factors that can lead to errors while transmitting information.

28. Which transmission medium discussed in this chapter provides the highest data rate?

DISCUSSION QUESTIONS

1. To what extent have you experienced interference or errors while using information and communications technologies such as cellular phones, radio, satellite television, a wireless remote control, or text messaging? Could you determine the sources of interference or errors?

2. What is the relationship between the bandwidth of a channel and the maximum data rate that it can reliably carry? How can channel capacity be increased?

3. Discuss the effect of M-ary signaling and the receiver's susceptibility to making errors in a digital communications system. Would communication systems that use M-ary signaling require more sophisticated error-control schemes than those that employ binary signaling (i.e., 2-ary signaling)?

4. How many different types of transmission media can you identify in your day-to-day use of information technologies? Do you use coaxial cable? Fiber-optic cable? Twisted pair cable? Wireless?

5. Visit the National Association for Amateur Radio Web site at *http://arrl.org*. Write a short paragraph describing how you might be able to obtain a license for becoming an amateur radio (also called HAM radio) operator. What kinds of licenses does the association offer?

CASE PROJECT

To help you understand the electromagnetic transmissions that surround you, make a list of all the wireless devices you use in the course of a week. Examples might include a garage door opener, a television remote control, cellular phones, handheld computing devices, satellite television, and wireless Internet access. Through research or manufacturer documentation, try to determine what frequency range each device uses. Finally, download and print a copy of the U.S. frequency allocation chart after finding it in an Internet search. Write the name of each device in the appropriate part of the chart.

INTRODUCTION TO FIBER OPTICS

LEARNING OBJECTIVES

In this chapter you will:

- Understand the importance of fiber-optic technologies in the information society

- Identify the fundamental components of a fiber-optic cable

- Understand the principles by which light travels within a fiber-optic cable, including refraction and total internal reflection

- Understand how a single fiber can carry multiple signals through wavelength division multiplexing

- Learn about the advantages and disadvantages of fiber optics as a transmission medium for various applications

- Gain exposure to cutting-edge fiber-optic approaches

INTRODUCTION

Fiber optics, the transmission of information via light over a glass or specialized plastic medium, has facilitated high-speed innovations ranging from the Internet's core transmission backbone and transoceanic technologies to cutting-edge medical imaging. Applications that demand transmission at high data rates ultimately resort to using light to carry information inside fiber-optic cables. Besides high bandwidth, some advantages of fiber-optic transmission include immunity to electromagnetic interference and low attenuation. This chapter describes the basic components of a fiber-optic cable as well as the physical and mathematical principles of transmitting light through fiber-optic cable. The chapter also describes the advantages and drawbacks of using fiber as a transmission medium, and introduces some of the novel technologies being used in state-of-the-art **optical communication** systems.

The notion of communicating with light is not new. People used light to convey information thousands of years ago by starting fires on high ground, such as hills and mountains, to send communication signals to distant locations. Today, boaters carry flares in case they need to signal an emergency and traffic lights indicate when drivers should stop, slow down, or proceed through an intersection. The subject of this chapter is fiber-optic communication, which involves trapping light inside an optical fiber, and how any form of information may be carried through it. Fiber is an **optical medium**, which means it is capable of transmitting light. Trapping and transmitting light within an optical medium for communication was first demonstrated by the Irish physicist John Tyndall in 1870. Tyndall's experiment showed that light could travel inside a curved stream of water. To conduct his experiment, Tyndall filled a bucket with water and pierced a small hole on one side (see Figure 9-1). He shone a light over the bucket and observed a phenomenon that caused some of the light to remain trapped inside the stream of water and to travel from the hole in the bucket to the end of the stream. This phenomenon, called **total internal reflection** (TIR), established the foundation for fiber-optic communications; essentially, fiber-optic cables replaced the stream of water for trapping light. TIR occurs when a beam of light traveling through one medium hits another medium and reflects back into the original medium.

Figure 9-1

Tyndall's experiment for trapping light inside a stream of water

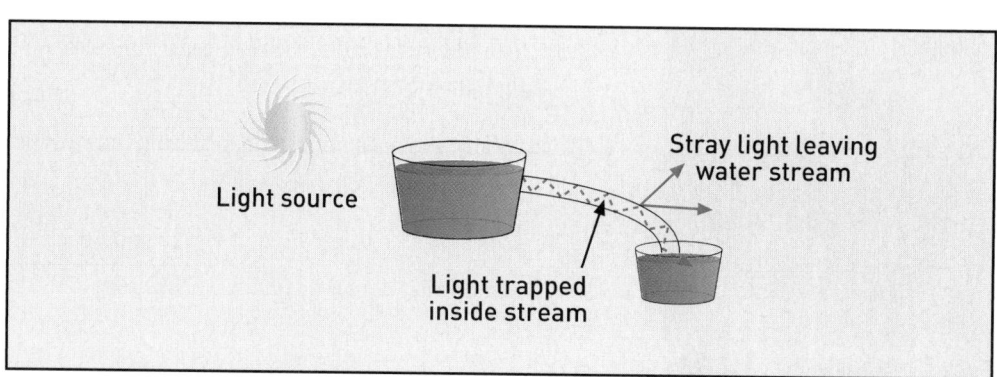

Later in the nineteenth century, Alexander Graham Bell constructed the **photophone**, a device that demonstrated how sound could be transmitted via an optical beam traveling through free space. These developments spurred more research on materials other than air or water that could effectively transport light beams to carry information. Scientists later developed thin strands of glass called optical fibers. These fibers were subject to high attenuation of light, meaning that they could only be used to communicate over short distances. Subsequent research led to successful demonstrations of fiber-optic communication systems with low-loss fibers, which were incorporated into the telephone and data networks of the 1970s, 1980s, and 1990s. These optical fibers remain a vital transmission medium in today's communication networks, and fiber-optic cables continue to be manufactured with progressively lower losses as technology advances.

STRUCTURE OF FIBER-OPTIC CABLES

The elementary structure of a fiber-optic cable is simple. A cylindrical material called the **core**, made of glass or specialized plastic, constitutes the central portion of the cable (see Figure 9-2). The core is surrounded by another glass or plastic material called the **cladding**. A third component, called the **coating** or **jacket**, surrounds the cladding to protect the fiber from its

Figure 9-2

General structure of a
fiber-optic cable

Transverse view

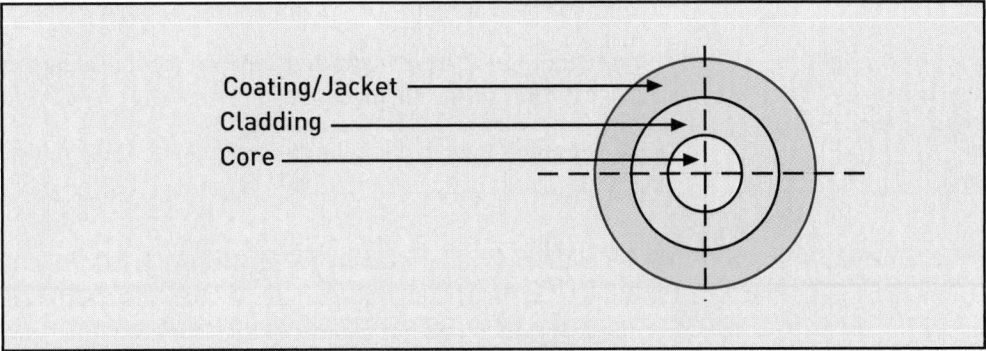

Longitudinal view

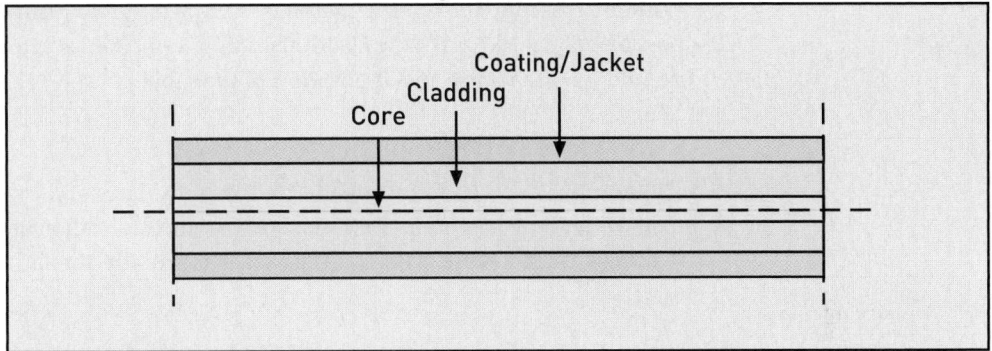

surrounding environment and to provide rigidity. This type of concentric configuration is often called a coaxial arrangement, similar to that of the center conductor and outer shield conductor in coax cables. Unlike coax cables, which transport electricity over copper wires, fiber-optic cables carry light through glass or plastic to convey information.

Each component of the cable serves an explicit purpose. The core and cladding pair is used to trap and guide the light inside the cable, while the coating/jacket protects the pair from the surrounding elements, such as moisture, dust, and other factors, and strengthens the fiber.

CORE

The core is the central part of the fiber; the light signal carrying the information essentially travels through the core. It is cylindrical or elliptical and made of either an optically transparent glass (silica) or plastic. The diameter of the core can range from a couple of micrometers (µm) to a couple of millimeters (mm), depending on the category of cable. A micrometer is equal to one millionth of a meter, and a millimeter is one thousandth of a meter.

The core material has an important property called the **refractive index** (RI), which typically is indicated by a number greater than 1. The RI, also called the index of refraction (IOR), is a measure of the speed of light through a material. Light travels fastest in a complete vacuum, so the RI of light through a vacuum is conventionally defined as 1. The RI of other materials is greater than one and is defined as follows:

Refractive index of an optical medium = Speed of light in a vacuum (300,000,000 meters per second)/speed of light in the optical medium

Air has an RI of just more than 1, whereas water has an RI of approximately 1.3. Glass has an RI of approximately 1.5, which means that the speed of light in a vacuum is about 1.3 times greater than the speed of light in water. Similarly, the speed of light in a vacuum is about 1.5 times greater than the speed of light in glass. RI is the property of an optical medium that also controls the amount of **refraction**, or bending of light, when a light ray crosses the boundary from one optical medium to another with different refractive indices. Because TIR can be considered an extreme case of refraction, the RI of the core and cladding is a critical property that determines how light will travel across the cable. TIR is explained further in the next section, "How Light Travels through a Fiber."

> **Problem**—Calculate the refractive index of an optical medium if light travels at a speed of 2.4×10^8 m/s inside the medium.
>
> RI = Speed of light in a vacuum/Speed of light in the optical medium = $3 \times 10^8 / 2.4 \times 10^8$ = 1.25

CLADDING

Cladding is the material that surrounds the core. The cladding is also made of glass or plastic, like the core, and always has a lower RI than the core. The combination of core and cladding materials with different refractive indices creates an environment in which TIR can occur. Without a secondary optical material with a lower RI surrounding the core, TIR is not possible. The thickness of the cladding in fiber-optic cables can vary greatly, just like the core, depending on the type and application of the cable.

COATING/JACKET

The coating/jacket surrounding the cladding insulates and protects the fiber from physical damage and environmental effects, such as moisture that might interfere with the inner workings of the cable. The coating/jacket is usually made of acrylic or another type of material.

HOW LIGHT TRAVELS THROUGH A FIBER

TIR is the basis of fiber-optic communication, and as mentioned previously, TIR may be considered an extreme case of refraction. When a light ray strikes a boundary of two materials with different RIs, it bends, or in other terms, refracts to an extent that depends on the ratio of the RIs of the two materials. Refraction is the phenomenon that causes a spoon inside a clear glass of water to appear shorter and bent to an observer looking from the outside. Light rays that strike the water/air boundary bend to create an image that is shorter than the actual height of the spoon in the water, because water has a higher RI than air. TIR is the phenomenon that makes the side of an aquarium act as a mirror when viewed at an appropriate angle. Light rays that strike the glass/air interface undergo TIR, and a person standing beside the aquarium can see the reflection of the inside of the tank.

In a fiber, the core and cladding are made of materials with different RIs, and the core has a greater RI than the cladding. Light rays that emerge from the core and strike the core/cladding boundary undergo refraction. However, if a ray strikes the core/cladding boundary at an angle that is greater than the **critical angle**, the ray undergoes TIR, where it is totally reflected from the boundary back into the core. Thus, if a light ray strikes a boundary with different refractive indices at an angle that is less than the critical angle, it undergoes refraction. However, if it strikes the boundary at an angle greater than the critical angle, it undergoes TIR.[1]

In Figure 9-3, two rays denoted by black and red lines strike the boundary at different angles of incidence. The angles of incidence are measured with respect to the vertical dashed reference line called the normal. The black ray emanating from Medium 1, which could be water or the fiber-optic core, strikes the boundary at an angle that is smaller than the critical angle (denoted

---

[1] Note that light rays that strike the core/cladding boundary at an angle less than the critical angle, besides undergoing refraction, are also partially reflected back into the core. If they strike the boundary at an angle that is greater than the critical angle, they are not partially reflected but are fully reflected. In other words, they undergo TIR.

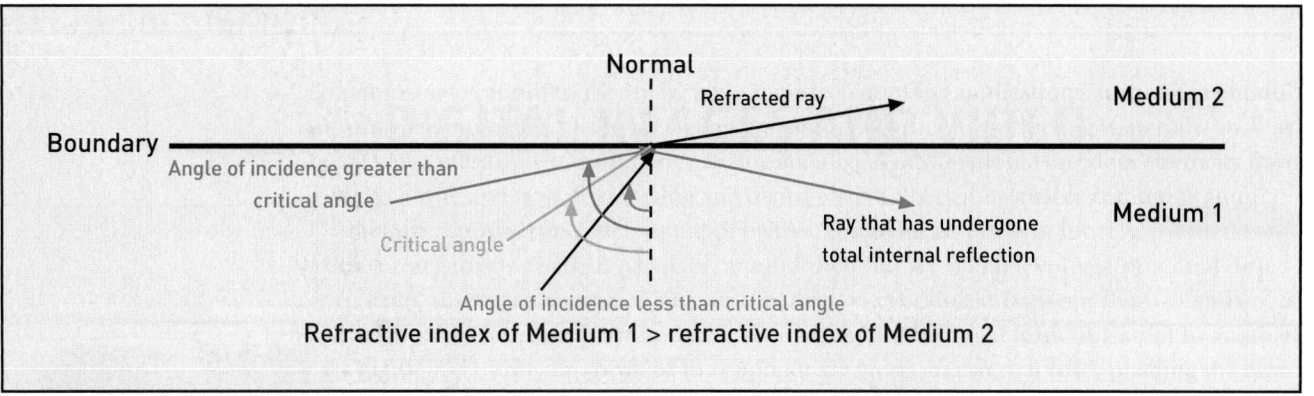

Normal

Boundary

Refracted ray

Medium 2

Angle of incidence greater than
critical angle

Critical angle

Medium 1

Ray that has undergone
total internal reflection

Angle of incidence less than critical angle

Refractive index of Medium 1 > refractive index of Medium 2

Figure 9-3

Principle of total internal
reflection

in green). The black ray is then bent away from the normal upon entering Medium 2, which could be air or the cladding. Thus, instead of the light ray continuing in the same path as it enters Medium 2, it bends in the direction away from the normal. The ray in red strikes the boundary at an angle that is greater than the critical angle, undergoes TIR, and is reflected back into Medium 1. TIR therefore causes the boundary to act as a mirror. By inserting light rays into the core at angles of incidence that are greater than the critical angle, a light ray can be continually trapped inside the core, as shown in Figure 9-3. The critical angle, above which TIR occurs, depends solely on the differences in RIs of the core and cladding materials. In general, the greater the ratio is between the cladding RI and the core RI, the smaller the critical angle. In Tyndall's experiment, the stream of water acted as the core and the surrounding air served as the cladding. Because water has a higher RI than air, TIR could occur within the stream of water for rays that struck the water-air boundary at an angle greater than the critical angle. For rays that struck the water-air boundary at an angle less than the critical angle, the rays were not trapped inside the stream, but left the water stream as stray light.

Based on this concept, Figure 9-4 illustrates how a ray of light that enters from one end of the fiber is trapped inside the core. The ray at an angle of incidence greater than the critical angle strikes the core-cladding boundary at point A and undergoes TIR. The reflected ray strikes the boundary at point B at the same angle of incidence as point A (which is greater than the critical angle), and is reflected back into the core. The ray travels in this manner throughout the length of the fiber. Any light ray that does not undergo TIR is refracted from the core-cladding boundary. This ray is usually lost in the cladding and does not reenter the core.

If a glass or plastic rod could always be laid out along a straight line and light could be introduced inside the rod with perfect horizontal alignment to the rod's longitudinal axis, as in the upper part of Figure 9-5, the ray would travel inside the rod in a straight line without having to bounce back and forth from the boundary, eliminating the need for the cladding and for TIR to occur. However, this setup is rarely possible in real life. TIR is therefore the solution to guiding light inside a cable with twists and bends, as shown in the lower part of Figure 9-5.[2]

Figure 9-4

A light ray trapped inside
the core undergoes
continual TIR

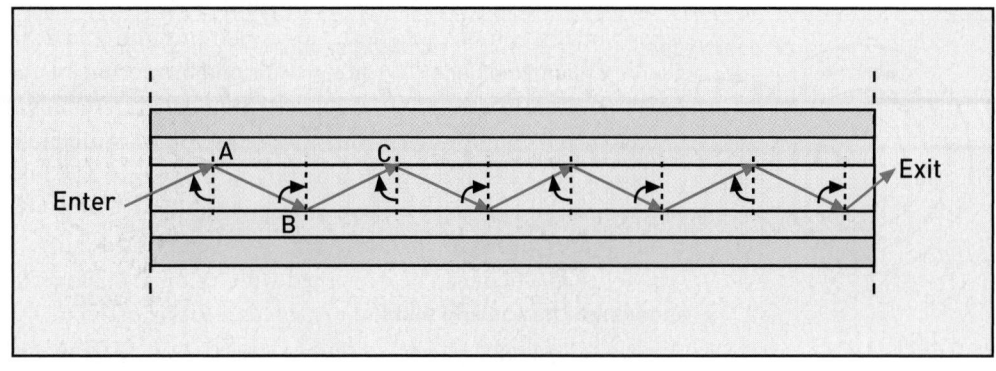

Enter

A

C

B

Exit

[2] Note that newer fiber-optic cables may rely on a different mechanism than TIR.

Figure 9-5

The importance of TIR in cables with twists and bends

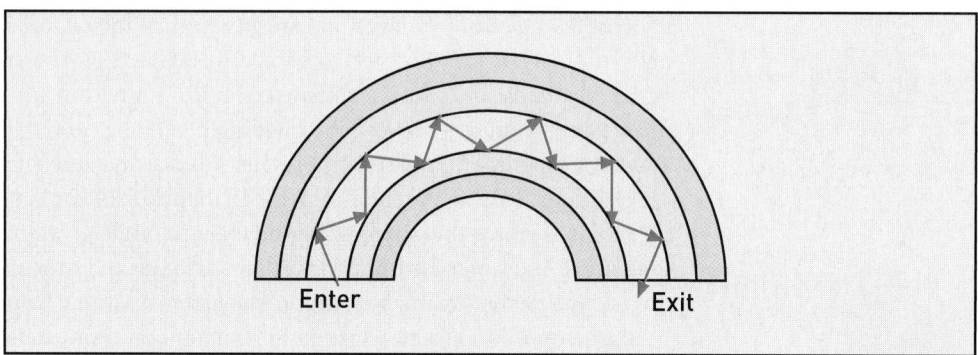

TYPES OF FIBER-OPTIC CABLES

Fiber-optic cables may be classified as either single-mode fibers or multi-mode fibers. Furthermore, single-mode fibers and multi-mode fibers may also be classified according to the refractive index profiles of their cores. These classifications are all explained in this section.

SINGLE-MODE FIBER

Single-mode fibers have a very small core diameter, on the order of a few micrometers (μm). This diameter is much less than that of a typical human hair, which is about 70 μm. The cladding diameter is tens of micrometers, making it much larger than the core. Single-mode fibers can transmit information at very high data rates across large distances, and at much higher rates than multi-mode fibers, because the small diameter restricts only a single mode of light to carry all of the information. In other words, the energy of the light traveling in a single-mode fiber is concentrated in a single mode, while the same energy of light in a multi-mode fiber is dispersed, or spread out, across multiple modes.

Modes in a fiber can be explained with a simple analogy. Imagine that each mode corresponds to a person and that light energy is a basket of apples. A single-mode fiber transmits a single mode of light, which is similar to one person carrying the basket of apples, and a multi-mode fiber transmits light in multiple modes, which is analogous to a group of people in which each carries a single apple from the basket. All the apples are transported, and each person gets to carry a single apple. Concentrating the light energy in a single mode results in very high data rates, on the order of tens of gigabits per second.

Because single-mode fibers require more expensive light sources and other equipment and are difficult to align with other components due to their small sizes, systems that use them are usually more costly to set up than those that use multi-mode fibers. Single-mode fibers are made of high-quality glass with minimal impurities (i.e., foreign materials within the raw material of the fiber), resulting in a low-attenuation fiber that is ideal for long-haul communication systems.

MULTI-MODE FIBER

Compared with single-mode fibers, **multi-mode fibers** have a much larger core diameter, ranging from tens to hundreds of micrometers. These types of fibers are usually used for lower-bandwidth, shorter-distance applications than single-mode fibers. Although multi-mode fibers are wider in diameter, their information-carrying capacity is less than that of single-mode fibers due to the physical effects they impose on the signals traveling through them. Multi-mode fibers can be made of either glass or plastic, depending on their application. They are also cheaper to use because they require very cheap light sources and are easy to align with other components due to their larger physical size.

STEP INDEX AND GRADED INDEX FIBERS

Although fiber-optic cables may be classified into single-mode or multi-mode, they may also be further classified according to how the refractive index varies throughout their core. A fiber-optic cable with a uniform refractive index throughout its core is classified as a **step index fiber**. In contrast, a cable whose core refractive index is nonuniform and varies gradually is classified as a **graded index fiber**. The value of the RI of a graded index fiber is highest in the center of the core and gradually diminishes toward the cladding. Both types of fiber are available in single-mode and multi-mode forms, such as **single-mode step index**, **multi-mode step index**, **single-mode graded index**, and **multi-mode graded index**. Figure 9-6 illustrates the refractive index profiles of step index and graded index fiber-optic cables. You can observe a sharp increase in the value of the RI at the boundary between the cladding and the core. The RI index of the core is then constant throughout the core. In contrast, the RI of a graded index fiber gradually increases in value starting at the cladding-core boundary, and reaches its highest value at the center of the core.

Both types of fiber have advantages and disadvantages. Figure 9-6 also illustrates the difference between how light is guided within a step index fiber and a graded index fiber. A step index fiber essentially guides light in the traditional way that was explained earlier; the path that a light ray travels within a graded index fiber is more sinusoidal in nature. Rather than being sharply reflected back from the core/cladding boundary, light within a graded index fiber bends gradually.

Figure 9-6

Refractive index profiles of step index and graded index fibers

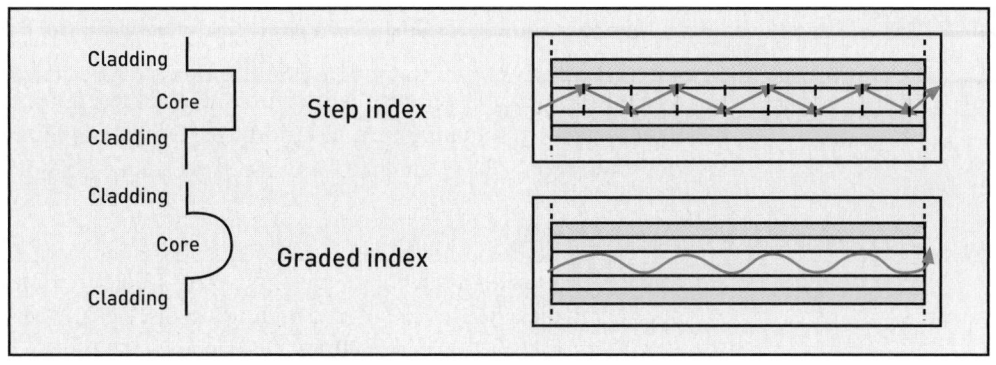

Figure 9-7 illustrates how light is used to carry bits of information within a basic fiber-optic communication system. Pulses of electricity corresponding to the bits arrive at the input **transducer**, a device that converts one form of energy to another. In fiber optics, devices similar to light bulbs are used as optical transducers at the input of fiber-optic cables to convert electricity into light. The most commonly used devices are **light emitting diodes** (LED) and **laser diodes** (LD); the latter light source is more expensive.

Electrical signals arriving at the input of these **optoelectronic devices** are used to modulate the light source, as discussed in Chapter 8. Just as radio waves can be modulated to carry information in radio-frequency communication systems, light waves emitted from sources such as lasers can be modulated by altering various properties of the light, including its wavelength, phase, and amplitude. The modulated optical signal is emitted by the source and coupled into the cable. Once the light is trapped inside the cable, it travels to the other end, where it is demodulated and an output transducer such as a **photodiode** (PD) or **phototransistor** converts the light back into electrical pulses.

Figure 9-7

A fiber-optic communications system

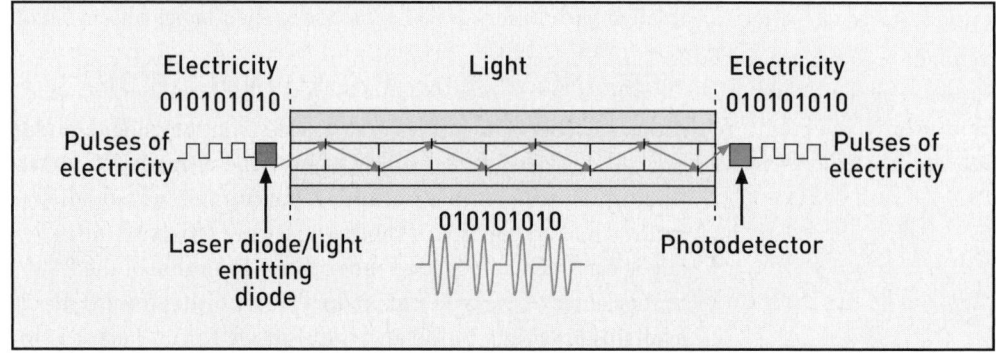

Although repeaters are not shown as part of the system in Figure 9-7, repeaters can be periodically inserted to receive weakened signals, amplify them, and retransmit them to the other end of the communications system.

Because LDs are generally more expensive than LEDs, and because they can be modulated at higher data rates, they are typically used for high data-rate applications, such as with single-mode fibers. In contrast, LEDs are commonly used with multi-mode fibers for shorter-haul systems at lower data rates.

Fiber-optic communication systems commonly operate in the infrared portion of the light spectrum. (The light spectrum was introduced in Chapter 8; refer back to it if necessary.) Light sources that operate in the infrared region usually emit at wavelengths of 1.3 µm and 1.55 µm, and the corresponding detectors are highly sensitive to light at these wavelengths. Fiber-optic cables are no exception; because most fiber-optic cables induce the lowest attenuation at 1.3 µm and 1.55 µm, light sources and detectors are commonly manufactured to operate at these wavelengths. Therefore, you can note that the attenuation of light within a fiber-optic cable is wavelength dependent.

Various **couplers** can be placed strategically between the light source and the fiber input and between the light detector and the fiber output. These devices efficiently couple light into and out of the cable so that light can be inserted at appropriate angles (greater than the critical angle) and to minimize insertion loss. Other components, such as **fiber connectors**, can connect one fiber segment to another with minimal loss of light.

WAVELENGTH DIVISION MULTIPLEXING

Figure 9-7 showed how an optical signal can carry a single channel of information from one end of a fiber to the other. Instead of transmitting just one signal, fiber-optic cables can carry multiple signals due to their large bandwidths, and hence carry multiple channels of information. Many optical signals can be effectively multiplexed into the same fiber, as opposed to transmitting just one signal. Fiber optics is consequently very popular in broadband applications where multiple channels are concurrently sent through a single fiber to carry video, voice, and computer data. The technology that makes such applications possible for optical communications is called **wavelength division multiplexing** (WDM). When a large number of channels are multiplexed onto a single fiber-optic cable, the approach is called **dense WDM** (DWDM). Multiplexing a smaller number of channels corresponds to **coarse WDM** (CWDM). A typical DWDM system might have 100 channels, and a typical CWDM system might have 10.

Chapter 8 covered two multiplexing techniques known as TDM and FDM. WDM is a similar approach, in which each information channel in the system is sent through the fiber at a separate wavelength. Because the frequency (f) and wavelength (λ) of an electromagnetic wave are related to each other ($\lambda = c/f$), WDM can be considered the optical equivalent of FDM. Note that c denotes the speed of light.

WDM may be implemented by using a separate light source, such as a laser, that emits at a different wavelength for each information channel (see Figure 9-8). For example, assume that two channels are to be multiplexed over a single fiber-optic cable. With WDM, two separate LDs that transmit at different wavelengths must be used to achieve the required wavelength separation.

Figure 9-8

A fiber-optic communication system employing WDM

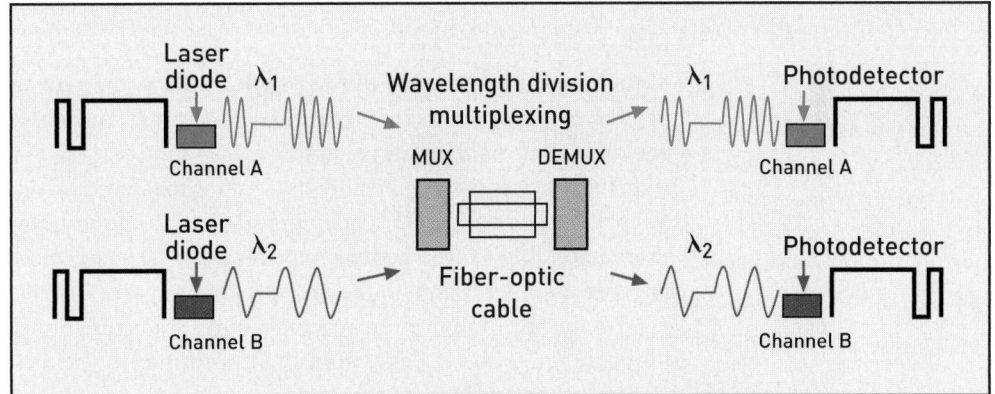

Each laser is ASK-modulated (a technique discussed in Chapter 8) with its corresponding bit stream and combined at the optical multiplexer, which combines all the light emitted from the lasers and inserts it into the cable. The received signal is then demultiplexed into its individual channels with a wavelength-selective filter that can separate light based on its wavelength.

In contrast to using multiple light sources to emit at different wavelengths, other sources are available, such as **tunable lasers** that can be adjusted to emit at a variety of different wavelengths. These devices are usually more expensive than single-wavelength sources and are not as popular in most fiber-optic communication systems.

As a final note, other multiplexing schemes such as TDM may be used in combination with WDM to transmit a very large number of channels across a single fiber-optic cable.

BENEFITS AND DRAWBACKS OF FIBER-OPTIC TECHNOLOGY

This section further explains the properties of fiber-optic cables that make them appealing for use in high-performance systems. This section also discusses their shortcomings.

HIGH TRANSMISSION RATE

One of the greatest advantages of optical fiber for communication is its capacity to support extremely high data transmission rates. Such rates are required for many applications, including video or music sharing, transmission of medical images, satellite imagery, cable TV broadcasting, networked video games, and multimedia communications. The Internet's transmission backbone primarily uses fiber-optic cable because it must support an enormous volume of information. It was recently demonstrated that a fiber-optic communication system with 160 WDM channels could transmit close to 26 terabits per second (i.e., 26,000,000,000,000).[3] This high data rate can only be achieved with fiber-optic cables by today's standards.

IMMUNITY TO ELECTROMAGNETIC INTERFERENCE

Because **light** rather than electricity is carried inside fiber-optic cables, they are immune to electromagnetic interference (EMI), a form of noise. The presence of EMI in a communication system directly affects the rate at which data can be transmitted across a wireless or copper-based communication channel. A channel with a large amount of interference has a small signal-to-noise ratio, which in turn affects the maximum data rate that can be transmitted, according to the Shannon-Hartley capacity theorem discussed in Chapter 8. Thus, the immunity to EMI enables the large capacity of fiber optics. Because copper cabling systems are affected by external sources of interference such as lightning, radio broadcasts, and power lines, fiber-optic cables are often used to replace copper in noisy environments. Because of their susceptibility to noise, some sound recording systems use fiber-optic cables during the digital recording process to minimize such noise. Fiber-optic cable is also a safer option than copper in an industrial environment, where the presence of an electrical signal could present a fire hazard.

LOW ATTENUATION

Another appealing property of fiber-optic cables is their low attenuation—on the order of fractions of a dB/km in some transmission systems. The glass used for fiber-optic cables is quite different from the glass we find in a glass mug or a car windshield. In essence, the purer the quality of glass used in the fiber-optic cable, the lower the signal degradation. Low attenuation translates directly to efficient long-distance communications, such as transatlantic or other suboceanic cable routes, because repeaters can be spaced further apart if low-attenuation cables are used, compared to their copper counterparts. Using a large number of repeaters in long-distance communication systems reduces transmission rates, increases costs, and leads to loss and large signal delay from the transmitter to the receiver because it takes time to regenerate the signal at each repeater.

HIGH SECURITY

Security is another important feature of fiber-optic technology. Unlike wireless or copper-based transmission systems, in which it is usually easy to tap into the communication link, fiber-optic systems are very difficult for outsiders to crack. You cannot tap into a fiber cable by merely slicing the cable jacket and clipping a crocodile clip on the cable, as you can with a copper cable. Even if someone gained access to the fiber and tapped into the transmission, several mechanisms can be used to detect eavesdroppers. Banks, embassies, and many governmental departments that require private communications use heavy-duty, robust fiber-optic cabling systems due to their high level of security.

SMALL WEIGHT AND SIZE

Because fiber cables are either made of glass or plastic, they are much more lightweight than a metal such as copper. A fiber as thin as a strand of hair has much more capacity than a thick, heavy, coaxial copper cable. In applications used on aircraft and satellite systems, where weight is a major issue, and in applications where space is a concern, fiber optics has obvious advantages.

LOW POWER CONSUMPTION

Because fiber-optic cables have low attenuation characteristics, the amount of power that must be supplied in the cable is much lower than in copper-based systems across the same

[3] *www.alcatel-lucent.com.*

distance. In addition, efficient LEDs and LDs, as well as efficient light detectors, are readily available to reduce the overall power consumption of an optical link.

HIGH INSTALLATION COST

One traditional disadvantage of fiber optics is its high planning, installation, and maintenance cost. To properly install, lay, and align these cables, personnel usually require extensive training, because the cables can be sensitive to stretching or other physical effects. Elaborate installations are often required for important applications to minimize these effects. Nonetheless, the cost of installing fiber-optic systems is decreasing because more efficient components and techniques are being developed to make installation easier.

DIFFICULTY IN SPLICING

Because fibers are made of either plastic or glass, they are more difficult to **splice** (i.e., bring together) than copper cables. Sophisticated splicing techniques, such as fusion or using a connector, are needed to bring two pieces of fiber together. Splicing fiber-optic cables is generally more difficult than soldering or crimping two pieces of copper wire together.

COMMERCIAL CABLES

The core and cladding constitute the basic components of a single optical fiber strand, as described in earlier sections. However, commercial fiber-optic cables are usually sold as multiple strands of fiber bundled together. In addition to the individual strands, commercial fiber-optic cables contain other components that serve various purposes, such as providing rigidity. Figure 9-9 shows commercial fiber-optic cables and their components. A single strand of optical fiber may be surrounded by thin strands of strengthening materials; these added materials protect the basic components from tension that may be imposed on the fiber during installation. A cable jacket, which can be made of PVC (polyvinyl chloride) or another material, provides stiffness and protection to the entire structure. Strands of fibers can also be bunched together to construct a thick cable system to carry multiple channels. Even more components can be used to surround the fibers for extra protection against tension and to ensure further privacy, as illustrated in Figure 9-9.

Figure 9-9

Structure of commercial fiber-optic cables

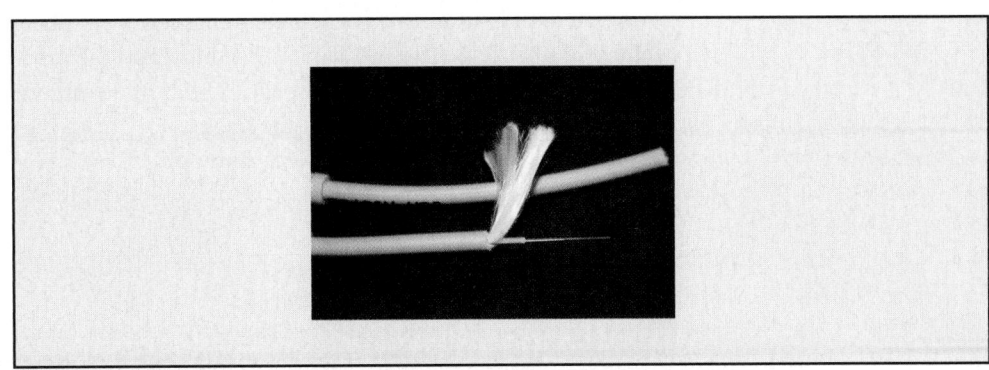

Dark fiber, or unlit fiber, is a common term in the telecommunications and IT industry to describe a fiber-optic infrastructure that has been established but is not yet being used. The term *dark* refers to the inactive or "unlit" nature of the fiber system. Many telecommunications companies have installed fiber-optic systems in anticipation of massive demand for broadband communications in the future. In addition, because the cost of installing fiber-optic systems after construction is prohibitive, the systems are often planned ahead and installed during construction of public and private buildings, malls, and commercial or residential areas. Until they are used, however, the cables are called dark fiber.

Fiber optics has become a significant area of research since its invention. It is used in a wide array of applications, including medicine, telecommunications, computing, and computer networking, among many others. One prominent use of these high-capacity transmission media, as mentioned previously, is in long-haul communication systems such as transoceanic cables to transport voice and other types of data. Cables that connect continents must have very low attenuation to minimize the number of repeaters, and fiber optics provides this feature.

Another significant application of fiber optics is in the Internet backbone. Most traffic across the Internet is carried over high-capacity backbone networks, as we discuss in forthcoming chapters. Most of these backbone networks use fiber-optic cabling due to its obvious advantages. Furthermore, fiber-optic technology is used in a wide array of high-speed local area networks, such as Gigabit Ethernet.

Besides computer networking, the computing field also benefits from fiber optics. For example, printed circuit boards such as the motherboard were discussed in Chapter 5. Chips on these boards are traditionally connected using metal traces on the boards; however, metallic interconnects are limited in terms of the density of interconnections and their speed. Furthermore, multiple boards are also traditionally connected to each other using metallic interconnects. Replacing metallic interconnects with fiber-optic cables in these applications has been proposed as an alternative; this subject has been an active area of research within the past decade or so.

An even more interesting area of research involves replacing metallic interconnects within the chips themselves with tiny fiber-optic strands. As chips become faster and more complex in terms of their interconnections, fiber optics has been proposed as a way to overcome the bottlenecks posed by traditional metallic interconnects, including high attenuation and noise.

Cable companies, such as those that provide cable TV or Internet service over coaxial cables, lay coaxial cables in neighborhoods to provide service to homes and businesses. These service providers, however, may not carry the signal from the subscriber to the provider's premises entirely over coaxial cables. The coaxial cable may just be employed to provide the signal to the home or office, and fiber-optic cables may be used at the curb to carry the signals from the curb to the cable company's facilities. This requires the signals from the coaxial cable to be converted into optical signals at the curb before they can be carried across the fiber-optic cable. This type of arrangement, called **hybrid fiber coaxial** (HFC), is frequently employed by cable service providers to overcome the higher attenuation, noise, and lower data capacity of coaxial cables.

The medical field has also benefited immensely from fiber-optic technology. One prominent use of fiber optics for medicine is in endoscopy, the process of inspecting organs of the body using a device called an endoscope. A fiber-optic cable as part of an endoscope can be used to provide light and transport the imagery of organs such as the stomach. Endoscopy was one of the first uses of fiber-optic technology.

CHAPTER SUMMARY

- Fiber-optic cables are made up of a few basic components, including the core, cladding, and coating or jacket.
- Light within a fiber-optic cable is carried from one end to the other based on a phenomenon called total internal reflection (TIR).
- The core and the cladding have different refractive indices, so that light that traverses the core and hits the core-cladding boundary undergoes refraction.
- Light rays that strike the core-cladding boundary at an angle greater than the critical angle undergo total internal reflection.
- Lasers and LEDs are the primary sources used to insert light into a fiber-optic cable.
- Photodiodes and phototransistors are transducers that convert light back into electricity.
- Fiber-optic cables may be classified as either single-mode or multi-mode. Single-mode fibers have a smaller core diameter and can support higher transmission rates across longer distances than multi-mode fibers.
- Fiber-optic cables may also be classified according to the nature of their core's refractive index. Step index fibers have a uniform refractive index throughout their core, whereas graded index fibers have a refractive index that varies gradually through their core.
- WDM is a technique used to multiplex multiple channels of information and transmit them across a single fiber-optic cable. The 1.3-μm and 1.55-μm wavelengths are the most popular due to the low attenuation that optical signals undergo at these wavelengths.
- Advantages of fiber-optic cables include high transmission rates, immunity to EMI, low attenuation, high security, small weight and size, and low power consumption. Disadvantages include higher installation and maintenance costs and difficulty in splicing.
- Commercial fiber-optic cables are sold with multiple strands of fibers packaged as a single cable with extra components that increase its strength.
- Established but unused fiber-optic systems are called dark fiber.
- Various commercial applications of fiber-optic cables include transoceanic communications, computer networking, computing, and medicine.

KEY TERMS

cladding

coarse WDM

coating

core

coupler

critical angle

dark fiber

dense WDM

fiber connector

fiber optics

graded index fiber

hybrid fiber coaxial

jacket

laser diode

light

light emitting diode

multi-mode fiber

multi-mode graded index

multi-mode step index

optical communication

optical medium

optoelectronic device

photodiode

photophone

phototransistor

refraction

refractive index

single-mode fiber

single-mode graded index

single-mode step index

splicing

step index fiber

total internal reflection

transducer

tunable laser

wavelength division multiplexing

REVIEW QUESTIONS

1. Give one example in which you might have used light to convey or receive information.
2. Name an everyday occurrence in which you can observe the phenomenon of refraction.
3. Briefly explain the two conditions that are needed for TIR to occur.
4. Of air and water, which has the higher refractive index?
5. List two devices that you use in everyday life to convert electricity to light.
6. List three advantages of using fiber-optic technology for communication.
7. Do you think that the advantages of fiber-optic technology far outweigh the disadvantages? Discuss in your own words.
8. Explain what is meant by the term *dark fiber*.
9. What is the purpose of the coating and jacket surrounding the cladding?
10. What materials are typically used for the core and cladding?
11. List two popular light sources that are used in fiber-optic communications.
12. List two popular light detectors that are used in fiber-optic communications.
13. Briefly explain how three channels of information can be multiplexed on a single fiber.
14. What is the main physical difference between single-mode fiber and multi-mode fiber?
15. What is the main difference between step index fiber and graded index fiber? How is light guided differently within a step index fiber than within a graded index fiber?
16. Which supports higher data rates: single-mode fiber or multi-mode fiber?
17. With the help of a diagram, what is a critical angle?
18. Calculate the refractive index of an optical medium if light travels at a speed of 2.2×10^8 m/s inside the medium.
19. Calculate the speed of light in a medium that has a refractive index of 1.34.

DISCUSSION QUESTIONS

1. Try to think of some application environments or implementation scenarios in which fiber-optic cable is preferable for reasons of safety or security.
2. Some corporate buildings, houses, and dormitories installed fiber-optic cable for high-speed Internet access before wireless LAN access became as popular as it is today. If you were building a new facility today, would you install fiber-optic cable or assume that high-speed wireless access would be more advantageous? Why or why not?
3. Should municipal governments invest in fiber-optic networks to help provide broadband Internet access to citizens? Give arguments for and against this investment.
4. Research the medical diagnostic procedure called endoscopy. How has fiber-optic technology made this procedure possible? If fiber-optic technology were not available, what would be an alternative diagnostic technology, and how would the patient's experience be different?

CASE PROJECT

Search online for companies that manufacture fiber-optic cable. See if any of the companies' Web sites provide information about their fiber manufacturing process. Try to answer the following questions: How pure does fiber-optic material have to be, and what analogies describe this purity? For example, if the ocean was made of fiber-optic material, would you say that you could almost see to the bottom? What are the major fiber-optic manufacturing companies?

What raw materials are used to produce fiber-optic cables? Describe some of the manufacturing processes that produce the purity required for fiber optics. What manufacturing improvements have occurred in optical technologies over the last few years? What other products are made by fiber-optic manufacturers? Do any of these products surprise you?

CHAPTER 10

WIRELESS COMMUNICATIONS

LEARNING OBJECTIVES

In this chapter you will:

- Understand the various applications that use radio waves

- Learn about satellite components and satellite orbits

- Understand the effects of the atmosphere on radio waves

- Learn about the underlying principles of the global positioning system

- Gain exposure to satellite-based Internet services and radio frequency identification

INTRODUCTION

Ever-increasing demands for mobile computing and communications applications keep wireless systems at the forefront of information technology. Most wireless systems transport information through the air via radio waves, which reside in the radio frequency portion of the electromagnetic spectrum and are modulated to carry both analog and digital information. Although other methods are used, radio waves provide the greatest flexibility, range, and mobility. Other wireless approaches, such as infrared or visible light, are used mainly under direct line-of-sight conditions, making them suitable primarily for short-distance communications. This chapter addresses radio-based wireless communication systems and provides an overview of some important radio applications, including global positioning systems, satellite Internet, and radio frequency identification systems.

Ever since Guglielmo Marconi first demonstrated how to communicate using long-distance radio waves, people have used them for information exchange. As explained in Chapter 8, radio waves are electromagnetic (EM) and occupy a large portion of the EM spectrum (see Figure 10-1). As illustrated in the figure, the frequency of radio waves ranges from a few hertz to hundreds of billions of hertz. Figure 10-1 illustrates microwaves (SHF and EHF waves—both radio frequency bands discussed in Chapter 8) as part of the radio spectrum, and we treat microwaves as such, although some sources classify radio waves and microwaves as separate entities.

Figure 10-1

The radio spectrum

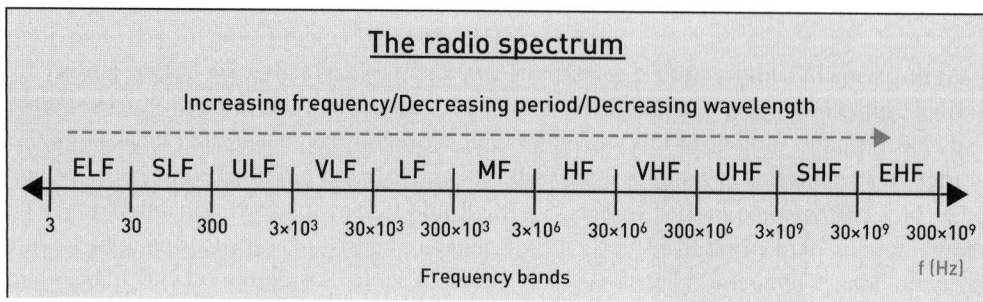

Most of our daily activities involve some form of wireless communication, such as listening to the radio or talking on a cellular phone. The mobility that radio waves provide, combined with the long distances they can travel and their capacity for carrying large amounts of information, make them indispensable to most information applications. Some of these applications are discussed in the following sections.

DIGITAL RADIO AND TELEVISION BROADCASTING

Most cities have radio stations that provide both AM and FM broadcasting over multiple channels. These analog modulation techniques were explained in detail in Chapter 8; they are used to carry music, speech, and data over radio waves. The traditional AM and FM techniques are still used extensively for radio broadcasting. Analog radio is vulnerable to unwanted effects such as interference from other sources, which cannot be removed effectively. Digital radio, which uses effective error-control techniques, was developed as a higher-quality alternative for radio listeners. Such services, which use digital signals to transmit information, are also called **digital audio broadcasting** (DAB) and **high-definition radio** (HD radio). These technologies offer high-fidelity music that is close to CD quality, as well as digital broadcasting that enables receivers to display moving pictures, other multimedia, and text.

Because terrestrial radio broadcasting has a limited reach, **satellite radio** has become an attractive alternative. Because satellite radio is digital, it overcomes the distance limitations of conventional radio broadcasts and provides continuous reception across a wide geographical area. By transmitting digital signals via satellite to terrestrial listeners, subscribers to satellite radio can listen without interruption, even when traveling coast to coast across the United States. Some car manufacturers and rental companies include these receivers as a standard feature in their vehicles.

The electromagnetic spectrum's VHF and UHF bands are designated for TV broadcasting, among many other applications. When terrestrial TV broadcasts first began, viewers could only receive a stable signal by using bulky antennas on rooftops or on their TV sets. Although this setup is still used in some parts of the world, cable and satellite television has replaced it in major cities. In cable TV, signals are typically broadcast to a cable company receiver via satellite, then sent over radio waves from the cable provider to the consumer via coaxial cable. Confining radio waves within a cable enables better picture and sound quality by reducing interference and provides hundreds of channels to the consumer.

Satellite television providers broadcast signals via satellite to customers over a large area. This technique requires the use of a dish antenna to receive broadcasts, and is more susceptible to interference from weather conditions than cable TV. In contrast, users can access many more stations than cable TV providers can offer.

Just as digital radio provides an alternative to analog radio, digital TV broadcasts are an alternative means of conveying TV signals. **Digital television** (DTV) broadcasting enables the transmission of many high-quality channels, high-fidelity audio, and such features as pay-per-view and video on demand. Many cable companies provide digital broadcast services to consumers in the form of digital cable. **High-definition television** (HDTV) is also in demand for its superior picture and sound quality. A subscription to HDTV services requires a high-end set that can display ultra high-resolution images to take full advantage of HDTV programming. Not all providers offer HDTV programming, and its popularity has been limited so far, but the reduced cost of HDTV sets and services is expected to help expand the market.

NATIONAL DEFENSE AND EMERGENCY SERVICES

The military relies heavily on wireless communications. Military ships, aircraft, and terrestrial units are in constant communication via designated radio frequencies. One major difference between civilian and military communications is the level of security. Because military communications are highly sensitive and radio waves can be intercepted by a third party, the communications are usually encrypted using sophisticated techniques to ensure secrecy.

Another critical application of radio waves is in emergency dispatch services. Radio networks are in place to dispatch first responders such as paramedics, police, and firefighters to fires, medical emergencies, accidents, and natural disasters.

AVIATION AND MARITIME COMMUNICATIONS

Radio waves are indispensable for communication in aviation and maritime applications. Private jets, commercial airliners, and military planes all use radio waves to communicate with airports and other ground stations en route to their destinations. Pilots receive navigational instructions from air traffic controllers and information about the weather, both of which are vital information for air safety. During flight, an aircraft constantly communicates with ground stations, either directly or via satellite. In addition, many airlines provide in-flight phone and Internet services to passengers. Both of these applications rely on radio waves.

The maritime industry also uses satellites for communication between ships at sea, between ships and stations on land, and even between ships and aircraft. Numerous maritime satellite systems such as **Inmarsat** (International Maritime Satellite Organization) are specifically designed to provide maritime communications. Given the growing number of commercial ships, private boats, and cruise liners, a reliable communications network is critical. Currently, this network is only possible using radio-wave transmissions.

COMPUTER NETWORKS

Computer networks employ radio waves to transmit digital signals between computing devices. With the growing trend in mobile computing, wireless connectivity is becoming an increasing necessity. Although a great deal of network traffic flows over copper or fiber-based transmission media, a significant portion also flows across wireless systems. Information on the Internet, for example, is often transmitted across great distances via satellites. In addition, remote users, such as those on ships or those with no access to DSL or cable modem connections, use satellites to connect to the Internet. Short-distance communication between printers and workstations, and between computers and other devices such as laptops and PDAs, can also be implemented in wireless mode.

One popular wireless networking technology is wireless fidelity (**Wi-Fi**), which uses radio waves at frequencies of 2.4 GHz and 5 GHz. Many business and home-based networks use Wi-Fi for mobile connectivity. Another important wireless networking standard, **Bluetooth**, is used to connect mouses, keyboards, printers, and digital cameras to computers. Bluetooth uses radio waves with a frequency of 2.45 GHz and generally provides a shorter range than Wi-Fi, because the radio waves are transmitted at lower power.

Another significant wireless technology that has gained attention is **WiMAX** (Worldwide Interoperability for Microwave Access). WiMAX is considered an alternative to DSL and cable modems for providing metropolitan wireless connectivity between users and Internet service providers. These technologies are discussed in greater detail in Chapters 11 and 12.

THE TELEPHONE NETWORK

Radio waves in telephony applications are most often encountered in the following areas:

- Between a cordless telephone and its base station
- Between a wireless headset and a mobile telephone
- Within the telephone network in areas that employ satellites
- Within the telephone network in areas that use long-distance terrestrial microwave links
- Between cellular telephones and base station antennas
- Between a satellite phone and the satellite

Cordless telephones can be found in most businesses and households. Short-distance radio communication between a cordless headset and the base is easy to achieve over a radio frequency, either in analog or digital format. Because low-power signals are transmitted between the headset and base, cordless phone use is usually limited to a closed building such as an office or residence.

As mentioned in the previous section, Bluetooth is a prominent standard used in wireless computer networking. It is also used to provide wireless connectivity between mobile telephones and their earpieces, which gives users hands-free access to mobile phones.

The telephone network also relies heavily on radio waves for transmitting signals across terrestrial and satellite links for long-distance communications. During a call from one continent to another, such as from Europe to Australia, the telephone signal is often relayed via a communications satellite. Without radio waves, long-distance communication would be impossible in many circumstances. Unfortunately, the time it takes a signal to be relayed across such long distances is significant and is often apparent as perceptible pauses in transcontinental telephone conversations via satellite.

Without radio waves, cellular telephones could not exist. Although most communication within the cellular network itself uses copper or fiber-optic lines, communication between the mobile phone and the base station occurs over the air. Some cellular telephone traffic also passes over satellites, which means that the system also uses radio waves beyond the mobile phone and base station. The final section of this book describes the technological underpinnings of wireless telephony.

Satellite telephones communicate directly with satellites, unlike cellular telephones, which communicate directly with terrestrial base stations. Satellite telephone services provide users with global coverage using a single telephone. Satellite telephones are useful in places where no landline phones or base stations are available, such as in mountainous or desert regions.

NAVIGATION

Radio waves are also used for navigational purposes. The global positioning system (**GPS**) uses a combination of numerous satellites, receivers, and ground stations to give positional information in three dimensions. By sending special radio signals from multiple satellites to a receiver, the latitude, longitude, and elevation can be calculated and displayed. Many cars and boats come equipped with GPS systems that can display maps and provide directions. Small handheld GPS devices are also popular, especially with mountain climbers and hikers. The technical principles of GPS operation are discussed later in this chapter.

SPACE EXPLORATION

Because electromagnetic waves can travel through a vacuum, they are used in deep space exploration. Space probes can communicate effectively with the Earth over radio waves, either directly or via satellite. For example, NASA's recent exploration of Mars used robots called *rovers* to analyze the planet's surface and send the information to the antennas of the **deep space network** (DSN) on Earth.[1] The Hubble space telescope, launched in 1990, is another application that has contributed to our understanding of the universe and faraway galaxies. Information gathered by the telescope is still transmitted to Earth periodically via radio waves.

REMOTE SENSING

Radio waves also enable remote sensing technologies that acquire information about an object, landscape, or person from a distance. One of the most common remote sensing techniques is **RADAR** (Radio Detection And Ranging). RADAR is used in numerous areas, including navigation, weather forecasting, mapping, imaging, space exploration, and a variety of military and research applications. RADAR works by emitting radio waves toward an object and analyzing the reflections of the waves from it. These results can then be used to take speed and distance measurements. A specialized form of RADAR, Synthetic Aperture RADAR (**SAR**), is primarily used to map the surface of the Earth; it is also used for geographical applications that involve large stationary objects.

RADIO FREQUENCY IDENTIFICATION

Radio frequency identification (RFID) is an innovative approach for identifying and tracking shipping containers, pets, and sometimes even people. It is in widespread use today and will probably gain even more popularity as the costs of RFID-related components decline. An RFID system is made up of small **RFID tags** that emit radio frequencies and readers that can detect and read the information stored on the tags. For example, these tags are widely used for pet identification. A small tag that holds information about the pet is embedded under the animal's skin, enabling it to be identified if lost. The technology is also commonly used in highway tollbooths to reduce traffic bottlenecks. An RFID tag on the windshield of the vehicle is read and the tag holder's account is charged, significantly reducing passage times through toll plazas.

RFID tags are classified as either active or passive. **Active tags** can emit radio waves through their antenna using energy from the battery when prompted by the tag reader, whereas **passive tags** are induced by the energy of the reader's radio waves to emit radio waves of their own. Active tags have a longer range than their passive counterparts and can store and transmit a greater amount of information. Passive tags are much smaller and cheaper, and are easy to manufacture.

In many applications, RFID tags compete with another type of tag called a bar code. Although both types of tags supply the reader with information on a specific product, bar code systems have the disadvantage of requiring direct line of sight between the laser source that reads the bar code and the bar code label. RFID tags, on the other hand, can be read remotely and do not require a line of sight with the reader, making them highly versatile. Also, RFID systems do not rely on lasers to read the tags. As is commonly known, a crumpled bar code is very difficult to read with a laser bar code scanner. This situation does not occur in RFID systems. Although RFID technology is increasing in popularity for many applications,

[1] *http://deepspace.jpl.nasa.gov/dsn.*

the lower cost of bar code systems and their wide deployment in existing applications will probably make them the preferred choice for the near future for numerous applications.

Several social considerations make RFID technology controversial, including questions of ethics and individual privacy concerning the possibility of implanting RFID identification in people. Other concerns address the potential privacy implications of tracking cars, credit cards, and other personal property.

PERSONAL AND COMMERCIAL

Many personal applications rely on radio waves, including baby monitors, remote keyless entry features on cars, microwave ovens, and garage door openers. Amateur radio, commonly called **ham radio**, also relies on radio waves to connect people all over the world, not unlike Internet chat rooms. Ham radio operators use a specifically assigned frequency band and must have a license. Citizens band (CB) radio, which does not require a license, enables short-distance audio communications and is used for personal conversations and emergencies. Radio waves are also used in microwave ovens to heat and cook food.

One prominent use of radio waves for commercial applications is to wirelessly transmit credit card information between sales points and banks or credit card companies. Stores can install special equipment called very small aperture terminals (**VSATs**) for communicating directly with credit card companies and banks via satellites. VSATs enable businesses to serve customers who want to make purchases with their credit cards.

Table 10-1 summarizes the applications discussed in this section and the frequencies they use.

Table 10-1

Applications of radio frequency

Application	Frequency
Wi-Fi	2.4 GHz, 5 GHz
Bluetooth	2.45 GHz
WiMAX	2.5 GHz/3.3 GHz/3.5 GHz/5 GHz
AM radio	535 KHz–1605 KHz
FM radio	88 MHz–108 MHz
Microwave ovens	2.4 GHz

SATELLITE SYSTEMS

Several of the applications mentioned in the previous section rely on satellite systems. Beginning with the launch of **Sputnik** in 1957, the space race between the former Soviet Union and the United States spurred rapid advances in satellite technology and significant developments in space travel and exploration. Satellites are crucial for many applications, including communications, imaging, surveillance, and exploration. Hundreds of satellites currently orbit the earth, and many are visible in the sky on a clear night.

SATELLITE COMPONENTS

A satellite is a complex piece of electronic equipment (see Figure 10-2), but it can be broken down into four main functional components (see Figure 10-3):

- Antennas
- Solar panels
- Communications payload
- Control system

Figure 10-2

A satellite

Figure 10-3

Components of a satellite

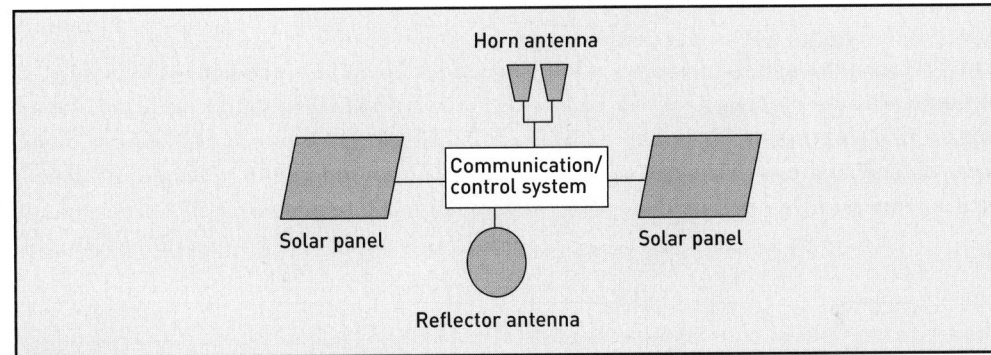

Satellite antennas come in many shapes and sizes, and are used for receiving and transmitting modulated radio waves that carry information (see Figure 10-4). These structures can be used for single-point transmission or for broadcasting to multiple users on the ground, such as in radio or TV broadcasting. Satellite antennas can be highly directional or can have a wide coverage area. Their appearance may resemble a dish (reflector antennas), a pole, a horn, or an array of smaller antennas. An antenna's size depends on the area of coverage as well as the wavelength of the electromagnetic wave used for communications.

Figure 10-4

Satellite antenna

Solar panels on the satellite consist of an array of transducers that convert sunlight into electricity, generating energy for the satellite. When the satellite is in the path of the sun's rays, its panels receive solar energy and convert them into electricity. The energy generated by these panels powers the various circuits of the satellite. Part of this electricity is also stored in rechargeable batteries that power the equipment when the sun's rays cannot reach the satellite's solar panels, due to its position. Although satellites carry fuel, it is only used to alter the satellite's location, such as when it is launched into orbit or when it has drifted from its designated orbit.

The communications system, sometimes called the communications payload, consists of various modulators and demodulators, transponders, encoders and decoders, multiplexers and demultiplexers, amplifiers, and other circuits. These components receive, process, and transmit information between the satellite and other stations. The communications system is divided into *uplink* and *downlink* components. Communications from the satellite to earth are the downlink, and the reverse communication is the uplink.

A satellite in orbit communicates with transmitters and receivers on land, the sea, or on other satellites. These terrestrial stations are commonly called *Earth stations* or *ground stations*. Some earth stations are mobile, some only receive broadcasts from the satellite, some only transmit to the satellite, and some only control the satellite. Although satellites are equipped with a complex list of autonomous functionality, they require the assistance of an Earth segment, called the *control station*, when they drift from their orbits and have to be repositioned via command signals.

The control system on the satellite constantly communicates with the control station to keep the satellite in good working condition. The control system sends information about the functionality and status of important components and the satellite's overall condition. The control station, in turn, sends appropriate signals to the satellite for troubleshooting. If the satellite drifts from its correct alignment due to various reasons, which include the gravitational forces of the moon, the sun, or the Earth, as well as other effects including *solar winds;* the signals from the ground station can help to realign the satellite.

SATELLITE ORBITS

A satellite orbit is the path the satellite follows while traveling around the Earth. Satellites are launched into space by rockets and positioned into a precise orbit that is either circular or elliptical. Once the satellite is positioned at a designated altitude above the Earth, it revolves around the Earth in this orbit. The altitude of the satellite usually classifies it into one of the following categories, although other categories exist:

- Low Earth orbit (**LEO**) satellite
- Medium Earth orbit (**MEO**) satellite
- Geostationary Earth orbit (**GEO**) satellite

A LEO satellite is positioned approximately hundreds of kilometers to a few thousand kilometers above the Earth's surface, and completes one revolution within its orbit about once every few hours or so, depending on altitude. The low altitude of these satellites makes them ideal for low-latency applications—in other words, applications that require the time it takes a signal to reach its destination to be very short. Such applications include voice communications as well as those for close observation of the Earth's surface and its atmosphere. LEO satellites are useful in low-latency applications because the closer a satellite is to the Earth, the less time it takes a signal to travel between the satellite and Earth transceivers. Signals from LEO satellites travel between the satellite and the Earth in a time that is roughly calculated by the following formula:

Time = Distance/speed

If we assume that the speed of EM waves is roughly 3×10^8 m/s, it is easy to calculate the time it takes for a signal to travel a given distance.

Because a satellite's maximum field of view, or area of coverage, is a function of its altitude, LEO satellites tend to cover only a small portion of the Earth's surface. To provide global coverage with LEO satellites, a large number are launched into low Earth orbit (see Figure 10-5). Because these satellites complete one revolution of the earth in a short amount of time, their area of coverage constantly changes. Earth-receiving stations often have to track the satellites' motion and reposition their antennas to track satellite signals. In addition, satellite signals have to be handed over constantly from one satellite to another, as one disappears from view and another comes into view. Nonetheless, many of the satellites in orbit are LEOs, because they are cheaper to launch. Launching cost is a strong function of orbital altitude.

Figure 10-5

Global coverage using
multiple satellites

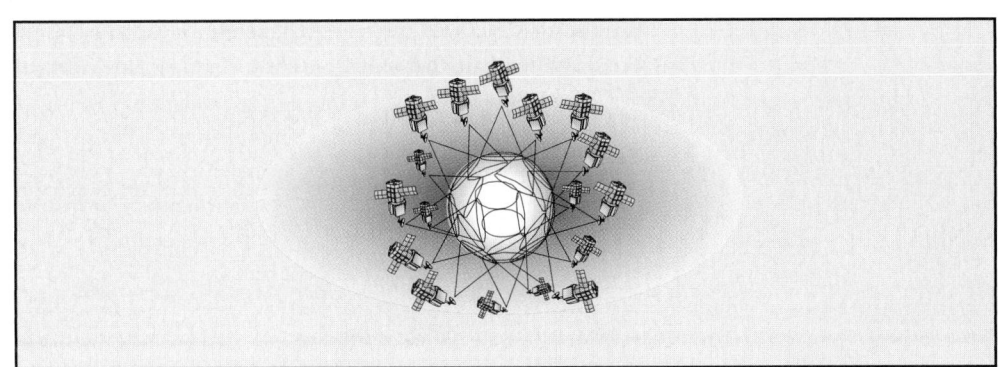

To achieve greater coverage, satellites can be positioned in higher orbit. For example, a satellite in a medium Earth orbit operates at an altitude of tens of thousands of kilometers above the Earth's surface. These satellites complete one revolution every couple of hours, so fewer of them are required to provide global coverage than LEO satellites. Because these satellites are in a higher orbit than LEOs, their latency is also higher. MEO satellites are commonly used for telecommunications, monitoring the weather, and navigation. For example, the global positioning system uses satellites in MEO orbit.

GEO satellites reside in an orbital altitude of approximately 36,000 km directly above the equator (see Figure 10-6). Because GEO satellites complete a single revolution in 24 hours, they appear to be stationary with respect to the Earth, which also completes a single rotation around its axis once every 24 hours. This is why the satellites are called geostationary.[2] GEO satellites have the largest area of coverage; three positioned in various orbital inclinations are enough to cover almost the entire surface of the Earth. Their high altitude results in a larger latency for real-time voice communications, because it takes a signal significantly longer to travel from Earth to the satellite and back, compared with signals to and from LEO or MEO satellites. GEO satellites are commonly used for telecommunications and broadcasting television and radio signals because Earth antennas can be fixed and do not need to be repositioned to track satellite signals. GEO satellites are also used for imaging cloud cover above the Earth's surface, for meteorological research, and to transmit signals between space probes and Earth stations.

Figure 10-6

GEO satellites positioned
above the equator

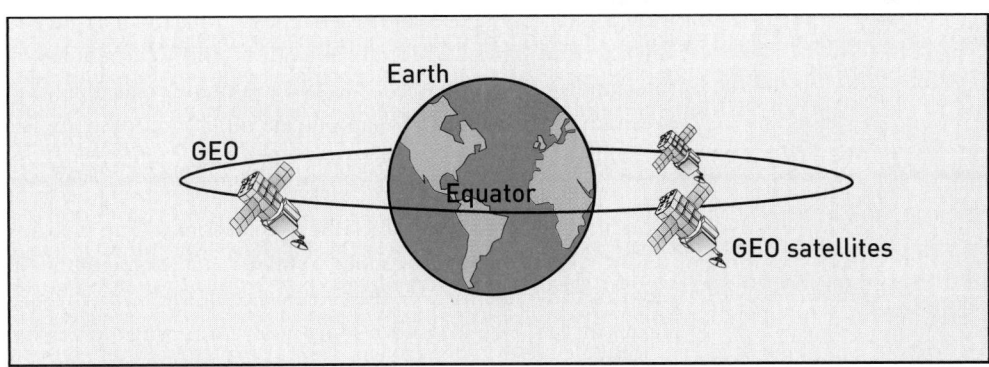

Problem—Calculate the time it takes for a radio frequency signal to travel between the Earth and a LEO satellite at an altitude of 1000 km from the Earth's surface.

1000 km = 1,000,000 m = 1×10^6 m
Speed of EM waves in vacuum = 3×10^8 m/s
Time = Distance/speed = $1 \times 10^6 / 3 \times 10^8 \approx 0.0033$ s = 3.3 ms

---

[2] Note that some GEO satellites are not geostationary but are geosynchronous. They complete one revolution around the Earth in 24 hours, but because they do not reside directly above the equator, they do not appear to be stationary with respect to the Earth.

Problem—Calculate the time it takes for a radio frequency signal to travel round-trip between the earth and a GEO satellite at an altitude of approximately 36,000 km.

36,000 km = 36,000,000 m = 36×10^6 m
Speed of EM waves in vacuum = 3×10^8 m/s
Time for one-way trip = Distance/speed = $36 \times 10^6 / 3 \times 10^8$ = 0.12 s = 120 ms
Round-trip time = 120 ms $\times$ 2 = 240 ms

EFFECTS OF THE ATMOSPHERIC CHANNEL ON RADIO WAVES

Systems that rely on radio waves over an open channel are extremely vulnerable to various effects, including interference from other radio sources; signal attenuation due to atmospheric effects, such as cloud cover, rain, ice, and reflections from buildings and other reflective surfaces. Much of the noise can be reduced, but it cannot be completely eliminated. Careful placement of receiver and transmitter stations can reduce reflections, and powerful amplifiers can amplify attenuated signals. Electronic filters weed out unwanted interference from other sources. Although such measures can increase reliability, unwanted effects cannot be eliminated completely, resulting in reduced quality.

Interestingly, not all radio wavelengths encounter the same effects in the atmosphere. Some wavelengths are more susceptible to its detrimental effects than others. For instance, high-frequency radio waves that have short wavelengths are more susceptible to **absorption** and **scattering**, which arise due to water vapor and other atmospheric particles, than radio waves with low frequency and large wavelengths. Absorption is the effect whereby the radio wave loses its energy due to materials in the atmosphere that absorb the wave's energy. Scattering occurs when a radio wave encounters an object that is close to the dimensions of its wavelength and the wave's energy is scattered in different directions.

THE GLOBAL POSITIONING SYSTEM (GPS)

GPS is used to provide positional information in three-dimensional space (latitude, longitude, and elevation) through a GPS receiver and a network of satellites. It was initially intended for military applications, but later became available to civilians. The system is currently controlled by the Department of Defense in the United States.

GPS receivers can be installed in cars, boats, and aircraft, and can also come in handheld form for individual use. The system is used for several applications, including navigation, surveying, time stamping, synchronization, and fleet management. By obtaining positional information, GPS provides navigation in many areas, including aviation, maritime, driving, and hiking. Surveying and mapping are other important applications of GPS. By scanning an area with a GPS receiver, you can create a map to determine topographical information. GPS is also used in farming to track large areas for crop planting and harvesting, in commercial containers and vehicles to track their whereabouts, and in police work to track stolen cars.

The GPS is a network of about 24 MEO satellites and numerous earth stations that constantly monitor these satellites. As the satellites orbit the earth, they periodically send out distinctive signals that the GPS receivers can detect. By processing these signals from three or more satellites, any GPS receiver can calculate its location in three dimensions. The basis of GPS is called **triangulation** (sometimes also called trilateration). Positional information can be obtained by means of triangulation, if the distance from at least three reference points is known.

Consider an airplane that is using GPS for navigation. First, the GPS receiver on the airplane must know its distance from three or more satellites. Assume that the airplane knows its distance from one of the satellites, and that this distance is x km. This means that the airplane is located on the surface of a sphere with a radius of x km and the satellite at the center, as in Figure 10-7. If the receiver also knows its distance from a second satellite, then it is located on another sphere with a radius of y km and the second satellite at the center. The resulting circle formed from the intersection of the two spheres would be the possible locations of the airplane (denoted in dotted lines in the figure). If yet a third distance measure is known—say, a radius of z km from a third satellite—then the number is narrowed down to two distinct points on the circle (denoted by the red dots in the figure). One of these points can usually be disregarded, because it is generally in an unfeasible location, such as near the center of the Earth, and the possible number of positions can be narrowed down to just one. In reality, however, four or more positions are used to pinpoint the location of objects. Geometry can then be used to deduce the position of the airplane, if the exact location of the three satellites with respect to the earth is known when the distance measurements are taken by the receiver.

In theory, the distance between the GPS receiver and at least three satellites must be known by the GPS receiver. However, this is not enough for the GPS receiver to derive its coordinates. The GPS receiver must also know the coordinates of at least three of the satellites when the measurements are taken.

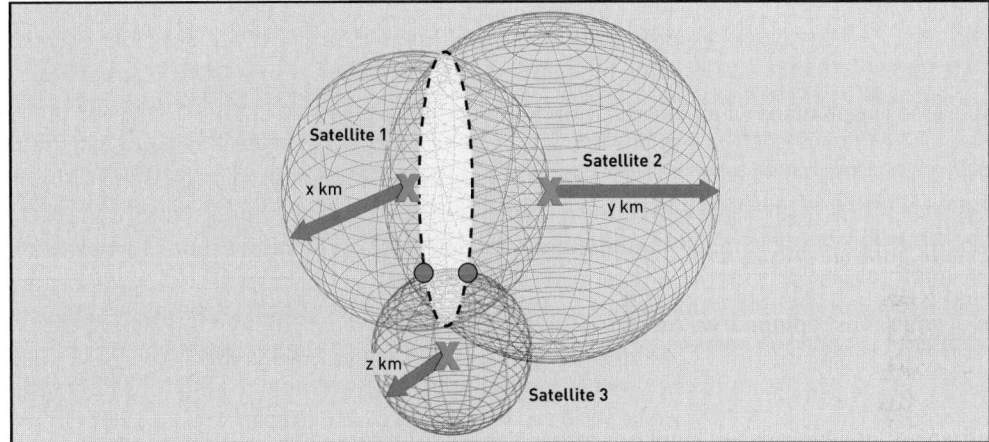

The distance from each satellite and the airplane is measured by generating a unique bit pattern for each satellite and synchronizing it with the GPS receiver. The pattern generated by the satellite is transmitted via the downlink, and just as the pattern is generated, the GPS receiver starts a timer. When the satellite signal is received, the timer is stopped and the delay between the generated bit pattern and the pattern from the satellite is recorded. This duration translates into the delay, or latency, encountered by the satellite signal to reach the receiver. Because the speed of electromagnetic waves within a vacuum is known (3×10^8 m/s), the distance between each satellite can be calculated using the formula presented earlier in this chapter. This procedure is implemented for all satellites, and the GPS receiver calculates distance information from each satellite. The next step is to use the positional information of the satellites to calculate the position of the receiver—in other words, the location of the airplane. Because satellites have predetermined orbits, their exact location at particular times can be stored in the GPS receiver beforehand or periodically transmitted to the receiver. Using these two pieces of information, the receiver can calculate the latitude, longitude, and elevation of the aircraft with respect to the Earth. This information is consequently used to navigate the aircraft to its correct destination.

The preceding explanation tries to simplify how the GPS system works. In reality, GPS is a highly complex system involving several other technical details that are beyond the scope of this textbook.

One important aspect of GPS is synchronizing satellite clocks with the clocks on the GPS receivers. Because satellites come equipped with *atomic clocks*, the synchronization between satellites is nearly perfect. However, because GPS receivers are not equipped with such precise and expensive clocks, a relatively simple method is needed to account for the offset in time between the satellites and the receivers. The clocks must be synchronized because the three spheres would otherwise not intersect. This situation would cause the GPS receiver to give erroneous positional information by several kilometers, rendering the system practically useless. To overcome this risk, more satellite measurements are taken into account, enabling the receiver clock to be synchronized with the atomic clocks of the satellites.

Although many satellite measurements are necessary for accurate estimates, it still might not produce exact results because of other effects acting on the system. For example, the speed of an electromagnetic wave in a vacuum is different from its speed when it penetrates the different layers of the atmosphere. This introduces errors into the distance measurements, which uses the speed of electromagnetic waves in a vacuum. However, these and other related errors can be virtually eliminated by using more than four satellites as well as differential GPS (**DGPS**), which enables almost perfect calculations by including another measurement taken by designated Earth stations acting as satellites. Furthermore, a recent approach called **wide area augmentation system** (WAAS) is used to provide better precision to aircraft and civilian users. WAAS employs a combination of ground-based stations and satellites rotating directly above the equator to provide for better accuracy in positioning.[3]

SATELLITE INTERNET

Broadband terrestrial Internet service is not available everywhere in the world. Even when cable modems or DSL are offered as competing alternatives, some people still prefer to use other means of connecting to the Internet, such as broadband satellite, due to the mobility it offers. In many circumstances, such as for ships at sea, satellite connection is the only alternative. The setup for a **satellite Internet** connection is very simple: it requires a subscription to a satellite Internet service; a small receiving/transmitting terminal, such as a VSAT; and a **satmodem**, the modem used for communicating directly with the satellite. The VSAT transmits and receives RF signals from the satellite, which then transponds the message to an Earth hub station that handles all the traffic in and out of the satellite. Satellite Internet subscriptions are generally offered as two alternatives. Some services offer only one-way access to the Internet (downloading only), and some support both downloads and uploads. Businesses might prefer a two-way service to upload data to a remote server, whereas individual users might prefer a one-way service. However, two-way connection is typically asymmetrical, and downloads are generally faster than uploads.

---

[3] *http://gps.faa.gov.*

CHAPTER SUMMARY

- Radio-wave communication provides mobility and flexibility in a wide range of applications, including radio and TV broadcasting, national defense and emergency services, aviation and maritime communications, computer and telephone networks, navigation, space exploration, remote sensing, radio frequency identification, and personal and commercial applications.
- Satellites receive radio frequency signals, process them, and relay them to their destination. A satellite's major components include antennas, solar panels, a communications payload, and a control system.
- Satellites may use a low Earth orbit (LEO), medium Earth orbit (MEO), or geostationary Earth orbit, among others. LEO satellites are closer to the surface of the Earth than MEOs and GEOs but have a smaller area of coverage. They complete a revolution of the Earth in much less time than MEOs and GEOs. GEOs revolve around the Earth in 24 hours, directly above the equator, which makes them appear stationary when seen from Earth.
- The atmosphere has several detrimental effects on radio waves, including absorption and scattering. These effects attenuate the strength of the waves, limiting the extent to which they can travel and their maximum data rate.
- GPS is used for a variety of applications ranging from navigation to surveying. The main principle of GPS is called triangulation, whereby multiple satellites are used to triangulate a position in three-dimensional space. Differential GPS is more accurate than regular GPS.
- Satellite Internet services are used primarily by customers who do not have access to a landline Internet connection, such as a cable modem or DSL.

KEY TERMS

absorption	RADAR
active tag	RFID tag
Bluetooth	SAR
deep space network	satellite Internet
DGPS	satellite radio
digital audio broadcasting	satellite television
digital television	satmodem
GEO	scattering
GPS	solar panel
ham radio	Sputnik
high-definition radio	triangulation
high-definition television	VSAT
Inmarsat	wide area augmentation system
LEO	Wi-Fi
MEO	WiMAX
passive tag	

REVIEW QUESTIONS

1. What is DAB, and what advantages does it have over conventional AM and FM broadcasting?
2. What is the advantage of subscribing to a satellite radio service?
3. List two applications of GPS.
4. Describe the function of a satellite antenna.
5. What is meant by the terms *uplink* and *downlink*?

6. Explain the reason for using satellites to communicate between Africa and Europe.

7. Explain why a satellite might occasionally shift from its orbit. Also, how can the satellite be repositioned into its orbit?

8. What purpose do a satellite's solar panels serve?

9. List the three main satellite orbits in order of decreasing altitude.

10. List the three main satellite orbits in order of increasing area of coverage.

11. Calculate how long it takes for an Earth station to receive a signal transmitted from a satellite that is 42,000 km above the Earth.

12. In what type of orbit would a satellite be positioned at an altitude of 12,000 km above the Earth?

13. How many GEO satellites would provide almost global coverage?

14. Why are there more LEO satellites than GEOs?

15. What techniques can counteract noise due to channel effects in wireless communications systems?

16. Create a list of the types of information that GPS receivers need to calculate a position in three dimensions.

17. Calculate the distance between a GPS receiver and a satellite if it took 30 ms for a signal to arrive from a satellite to the receiver.

18. In what type of orbit would a satellite be positioned if there was a 40-ms delay between the satellite and an Earth station directly below it?

19. Explain the reason for using as many satellite measurements as possible in GPS applications.

20. Explain why a distance measurement taken between an Earth station and a satellite, as with DGPS, would reduce timing errors.

21. What is RFID, and what sorts of applications does it have?

22. Explain why a satellite Internet subscription is sometimes preferred over other means of online access.

DISCUSSION QUESTIONS

1. Think of some ideas for using RFID besides those you studied in this chapter. From a social, political, and economic standpoint, make arguments for and against the use of RFIDs in human beings.

2. Why do you think civilian GPS is less accurate than military GPS? What types of events could disable the GPS system worldwide?

3. Do research on the Internet to determine the average lifetime of a satellite. What limits its lifetime? How can its lifetime be extended? What happens to a satellite once it is no longer useful?

4. Discuss some reasons that LEO or GEO satellites are not preferred over MEO satellites for GPS. Would more satellites be required using LEO satellites with GPS?

5. Visit www.heavens-above.com to obtain information about the various satellites that are currently observing your hometown. Write down the names and coordinates of at least three satellites.

6. Visit http://science.nasa.gov/realtime/GIFTrack/giftrack_frames /frameset.htm to obtain information about the altitude and coordinates of the Hubble space telescope at the time you visit the site.

CASE PROJECT

Do some research on the Internet to identify service providers that offer Internet access via satellite. What type of equipment is required to access the Internet via a satellite from home? How much would it cost to set up the service? What upload and download speeds are typically offered by service providers? Which frequencies are used for uploading and which are used for downloading? Do you think weather conditions affect upload and download speeds?

PART

5

INTRODUCTION TO COMPUTER NETWORKING

CHAPTER 11

LOCAL AREA NETWORKS

LEARNING OBJECTIVES

In this chapter you will:

- Gain familiarity with the most popular types of local area networks (LANs), with a focus on Ethernet

- Understand LAN design characteristics, including topology, access mechanisms, physical transmission media, and equipment

- Explain what a frame format is and what purpose it serves in LANs

- Define the functions of LAN operating systems

- Understand the technical architecture of Wi-Fi

INTRODUCTION

The ability to easily share information over a network has helped to define modern computing. A network is a set of hardware and software designed to interconnect multiple devices for processing, storing, and exchanging information. In the IT industry, networks are usually categorized by geographical scope into local area networks (LANs) and wide area networks (WANs). However, many other network categories exist, including personal area networks (PANs), metropolitan area networks (MANs), campus area networks (CANs), and storage area networks (SANs). This chapter introduces LANs, including their underlying logical and physical topologies, access control mechanisms, physical media and equipment, and LAN operating systems. The chapter details the evolution and characteristics of the most popular type of LAN—Ethernet—and explains how wireless local area networks (WLANs) work.

There are many arbitrary ways to categorize networks. Networking industry classifications have historical roots in the evolution of network technologies and are like any classificatory system: they divide knowledge into manageable units for people to address, control, and understand. To illustrate this phenomenon, consider the question of how many oceans are on earth. The answer depends on whom you ask and in which country they reside. At a minimum, we conventionally divide oceans into the Pacific, the Atlantic, the Arctic, and the Indian. These labels serve to make the topic easier to discuss, but they are scientifically a misnomer because there is only one completely interconnected ocean on earth, as any satellite photo illustrates. Categorizing networks is analogous; while a network is one interconnected entity, classifications help us to understand, discuss, and design technical architecture.

The IT industry almost always distinguishes between a **LAN**—a network that spans a confined geographical distance, such as a building or home—and a **WAN**, which spans a larger geographical area, such as a city, nation, or the world at large. Geographically expansive WANs, especially the Internet, have become synonymous with networking, but a sizable percentage of networked communications occurs within a corporation or other limited area.

A computer user in a professional office requires shared access to a variety of local resources, including printers, electronic storage, database management systems, electronic mail servers, shared firewalls regulating Internet access, or any number of internal Web server applications, as shown in Figure 11-1. These services can reside anywhere on the network, but they are often located in the same building as the users that access them. A network performs the critical function of providing users with high-speed, controlled access to its resources.

Figure 11-1

Resources accessed via a LAN

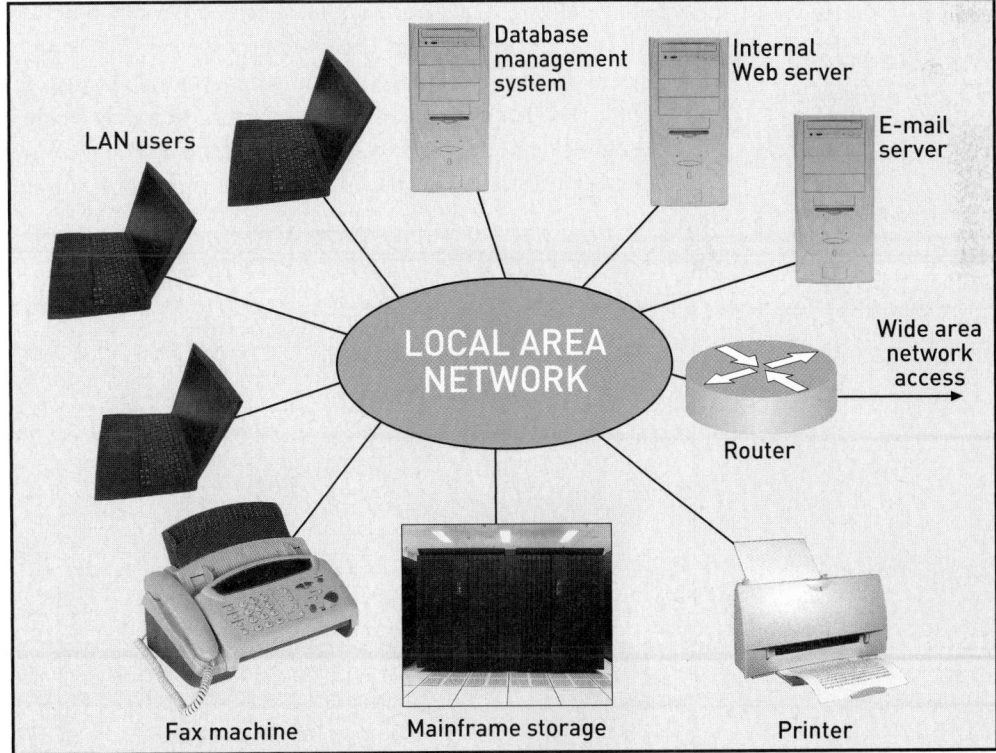

LANs can support between two and more than 1000 devices. Some LANs are privately owned and used by individuals, businesses, universities, or other private organizations, while other LANs are publicly available for access via the Internet. LAN users are typically employees or other authorized clients who use computers in a private enterprise, or home users and mobile users who rely on wireless LAN access to the Internet. For example, users often establish small LANs in their homes to interconnect multiple computers to a printer, a shared file server, and an Internet access device such as a cable modem. Private, home LANs are usually wireless or operate over high-grade twisted pair or fiber-optic cable that is hard-wired in the house.

Public LAN services provide shared Internet access to the public. Such services include high-speed wireless LANs on university campuses, in coffee shops, in Internet cafes, or in neighborhoods.

The following sections describe technical characteristics that historically have defined and differentiated types of LANs.

LAN DESIGN CHARACTERISTICS

Understanding the individual types of LANs, such as switched Ethernet and Wi-Fi, requires some background in network topology, access control, physical media and equipment, and LAN operating systems. The following section describes these design characteristics.

PHYSICAL AND LOGICAL TOPOLOGY

One distinguishing characteristic of a LAN is its **topology**, or configuration. First, it is important to distinguish between logical topology and physical topology. A LAN's **physical topology** refers to how the multiple devices, often called network **nodes**, are physically connected to each other. Many LANs connect devices in a **star** configuration, with cables or wireless links connecting each computer to a central wiring hub or LAN switch, as illustrated in Figure 11-2. For example, the LAN switch or hub may reside in a wiring closet of an office building, while nodes in offices throughout the floor are connected to the central switch via single cable runs called **drops**. These cable runs resemble the spokes of a wheel.

Figure 11-2

Star topology

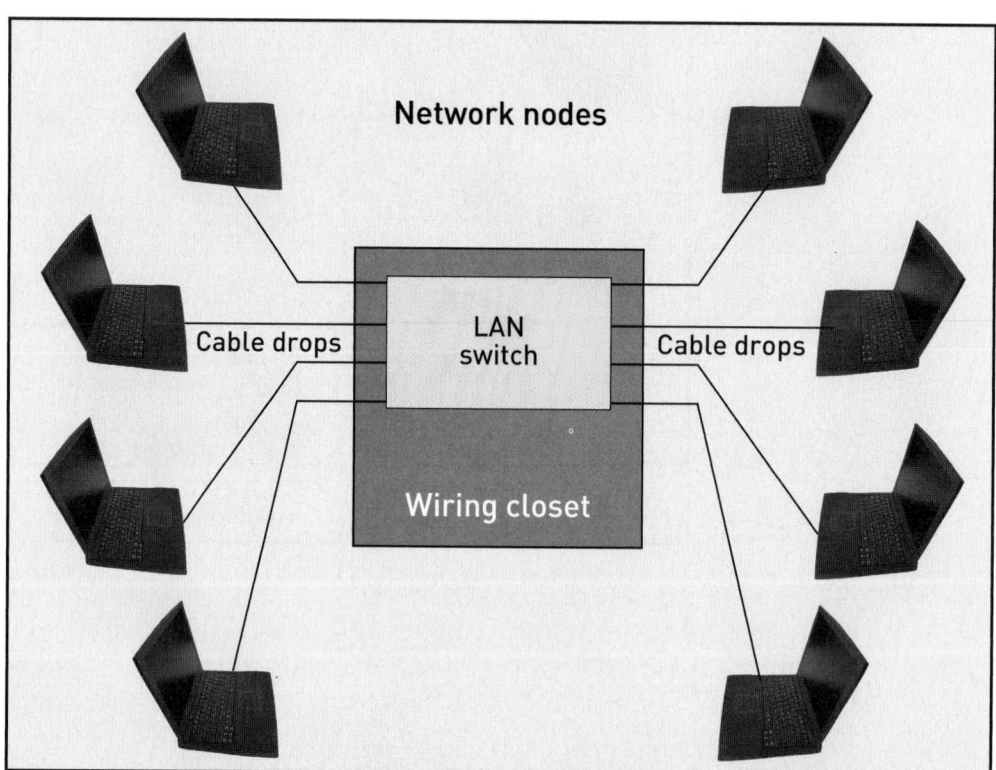

One advantage of a star configuration is that if one cable connection fails, the other connections continue operating. If a LAN was physically wired as a ring, in which cables run between subsequent devices rather than to a central hub, a failure in any cable connection would adversely affect the entire network. Of course, one disadvantage of a star configuration is that if the LAN hub or switch fails, service to all the nodes is disrupted.

While the physical topology describes how network nodes are connected within the LAN, the **logical topology** dictates how information flows among the nodes. Common logical topologies for LANs are the **bus**, the **ring**, and the star. In a logical ring topology, shown in Figure 11-3, data flows sequentially from one computer to the next in an orderly pattern around the ring.

Figure 11-3

Ring topology

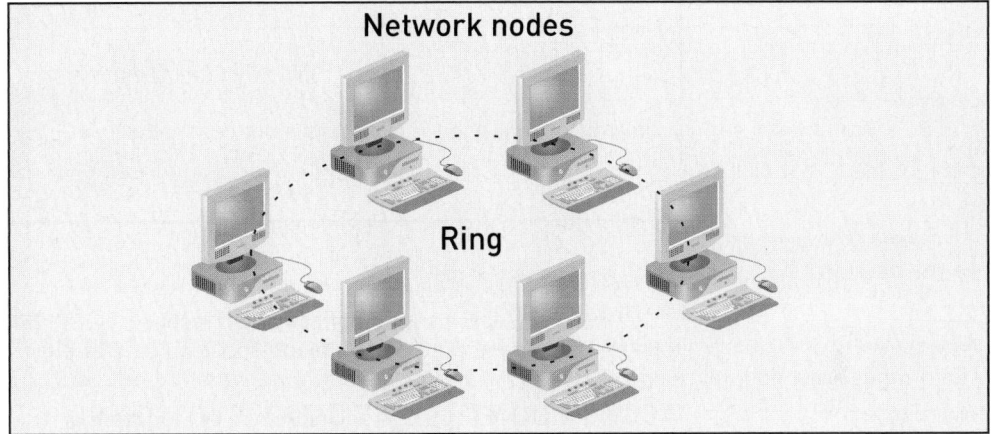

If the LAN is physically wired as a star but uses a logical ring topology, information still flows from node to node, originating at one node, passing through the central hub, and then going to the next sequential node on the ring. This type of LAN configuration is often called a **star-wired ring** (see Figure 11-4).

Figure 11-4

Star-wired ring

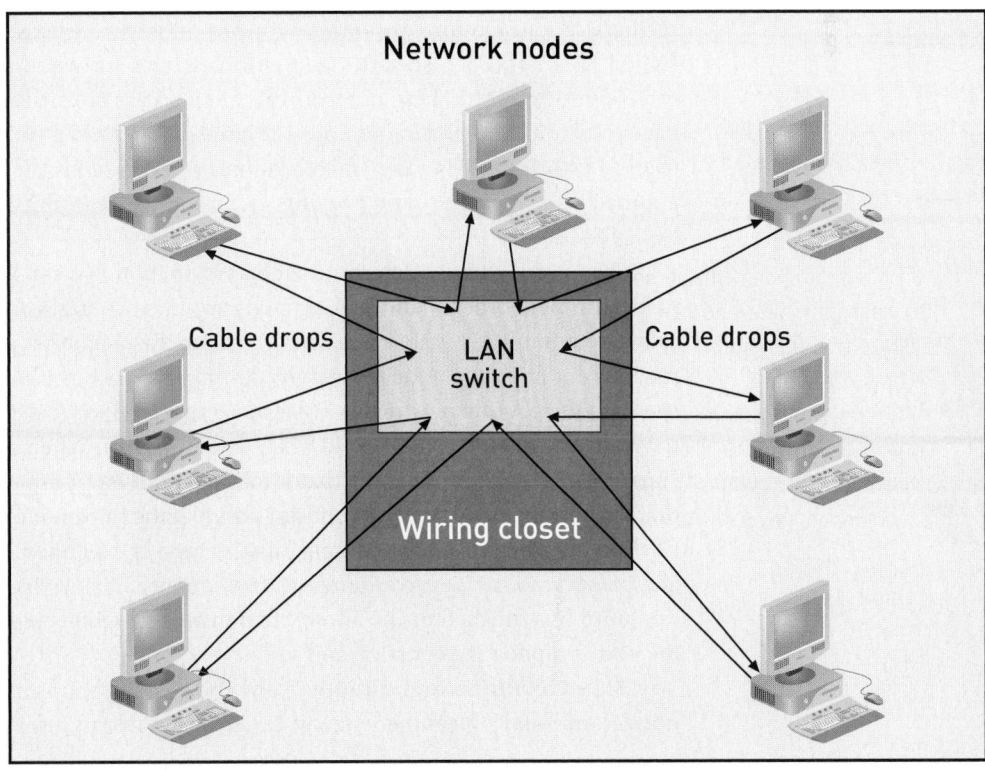

A bus topology interconnects all computing nodes via a continuous cable, as shown in Figure 11-5. In a logical bus configuration, information is transmitted simultaneously to all networked devices, rather than flowing in an orderly sequence from node to node.

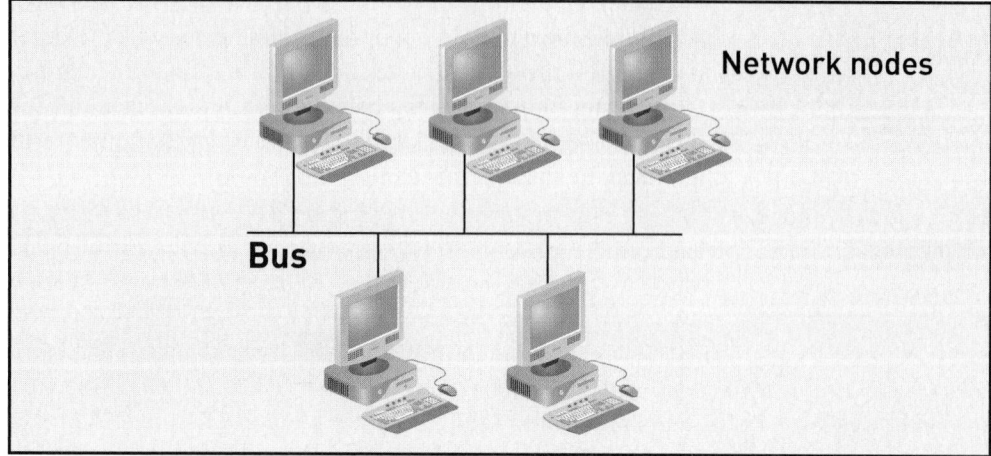

Finally, in a logical star topology, signals flow from a computer to a centralized node and then to another computer in a hub-and-spoke pattern, as illustrated in Figure 11-2.

ACCESS CONTROL AND LAN SWITCHING

In some ways, computers on a network communicate with each other like human beings. If 20 people at a party talked at the same time, the result would be an incomprehensible cacophony of sound. Similarly, with so many devices concurrently transmitting or receiving information in a LAN, signals can collide unless they have proper access rules or switching configurations. Understanding how nodes access and transmit data requires knowing the difference between shared LAN approaches and the more modern LAN switching approach.

Historically, devices on a LAN have shared a common segment of the LAN's transmission medium. Because numerous devices concurrently try to access the same network segment, a shared LAN approach specifies and enforces rules for when each device may transmit or receive information. These procedures, called **access control mechanisms**, seek to prevent data collisions and ensure an access response time that seems instantaneous to computer users. Historically, the two most common access control procedures for shared LAN approaches have been **token passing** and carrier sense multiple access with collision detection (**CSMA/CD**).

Token passing uses a **deterministic** access method, whereby each network node is given a predetermined, orderly, and sequential opportunity to transmit information. The method is called deterministic because you can determine the maximum amount of time it takes information sent from any node to reach its destination. Each node communicates only during its specified turn. A **token**, which is just a signal that comprises a specific bit pattern, is transmitted from one device to the next in a sequential pattern. When a device receives this signal, it "possesses the token," meaning that it may transmit its information over the network. When the transmitting node receives confirmation that the information successfully reached its destination, or when its allocated transmission time is expended, the device releases the token and passes it to the next computer in the sequence. The device may then transmit information until it is finished or the allowable transmission time elapses. The token then passes to the next computer in sequence, and so on.

CSMA/CD differs significantly from the token passing access method. CSMA/CD is **nondeterministic**, meaning that any device may transmit information at any given moment, provided that no signals are already being transmitted over the LAN. The method is called nondeterministic because you cannot determine beforehand how long data may take to reach its destination. In LANs that use CSMA/CD, a node preparing to transmit information first "listens" to the network to determine whether transmissions are occurring. The node starts to transmit only if it detects that the network is free of transmissions. With CSMA/CD, information is transmitted simultaneously to all nodes of the network. Unlike token passing, the devices do not require possession of a token to transmit.

What if two devices simultaneously determine that the LAN is available and both begin to transmit information? CSMA/CD includes a "collision detection" feature to handle this problem. Devices can detect collisions on the LAN, primarily through variations in detected signal energy levels, and retransmit the information after a random period of time. To forestall a possible second collision from two devices waiting the same interval of time and then retransmitting, each device has a unique waiting time before retransmission.

An access mechanism that is closely related to CSMA/CD is carrier sense multiple access with collision avoidance (**CSMA/CA**). In some LANs, such as wireless LANs, collisions cannot be detected. Using CSMA/CA, a node that has information to transmit first listens to the network to see if another node is transmitting. If the node finds that the network is idle, it sends a "request to send" (RTS) packet to other nodes. The other nodes can then send a "clear to send" (CTS) packet, alerting the transmitting node that it is free to send data.

If the transmitting node detects that the network is not idle, it waits a brief time before starting the procedure again.

Assuming that a LAN uses the CSMA/CD access method, what happens as more and more devices are attached to the network? Using CSMA/CD, devices on a common LAN transmit only if the network is clear. Collisions occur when two or more nodes transmit simultaneously. As more nodes are added to the network, collisions will occur so frequently that network performance will degrade. Most commonly, the network will experience high **latency**, meaning that information takes longer to transmit to its destination, like a car on a congested highway. Regardless of how much bandwidth the network may have, latency becomes a problem in a shared CSMA/CD network if it has too many networked devices.

Because of this problem, switched LANs are preferable to shared LANs. LAN switches segment a network into pieces (see Figure 11-6). Rather than each computer broadcasting a message to all nodes on a network, a computer only transmits to a centralized LAN switch, which detects the destination of the data and sends it directly to the associated device(s). As a result, each device connected to the LAN switch enjoys full network bandwidth rather than sharing it with other devices. LAN switching is especially helpful for connecting servers and other frequently accessed computers. Networks sometimes use LAN switches exclusively, providing full bandwidth to each device, and sometimes use a combination of LAN switching and shared network segments, as shown in Figure 11-7.

LAN PHYSICAL MEDIA AND EQUIPMENT

A LAN contains a range of hardware and physical media, including a network interface controller, transmission media, wiring hubs, high-speed switches, routers, and servers.

To connect to a LAN, a computer must have a **network interface controller** (NIC), which is usually built into the computer's motherboard. A NIC is basically a small card (printed circuit board) with a couple of integrated circuits and other electronic components. The NIC provides the physical interface to a network medium or wireless LAN and supports an addressing system that is critical to the LAN's operation. Each NIC, which is sometimes called a network adapter or network card, has a unique physical address permanently assigned to it under one of several address spaces. These unique addresses are similar to a mailing address, and serve the same purpose: they help information reach its destination.

The address spaces are known as the MAC-48, EUI-48, and EUI-64, where **MAC** stands for Media Access Control and **EUI** stands for extended unique identifier. For example, Ethernet,

Figure 11-6

Gigabit LAN switches

Figure 11-7

LAN switching

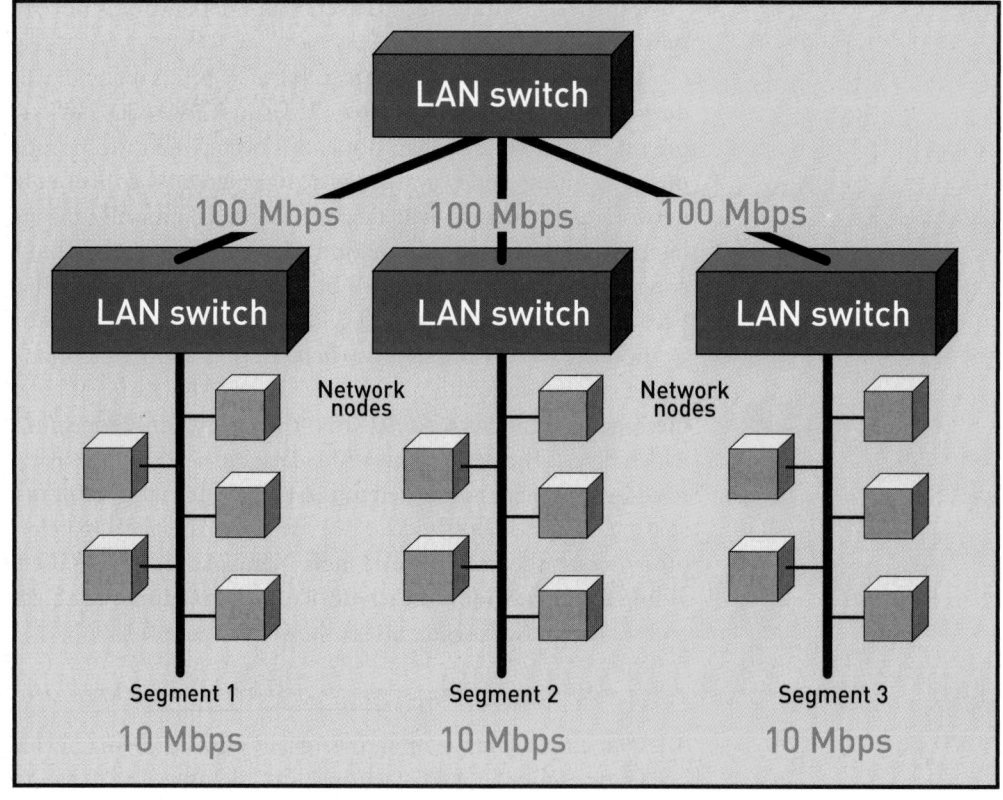

the most popular type of LAN, has long used Ethernet cards with 6-byte, 48-bit MAC addresses. The first three bytes are called an organizationally unique identifier (OUI), and are assigned to a specific NIC manufacturer. The last three bytes are a unique number assigned by the manufacturer to individual NIC devices. The 48-bit address is usually written in hexadecimal to make it easier for people to read. Recall that the hexadecimal numbering system uses

16 characters (0, 1, 2, 3, 4, 5, 6, 7, 8, 9, A, B, C, D, E, F) and that each character can conveniently act as a shorthand for four binary digits, as follows:

Hex	Binary	Hex	Binary
0	0000	8	1000
1	0001	9	1001
2	0010	A	1010
3	0011	B	1011
4	0100	C	1100
5	0101	D	1101
6	0110	E	1110
7	0111	F	1111

Twelve hexadecimal characters can act as shorthand for a 48-bit Ethernet address.

Problem—Convert a 48-bit Ethernet address into a human-readable hex address.

The 48-bit Ethernet address is 000000010101000110101011110011001010111101100001.

The address is usually written in hexadecimal by dividing it into groups of four bits and converting each group into its hexadecimal equivalent.
The hexadecimal equivalent is 0151ABCCAF61.

The purpose of hexadecimal notation is to make the address easier and shorter for people to read and process. Digital technologies only read ones and zeros, but IT professionals and users find it easier to read hex.

Because each 48-bit address must be unique, the Institute of Electrical and Electronics Engineers (IEEE) acts as a standards body and allocates a unique identifier to each manufacturer. The NIC manufacturer then assigns the last three bytes to each device it produces.

In addition to addressing, the NIC converts information provided by the computer into a format that the LAN can transmit and that another network device can read. To format this information, the NIC usually breaks it into small groups of bits, called **frames**, and includes such data as a source and destination address and a variety of control parameters. Frames are explained in more detail later in this chapter.

Information formatted by a NIC is then sent over a physical transmission medium. The most common transmission media used by LANs are free space, fiber-optic cable, twisted pair, or some combination of these. For example, many LANs connect devices using a combination of twisted pair cable and fiber-optic cable. On a floor of an office building, twisted pair cable often extends horizontally from a wiring closet to each office, while higher-speed fiber-optic cable (called riser cable) extends vertically between floors to interconnect equipment in wiring closets.

Recall that twisted pair cable consists of two plastic-coated copper wires twisted around each other. The quality of twisted pair varies according to the number of twists per foot of cable. The greater the number of twists, the more the cable is shielded from interference. However, a high number of twists also increases attenuation. The varieties of twisted pair that are most widely used in LANs today are Category 5 (CAT 5) or Category 6 (CAT 6) cable. These "categories" are actually common specifications developed by standards institutions to describe features such as wire gauge.

Fiber-optic cable, which consists of extremely pure strands of glass or plastic that transmit light, provides extremely high bandwidth and is used for LANs operating at speeds of gigabits per second and higher. Fiber-optic cable is also advantageous in environments that are subject to high levels of electromagnetic interference (EMI): fiber, which transmits light, is not affected by EMI. Fiber is also more difficult to tap into, and thus is desirable in high-security implementations.

The other common transmission medium for LANs is simply free space. WLANs use radio frequency communications rather than fiber-optic cable or copper cable.

LAN OPERATING SYSTEMS

Another important LAN component is software. An operating system that provides network services over a LAN is sometimes called a network operating system (NOS) or a server operating system, depending on the functionality provided. Some computer operating systems, such as UNIX, Linux, and Windows NT (discussed in Chapter 5), have built-in networking features, while other software, such as Novell's Netware, has specifically been designed as a **LAN operating system**.

One important function of a NOS is to manage and control networked access to LAN resources such as printers, files, applications, and messaging services. Network operating systems also provide security by managing user directories, monitoring remote LAN access, and incorporating encryption and other security features. They also provide network management, including diagnostic tools that monitor and analyze network traffic across the LAN and to and from important servers.

THE EVOLUTION OF LAN TYPES

Two devices on a network cannot exchange information without common standards that specify how data should be formatted, segmented, and transmitted. Several types of LANs have risen to prominence throughout the years, but **Ethernet** (both wired and wireless) has become the most popular by far. Other types of LANs have included **token ring** and Fiber Distributed Data Interface (**FDDI**). LAN standards differ based on numerous characteristics:

- Shared or switched access method
- Topology (ring, bus, star)
- Medium (twisted pair, free space, fiber)
- Speeds (10 Mbps, 16 Mbps, 100 Mbps, 1 Gbps, and higher)
- Distance
- Cost
- Performance
- Mobility
- Manageability
- Number of devices supported
- Frame format

ETHERNET LANS

Ethernet, a widely implemented LAN standard, was originally developed in 1976 by Bob Metcalfe at Xerox's Palo Alto Research Center (PARC). The IEEE eventually assumed responsibility for establishing Ethernet standards; one of the first became known as the **IEEE 802.3 standard**. The original transmission medium that supported Ethernet was coaxial cable, but now it is commonly implemented over free space, twisted pair cable, or fiber-optic cable.

Shared wired Ethernet LANs use the nondeterministic access method, CSMA/CD, and operate at speeds in the Gbps range and higher over a bus logical topology. Wireless Ethernet LANs employ CSMA/CA and operate at speeds ranging from a couple of Mbps to 100 Mbps. Many wired Ethernet LANs use **Ethernet switching**, described earlier as LAN switching, which divides the LAN into small segments. Figure 11-8 shows an Ethernet switch. Instead of having the limited capacity of a shared Ethernet LAN (10 Mbps or 100 Mbps), each device or group of networked devices is assigned a dedicated segment that is directly connected into the Ethernet switch, which then switches traffic to the receiving computer based on address.

Figure 11-8

10/100-Mbps switch

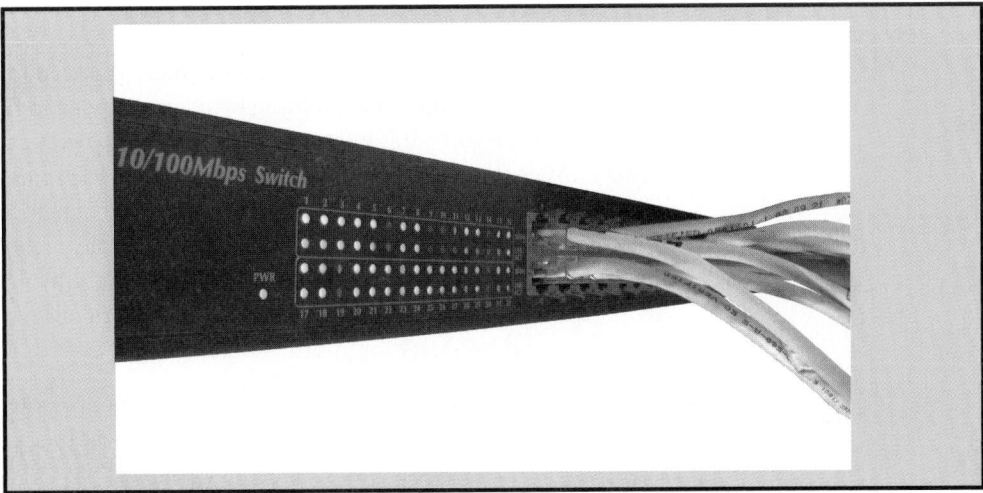

A major difference between LAN standards is how they format information for transmission. Information must be formatted before being transmitted across a network so that the images, numbers, and text in the data can be sent as "frames" or "packets." A frame is a string of ones and zeros that contains various components, as shown in Figure 11-9. Information is broken down into smaller groups and "packaged" with source and destination addresses and other components that are necessary for correct transmission. These packets are then reassembled at the receiver and presented to the user.

Figure 11-9

Ethernet frame format

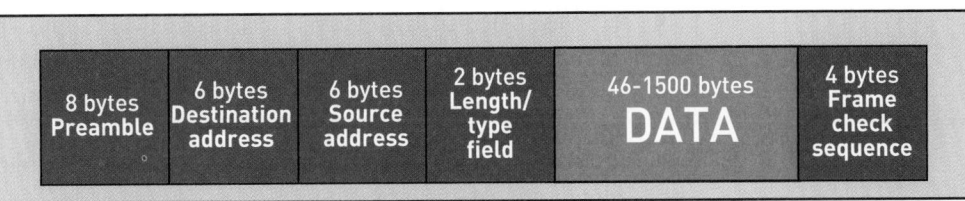

The standard Ethernet frame format includes the following components:

Preamble—The preamble is a series of eight bytes that identifies (or flags) the transmission as an Ethernet transmission, helps achieve synchronization between devices, and indicates the start of transmission. The Ethernet preamble is the byte 10101010 repeated seven times, followed by 10101011.

In its entirety, then, the Ethernet preamble is 10101010 10101010 10101010 10101010 10101010 10101010 10101010 10101011. Note that the last bit in the last byte of the preamble is different from the last bit in the other preamble bytes, to indicate that the following bit is the first one in the destination address.

Destination address—An Ethernet address is 48 bits, or 6 bytes long, and is the MAC address of the Ethernet frame's recipient.

Source address—The Ethernet source address is 48 bits, or 6 bytes long, and is physically associated with the network interface card of the transmitting device.

Length/type field—This field is a 2-byte designation with the important function of indicating the length of the forthcoming data payload. In some cases, the field also indicates the type of protocol used by the LAN.

Data—The actual information to transmit is a variable-length payload of at least 46 bytes and a maximum of 1500 bytes.

Frame check sequence (FCS)—The Ethernet frame concludes with a 4-byte frame check sequence designed to provide error control.

The information content of an Ethernet frame is expected to be at least 46 bytes long. If it is less than 46 bytes, additional **padding** bytes are included to fill out the frame. For example, if the information content is only 40 bytes long, then 6 bytes of padding are added.

Also notice the amount of **overhead** included in each frame. Overhead includes everything transmitted that is superfluous to the bytes of actual content. Each Ethernet frame contains 26 total bytes of overhead: the 8-byte preamble, the 6-byte destination address, the 6-byte source address, the 2-byte type field, and the 4-byte frame check sequence. Even though the overhead is extraneous to the actual information intended for transmission on the network, it is necessary for information exchange.

> **Problem**—If a computer has 140 bytes of information content to transmit via an Ethernet LAN, what percentage of the transmission is overhead?
>
> Total overhead of 26 bytes + total data payload of 140 bytes = 166 total bytes transmitted
>
> Percentage of overhead = 26/166 = 15.7%

The original forms of Ethernet were called Thicknet and Thinnet, but they are no longer used. They operated over coaxial cable and provided a maximum data rate of a few Mbps. Several varieties of Ethernet LANs are currently available; they have various speeds, media, and other characteristics:

- 10Base-T uses unshielded twisted pair (UTP) wiring and operates at 10 Mbps.
- Fast Ethernet operates at 100 Mbps over UTP or fiber.
- Gigabit Ethernet reaches speeds of 1 Gbps and higher, as the name suggests.
- Switched Ethernet, as opposed to a shared network approach, allocates the full bandwidth to single devices, as discussed in LAN switching earlier in this chapter.
- The other important variant of Ethernet is wireless Ethernet, which we discuss in detail later in this chapter.
- A faster form of wired Ethernet that has recently emerged is 10-Gigabit Ethernet. Its adoption is gaining momentum.

TOKEN RING AND FDDI

Two other types of LANs, token ring and FDDI, have waned in popularity compared with Ethernet and wireless LANs. Token ring is historically associated with IBM and defined as a standard in the IEEE's 802.5 specification. As its name suggests, token ring uses the token passing access method, meaning that a token (or group of bits) continually passes from node to node around the LAN in a ring logical topology. The token that token ring LANs use is actually a group of 3 bytes (or 24 bits). It includes a start byte, an access control byte, and an end delimiter byte, as shown in Figure 11-10.

Figure 11-10

The structure of a token

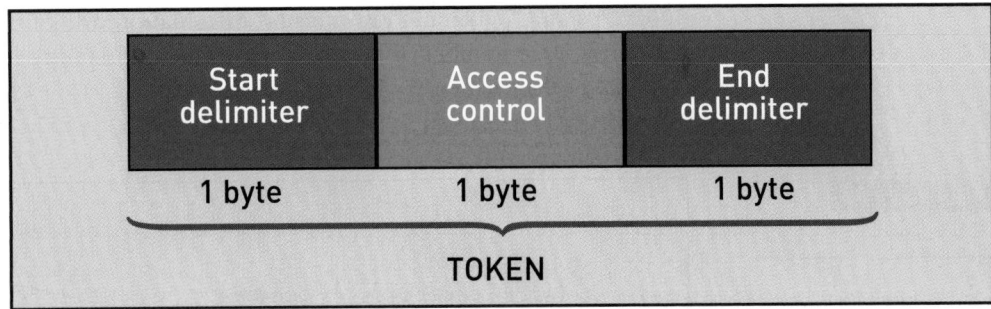

Start delimiter	Access control	End delimiter
1 byte	1 byte	1 byte

TOKEN

When a device possesses the token, it may transmit information, which circulates in an orderly manner from node to node until it reaches its intended destination, as shown in Figure 11-11.

Figure 11-11

Token ring network

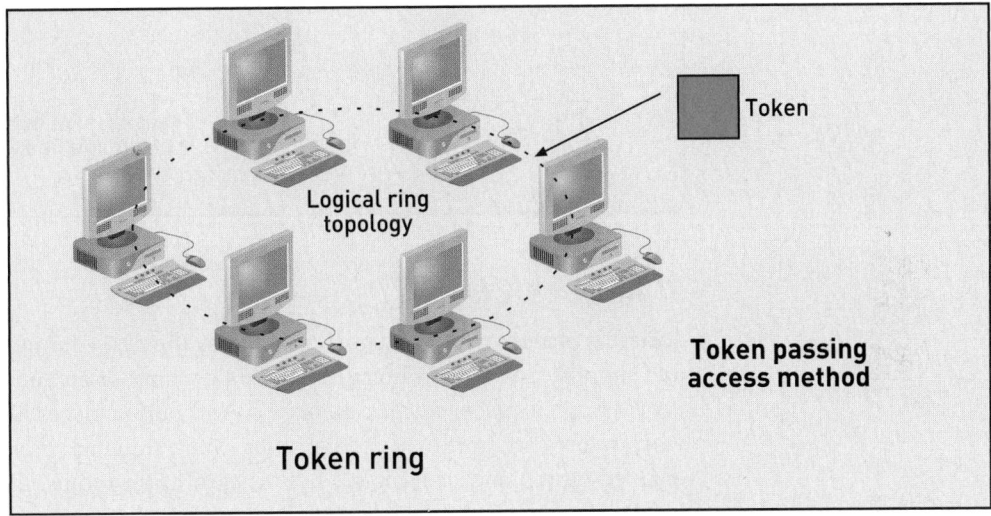

Another characteristic of token ring LANs is its frame format, which is similar to the standard format used to transmit information over an Ethernet LAN. The token ring frame format begins with the 3-byte token and is followed by a 6-byte destination address, a 6-byte source address, the actual information content to be transmitted, a 4-byte frame check sequence, a 1-byte delimiter, and finally a frame status byte. Figure 11-12 illustrates the standard token ring frame format. In contrast to Ethernet, a token ring frame does not have to be a minimum of 46 bytes.

Figure 11-12

Token ring frame format

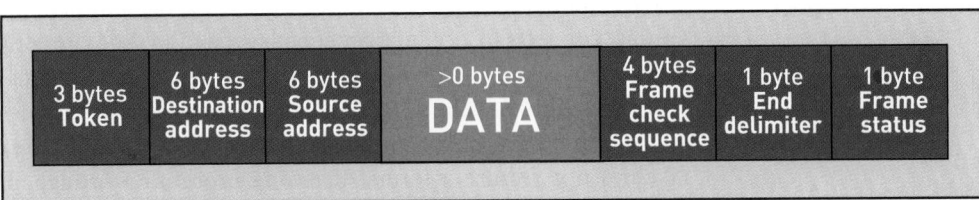

3 bytes Token	6 bytes Destination address	6 bytes Source address	>0 bytes DATA	4 bytes Frame check sequence	1 byte End delimiter	1 byte Frame status

An FDDI LAN is similar to token ring, in that it uses the token passing access method and a ring topology (see Figure 11-13). One typical difference is that FDDI's logical ring is actually a "dual ring," which provides redundancy. Having dual rings increases capacity and provides for fault tolerance, so that if one ring fails, the other can support traffic over the network. As its name suggests, FDDI operates over fiber-optic cable, which means it can achieve speeds of

100 Mbps and higher and can transmit information over fairly long distances. It can also support a large number of nodes. However, the availability of higher-speed Ethernet has diminished the need for the FDDI standard.

Figure 11-13

FDDI LAN

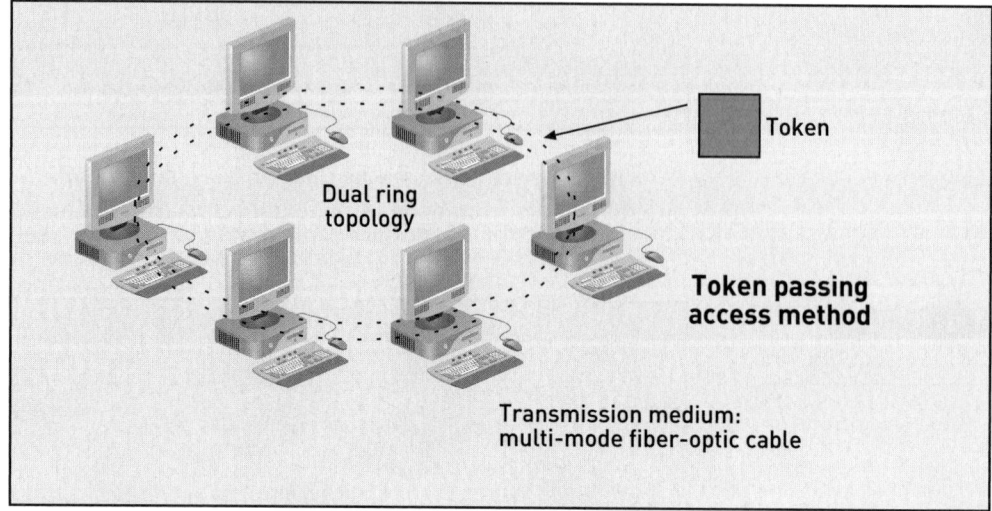

Dual ring topology

Token

Token passing access method

Transmission medium: multi-mode fiber-optic cable

WIRELESS LANS

Society is increasingly "unplugged." Driven by the desire for mobility and ubiquitous access, and supported by the proliferation of wireless laptops and personal handheld devices, the cord has been cut in businesses, homes, and public places. Business travelers depend on wireless network access to do their jobs. Most students arrive at college with a wireless-enabled laptop and use wireless networks in dorm rooms, libraries, and student centers. Public Internet users rely on wireless access points in coffee shops, bookstores, and other public venues. Businesses are increasingly supplementing LANs with wireless add-ons to meet demands for convenience, cost savings, and rapid implementation. Some workplaces such as hospitals need mobile computing access rather than access at a fixed station. A sizable percentage of home Internet users have established wireless LANs in their homes, using a wireless-enabled laptop to connect to a high-speed Internet connection, a printer, or another home computer.

WLAN technology is enabling this mobility; it is a phenomenon like the rise of cellular telephones decades earlier. WLANs use radio-frequency signals, rather than light or electricity over cables, to connect users within a limited geographical range, similar to the way a cellular phone connects to a cellular antenna base station. The currently predominant WLAN implementation is known as **Wi-Fi** (wireless fidelity), a catchy designation indicating that a product complies with the IEEE's 802.11 wireless Ethernet specifications. Wi-Fi uses a carrier radio frequency in the 2.4-GHz or 5-GHz range. This frequency is in the unlicensed category, meaning it does not require a formal licensing procedure through the Federal Communications Commission (FCC). Wi-Fi access points, called hot spots, have cropped up across the United States in coffee shops, airports, neighborhoods, and individual homes.

WLAN Standards

Wi-Fi commonly employs the 802.11b or 802.11g wireless technology standards, which are currently the dominant approaches. The various standards differ based on operating frequency, maximum data rate, distance, encoding schemes, and degree of adoption. The following list describes some of the major efforts to establish wireless technology standards:

IEEE 802.11a—Wireless technology operating at a frequency of 5 GHz with a maximum data rate of 54 Mbps.

IEEE 802.11b—Wi-Fi wireless technology operating at a frequency of 2.4 GHz with a maximum data rate of 11 Mbps.

IEEE 802.11g—Backward compatible with 802.11b, operating at a frequency of 2.4 GHz and a maximum data rate of 54 Mbps.

IEEE 802.11i—Standards for wireless security mechanisms.

IEEE 802.15—Known as WPAN, or wireless personal area networks.

IEEE 802.16—WiMAX, an emerging technology that seeks to provide high-speed wireless access over much longer distances than implementations of the IEEE 802.11 standards.

Personal WLANs

A home-based wireless LAN uses a **wireless access point** (WAP), a device that connects wireless computers to a wired network to enable high-speed Internet access and other services. Laptops with an installed wireless adapter then communicate with the WAP over the 2.4-GHz or 5-GHz frequency range, as shown in Figure 11-14.

Figure 11-14

Home WLAN implementation

Business WLANs

Business LAN environments often use a mix of wired and wireless technologies, connecting 802.11-compliant devices to wire-based Ethernet LANs via an access point that acts as a bridge between the wireless and wired realms, as shown in Figure 11-15.

Figure 11-15

Business WLAN implementation

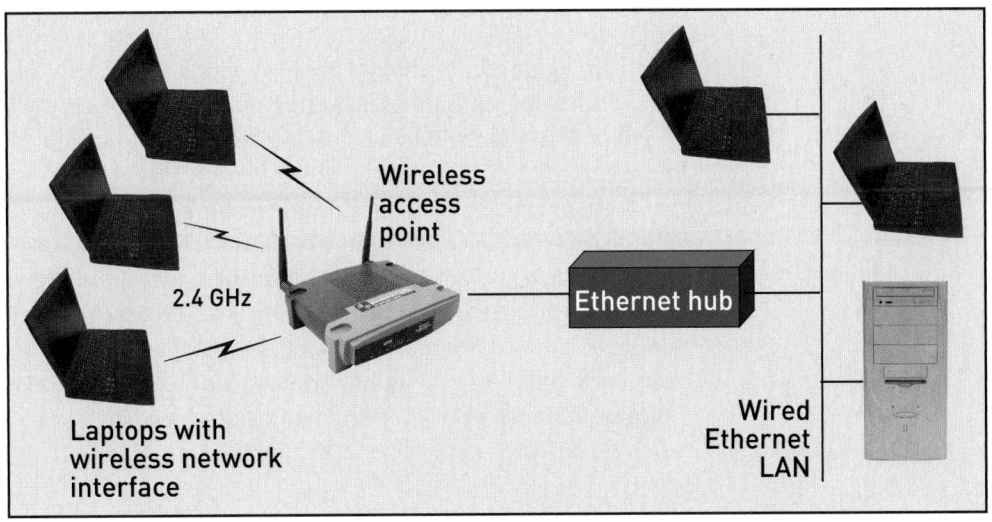

Advantages and Disadvantages of Wireless LANs

The main advantages of wireless LANs are mobility and flexibility. A computer user can access a network almost anywhere rather than having network access tied to a home, business, or university computer terminal. This ubiquitous access increases the productivity of workers who travel regularly. WLANs also help businesses by providing convenient, cost-effective network access without requiring them to hard-wire cables to each office. Adding and deleting network users on an office WLAN is easier than on a wired network.

Challenges to wireless LAN implementations include security, range limitations, bandwidth, and manageability, although there are efforts to address these limitations. For example, experimental development is progressing for WiMAX, a technology specified by IEEE 802.16. WiMAX could improve on Wi-Fi by providing longer-range broadband wireless services to compete with access mechanisms such as cable modem and digital subscriber line (DSL). WiMAX is discussed in more detail in Chapter 12.

One of the greatest disadvantages of wireless LANs is their lack of security. Information is transmitted "over the air" rather than via cables, which makes it easier to intercept signals and gain unauthorized access to transmitted information. As long as hackers are in range of a wireless access point, they can intercept signals that are not adequately encrypted. Another type of vulnerability occurs when an unauthorized user places an unauthorized device on the network, either by connecting a laptop to a wireless access point or by inserting a rogue **network access point**. Even authorized network access points present a challenge, as they are often shipped with a default configuration that is relatively unsecure. These access points must be properly configured for user authentication and other security measures. Also, a hacker who wants to disrupt network communications for malicious purposes can launch a "denial of service" attack, which overwhelms a network with an inordinate amount of requests or jams wireless signals.

Public locations that provide wireless Internet access also serve as a magnet for hackers who want to obtain sensitive or confidential information, such as user IDs, passwords, bank account numbers, or credit card numbers. One hacking technique is to set up a wireless hot spot in a public location such as an airport or hotel. This rogue hot spot appears on a computer's list of available wireless networks, and because the service is "free," it can lure unsuspecting users to use the network to connect to the Internet. Chapter 15 provides a detailed description of wireless security vulnerabilities and preventive security measures.

Municipal Broadband Wireless

The relatively low cost and wide availability of wireless LAN technologies have allowed some cities and municipalities to offer public Internet access via Wi-Fi network access points in parks, libraries, and other public locations. These networks are driven by economic, political, and technical factors. For example, ubiquitous wireless Internet access can help to economically revitalize cities that want to attract business growth and professionals. In the political realm, some cities see wireless Internet access as a way to serve the public and to address the so-called digital divide: the gap between people who have access to digital technologies and people who do not. Laptops and other mobile devices usually have built-in Wi-Fi capability, so public access to the Internet does not require a city to expend money or other resources on the client side.

These networks might also be able to support municipal functions. For example, many cities have experience with wireless connectivity through the operation of wireless public safety networks used by first responders. These networks have traditionally operated over frequency ranges that are licensed from the FCC. Because many city workers now require more multimedia-based information than voice only, and because the unlicensed frequency bands associated with Wi-Fi are inexpensive, some cities are interested in switching their traditional voice applications to Wi-Fi-based multimedia networks.

CHAPTER SUMMARY

- A network shares information using a set of hardware and software designed to interconnect multiple computers, communication devices, and other types of devices.
- Categorizing networks is somewhat arbitrary, but in the networking industry, the most common categories are local area networks (LANs) and wide area networks (WANs).
- LANs are networks that span a relatively small geographical area, such as the floor of an office or a house. LANs interconnect local computing resources and provide WAN access.
- WANs are networks that span a large geographical area such as a city, a country, or the world at large. The Internet is the most important example of a WAN. Chapter 12 discusses WANs in detail.
- Some design characteristics of LANs include topology (such as bus, star, and ring), access method (such as CSMA/CD, CSMA/CA, or token passing), and type of medium (wireless, fiber, twisted pair).
- Ethernet, including switched Ethernet, has become the most prevalent type of local area network.
- Some wireless local area networks (WLANs) include the IEEE 802.11b and IEEE 80211g standards, more commonly called Wi-Fi. WLANs have great advantages, including mobility and flexibility, but they also present security challenges.

KEY TERMS

access control mechanism
bus
CSMA/CA
CSMA/CD
deterministic
drops
EUI
Ethernet
Ethernet switching
FDDI
frames
IEEE 802.3 standard
LAN
LAN operating system
latency
logical topology
MAC
network access point

network interface controller
nodes
nondeterministic
overhead
padding
physical topology
ring
star
star-wired ring
token
token passing
token ring
topology
WAN
WLAN
Wi-Fi
wireless access point

REVIEW QUESTIONS

1. Describe three devices or services a user might access via a LAN.
2. Draw a picture of the following LAN topologies: star, ring, and bus.
3. Describe the difference between logical and physical topology.
4. What is the purpose of an access control method?
5. Explain the difference between deterministic and nondeterministic LAN access, and name an example of each approach.
6. What access method does traditional wired Ethernet use? What method does wireless Ethernet use?
7. What piece of networking equipment acts as the physical interface between a computer and the network?
8. Why is a unique MAC address assigned to each manufactured NIC?
9. Why is a NIC's 48-bit physical address printed in hexadecimal form on the card?
10. What professional organization allocates unique addresses to NIC manufacturers?

11. What is riser cable?
12. What problem occurs when increasing numbers of nodes are added to a network that uses the CSMA/CD access method? What network equipment addresses this problem?
13. Name two examples of a LAN operating system.
14. Name an important function of a LAN operating system.
15. What is the most popular type of LAN?
16. Name several types of transmission media over which Ethernet operates.
17. What is a frame?
18. The Ethernet frame format expects a minimum of 46 bytes of data. What happens if the data content to be transmitted falls below this minimum?
19. If a computer transmits 100 bytes of information content over Ethernet, what percentage of the transmitted frame is overhead?
20. What is the popular name for the IEEE's 802.11 wireless Ethernet specification?
21. Name three public places in your area that are Wi-Fi hot spots.
22. Name one advantage and one disadvantage of wireless LANs.

DISCUSSION QUESTIONS

1. As a business owner starting a new office, would you invest money in wired or wireless LANs? Explain.
2. What is the technical difference between deterministic and nondeterministic network access? What are some examples of applications that would be better served by each access method?
3. Should public Wi-Fi access be a service that governments provide? What is the role of local government, and does public Internet service fit into this role?
4. What security challenges are posed by wireless LANs? What preventive measures do you or should you take when accessing the Internet wirelessly?

CASE PROJECT

Design a municipal wireless network with Wi-Fi Internet access.

You are the top IT official of a small city, and you are asked to present a strategic plan to the mayor for implementing citywide Internet access via Wi-Fi. Develop a presentation that addresses the following five groups of questions.

- What would be the justification for installing public Wi-Fi access? Provide a detailed explanation of the economic, technical, and political considerations behind your recommendation.
- What are the greatest challenges and obstacles to implementing the network?
- How would this service affect commercial network providers in the area? Would it help or hurt their businesses?
- What have other cities done to provide Wi-Fi access, and why? Present examples to the mayor of what other municipalities have done, their rationales, and the challenges they faced.
- Should city IT workers manage the network, or should it be outsourced to a commercial service provider?

CHAPTER
12

WIDE AREA NETWORKS

LEARNING OBJECTIVES

In this chapter you will:

- Understand the concept of a wide area network

- Identify the main technical components of a wide area network

- Distinguish between packet switching and circuit switching

- Understand virtual private networks (VPNs)

- Gain familiarity with the most important commercial WAN services

- Understand WAN access technologies, including dedicated lines, xDSL, cable modem access, and WiMAX

- Identify important network management functions

INTRODUCTION

Wide area networks (WANs) span a large geographical area, in contrast to LANs, which operate in more confined areas. One type of wide area networking is called a metropolitan area network (MAN), which provides information exchange or network access over any city or suburb within a 10- to 50-mile range. Other examples of geographically expansive networks include cellular telephony, satellite television services, the Internet, and commercial networks geared toward business applications. For historical reasons, WAN services have evolved from data communication technologies that cater to large businesses.

This chapter describes the distinguishing technical characteristics of WAN services, including packet switching, virtual private networking, WAN equipment, and WAN protocols. The chapter also presents an overview of popular WAN services, including Internet services, frame relay, Asynchronous Transfer Mode (ATM), multiprotocol label switching (MPLS), and private networks, and describes alternatives for accessing WANs. The discussion concludes with an overview of network management functions that are vital to enterprise wide area networks.

Network categories, as mentioned previously, are somewhat arbitrary. Using geographical distance as a sole differentiator is not precise or even relevant for many technologies. Nevertheless, industry convention describes a WAN as a network that spans a large geographical distance. The largest example of a WAN is the public Internet, but many other types of WANs exist, particularly to serve large organizations that exchange information over networks. Therefore, wide area networking is sometimes referred to as **enterprise networking**, and is often geared toward business information exchange among companies, customers, and suppliers.

Wide area networking developed from the way businesses first used computer networks to exchange information internally, beginning in the mid-1970s. The information exchanged within business computer networks was entirely alphanumeric (numbers and text), and these data networks were completely separate from telephone networks. Sending video, images, and multimedia applications over the network was not yet possible. Business computers were enormous but had limited storage capacity. A major advancement at the time was the introduction of IBM's Systems Network Architecture (SNA) and Digital Equipment Corporation's DECnet. These networking architectures provided a set of protocols (standard rules for formatting and exchanging information) and used computers and other networked devices manufactured by a single vendor. At the time, there was little interoperability between computing devices made by different manufacturers. The main components of a network were an enormous mainframe computer connected through intermediary devices to computer terminals. These "dumb terminals" allowed users of the private networks to input and output data to and from the mainframe, but they had limited intrinsic functionality or processing ability.

Several architectural features distinguished these networks from modern WANs. They primarily enabled information exchange within a single enterprise rather than allowing for communications with external customers and suppliers, as modern networks provide today. The networks almost exclusively used dedicated private telecommunication lines, meaning that businesses would lease point-to-point private lines (at low transmission rates such as 9.6 Kbps) that extended end to end from one networked location to another. These point-to-point lines were provided by telecommunications carriers and were available only to the company that leased the lines. This approach differs from modern public WAN services, in which customers often share common transmission media and services through virtual private networking. The older networks did not necessarily offer integrated voice, data, and multimedia information exchange. Most strikingly, these networks used proprietary protocols, meaning that a business using IBM's SNA could not communicate with a business using DECnet or another proprietary protocol. WANs now support voice, data, and multimedia information, use open network protocols, and often are offered over a public network such as the Internet.

PACKET SWITCHING

Packet switching is an important general concept of computer networking and an approach used by the most popular WAN: the Internet. Information sent over the Internet is broken into small segments called **packets**. Each packet contains the actual information content to be transmitted, such as an electronic mail message, as well as supplementary overhead information, such as the order of the packet, the sender's binary address (called the source address) and the binary address of the packet's destination (called the destination address). In common industry usage, a packet is similar to a frame, a term associated with some types of WANs (like frame relay) and with LAN technology, as explained in Chapter 11. As shown in Figure 12-1, the path that one packet traverses over a network from source to destination may be different from the next packet's path, depending on network congestion or other conditions. In this type of networking approach, known as connectionless **packet switching**, no dedicated end-to-end physical connection is established for the duration of data transmission. On the Internet, network devices called routers read the destination address and determine how to expeditiously route packets through the networks, based on routing algorithms that are designed to minimize latency—the delay that a packet undergoes from source to destination. Routers are also designed to minimize **hops**, the number of times a packet traverses various routers as it is transmitted over a network. Once all the packets from a given transmission reach their destination, they are reassembled in correct order. The receiving device knows the proper order because each packet has its own sequence number that is included as overhead.

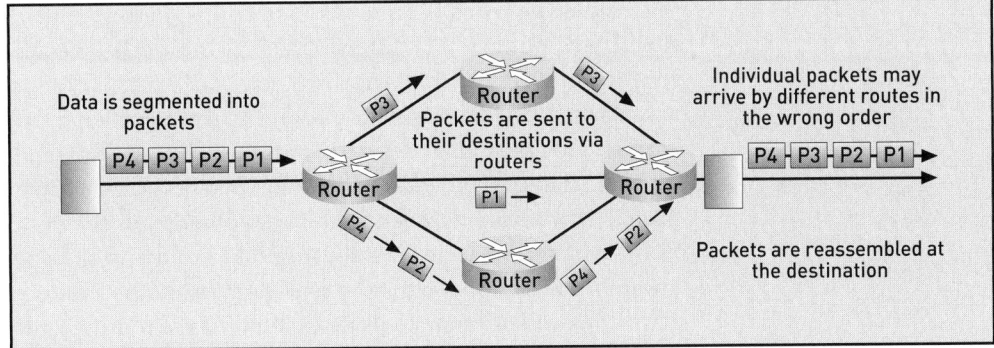

The packet-switching approach contrasts with the **circuit-switching** approach of the traditional telephone network, which is described in Chapter 16. The circuit-switching approach establishes a physical, dedicated end-to-end path through the network between a caller and receiver, and maintains the path for the entirety of the call. This path cannot be used by another party. In contrast, a path is only occupied between two points in a packet-switched network (such as between two routers) when the packet is actually being transmitted from one point to another. Once the packet reaches its destination, the path is free to be used by other parties. The packet-switching approach uses resources more efficiently because no single user can exclusively occupy a given path.

In the United States, the development of the packet-switching approach was influenced by concerns about survivable communication networks during the Cold War. Traditional networks at the time were much more centralized and hierarchical, which made them more susceptible to widespread communication failures in the event of a strategic ballistic missile attack. The distributed packet-switching architecture was motivated in part by the knowledge that the network could continue to operate over surviving nodes even if some of the network nodes were destroyed. Packet switching was independently developed by Paul Baran at RAND Corporation in the United States and by Donald Davies at the National Physical Laboratory in the United Kingdom. As we explain later in this chapter, the development of packet switching helped to make the Internet possible.

NETWORK PROTOCOLS

Network protocols are another important technical enabler of WANs. Unlike older types of networks, such as the SNA and DECnet systems mentioned earlier, modern wide area networking has made the transition to more open network protocols. Protocols are the rules that enable information exchange over a network, including how many bits make up a binary network address, how to place information in standard formats that anyone can read, and rules for performing error checking over the network. Earlier protocols such as SNA and DECnet were **proprietary protocols**—only organizations that used the same closed standards and associated technologies could exchange information. The dominant family of network protocols in modern architectures is TCP/IP, or **Transmission Control Protocol/Internet Protocol**. TCP/IP is an **open protocol**—it is not controlled by a single vendor, it is available for any manufacturer to use, and it enables universal access to the Internet. Network protocols and TCP/IP are critical to network functionality, and are the topic of Chapter 13.

WAN ARCHITECTURAL COMPONENTS

Routers are one of the most important components in wide area networking. A router (see Figure 12-2) is an intelligent switching device that determines how to direct (or route) a packet across a network, based on the packet's destination address and network conditions. An enormous, interconnected web of thousands of routers make up the backbone of the Internet. This web of routers provides a diverse array of possible paths to accommodate a transmission from any given source to a destination. The addresses that routers analyze over the Internet are called IP (Internet Protocol) addresses; routers make packet forwarding decisions based on these IP addresses. A single networked device often has multiple addresses serving different functions. As discussed in Chapter 11, a computing device on a local area network has a Media Access Control (MAC) address that is physically associated with a LAN adapter, such as an Ethernet card. It also usually has an IP address associated with its virtual location relative to other devices on the overall Internet. When an information packet arrives from a network at any given router, the router reads the destination IP address. The router "looks up" information in a **routing table** to direct the packet in the most efficient way toward its destination, either by sending it to another router or by leaving the packet on its current network to route to its destination. These routing tables are constantly changing and automatically updated as routers probe their network environments, exchange information with other routers, and dynamically update their routing tables.

Figure 12-2

IT equipment rack with routers

Customers who access the Internet or another type of WAN are connected to an **edge router**, a router that interconnects one or more LANs at a customer location to the wide area network. The customer location often connects to a WAN service provider's network via a terrestrially based digital transmission line. A device called a **CSU/DSU** (Channel Service Unit/Data Service Unit) is an important network component between the edge router and the dedicated transmission line, as shown in Figure 12-3. The CSU/DSU serves as a physical interface to the digital transmission circuit and performs important functions such as signal formatting, timing, and some error control. The digital circuit also terminates at the service provider's network in a CSU/DSU. In other words, there is a CSU/DSU on each side of the network. The equipment at a customer location that serves as the demarcation point between a local network and a service provider's WAN is often called customer premises equipment (CPE). Other examples of CPE include DSL modems and cable modems, as described later in this chapter.

Figure 12-3

WAN architectural
components

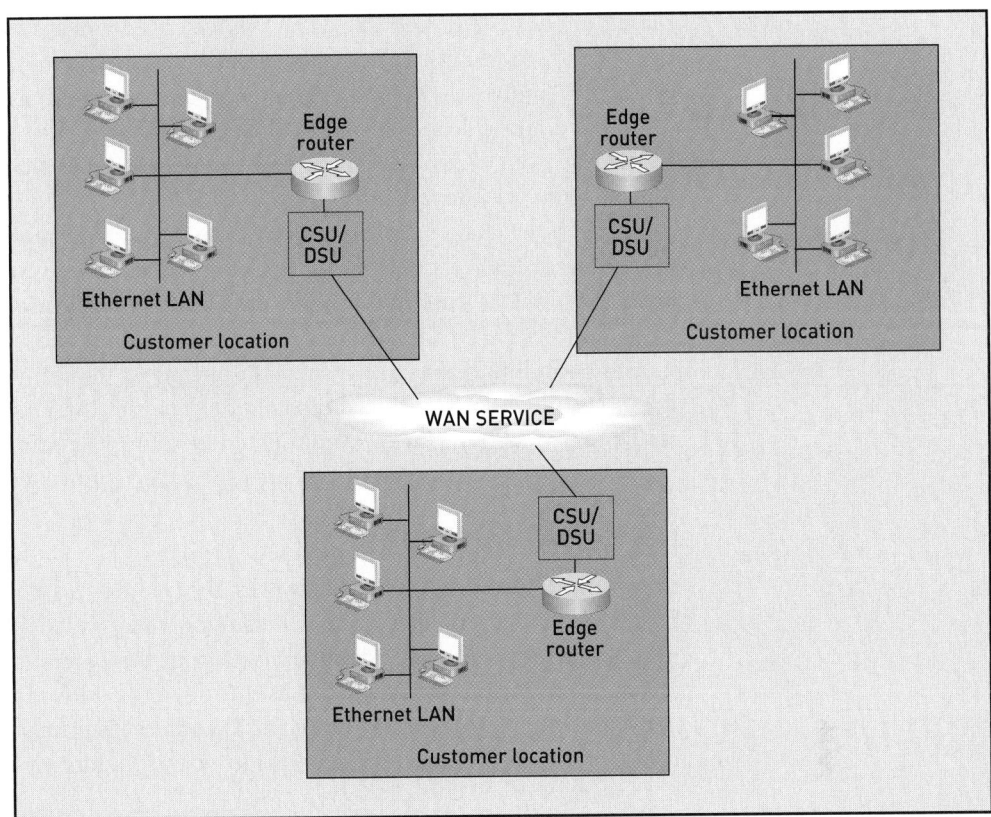

WAN ALTERNATIVES

Enterprises that require networked services over a large geographical area have many options.

PRIVATE NETWORKS

If a company such as a bank wants a dedicated transmission path between its branches for private digital communication lines, the company has the option of leasing dedicated private lines from a network provider. These private lines run end to end between locations and regularly exchange information from data, voice, image, and video applications, as shown in Figure 12-4. A private line is not shared with other customers, so it can guarantee performance and availability to some degree. Private networks are also an excellent alternative for transmitting sensitive information that requires more security than a public network service

Figure 12-4

Dedicated private
network

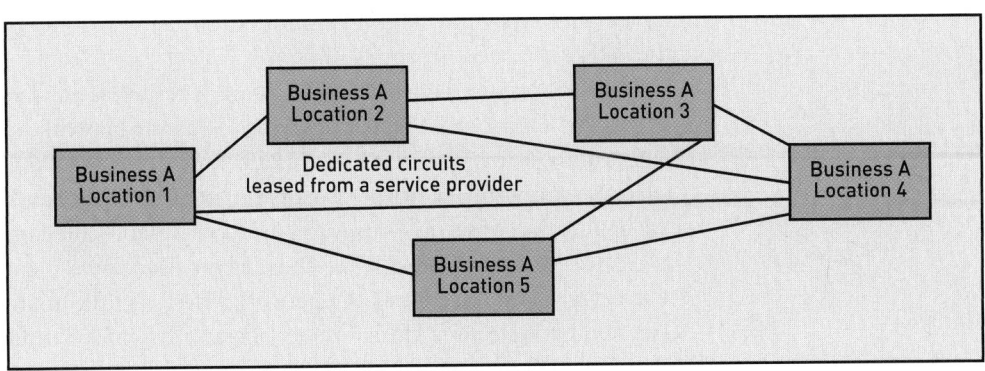

can offer. These dedicated lines may run terrestrially over fiber-optic cables or high-grade twisted pair, via undersea fiber-optic cables, or even via satellite for international communications or remote areas that do not have adequate terrestrial telecommunications.

At one time, wide area networks almost always used these end-to-end private communication lines, but many businesses now use public WANs that are shared by multiple customers. However, the transition from dedicated private networks to more public networks presented some challenges to businesses. The dedicated communication links, which range in speed from 1.544 Mbps to gigabits per second, have some performance and security advantages. Each link is leased monthly from a telecommunications carrier, which reserves the entire bandwidth for the business that leases the lines and prevents any other traffic. The transmission link between two sites can operate at the full transmission rate at any time. Because only one institution's information traverses these links, the information is more secure than data transmitted over a shared medium. The primary disadvantage of a private network is cost, because a single enterprise bears the entire cost of the lines.

INTERNET SERVICES

In contrast to dedicated private networks, the Internet is a public network shared by many businesses and users. Figure 12-5 illustrates a simplified WAN configuration in which multiple institutions and individual users from the general public share a common WAN service. Many businesses use the infrastructure of the public Internet to communicate internally and with customers and suppliers. The Internet's primary technological foundation is its use of the Internet Protocol, a packet format and addressing standard that we describe in the next chapter.

Figure 12-5

Shared public WAN

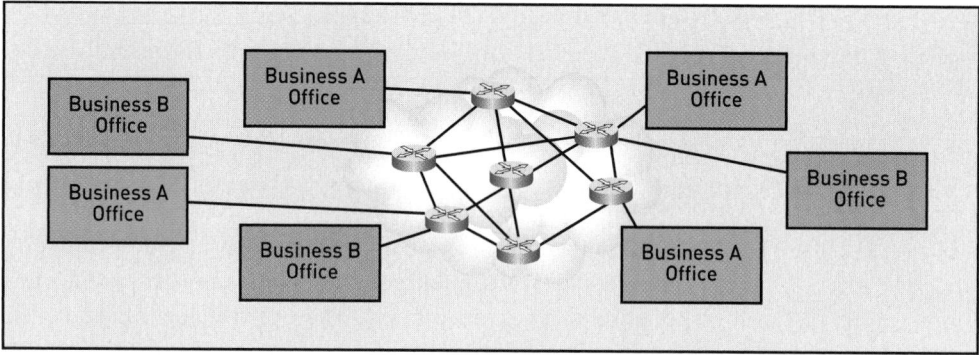

Using the Internet and other shared public WANs is much more cost effective than using dedicated private networks. However, this arrangement creates two concerns: information is less secure than on a private dedicated network, and the network's performance and availability are affected by the volume of information sent by other users. As traffic increases across the network, its performance suffers.

Virtual private networks (VPNs) are a response to these performance and security concerns. VPNs combine the cost savings of public network services with performance guarantees and security services. A VPN is a private network arrangement that runs over a public telecommunications network, usually the Internet. Privacy and security are achieved through the application of security measures and a technique called **tunneling**. The usual technique to create security through a WAN is to encrypt all traffic as it enters the public shared network and decrypt it at the edge of the network where the information leaves the public domain. This approach creates a virtual tunnel of encrypted information that is difficult for hackers to view. Through the use of encryption and firewalls (discussed in Chapter 15), a VPN that runs over the public Internet is nearly as secure as information on private dedicated lines, and cheaper than using private networks. The other important characteristic of VPNs is performance. VPN services offer quality-of-service (**QoS**) guarantees to its customers to cover such factors as latency of packets, dropped packets that cannot be routed to their destination, and network availability. VPN customers usually obtain service level agreements (**SLAs**) from the VPN provider. These SLAs contractually guarantee that the service provider will fulfill the QoS guarantees; if not, the provider will pay penalties such as fee reductions.

FRAME RELAY SERVICE

Frame relay is another WAN option offered by service providers to businesses and other institutions. The frame relay network is owned and operated by the service provider but is used by the customer. This service uses packets, called frames, with a variable number of bits that are switched throughout the WAN until they reach their destination. Frame relay is really a standard for how these variable-length packets, or frames, are formatted and transmitted across the network.

The frame relay standard uses a different format from other WAN standards. A software-defined (virtual) path is set between two devices on the network, and the devices are expected to exchange information. There are two types of software-defined paths: switched virtual circuits (SVCs), which are created for each information transfer and then terminated once the information transfer is complete; and permanent virtual circuits (PVCs), which are permanent paths that do not terminate when information transmission ceases. Frame relay services are beneficial to businesses because they support fast transmission of packets.

Frame relay service providers guarantee their customers a minimum bandwidth known as the committed information rate (CIR), along with a certain allotment of ad hoc, higher bandwidth called the extended information rate (EIR). Frames that exceed the CIR bandwidth are indicated as "discard eligible" and dropped by the network if it becomes too congested.

Historically, frame relay originated as an improvement over an older WAN technology known as **X.25**. This packet-switching WAN standard, which was common in the 1980s and into the 1990s, was designed for transmitting information over error-prone analog telecommunications systems. X.25 used a similar approach as frame relay in creating virtual circuits, but it also included extensive error-correction techniques. Frame relay was designed for higher-quality digital links. Because these digital facilities were much less susceptible to errors, the frame relay approach could spend less overhead performing error correction and therefore could achieve higher transmission speeds.

ASYNCHRONOUS TRANSFER MODE (ATM)

WAN service providers also offer **Asynchronous Transfer Mode** (ATM), a network alternative that formats information into fixed-length packets. These packets are normally called cells in the context of ATM. ATM cells have a total length of 53 bytes: five bytes for overhead and 48 bytes reserved for the actual data being transmitted. ATM is a connection-oriented WAN approach; it establishes an end-to-end virtual connection between the transmission device and destination device before transmission begins. Some applications cannot tolerate long delays when transmitting large packets to their destinations. Digitized voice is one example of a traffic type that would not perform adequately using large, high-latency packets as opposed to smaller cells relayed with relatively low latency. Any delay in voice transmissions is detectable to the user. ATM's fixed transmission delays, virtual circuits, and fixed cell size are beneficial for such applications.

MULTIPROTOCOL LABEL SWITCHING (MPLS)

The types of WAN services described thus far, such as frame relay, ATM, and Internet services, use distinctly different approaches, protocols, frame formats, and other distinguishing characteristics. A more practical type of WAN service, called multiprotocol label switching (**MPLS**), is designed to simultaneously support many types of WAN traffic. In other words, the MPLS service can transport IP packets, frame relay frames, and ATM traffic. Most companies would find this service useful, because they transmit a significant amount of IP packets and might use frame relay or ATM.

MPLS service can handle variable-length packets, in contrast with ATM, which transmits fixed-length cells. One major difference between MPLS and traditional IP packet switching is that MPLS is connection oriented. Traditional packet switching is often **connectionless**, meaning that a dedicated end-to-end connection is not established for the duration of transmission. Unlike this approach, MPLS routes packets along preconfigured paths.

WAN ACCESS ALTERNATIVES

The WAN access mechanism is the approach for connecting a local area network or an individual handheld device or computer to a wide area network. The access mechanism selected depends on several variables. For example, the bandwidths of WAN access mechanisms vary considerably, so a user's required transmission capacity can determine the selection. If users require mobile access to a WAN, alternatives narrow significantly. Another factor is security. For example, a dedicated line from an office building into a WAN provides greater information security than wireless connectivity or a shared transmission medium. Finally, WAN access choices are sometimes limited because not all alternatives are ubiquitously available. The following sections describe several WAN access alternatives.

LEASED PRIVATE LINES

Businesses and other enterprises that have more than a dozen employees and that need WAN access usually lease a dedicated private line to a commercial WAN service (see Figure 12-6). The line is rented on a monthly basis from a telecommunications service provider, and comes in the following transmission speeds:

- 768 Kbps—Sub-T1 link (also called a fractional T1 line)
- T-1 link—1.544 Mbps (also called a dedicated T1 line)
- T-3 link—45 Mbps
- OC-3—155 Mbps
- OC-12—622 Mbps
- OC-48—2.488 Gbps

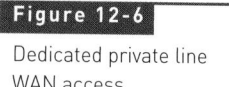

Figure 12-6

Dedicated private line
WAN access

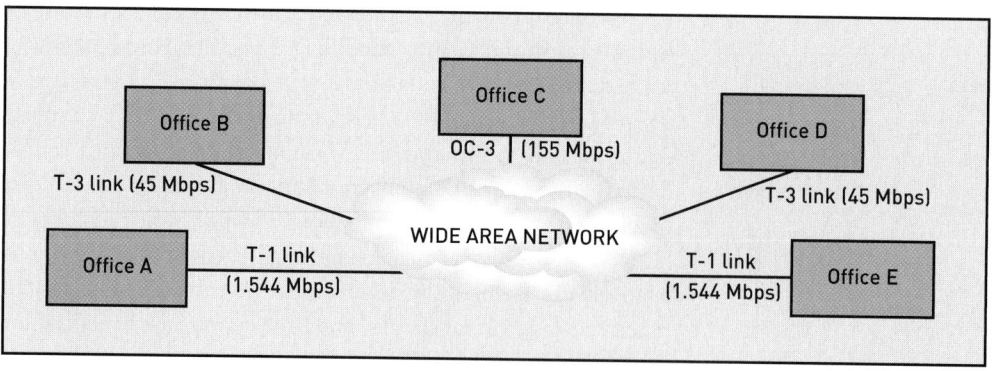

A medium-sized or large business that uses a WAN for voice communications, multimedia applications, and Internet access usually subscribes to a fractional T1 line or a dedicated T-1 or T-3 link into a commercial service provider's network. Supporting a bandwidth-intensive application such as Web hosting would necessitate a high-speed, dedicated-access link. Examples include an OC-12, which transmits at 622 Mbps, or an OC-48, which transmits at 2.488 Gbps.

DIGITAL SUBSCRIBER LINE

Small enterprises and individual households that access the Internet or other WANs can choose **digital subscriber line** (DSL) service if it is available in their area. DSL is a WAN access alternative that connects a user's DSL modem (see Figure 12-7) to the twisted pair cables installed as part of the traditional telephone network. DSL allows for the simultaneous transmission of voice data and more bandwidth-intensive multimedia information such as Internet traffic. DSL splits the existing phone line into three channels: a channel for transmitting voice, and upstream and downstream channels that carry traffic to and from the WAN, presumably the Internet.

Figure 12-7

DSL modem

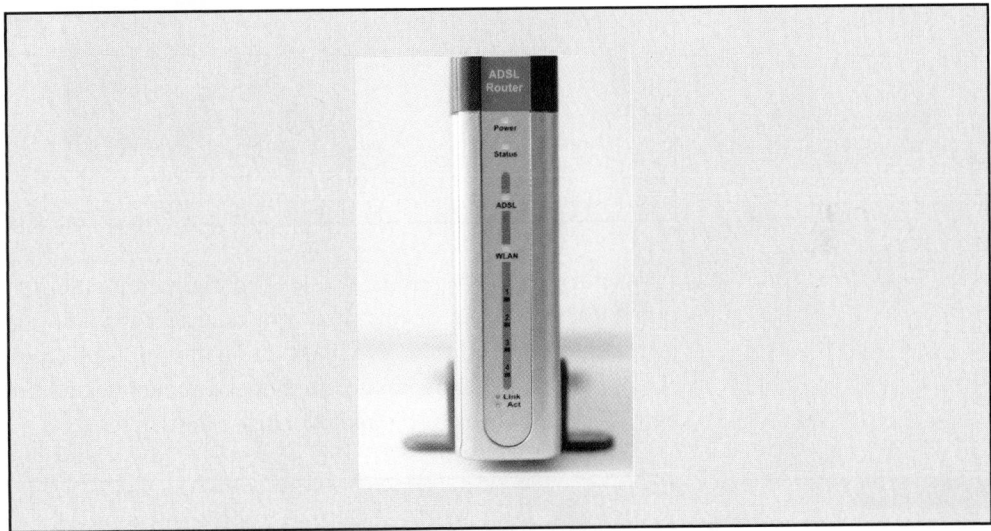

Many types of DSL technology are available, such as **symmetric digital subscriber line** (SDSL), **asymmetric digital subscriber line** (ADSL), and **high data rate digital subscriber line** (HDSL). References to the family of DSL variations are sometimes called "xDSL." DSL is not a new technology; it was designed in the late 1980s by Bellcore, a research center for regional phone companies, to allow high-bandwidth digital signals to run over twisted pair wires that were already carrying analog voice signals.

Because DSL uses the installed base of phone lines to residences and office buildings, it is usually offered by a local telecommunications carrier. These service providers have offered DSL, usually ADSL, in response to competition from cable companies that offer voice and high-speed Internet service over their installed base of coaxial cable.

DSL service offers some clear advantages, as described earlier in this section. However, this service is not ubiquitously available, and its quality depends on the distance between the user site and the telephone company (the "central office") that houses the DSL termination equipment. This equipment is known as the DSLAM, or DSL access multiplexer (see Figure 12-8). The distance limitation between the central office and a DSL user is approximately 18,000 feet.

Another disadvantage is the inherent asymmetry in some DSL service. The downstream transmission rate, which is the connection speed from the WAN to the end-user, is sometimes up to three times faster than the upstream rate (the connection in reverse). This consideration might be important for small businesses or individual users who upload large amounts of information to a WAN. Under ideal circumstances, DSL service can offer a transmission rate of up to 7.1 Mbps downstream and approximately 768 Kbps upstream.

Figure 12-8

DSL access multiplexer

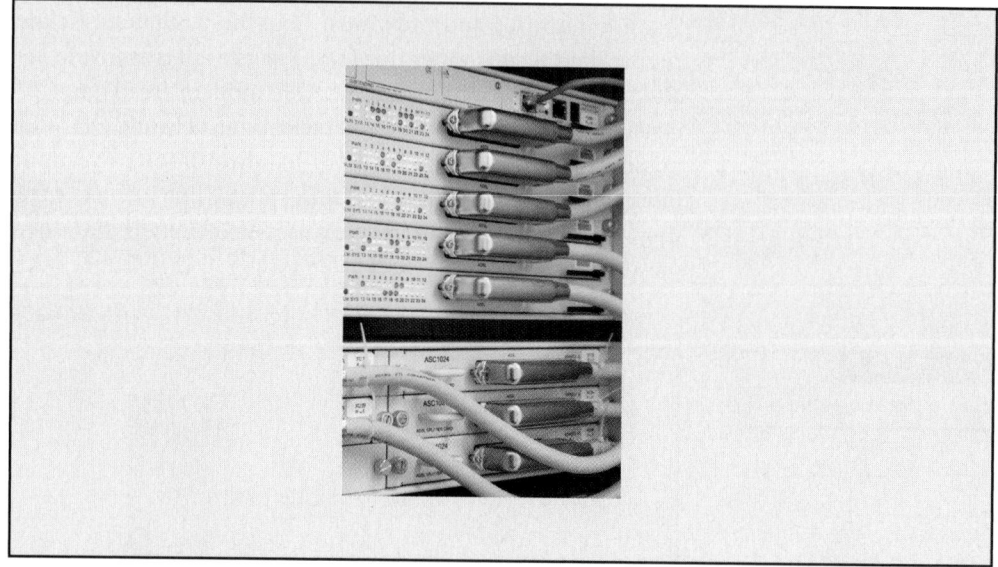

As illustrated in Figure 12-9, a DSL configuration requires a DSL transceiver (commonly called a DSL modem or router) at the customer's location and the DSLAM at the service provider's central office. The DSLAM terminates multiple lines from customer locations and routes information either to the Internet or through the traditional public switched telephone network. This network is discussed in detail in Chapter 16.

Figure 12-9

DSL components

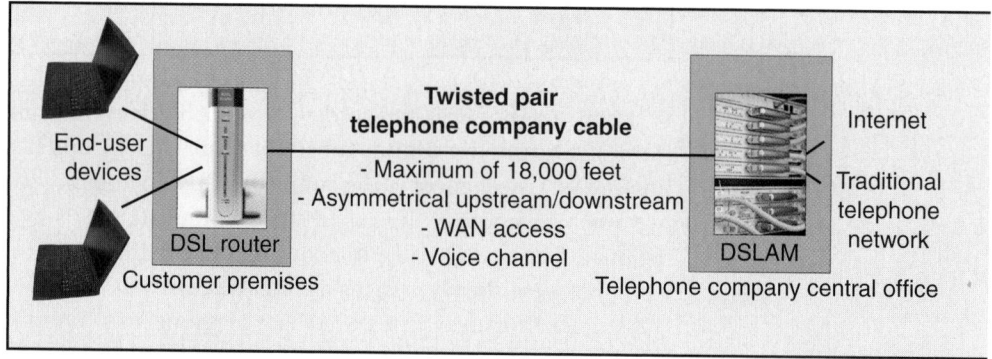

CABLE MODEM ACCESS

High-speed Internet access is also commonly provided through cable modems (see Figure 12-10) that connect customer premises equipment to the Internet using existing cable television infrastructure and coaxial cable. Internet access via cable offers transmission rates in the megabit per second range—approximately 30 to 50 Mbps downstream and 5 Mbps upstream in some areas. Some cable companies offer tiered services and charge a premium for higher transmission rates.

Cable WAN access is geared primarily to residential customers who already subscribe to cable television service and is offered as part of a package with cable television, high-speed Internet access, and Voice over IP service (see Chapter 17). Residential users who access the Internet over cable actually share a coaxial line, meaning that transmission speeds depend on the number of neighborhood residents who are concurrently using the service.

Figure 12-10

A cable modem

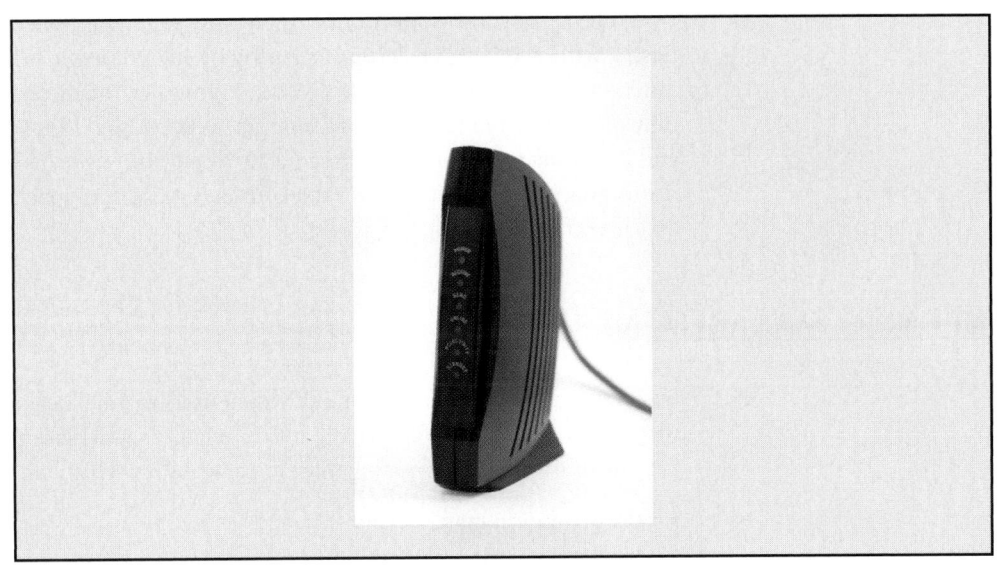

WiMAX

High-speed wireless broadband access to the Internet is commonplace over small geographical areas through technologies such as Wi-Fi, as described in Chapter 11. A significant technological requirement is for products and services to provide the same type of wireless WAN access over greater distances, such as across a city using a metropolitan area network (see Figure 12-11). **WiMAX** is an emerging broadband technology that could respond to this requirement. WiMAX (Worldwide Interoperability for Microwave Access) is actually another name for IEEE 802.16, a formal set of networking standards for wireless metropolitan area networks developed by the Institute of Electrical and Electronics Engineers (IEEE).

Metropolitan broadband wireless has many applications. It is a competitive alternative to land-based WAN access options such as DSL, cable modem access, or dedicated private lines. Rather than accessing the WAN via the infrastructure of a cable company or telephone company, users would access the WAN wirelessly. In addition to WAN access, it could also be used to interconnect fixed Wi-Fi hot spots or to provide businesses with a backup network infrastructure in the event of an outage in land-based infrastructures. There is also interest in WiMAX as a low-cost approach to establishing Internet access in areas such as developing countries, which do not necessarily have installed telecommunications infrastructures.

Figure 12-11

A wireless metropolitan area network antenna

WiMAX networks would consist of WiMAX antennas transmitting to residences and businesses with a WiMAX receiver, to laptops with WiMAX cards, or to Wi-Fi hot spots. The antenna would connect to other WiMAX antennas or directly into a telephone company network connected to the global Internet. WiMAX provides a theoretical maximum uplink and downlink speed of approximately 70 Mbps, although this rate can be achieved only over short ranges. Performance varies depending on distance, spectrum allocated, and number of concurrent users sharing bandwidth.

NETWORK MANAGEMENT SYSTEMS

Businesses, universities, and even individual users require local and wide area network services that are always available and that perform reliably. To achieve these requirements, users must provide or outsource network management activities, which fall into four categories:

- Configuration management
- Performance management
- Fault management
- Security management

CONFIGURATION MANAGEMENT

A computer network is in a constant state of flux. Network topologies change as new servers, routers, or wireless nodes are added and as cables are reconfigured (see Figure 12-12). In large corporations, the number of network users changes on a daily basis as employees arrive and depart. Each of these users may require access to a different set of software, hardware, and network services. For example, each desktop computer uses a certain operating system, set of software applications, LAN interface card, network protocol software, network operating system, and network management software.

Figure 12-12

Network manager reconfiguring cables

The function of **configuration management** is to track and manage all the hardware and software associated with the computer network, and to manage any changes that occur to these architectural elements. This information is usually tracked in a distributed database management system that automatically surveys network devices and provides updated, real-time reports. The system also provides graphics of network topology, computing devices, and associated software and hardware.

PERFORMANCE MANAGEMENT

Another function that network managers must provide is **performance management**, which ensures that the network is performing adequately for the applications and users it supports. Performance management software tests the network to make sure it is providing acceptable information transmission speeds, transmitting information with an acceptable level of dropped packets and errors, and providing acceptable throughput and utilization levels. The network should be large enough to accommodate current and planned traffic patterns, but not so large or overdesigned that it is not cost effective.

Performance management software automates much of this process. At regular intervals, a performance management tool gathers network statistics about various performance parameters, such as response time, dropped packets, or utilization. Next, it compares the data to a baseline or normal range of statistics called a performance threshold. If the collected data exceeds a performance threshold level, the software generates an alert for the network manager.

FAULT MANAGEMENT

One of the most important functions of network management is **fault management**—solving a network outage or performance problem when it occurs. An effective fault management system can automatically detect network degradation or a service interruption, isolate the source of the problem, and sometimes even diagnose and fix the problem. If the system cannot fix the problem, it notifies appropriate personnel. A fault management system also keeps a fault log that maintains records of network problems and outages and how the problems were resolved.

SECURITY MANAGEMENT

Network security is a vital management function that is often handled by a dedicated security team. Some of these **security management** functions include network access control, user authentication, firewall management, and critical infrastructure protection. Some security threats to enterprise networks include worms and viruses, spam, unauthorized access attempts, and data interception. Those involved in security management implement technologies to prevent these attacks and monitor networks in real time to detect and solve problems. Security management techniques include encrypting information to provide private information exchange, implementing authentication technologies to verify that authorized users are actually who they say they are, and access control—the process of controlling and monitoring access to information resources through technologies such as firewalls. Chapter 15 describes network security functions in greater detail.

CHAPTER SUMMARY

- A WAN is a network that spans a large geographical distance and can transmit voice, data, and multimedia information.
- The Internet is the most prominent example of a WAN.
- Most WAN services are run over a public network rather than private lines due to cost concerns.
- Most WANs, including the Internet, are based on a network approach known as packet switching, which breaks information into small segments called packets prior to transmission. These packets are sent over a network, possibly routed along different paths, and reassembled at their destination.
- A virtual private network (VPN) runs over a public network like the Internet, but it emulates a private network's higher performance and security by applying security measures like encryption and by offering quality of service (QoS) guarantees.
- WANs require network protocols, the standard rules that allow information to be exchanged over the network.
- Important types of WANs are Internet services, frame relay, Asynchronous Transfer Mode, and multiprotocol label switching.
- Most businesses and large institutions access WANs via leased lines, while smaller businesses and individual users access WANs via wireless technologies, DSL, and cable modems.
- Ensuring adequate WAN performance and reliability requires network management functions, including configuration management, performance management, fault management, and security management.

KEY TERMS

asymmetric digital subscriber line	packet
Asynchronous Transfer Mode	packet switching
circuit switching	performance management
configuration management	proprietary protocol
connectionless	QoS
CSU/DSU	router
digital subscriber line	routing table
edge router	security management
enterprise networking	SLA
fault management	symmetric digital subscriber line
frame relay	Transmission Control Protocol/Internet Protocol
high data rate digital subscriber line	tunneling
hop	virtual private network
MPLS	WiMAX
network protocol	X.25
open protocol	

REVIEW QUESTIONS

1. What is packet switching, and how is it different from circuit switching?

2. What historical circumstance influenced the development of packet switching in the United States?

3. Define connectionless network services, and give an example of a type of WAN that uses this approach.

4. Describe the difference between a proprietary protocol and an open protocol, and give one example of each.

5. What is the primary function of a router?

6. What is the function of a CSU/DSU, and where does it reside on a WAN?

7. Draw a diagram of a dedicated private network that serves eight locations. Build in redundant paths in case of transmission line outages.

8. What are the inherent advantages of a private network, and what is the main disadvantage?

9. What is a virtual private network, and what need does it address?

10. Explain the concept of tunneling.
11. What types of performance metrics might a QoS agreement specify?
12. Is there any difference between the terms *frame*, *packet*, and *cell*?
13. In frame relay networks, what is the difference between PVCs and SVCs?
14. Are ATM cells fixed length or variable length?
15. What does MPLS stand for?
16. What is the chief advantage of MPLS networks?

17. Give some examples of transmission speeds for alternative leased-line Internet access options.
18. Is DSL offered by cable companies or by telephone companies?
19. What transmission medium does DSL use?
20. What transmission medium does cable modem Internet access use?
21. What are the four categories of network management, and what do they describe?

DISCUSSION QUESTIONS

1. What market factors might have driven the evolution in businesses from using proprietary WANs such as IBM's SNA and DEC's DECnet to using more open WANs such as Internet services?
2. What challenges do businesses face for exchanging information and communicating with customers and suppliers if they use public networks? What are some possible solutions to these challenges?

3. If the emerging WiMAX service becomes a widespread alternative for broadband wireless Internet access in a metropolitan area, what would the implications be for users and for service providers such as cable companies and telephone companies?

CASE PROJECT

Develop a strategic plan for a small business wide area network.

Imagine that you are the head of IT for a new manufacturing company that is building three factories and a corporate headquarters in the United States. You are responsible for defining network requirements and for submitting a request for proposals (RFP) to three WAN service providers. Write a one-page RFP that includes the following information:

- *Locations*—Where will the three factories and the corporate headquarters be located?
- *Number of users*—How many employees will require network access at each location?
- *Application requirements*—What types of information will be exchanged between these locations? For example, does the information include voice, image, or video applications? Does it include factory data? Between which locations will these applications need to be networked? Will there be any bandwidth-intensive applications?
- *External connectivity*—Will any external users (such as customers and suppliers) require network access to support these applications?

- *Performance requirements*—How critical is network performance and reliability? Will you want a guaranteed service agreement with the network provider?
- *Security requirements*—How sensitive will the transmitted information be? Will you require networked security services such as firewall management?
- *Future growth*—Do you expect to add employees, application traffic, or more locations in the next five years?
- *Additional services*—Do you plan to outsource any network management services to the network provider or a third party?

CHAPTER 13

COMMUNICATION PROTOCOLS

LEARNING OBJECTIVES

In this chapter you will:

- Understand the purpose and importance of communication protocols

- Gain familiarity with the seven-layer OSI reference model

- Learn about TCP/IP, the protocols that enable communications over the Internet

- Understand the role of standards organizations in the development of protocols

- Identify the economic and political implications of IT standards

INTRODUCTION

People can communicate and interact efficiently if they follow certain rules, or protocols. These protocols include using a shared alphabet, shaking hands, or addressing an envelope in a standard format with name, address, city, state, and zip code. Similarly, digital devices can share information only if they adhere to standard formats for structuring the underlying binary code. This chapter describes the important role of communication protocols, also called network protocols or standards, in enabling digital information exchange over computer networks. It also describes the OSI (Open Systems Interconnection) reference model, a framework for organizing and understanding groups of protocols. The chapter examines a set of protocols called TCP/IP, the universal and open standards that allow devices to exchange information over the global Internet. Several important organizations are involved in establishing protocols through a complex array of processes for setting standards, including the Internet Engineering Task Force (IETF) and the World Wide Web Consortium (W3C). The final section describes how these protocols are not only technological design decisions, but how they can help to make economic and political choices mediating between competing economic interests, global trade interests, and social issues such as law enforcement versus individual privacy.

Computers and communication devices can seamlessly share and exchange information only if they "speak the same language," adhering to a common system of rules known as protocols. To understand how and why computers use protocols, consider how people rely on protocols in everyday life, as shown in Figure 13-1. Protocols govern any human interaction or exchange of information and are so pervasive that they are almost invisible. For example, people in various cultures understand the protocols used to greet each other, such as shaking hands or bowing. Similarly, protocols govern routine interactions such as stopping your car at a red light.

Figure 13-1

Protocols in real life

Greeting   Communicating   Interacting

Protocols vary from society to society. For example, automobile drivers in some countries use the left side of the street rather than the right. Rules for written communication also vary considerably among cultures—different languages employ different alphabets, words, and conventions for reading from left to right or right to left. The everyday communication protocol that is most analogous to IT protocols is the process of sending a letter. It requires adherence to strict conventions such as grouping characters into words, separating words into sentences using punctuation, using a spell checker, and placing the letter (the information content) in an envelope bearing the recipient's address in a predetermined format.

The example of letter writing is often used to help explain IT protocols because its social conventions are so similar to the protocols that enable information exchange between digital computing devices over networks. However, rather than providing order to the alphabets people use, IT protocols provide order to the binary streams used by digital devices. **Protocols** are predetermined, agreed-upon rules that specify how information is formatted, exchanged, and interpreted between nodes over a network. Protocols are simply the standards that provide order to information exchange over a communications network.

The purpose of communication protocols, which are also called computer networking protocols or standards, is to ensure **interoperability**—the ability of equipment and software made by different manufacturers to work with each other. Just as certain verbal languages are not interoperable, there was no way to connect different computing environments over a common network before network protocol standards were available. Instead of using common, interoperable protocols, networks consisted of single-vendor, proprietary (closed) network protocols such as IBM's Systems Network Architecture (SNA) or Digital Equipment Corporation's DECnet, as described in the previous chapter. Unlike proprietary protocols, **open standards** allow diverse computing environments to interoperate by providing published guidelines for formatting and exchanging information. Protocols establish guidelines for how to perform the following functions (see Figure 13-2):

- Formatting information in binary code
- Breaking up information into manageable units prior to transmission (these units include packets, frames, and cells, as discussed in the previous chapter)

- Detecting the presence of another node on the network
- Specifying what kind of standard connector an Ethernet cable should have
- Initiating or terminating a connection
- Appending a source and destination address to the bits in a standardized format
- Applying error detection and correction methods (for example, adding a parity bit)

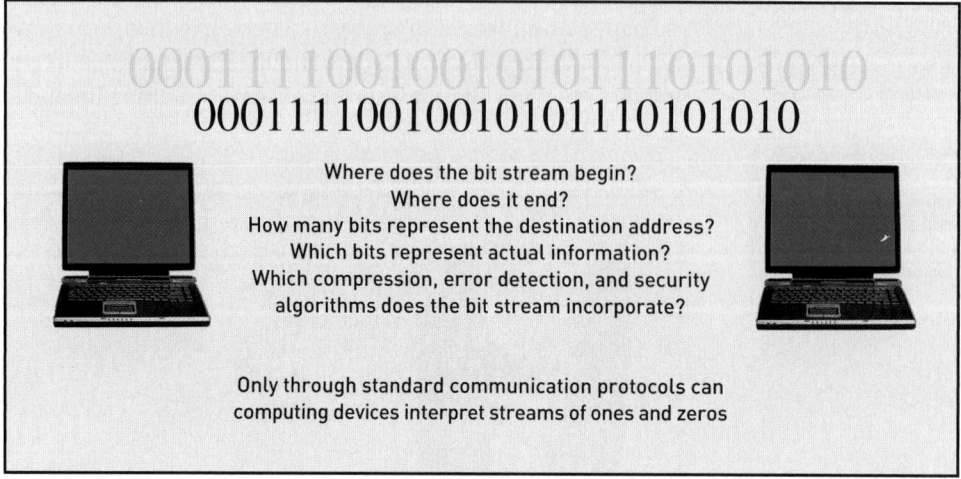

Figure 13-2

The role of communication protocols

Recall from previous chapters that each LAN standard uses a different protocol for formatting information within a packet, and that various protocols such as frame relay, MPLS, and ATM dictate the format and other specifications for WAN transmission. Recall also that ASCII is a standard for formatting information, and that Binary Coded Decimal is a technique for encoding information in binary. Information technology protocols, of which communication protocols are a subset, permeate every aspect of IT.

PROTOCOL SUITES

With so many protocols simultaneously acting on the same information, it can be challenging to understand their individual responsibilities. A taxonomic approach of grouping protocols into families, called network **protocol suites**, is useful. Protocols are hierarchical, in that any given protocol depends on or builds upon protocols that have already been applied. This hierarchical quality makes protocol families conducive to divisions into "layers," with each layer defining a specific function. The terminology can quickly become confusing. For example, what's the difference between a "layer," a "protocol standard," and an actual protocol implementation? Protocol **layers** are just conceptual tools for grouping and dividing protocol functions, while protocol standards are the actual, agreed-upon specifications and rules for protocol functions. A protocol implementation (sometimes called a protocol **stack**) is the vendor-specific software or hardware that adheres to these standards.

THE OSI REFERENCE MODEL

There are literally thousands of IT protocols. To help organize, discuss, and understand the many functions required to exchange information between computing devices, the computer networking field often relies on the Open Systems Interconnection (OSI) reference model. The **OSI reference model**, developed in the 1980s by the International Organization for Standardization (ISO), divides protocol functions into seven categories, or layers. As shown in Figure 13-3, the seven layers are the Application layer, the Presentation layer, the Session layer, the Transport layer, the Network layer, the Data Link layer, and the Physical layer.

Keep in mind that the OSI model is a conceptual tool to understand protocol functions and help organize the large number of protocols. The actual protocols achieve various network functions and enable interoperability between computers and associated devices. The conceptual framework helps organize and describe how various protocols contribute to information exchange between computing devices. The ability of two devices to exchange information requires numerous protocols, not just one. A protocol that provides a specific function depends on other protocols providing different functions.

Figure 13-3

The OSI reference model

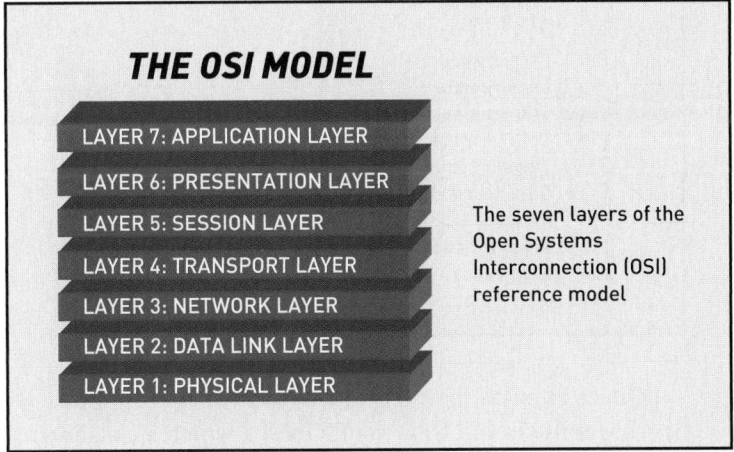

The following sections describe the layers of the OSI model. Each description includes real-world examples of protocols that are considered part of each layer.

The Physical Layer

Layer 1, the **Physical layer**, specifies the functions of protocols that define the electrical, optical, and mechanical specifications for the interface between a device and a transmission medium, such as fiber-optic cable, coaxial cable, or twisted pair. For example, voltage levels, transmission rates, signal timing, connection specifications, and other characteristics are addressed at this layer. An example of a standard specification that operates at this layer is the TIA/EIA-232 serial interface specification (formerly known as RS-232), shown in Figure 13-4. To provide a simple example of the kind of information that a Physical layer standard specifies, Figure 13-5 depicts the standard pin configuration on the 9-pin TIA/EIA-232 interface. Note that pins 2 and 3 are designated for transmitting and receiving information. Other pins have specific functions such as issuing a request to send information or indicating that it is clear to send.

Figure 13-4

TIA/EIA-232 9-pin interface

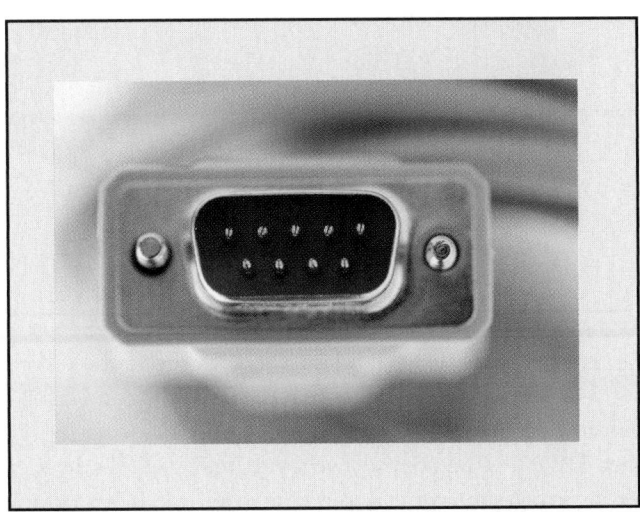

Figure 13-5

TIA/EIA-232 9-pin
specification

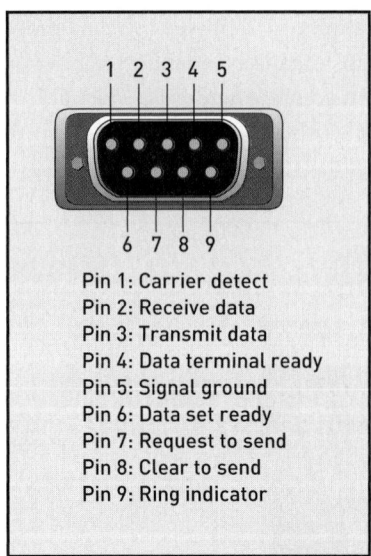

Pin 1: Carrier detect
Pin 2: Receive data
Pin 3: Transmit data
Pin 4: Data terminal ready
Pin 5: Signal ground
Pin 6: Data set ready
Pin 7: Request to send
Pin 8: Clear to send
Pin 9: Ring indicator

Other Physical layer protocols include Synchronous Digital Hierarchy (SDH) and Synchronous Optical Network (SONET), which are standards for transferring information onto a fiber-optic network. These standards specify important features such as the bit rates for optical signals.

The Data Link Layer

Layer 2, the **Data Link layer**, specifies a number of important protocol functions. For example, some Data Link layer specifications define physical addressing, such as Ethernet addresses for a network interface card (see Figure 13-6) that connects a computing device to a local area network. Other important network functions at this layer include topology (such as ring or bus) and frame formats. The Data Link layer is closely associated with specific LAN standards. For example, Ethernet and token ring operate at this layer, as do some types of wide area networks, such as frame relay.

Figure 13-6

Network interface card

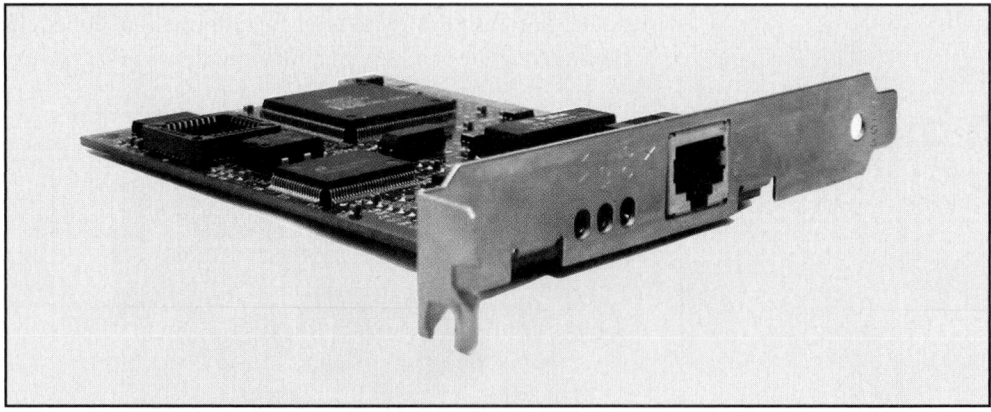

Part of the frame relay standard specifies the frame relay frame structure, as shown in Figure 13-7. The standard dictates that the first 8 bits and last 8 bits of each frame are called a flag, indicating the beginning and end of each frame. The 8-bit flag is always the binary number 01111110. The second two bytes make up the header, which consists of physical addressing, congestion control, and other overhead information. The next section of the frame is of variable length and contains the actual information, or data, to be transmitted. The next two bytes make up the frame check sequence (FCS) field, a value assigned by the transmitting device and verified by the receiving device to ensure transmission integrity. This brief example helps to explain what a network protocol specification does at the Data Link layer.

Figure 13-7

Standard frame relay
frame structure

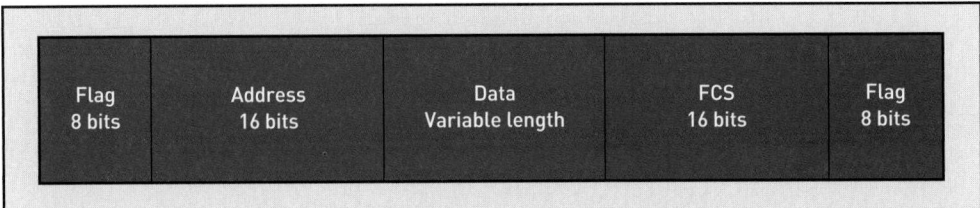

Flag 8 bits	Address 16 bits	Data Variable length	FCS 16 bits	Flag 8 bits

The Network Layer

Layer 3, the **Network layer**, is responsible for how packets should be routed and switched through a network. The Network layer handles the logical (software-defined) network address of the sender and destination. This logical address is not the same as the physical addresses defined in Layer 2. The specifications defined at this layer are critical for the successful functioning of routers (see Figure 13-8), which use logical source and destination addresses to determine how to forward information packets across a network. One of the best-known network protocols operating at this layer is the Internet Protocol (IP), the standard Network layer protocol used to route information over the Internet.

Figure 13-8

Network router

Another example of a Network layer protocol is the Border Gateway Protocol (BGP), designed to exchange network reachability information among BGP systems. It maintains information about how to route packets based on destination IP addresses. To gain a sense of the complexity of any given network protocol, you can go to *http://tools.ietf.org/html/rfc4271* and examine the 103-page document that describes BGP. The following is a brief excerpt from this specification:

```
Message Header Format
Each message has a fixed-size header. There may or may not be a data
portion following the header, depending on the message type. The
layout of these fields is shown below:
    0                   1                   2                   3
    0 1 2 3 4 5 6 7 8 9 0 1 2 3 4 5 6 7 8 9 0 1 2 3 4 5 6 7 8 9 0 1
   +-+-+-+-+-+-+-+-+-+-+-+-+-+-+-+-+-+-+-+-+-+-+-+-+-+-+-+-+-+-+-+-+
   |                                                               |
   +                                                               +
   |                                                               |
   +                                                               +
   |                           Marker                              |
   +                                                               +
   |                                                               |
   +-+-+-+-+-+-+-+-+-+-+-+-+-+-+-+-+-+-+-+-+-+-+-+-+-+-+-+-+-+-+-+-+
   |          Length               |      Type     |
   +-+-+-+-+-+-+-+-+-+-+-+-+-+-+-+-+-+-+-+-+-+-+-+-+-+

Marker:

     This 16-octet field is included for compatibility; it MUST be
     set to all ones.

Length:

     This 2-octet unsigned integer indicates the total length of the
     message, including the header in octets. Thus, it allows one to
     locate the (Marker field of the) next message in the TCP
     stream.  The value of the Length field MUST always be at least
     19 and no greater than 4096, and MAY be further constrained,
     depending on the message type. "Padding" of extra data after
     the message is not allowed. Therefore, the Length field MUST
     have the smallest value required, given the rest of the
     message.

Type:

     This 1-octet unsigned integer indicates the type code of the
     message. This document defines the following type codes:

                        1 - OPEN
                        2 - UPDATE
                        3 - NOTIFICATION
                        4 - KEEPALIVE
```

This short excerpt from the BGP standard is taken from RFC 4271. RFC refers to the "Request for Comments" series documenting Internet standards. RFCs are discussed later in this chapter.

The Transport Layer

The protocols at Layer 4, the **Transport layer**, ensure that information has successfully moved between two points on a network. Protocols at this layer have many functions, including segmenting data for transmission over a network and ensuring that data arrives at its destination in the proper order. Transport layer protocols perform the important functions of detecting any errors that may have occurred in transmission and correcting these errors through a number of remedies, including retransmitting the information. Two common examples of Transport layer protocols are Transmission Control Protocol (TCP) and User Datagram Protocol (UDP).

As another example of the complexity of network protocols, TCP is specified in *multiple* documents. TCP, a core protocol of the global Internet, was first described in 1981 (see *http://tools.ietf.org/html/rfc0793*). Other standards that define elements of TCP specify how to perform congestion control (see *http://tools.ietf.org/html/rfc2581*) and how to apply algorithms that compute retransmission timers, which sending devices use to retransmit information when the receiver has not provided any feedback (see *www.ietf.org/rfc/rfc2988.txt*).

To provide an idea of what these detailed standards define, the following is an abridged segment of the standard that specifies how to calculate the TCP retransmission timer:

```
The Basic Algorithm

    To compute the current RTO, a TCP sender maintains two state
    variables, SRTT (smoothed round-trip time) and RTTVAR (round-trip
    time variation). In addition, we assume a clock granularity of
    G seconds.

    The rules governing the computation of SRTT, RTTVAR, and RTO are
    as follows:

    (2.1) Until a round-trip time (RTT) measurement has been made for
          a segment sent between the sender and receiver, the sender
          SHOULD set RTO <- 3 seconds (per RFC 1122 [Bra89]), though
          the "backing off" on repeated retransmission discussed in
          (5.5) still applies.

    (2.2) When the first RTT measurement R is made, the host MUST set

              SRTT <- R
              RTTVAR <- R/2
              RTO <- SRTT + max (G, K*RTTVAR)

          where K = 4.

    (2.3) When a subsequent RTT measurement R' is made, a host MUST set

              RTTVAR <- (1 - beta) * RTTVAR + beta * |SRTT - R'|
              SRTT <- (1 - alpha) * SRTT + alpha * R'

          The value of SRTT used in the update to RTTVAR is its value
          before updating SRTT itself using the second assignment. That
          is, updating RTTVAR and SRTT MUST be computed in the above
          order.

          The above SHOULD be computed using alpha=1/8 and beta=1/4 (as
          suggested in [JK88]).

          After the computation, a host MUST update
          RTO <- SRTT + max (G, K*RTTVAR)
```

This short excerpt is taken from RFC 2988.

The Session Layer

The **Session layer**, as the name suggests, describes protocols that establish, maintain, and terminate sessions between two computing devices that are exchanging information. The term *session* refers to the logical connection between two devices from the establishment of the connection to the termination of the session. Some examples of real-world Session layer protocols are Remote Procedure Call (RPC) and H.323, a protocol that provides audio and video sessions over a packet network. H.323 is considered a VoIP-related protocol, and is described in Chapter 17.

The Presentation Layer

For different computing environments to share information across a network, the information must be encoded and formatted in a common standard. Layer 6, the **Presentation layer**, formats and encodes information from an application so that other computing devices using the same standard will understand it. The Presentation layer is aptly named because it is responsible for the presentation of information. Standards at this layer convert data to standard formats and provide important roles such as compressing information or applying encryption.

Well-known Presentation layer standards include Motion Picture Experts Group (MPEG), a standard method for compressing digital video information and Joint Photographic Experts Group (JPEG), a standard for encoding and formatting digital images. MPEG and JPEG were described in Chapter 7. ASCII, EBCDIC, and UNICODE, all explained in Chapter 3, are well-known Presentation layer standards for encoding textual information. Figure 13-9 shows part of the ASCII table that serves as a standard for converting textual characters to binary and vice versa. An example of a Presentation layer encryption standard is Secure Sockets Layer (SSL).

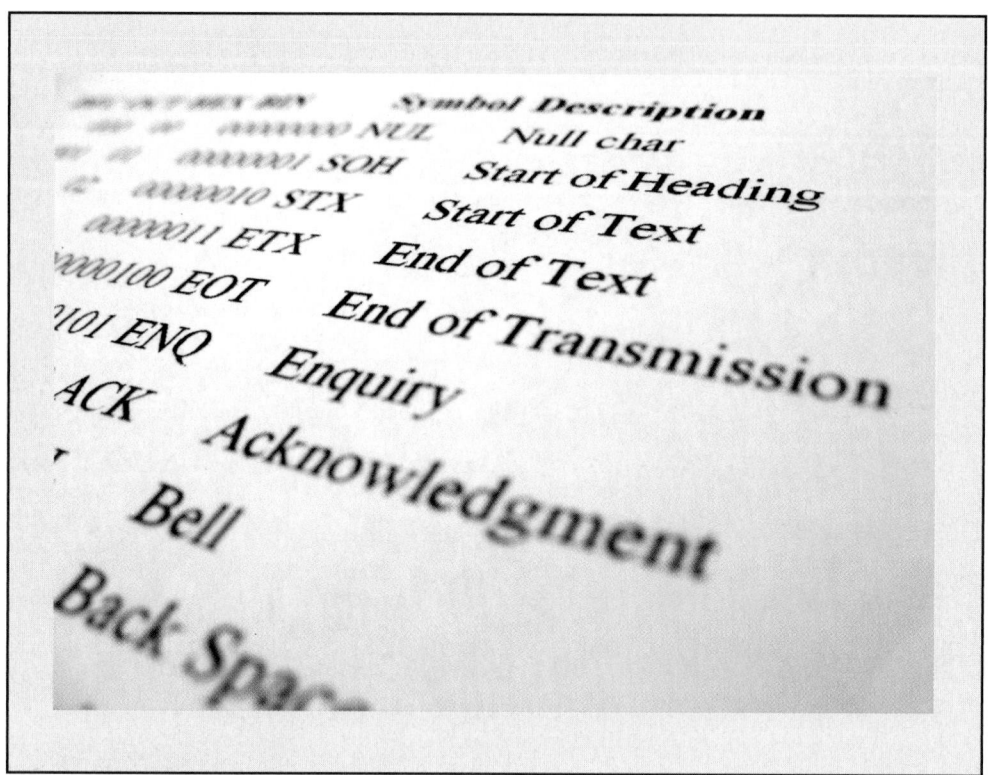

The Application Layer

A user interacts with a network primarily through applications running on a computer. These applications, in turn, interact with **Application layer** protocols that initiate the communications process and coordinate the exchange of information between two networked devices. Common network protocols that operate at the Application layer include Hypertext Transfer Protocol (**HTTP**) for Web communications, Simple Mail Transfer Protocol (SMTP), and File Transfer Protocol (FTP). HTTP is an important Application layer protocol because it enables the exchange of information between a Web browser and a Web server. The "http" that often appears at the beginning of a uniform resource locator (URL) is a reference to the HTTP protocol (see Figure 13-10).

Figure 13-10

URL specifying the HTTP
protocol

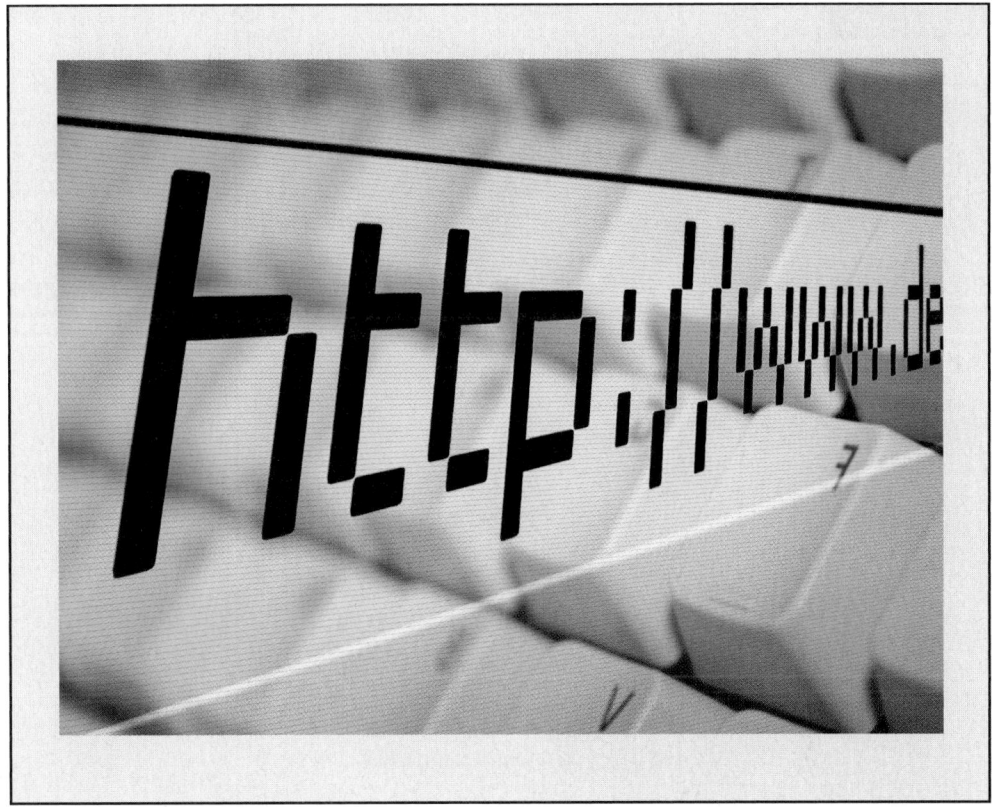

HTTP is a higher-level protocol that runs on top of other protocols, usually TCP and IP. One area specified in the HTTP standard is the syntax and semantics of a URL. As more evidence of how complex and detailed protocol parameters are, the HTTP specification alone is 175 pages long. Anyone can read and review such specifications online. For example, much of the HTTP standard is documented in RFC 2616 (see *www.ietf.org/rfc/rfc2616.txt*).

THE TCP/IP PROTOCOL SUITE

An important group of protocols is the **TCP/IP** protocol suite. These networking protocols have become the universal set of rules for information exchange between computers and associated devices over the Internet. To communicate over the Internet, a computing device must "speak" TCP/IP. By strict definition, TCP/IP is two protocols—TCP (**Transmission Control Protocol**) and IP (**Internet Protocol**). Each protocol performs a distinct function. However, TCP/IP customarily describes an entire family of protocols known as the TCP/IP protocol suite. For example, it includes protocols for tasks such as file transfer (FTP), electronic mail (SMTP), and a host of routing protocols. TCP/IP, by convention, is the group of protocols that work together to facilitate information sharing over the Internet.

Unlike the seven-layer OSI framework, the protocols of what is usually called the TCP/IP protocol suite are grouped into four layers: the Network Interface layer, the Internet layer, the Transport layer, and the Application layer, as shown in Figure 13-11. Each layer specifies a different function and depends on a preceding layer. Extensive information about TCP/IP and related topics, such as Internet addressing, is presented in Chapter 14.

Figure 13-11

The TCP/IP protocol suite

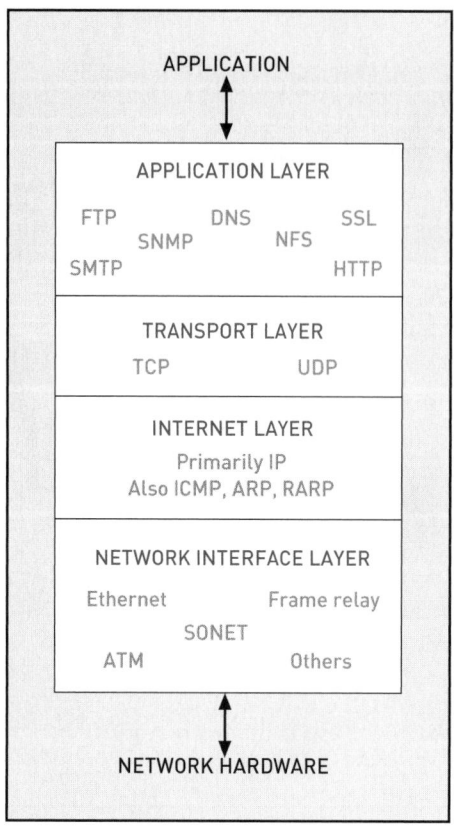

PUTTING IT ALL TOGETHER

Again, communication between two devices requires multiple protocols. To understand the functionality of various protocols, consider the following oversimplified example. Within a computer, data passes from a software application such as a Web browser to software that implements TCP. Data is divided into manageable pieces, and formats and routines are applied to ensure that data arrives correctly at its destination. The next step involves the Internet Protocol (IP), which applies information to logically address and route the data to its destination. Then, an Ethernet protocol or other network interface protocol may append a physical address of a piece of networking hardware. The network interface protocol also transforms the data into an appropriate format for a stream of bits to pass physically from a computer to a network. The connection relies on standards for the physical Ethernet jack, which is used to help interface a computing device to an Ethernet LAN. Depending on the information transmitted, the exchange might also rely on the ASCII standard for formatting text, JPEG for formatting images, or the MP3 standard for formatting and compressing audio.

ORGANIZATIONS THAT SET STANDARDS

If protocols are agreed-upon rules, who has to agree? Someone has to design, publish, and maintain these network protocols. No single organization, government, or company assumes responsibility for establishing the enormous number of standards required for network communications. Standards are sometimes referred to as either *de jure* standards or *de facto* standards. *De facto* ("from the fact" in Latin) standards rise to dominance not through a formal process of collaborative effort, but are adopted over time by gaining momentum or because they are considered best industry practices. *De jure* ("from the law" in Latin) standards develop through a formal, premeditated process, such as a standards-setting organization that collaborates to develop a protocol standard that addresses a specific requirement.

A complex matrix of organizations, sometimes in competition with each other, assumes responsibility for setting standards in various areas. These groups might be national, international, professional organizations, consortia of technology companies, or completely open to public participation. How, then, do they establish standards, given such diversity? Each organization has its own formal procedures for developing standards, which include the participation of individual members. These members are often employees of businesses with a stake in the outcome of standards development. Some important standards organizations are introduced in the following sections.

AMERICAN NATIONAL STANDARDS INSTITUTE (ANSI)

Founded in 1918, ANSI is one of the oldest standards institutions in the world. **ANSI** is a private, nonprofit organization with participating members from businesses, organizations, academic institutions, and governmental agencies. Its mission is to promote and coordinate standards that are in the economic interest of the United States. ANSI is primarily a coordinating umbrella organization for voluntary standards development. Some ANSI-approved standards include Fiber Distributed Data Interface (FDDI) for LANs and standards associated with ADSL (Asymmetric Digital Subscriber Line). ANSI also serves as the official U.S. representative to the International Organization for Standardization (ISO). More information about ANSI is available at *www.ansi.org*.

ELECTRONIC INDUSTRIES ALLIANCE (EIA)

The Electronic Industries Alliance (**EIA**) is a trade organization of more than 2500 companies involved in manufacturing electronic systems and components. The EIA specifies electrical standards for a variety of networking components, including TIA/EIA-232 connectors and Category 6 cable under TIA/EIA-568-B.2-1. In these specifications, TIA refers to a British standards organization called the Telecommunications Industry Association, now known as the CITA (Communications and Information Technology Association). For more information about these organizations, see *www.eia.org* and *www.tia.org.uk*.

INTERNATIONAL ORGANIZATION FOR STANDARDIZATION (ISO)

The **ISO** is the international standards-setting organization made up of national standards bodies from more than 150 countries. For example, ANSI is the standards body representing the United States in the ISO. The member organizations each have one vote and are not necessarily official delegates of national governments, although in some cases they are. The ISO is actually a nongovernmental entity spanning public and private groups. The group of MPEG standards that define the encoding and compression of movies is an example of one of the ISO's standards affiliations—in this case, through a joint technical committee (ISO/IEC JTC 1) known as the Motion Picture Expert Group. For more information, see *www.iso.org*.

INSTITUTE OF ELECTRICAL AND ELECTRONICS ENGINEERS (IEEE)

The **IEEE** (*www.ieee.org*), a nonprofit professional organization of engineers, has contributed important network protocol standards. One prominent example is the IEEE 802 series of local area networking standards, which the IEEE began specifying in the 1980s. Important IEEE specifications include Ethernet, token ring, and wireless LAN standards. The following are examples of IEEE working groups that are developing network standards:

- IEEE 802.3 CSMA/CD (for Ethernet LANs) Working Group
- IEEE 802.5 Token Ring Working Group

- IEEE 802.11 Wireless LAN (WLAN) Working Group
- IEEE 802.15 Wireless Personal Area Network (WPAN) Working Group
- IEEE 802.16 Broadband Wireless Access (BBWA) Working Group

INTERNET ENGINEERING TASK FORCE (IETF)

The standardization process for Internet protocols involves several organizations that fall loosely under the umbrella of the Internet Society (ISOC). One of these organizations is the Internet Engineering Task Force (**IETF**), the group most directly responsible for developing Internet protocol standards. The IETF (*www.ietf.org*) has no formal membership, but has more than 100 "working groups" that advance Internet standards. The Internet standards process involves proposing a draft standards specification, followed by a period of iterative revisions by a working group. Since 1969, an electronically published archive known as the "request for comments," or **RFC** series, has documented the Internet standards process. Even if a proposed draft becomes an official Internet standard, it retains this RFC nomenclature (for example, RFC 1142). Thousands of RFC exist, which should demonstrate the complexity of Internet specifications. You can view all RFC documents dating back to 1969 at *www.rfc-editor.org*.

INTERNATIONAL TELECOMMUNICATION UNION (ITU)

The **ITU**'s Telecommunication Standardization Sector (ITU-T), formerly known as the International Telegraph and Telephone Consultative Committee (CCITT), is an international telecommunications standards organization operating under the auspices of the United Nations (*www.itu.org*). The ITU-T has 13 study groups that are responsible for standardization in widely ranging technical areas, including network management systems, signaling standards, numbering systems, and transmission systems.

WORLD WIDE WEB CONSORTIUM (W3C)

The World Wide Web Consortium, or **W3C** (*www.w3c.org*), was founded in 1994 with the mission of promoting continued development of Web technologies and ensuring interoperable, vendor-independent Web standards. For example, the W3C developed and continues to maintain specifications for Extensible Markup Language (XML) through the XML core working group. Other examples of W3C working groups include the voice browser working group and the XML encryption working group.

NATIONAL INSTITUTE OF STANDARDS AND TECHNOLOGY (NIST)

In the United States, the National Institute of Standards and Technology (**NIST**) was formed to develop standards and technologies that promote U.S. economic competitiveness and innovation. Dating back to 1901, NIST is a nonregulatory agency of the U.S. government located in Gaithersburg, Maryland and Boulder, Colorado. NIST is involved in setting standards in a diverse selection of areas ranging from measurements to automobile manufacturing to health care. As an example of the agency's efforts in the area of information technology, NIST has contributed to encryption protocols such as Advanced Encryption Standard (AES). For information about NIST, see *www.nist.gov*.

ECONOMICS AND POLITICS OF PROTOCOLS

One interesting characteristic of communication protocols is that they are invisible to many users. A user who accesses a Web site on the Internet may not be aware that the connection uses an Internet address specified by IP, that information between the Web site and Web browser is being exchanged using HTTP, or that the actual information is formatted and developed using HTML or XML.

Users who are aware of the underlying protocols might assume that they are "just a technical design decision" and do not necessarily have social or economic implications. However, this assumption is not quite accurate. Because protocols represent a source of control over technology, they can serve as points of economic competition and political influence. The following sections describe some examples of the economic and political nature of protocols.

PROTOCOLS AND THE PUBLIC INTEREST

Communication protocols, like most technical standards, are not established by legislatures or elected officials, but by standards organizations. These organizations are usually made up of people who work for corporations that have an interest in the outcome of the standards process. Though not established by elected officials, some of the design decisions nevertheless establish public policy. For example, when a user purchases a song over a network, a proprietary standard that provides digital rights management (DRM) is downloaded with the actual song. DRM features are designed to prevent piracy of songs and other digital property.

These standards have implications. For example, the proprietary standard appended by one music seller is different from the standard applied by another music seller. As a result, music downloaded via Apple's iTunes service can be played on an iPod but not on another type of digital music player. This lack of interoperability was not acceptable when consumers played record albums, cassette tapes, or CDs. Any CD could be played on any CD player, but DRM standards do not make similar demands of networked music purchases.

Encryption standards are another type of network standard that has always intersected with public policy issues. Encryption policy has to balance the need for individual privacy with the need for law enforcement and surveillance. To illustrate how politically charged encryption is, note that the U.S. government until 1996 categorized encryption products as munitions under the U.S. International Traffic in Arms Regulations (ITARS). In other words, businesses or individuals who manufactured encryption products fell within the category of arms dealers. Like firearms, encryption products could not be exported without a license and could not be exported to some countries under any conditions. Later U.S. regulatory modifications eased export restrictions considerably, but products with strong encryption standards have encountered a patchwork of laws, with some countries prohibiting any encryption, others requiring licenses, and others restricting encryption based on encryption strength.

The issue of wiretapping standards, which are negotiated between governments and standards organizations, is another interesting issue. So are personal identifiers built into network standards that can identify the whereabouts of a person using a network. This latter issue is called the location privacy issue and is often brought up in the context of wireless networks.

If standards create issues for public policy, the related question is how the public interest is represented in the standards-setting process. As mentioned previously, standards are not established by typical network users. With a few exceptions, they are not established by courts, states, or any body of elected representatives. Hundreds of standards organizations exist, all with their own policies for who can be involved in setting standards. Some of these organizations exclude nonmembers from the decision-making process, and membership sometimes requires paying high fees. Other standards organizations do not allow open public participation, nor do they provide transparent access to standards specifications.

Other standards-setting organizations, including the IETF and the W3C, provide free and open membership and provide standards documentation to the public for free. Even in these cases, though, it can be challenging for the public to become involved in setting standards. Without financial backing from corporate employers with a stake in the outcome, participating in the standards process is uncompensated activity, making it a luxury that most members of the public cannot afford. Involvement in standards work also might require detailed knowledge of very esoteric technologies. Furthermore, there are hundreds of network standards.

ECONOMIC IMPLICATIONS OF PROTOCOLS

The economic stakes of network standards and all other technical standards are easy to understand, because having influence over the outcome of standards selections can produce great economic advantages. If a company's technological specification is selected as a standard, the company benefits economically because it has already invested in the specification. Furthermore, if a company's specification is selected as a standard and it holds patents on the underlying technology, any other company that wants to use the standard in its own product might have to pay royalties to the patent holder.

Economic issues in setting standards are closely related to the issue of innovation. For example, the argument for allowing patents in standards is that the developer of a technology will have no incentive to innovate if it cannot protect the intellectual property underlying the invention. Conversely, some economists argue that product innovation is maximized when standards are made available on a royalty-free basis or under reasonable conditions. In this way, all product developers can use the standard to make innovative and competing products.

Another economic issue for standards is more global, and concerns nations that can use regulations and standards as global barriers to trade. One well-publicized standards conflict escalated to the highest levels of the Chinese and U.S. governments. In 2003, the standards administration of China announced that a proprietary protocol for wireless local area network encryption would be nationally mandated. Furthermore, foreign equipment manufacturers that wanted to do business in China would have to license the proprietary protocol from one of China's network equipment manufacturers and, perhaps, even would have to reveal underlying technical designs to the Chinese manufacturers. This policy effectively would create a trade barrier for foreign manufacturers that wanted to sell wireless LAN products in China.

Clearly, communication protocols are important not only from the perspective of technical interoperability. They can have implications for public policy, global trade, economic competitiveness, and innovation policy.

CHAPTER SUMMARY

- Computing devices can exchange information over a network if they adhere to standard rules called protocols. Protocols perform a number of functions, including how to format information in binary code, how to indicate the initiation or termination of a connection, and how to append standard addressing information prior to transmission.
- The OSI reference model is a framework that groups the many protocols and protocol functions required to exchange information over a network. The seven layers of the OSI model are the Physical layer, the Data Link layer, the Network layer, the Transport layer, the Session layer, the Presentation layer, and the Application layer.
- The TCP/IP protocol suite is the group of communications protocols that collectively enable information exchange over the public Internet (and private TCP/IP networks). By strict definition, TCP/IP is two protocols (TCP and IP), but it is usually interpreted to include many other protocols used in Internet communications, such as FTP, SMTP, and HTTP.
- Many standards-setting organizations establish communication protocols. Many members of these organizations work for companies that have a stake in the outcome of protocol design decisions. These organizations include ANSI, EIA, NIST, the IEEE, the IETF, the ITU, the ISO, and the W3C.
- Protocols have economic implications because they represent a means of control over technology, mediate between competing equipment and software manufacturers, determine how innovative and competitive markets can be, and sometimes even serve as barriers to trade in global IT markets.

KEY TERMS

ANSI	NIST
Application layer	open standard
Data Link layer	OSI reference model
de facto standard	Physical layer
de jure standard	Presentation layer
EIA	protocol
HTTP	protocol suite
IEEE	RFC
IETF	Session layer
Internet Protocol	stack
interoperability	TCP/IP
ISO	Transmission Control Protocol
ITU	Transport layer
layer	W3C
Network layer	

REVIEW QUESTIONS

1. Explain why protocols are necessary for communication between people in everyday life.
2. Describe an everyday protocol you used today to interact with people around you.
3. Explain why protocols are necessary for communication between computing devices.
4. Name several specific functions performed by communication protocols.
5. What is a protocol suite, and what is the purpose of the OSI reference model?
6. Name the seven layers of the OSI reference model.
7. Name a protocol that operates at each layer of the OSI reference model.
8. What is the function of Presentation layer protocols?
9. What is the function of Network layer protocols?
10. Name some communication protocols that you use when you access the Internet.
11. Name an important standard developed by the W3C.
12. What organization establishes standards for most Internet protocols?
13. What is the request for comments (RFC) document series?
14. Explain the difference between *de jure* and *de facto* standards.
15. Which standards body operates under the umbrella of the United Nations?

DISCUSSION QUESTIONS

1. Why does the OSI reference model have seven layers, while the TCP/IP protocols are usually organized into four layers? What is the difference between the OSI model, a layer, a protocol standard, and an actual protocol implementation? Do we really need all these different categories?

2. Who is in charge of standards? Make arguments both for and against the involvement of governments either in regulating standards, establishing standards, or influencing standards through procurement policies.

3. On the surface, technical standards appear to be just technical design choices. In what ways can standards also have economic implications? How can standards influence economic competition, innovation, and global trade policy?

4. Should standards organizations charge membership fees, or should they be open to general public participation? Make arguments for and against open membership.

5. Name some examples of information and communication technologies you use in everyday life, and try to identify some of the standards that enable them to interoperate with other technologies.

CASE PROJECT

You work for a start-up company that has invented a new approach for high-speed metropolitan wireless communications. The chief executive officer has asked you to start and participate in a standards organization to establish common communication protocols with your business partners, suppliers, and competitors in this market. Check out the Web sites of some standards organizations (such as the IETF, the W3C, and the IEEE) for policy ideas. Answer the following questions about the policies you would set for the new standards body:

- *Membership*—Who should or could become members of the organization? Will you try to exclude competitors or include them? What is the rationale for your decision? Will the public be able to participate in the new organization?

- *Objectives*—What is the purpose of having a standards organization in this area?

- *Documents*—Will the specifications developed by your standards organization be available for the public to view, be kept proprietary, or be available for a fee?

- *Process*—How will the standards organization be run? Will interactions between participants be in person or electronic? How will disagreements be handled?

- *Intellectual property considerations*—Who, if anyone, will own the rights to the protocol specifications? Will participants have to disclose any patents they hold? Will other companies have to pay to implement the specification in their products, or will it be available on a royalty-free basis?

LEARNING OBJECTIVES

In this chapter you will:

- Become familiar with important Internet technology milestones

- Understand fundamental Internet architectural features such as Internet exchange points, the Domain Name System, IP addresses, and Uniform Resource Locators

- Understand the technology underlying popular Internet applications

- Examine the centralized administrative functions that keep the Internet running, including management of domain names and Internet addresses

- Contemplate economic and social issues associated with the Internet

INTRODUCTION

The Internet is not a single technology or application, but a complex set of systems developed over time within different social, political, and economic contexts. The Internet uses common protocols and architectures that enable information sharing among a diverse array of systems and applications. Many of the Internet's underlying technological approaches have been addressed already in this book—the Internet transmits images, data, video, and audio, and involves storage, transmission technologies, and computer networking. This chapter expands on the information presented in earlier chapters, first by presenting a brief history of the Internet and its predecessor networks. It then reiterates, or introduces information for some of the key technologies required for the Internet to operate, such as the TCP/IP protocol suite, Internet routers, packet switching, Internet exchange points (IXPs), and the Domain Name System (DNS). The chapter also examines some important Internet applications that improve business productivity and individual communications, and some central administrative aspects of the Internet such as the administration of domain names and IP addresses. Finally, the chapter explores unique economic and social issues that have emerged with the evolution of the Internet.

The history of the Internet can perhaps be traced as far back as the Cold War. In 1957, the former Soviet Union launched the world's first satellite, Sputnik. In response, the U.S. government founded ARPA, the Advanced Research Projects Agency, a new agency within the Department of Defense (DoD). ARPA's task was to keep the United States on the cutting edge of military science and technology applications. In the late 1960s, it was responsible for developing ARPANET, a computer network and research and development project that became the precursor to the modern Internet. In the late 1960s and early 1970s, computers were not widely used, and there was almost no way for different types of computers to communicate with each other or to share information electronically. The original purpose of ARPANET was to interconnect geographically dispersed and technically disparate computers at various research locations.

As described in Chapter 12, "survivable communications" were also a concern during the Cold War. Would communication systems survive and continue operating in the event of a major military attack on the United States? The hierarchical, centralized nature of communication systems such as the traditional telephone network made them more susceptible to severe disruption, because an outage in a single location could disable the entire system. This traditional approach is called circuit switching, and it is still used in today's legacy telephone networks. Circuit switching establishes a physical path through the network between a caller and receiver and maintains the path for the entirety of the call.

One Internet milestone that emerged from the requirement for a more reliable communications scheme was **packet switching**, which was developed in the United States by RAND Corporation engineer Paul Baran in the late 1960s. In contrast to circuit switching, packet switching breaks information into smaller segments called packets and routes each packet individually through the network, usually over the most efficient or most available path. The packets arrive at the final destination and are reassembled into the original information. As Chapter 12 described, packet switching is a decentralized, distributed approach in which multiple paths are available for transmission. A particular path does not need to be established beforehand and maintained for the duration of the transmission, which provides a more resilient and survivable system. The destruction of a single location would have minimal impact on communications because information could be routed around the problem over an alternative path. Packet switching eventually became the technique underlying the Internet.

Another major milestone came in the late 1970s, when **TCP/IP** (Transmission Control Protocol/Internet Protocol) was developed as the common communications standard for ARPANET. Two Internet luminaries, Vinton Cerf and Robert Kahn, established the technical groundwork for TCP/IP. At the time, ARPANET went through a series of structural and institutional changes. Part of the network was split into MILNET, which focused on military issues. ARPANET was transformed again in the 1980s when the National Science Foundation took control of part of ARPANET, called NSFNET. Commercial, public TCP/IP networks emerged in the late 1980s and 1990s, interconnecting with each other and government-funded backbones to become the growing and publicly expanding network we know as the Internet. Control and development expanded into the commercial sector, and a myriad of Internet service providers and other private companies began operating and managing large segments of the Internet.

Even by 1990, Internet users were still a relatively small group of mostly academic, research, and federal government communities. It took the major technological breakthrough of the World Wide Web to begin transforming the Internet into a useful mass communications medium. The development of the Web is attributed to Oxford graduate Tim Berners-Lee of the European Organization for Nuclear Research (CERN) in Geneva, Switzerland. Berners-Lee and others never anticipated the eventual Web technology explosion. Instead, they envisioned the Web as a specific solution to a specific problem—an application to enable communications among computers of nuclear physicists throughout the world, including those at CERN and Los Alamos National Laboratory in the United States. An accompanying breakthrough in the early 1990s was the development of a Web browser, Mosaic, which allowed user-friendly access to Web sites. The lead programmer of Mosaic was Marc Andreessen of the National Center for Supercomputing Applications (NCSA) at the University of Illinois. Andreessen also founded Netscape (originally called Mosaic Communications Corporation), which was eventually bought by America Online, now part of Time Warner.

As shown in the partial historical progression in Figure 14-1, the Internet is not a recent, singular, or sudden phenomenon. Its early technological roots trace back to a series of advancements in the 1970s, 1960s, and beyond. Technological advancements, some anticipated and some unexpected, have continued to change the character of the Internet. For example, one technological change was the mass untethering of Internet connectivity into the realm of wireless access. Laptop computers, cellular phones, and handheld devices that wirelessly access the Internet contributed to greater ubiquity, convenience, and mobility. If history is any indication, new technological breakthroughs, many of them unforeseen, will continue to transform the Internet.

Figure 14-1

Some Internet history milestones

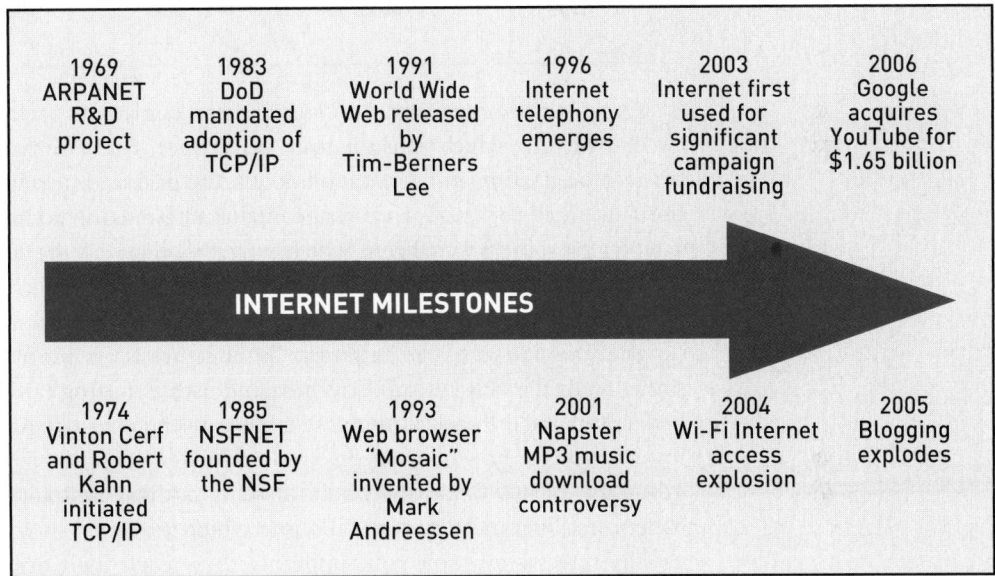

INTERNET ARCHITECTURAL COMPONENTS

This section examines some of the main architectural components that enable individual networks to form the global Internet. The Internet includes the following important technological systems and components:

- Internet backbone and routers
- Internet exchange points (IXPs)
- The Internet Protocol (IP)
- The Domain Name System (DNS)
- Uniform Resource Locators (URLs)

INTERNET BACKBONE AND ROUTERS

No single entity owns the global Internet or even a significant portion of it. As illustrated in Figure 14-2, the **Internet backbone**—the global collection of high-capacity trunks (transmission lines) that carry the bulk of Internet traffic—is not owned and operated by any single company or government. Rather, it is a collection of high-speed, interconnected networks run by large **network service providers** such as AT&T, British Telecom, France Telecom, Qwest, and Verizon, to name just a few.

Figure 14-2

Abstract conceptualization of Internet backbone

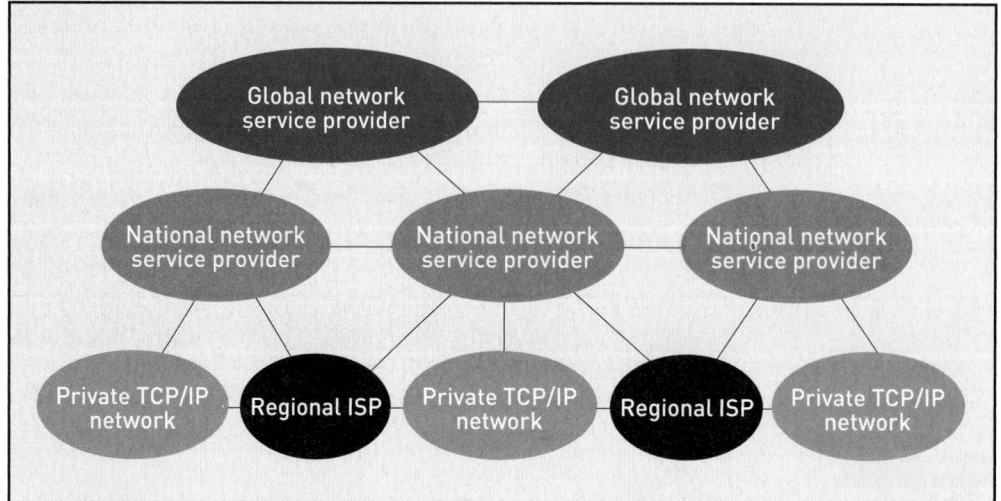

As explained previously, the Internet is an enormous packet-switching network. Unlike circuit switching, which transmits a stream of information over an established end-to-end path between origination and destination points, the packet-switching approach segments information into small packets. Each packet contains a destination address and is transmitted over the Internet via **routers**, intelligent switching devices that use the addresses to direct packets over the Internet to their destination. The individual packets, which can be routed over different paths, are then reassembled at the destination into the original information stream.

The foundation of the Internet's architecture is an enormous number of routers. The router reads the destination IP address and uses a **routing table** to look up information for how to forward the packet. A routing table is essentially a database on the router that provides information for how destinations can be reached most efficiently. Routing tables contain a minimum of two fields: the prefix of an IP address destination and the next destination (often another router), or destinations to which the router should forward the packet. Packets usually traverse multiple routers before they reach their destinations. Each step along a packet's route is called a **hop**. In many networks, a single routing table field can specify several next hops.

The routers in one service provider's networks can communicate with routers in other such networks because they adhere to the same **routing protocols**. These protocols enable routers to share network changes that are reflected in updates to router tables. An example of a routing protocol that provides this service is **Border Gateway Protocol** (BGP).

INTERNET EXCHANGE POINTS (IXPs)

Countless individual networks interconnect to create the global Internet. Some of these networks are owned by national and international telecommunications providers, some are operated by Internet service providers (ISPs), some are operated by regional cable or telephone companies, and some are privately owned by corporations, universities, and individuals. Traffic from one network flows seamlessly to other networks across the Internet through interconnection locations called **Internet exchange points** (IXPs). These IXPs contain shared switching equipment, as shown in Figure 14-3. Each network is connected to one of numerous IXPs (formerly called network access points) and has a high-speed, fiber-optic link that interfaces to the access point. The exchange point serves as a juncture at which packets from different networks are exchanged and routed toward their appropriate destinations.

IXPs are the physical location where disparate backbone trunks from service providers interconnect and exchange Internet traffic. Physically, these transmission trunks terminate at rates of gigabits per second into ports on high-speed switches and routers.

Making these connections obviously requires physical infrastructure, but it also requires agreements about cost and performance. These **peering agreements** allow service providers to share the costs of shared exchange points and provide service-level agreements for characteristics such as reliability and latency, the delay that packets undergo en route to a destination.

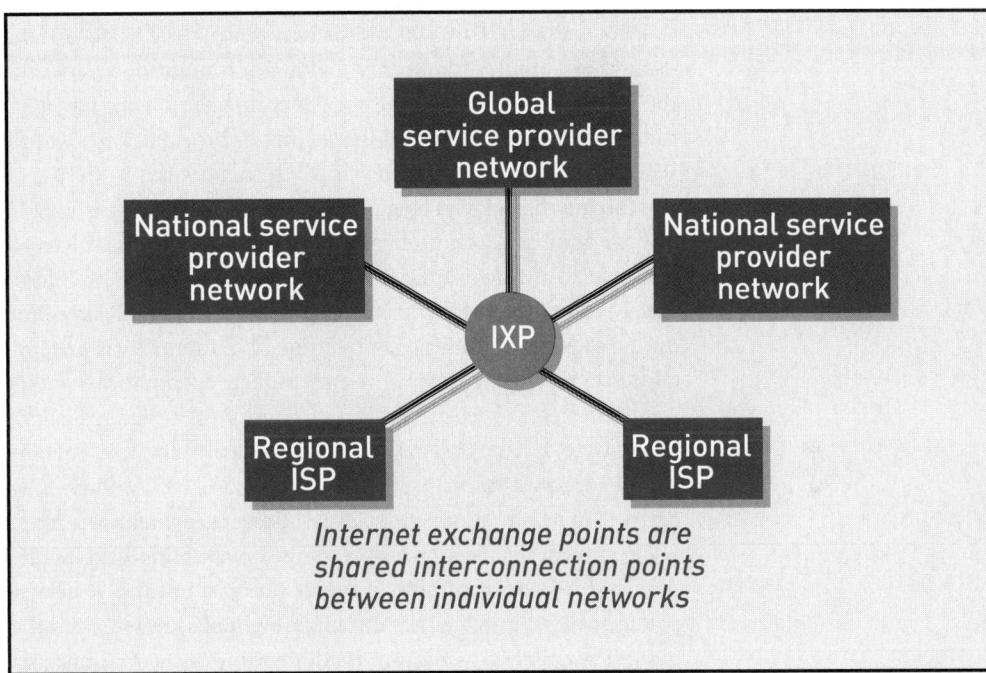

Internet exchange points are shared interconnection points between individual networks

THE INTERNET PROTOCOL

To communicate over the Internet, computing devices must use the TCP/IP family of protocols. TCP/IP is a network protocol suite. The **Internet Protocol** (IP) is a critical part of TCP/IP and the circulatory system of the Internet in many ways. IP is the one protocol needed in almost every instance of information sharing over the Internet. As a result, one common definition of "being on the Internet" is the use of IP. This definition is not completely accurate because some businesses use private IP networks that are not connected to the public Internet. The function of IP is to route blocks of information from a source to a destination over a complex network. To perform this routing, IP uses a hierarchical addressing scheme that assigns a hardware-independent (logical rather than physical) address to every device connected to the Internet. Recall that the IP address is software defined; it is distinct from a MAC address that is physically associated with a LAN adapter such as an Ethernet card.

IP Addresses

Each device that communicates over the Internet must use a unique address known as an **IP address**. To understand the need for an IP address, consider an analogous information delivery system—the postal system. For a postal carrier to deliver a letter, it must contain a destination address. This address must also be unique to the recipient. If two houses had the same address, the postal service could not determine how to deliver the letter. Similarly, each Internet device connection requires a unique address. However, instead of a name, street address, and zip code, an IP address is a unique binary number.

The traditional standard for IP addresses, called **IPv4** (IP Version 4), specifies 32 bits for each address. This standard is more than 20 years old, which means it predates the introduction of the Web. IPv4 is actually the original version of IP, despite the "4" in its name.

An IP address is a combination of 32 ones and zeros such as the following:

01011110000101001100001111011100

This 32-bit address is understood by a computer but is awkward for people to use. Therefore, industry convention dictates a shorthand method, **dotted decimal format**, for discussing and managing IP addresses. For example, an IP address in dotted decimal format might be 94.20.195.220—a format many Internet users may recognize. For a refresher course on how to convert a digital bit stream of 32 ones and zeros to dotted decimal format, see Chapter 3.

The dotted decimal format is easy for people to comprehend and manage, but it is not useful to computers and other digital equipment in a network. The successful functioning of the Internet is predicated upon each connected device having a unique IP address, the ability of routers to "read" these addresses and appropriately forward IP packets, and having enough IP addresses for anyone who wants Internet access.

The Internet address length of 32 bits theoretically provides 4,294,967,296 (calculated as 2^{32}) unique addresses. A number of circumstances prompted concerns that four billion Internet addresses would be insufficient to meet global demands for Internet connectivity. First, in the early stages of Internet development, U.S. institutions received large blocks of addresses—in some cases, on the order of 16 million. At the time, the Internet was primarily an American enterprise, and addresses were generously allocated to large corporations and government agencies. Addresses were allocated in fixed-sized blocks for technical design reasons related to router efficiency. These blocks were called Class A, B, or C address blocks, and they contained roughly 16,000,000, 65,000, or 254 addresses, respectively. There were also Class D addresses, which were reserved for multicast delivery of information, and Class E addresses, which were reserved for experimental purposes. The technical details are complex, but the simplified explanation is that this class system was designed in part to minimize router processing time. It also resulted in address distribution patterns whereby a single organization might possess more than 16 million addresses but use only 5000. The other 15 million-plus addresses would remain unused but would be unavailable for other organizations to use. As the Internet grew internationally and new applications such as wireless Internet access and Internet telephony emerged, the Internet Engineering Task Force (IETF) identified the possibility that the reserve of Internet addresses might be exhausted. The need for more global Internet addresses was recognized in the early 1990s.

The IETF engineered two initial technical approaches to conserving Internet addresses: Classless Interdomain Routing (CIDR), which eliminated the Class A, B, and C distinctions, and Network Address Translation (NAT), a technique that allowed a network device such as a router to share a limited number of public IP addresses among many devices on a private network. When a computing device on a private network accesses the Internet, NAT dynamically allocates a globally unique, temporary, public IP address for transmission over the public Internet (see Figure 14-4).

Figure 14-4

Network address translation

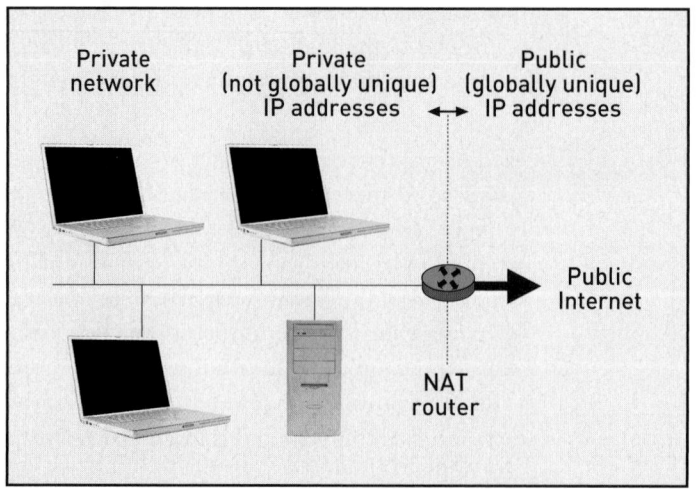

IPv6

In addition to these address conservation strategies, the IETF selected a new standard, now called **IPv6** (Internet Protocol Version 6), to exponentially expand the number of globally unique addresses. IPv6 expanded the address length from 32 to 128 bits, supplying 2^{128}, or 340,232,366,920,938,463,463,374,607,431,768,211,456 (340 undecillion) unique addresses.

If we need a shorthand notation to talk about 32-bit IPv4 addresses, then we clearly need one for 128-bit IPv6 addresses. It is not practical to say, "My IP address is 01110100100111011000011010101011101111010001100100110010010010011101001001110110000110101011101111010001100100110010010010101110000." Just as dotted decimal format is used to represent an IPv4 address, IPv6 has its own shorthand representation:

X:X:X:X:X:X:X:X

In this representation, each X is equal to the hexadecimal representation of 16 bits. Thus, the convention for IPv6 notation is to use the hexadecimal numbering system. The following is a random example of an IPv6 address in shorthand notation:

FDDC:AC10:8132:BA32:4F12:1070:DD13:6921

This representation consists of eight groups of four hexadecimal numbers, separated by colons. Recall from previous chapters that each hexadecimal number represents four binary numbers. For example, the hexadecimal number 0 equals the binary number 0000, the hexadecimal number 5 equals the binary number 0101, and so on.

Therefore, the preceding representation of an IPv6 address can be translated as follows:

```
FDDC = 1111110111011100
AC10  = 1010110000010000
8132  = 1000000100110010
BA32  = 1011101000110010
4F12  = 0100111100010010
1070  = 0001000001110000
DD13  = 1101110100010011
6921  = 0110100100100001
```

Putting it all together, the following human-readable address:

FDDC:AC10:8132:BA32:4F12:1070:DD13:6921

is equivalent to the machine-readable IPv6 address that follows (with spaces added).

1111 1101 1101 1100 1010 1100 0001 0000 1000 0001 0011 0010 1011 1010 0011 0010 0100 1111 0001 0010 0001 0000 0111 0000 1101 1101 0001 0011 0110 1001 0010 0001

As cumbersome as the hexadecimal version appears, it is obviously better than writing out the entire 128-bit string of ones and zeros. Even if you used decimal notation, it would require more digits than hexadecimal to represent the same string of numbers.

IPv6 deployment is occurring more rapidly in Asia and other countries than in the United States. Both protocols, IPv4 and IPv6, probably will coexist for the foreseeable future in several ways, including "dual stack" approaches that require the installation of both IPv4- and IPv6-compliant software on computing devices; "tunneling" approaches in which IPv6-formatted packets are encapsulated within IPv4 packets or vice versa; and "translation," whereby IPv4 packets are translated entirely into IPv6 packets or vice versa.

THE DOMAIN NAME SYSTEM

Another critical component of the Internet is the **Domain Name System** (DNS). Even the shorthand dotted decimal format, which was designed to make IP addresses less unwieldy, is difficult to remember and use. Fortunately, Internet users do not have to remember numeric

IP addresses while using the Internet. Instead, users can employ alphanumeric names that are easy to remember, such as *www.yale.edu*. These are known as domain names. All Internet users rely on domain names to send and receive e-mail or to surf the Web.

Each domain name has an associated IP address. The following is one example of a domain name and an associated IP address:

cnn.com   64.236.29.120

Web browsers will accept either the domain name or the IP address. Try opening a Web browser and typing *www.cnn.com*. Then type 64.236.29.120 in the address bar. Either approach directs you to the CNN Web site.

The DNS is like a hierarchical tree. The suffix, which is the component at the far right of any domain name, is called the top-level domain (**TLD**). There were originally seven generic TLDs, but the rapid growth of the Internet necessitated new TLDs. The original seven TLDs included:

- .com (for commercial businesses)
- .org (for nonprofit organizations)
- .edu (for educational institutions)
- .gov (for the U.S. government)
- .mil (for the U.S. military)
- .net (for a network)
- .int (for international entities)

Some newer TLDs are .biz, .info, .coop, .museum, .name, .pro, and .aero. Additionally, hundreds of two-letter TLDs represent particular countries, such as .jp for Japan and .uk for the United Kingdom. The following examples of top-level domains are called "country codes," or ccTLDs:

.af	Afghanistan	.bd	Bangladesh	.be	Belgium
.br	Brazil	.ca	Canada	.cn	China
.eg	Egypt	.fr	France	.gr	Greece
.hk	Hong Kong	.in	India	.il	Israel
.mg	Madagascar	.mx	Mexico	.nl	Netherlands
.pk	Pakistan	.sa	Saudi Arabia	.tr	Turkey
.va	Vatican City	.ve	Venezuela	.vn	Vietnam
.za	South Africa	.zm	Zambia	.zw	Zimbabwe

Within a domain name, the word to the left of the top-level domain is called a second-level domain. For example, "whitehouse" is the second-level domain in *www.whitehouse.gov*. Domain names can also have third- and fourth-level domains. Each domain name is readable to people but not to routers and other computing devices, which read the associated numeric IP address. A method is needed to translate between alphanumeric domain names and the associated IP addresses required for routing information across the Internet. This translation is called **address resolution** and is performed by the DNS. The DNS is an enormous database management system that is distributed on numerous servers around the world. Functionally, it tracks domain names and their associated numeric IP addresses.

The DNS not only allows users to locate the IP address of any Web site, it allows them to find an e-mail address or the address of other Internet applications such as FTP files.

An important architectural component of the Internet is its collection of **root name servers**, which are usually just called root servers. These servers maintain a master file, called the **root zone file**, that lists the names and IP addresses of the official DNS servers for all TLDs. In other words, the root zone file is the top of the hierarchical DNS namespace, as illustrated in simplified form in Figure 14-5. The root servers send updates of the root zone file to other DNS servers.

Figure 14-5

The DNS namespace

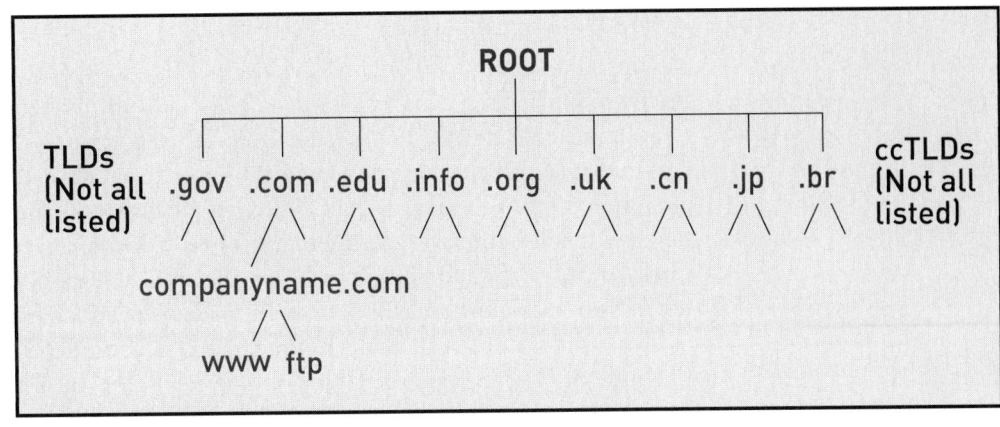

UNIFORM RESOURCE LOCATORS

Besides domain names, resources on the Internet are further identified by a URL, or **Uniform Resource Locator**. A URL is a string of characters associated with a specific information resource, such as *www.ebay.com*, *www.gmu.edu*, and so on. A full URL conveys a great deal of information. Consider the following URL:

http://www.university.edu/it101/section2/syllabus.html

As indicated in Figure 14-6, the prefix "http" signifies that the Hypertext Transfer Protocol is the network protocol used to access the information resource. The next part of the URL, "www.university.edu," represents the domain name of the server that houses the requested information. The supplemental information "it101/section2/" denotes the path or directory to the requested file. Finally, "syllabus" is the actual file, and "html" denotes that the file format uses Hypertext Markup Language.

Figure 14-6

Uniform Resource Locator

Many URLs relate to Web access via HTTP, but note that URLs also apply to many other Internet protocols and information resources. For example, instead of "http," the first part of a URL could include "ftp" for File Transfer Protocol, "news" for Usenet news, or other Internet resource types.

INTERNET APPLICATIONS

Uses of the Internet are constantly changing. At one point, the Internet primarily allowed file sharing and electronic mail. Over time, however, Internet applications have expanded to the World Wide Web, text messaging, Internet telephony, multimedia file sharing, and much more. Business productivity has risen with the development of applications that enable electronic commerce over the public Internet and private, internal networks called intranets. The Internet has also improved the ability for people to communicate and perform day-to-day functions such as registering for classes, researching information, making airline reservations, and paying bills. The following sections discuss some important Internet applications that have facilitated these improvements.

E-MAIL

Although the Internet was originally envisioned as a tool for resource and file sharing, one of its first popular uses was electronic mail. E-mail quickly became a significant percentage of Internet traffic, and continues to be. Unfortunately, spam is an unintended consequence of electronic mail—it plagues many Internet users with a daily barrage of unwanted, unsolicited, and often offensive messages. Some Internet analysts estimate that spam accounts for fully half the e-mail traffic over the Internet. Not only does spam waste users' time, it consumes Internet bandwidth and processing power. Nevertheless, traditional e-mail obviously continues to thrive; it is now difficult to imagine business or social life without it.

E-mail is a **store and forward system** that does not require the simultaneous online presence of senders and receivers (see Figure 14-7). In a typical transaction, a sender composes a message using an e-mail application (also called a client), then sends the message to a local mail server. The local server transmits the message to a remote server, which stores the e-mail until the recipient accesses it. The Internet uses common messaging protocols and formats to enable this interoperability. The *de facto* messaging protocol that historically has supported Internet e-mail is **SMTP**, or Simple Mail Transfer Protocol. SMTP is a well-documented, universal, Application-layer Internet mail standard originally specified in the early 1980s by the IETF. E-mail messages primarily consisted of text when the Internet community developed SMTP, but today messages incorporate multimedia and include attachments, thanks to newer messaging formats such as **MIME** (Multipurpose Internet Mail Extensions).

Figure 14-7

E-mail store and forward delivery

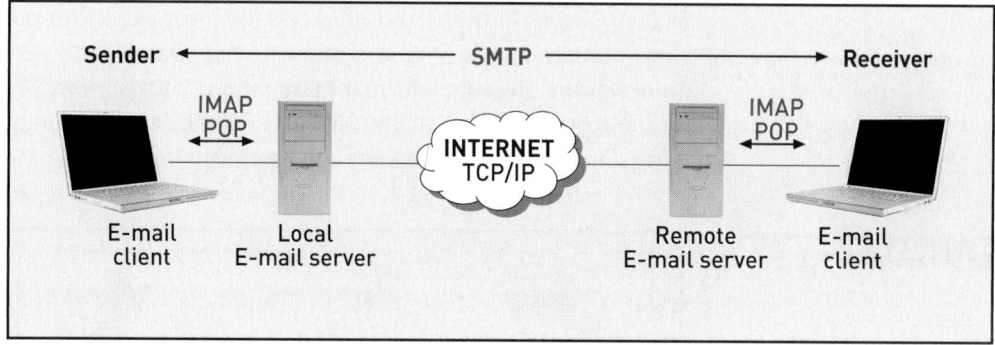

SMTP and other messaging formats involve the delivery of electronic mail between e-mail servers, not between the recipient and the sender. The arrival of e-mail at a local server and its transmission from the remote server to the recipient are separate transactions that use different sets of communications protocols, known as mail retrieval protocols. Examples include **Post Office Protocol** (POP) and **Internet Message Access Protocol** (IMAP).

MESSAGING

A more real-time messaging approach than e-mail is text messaging (also called instant messaging). Unlike e-mail, text messaging generally requires both users to be online or on their mobile phones simultaneously. Text messaging (see Figure 14-8) is explosively popular in social and business environments because it represents a much faster alternative to e-mail. When you send a text message to a recipient, a window opens and displays the message on the recipient's computing device. Each party can type messages that appear in the text window, generating a real-time, text-based conversation. This messaging approach is widely used in business environments and for social interactions between friends and families. Besides instant conversations, real-time messaging enables the immediate sharing of Web links, pictures, video, files, and other information.

Figure 14-8

Text messaging

THE WORLD WIDE WEB

The advent and rise of the World Wide Web in the early 1990s spurred explosive growth of the Internet. Even though the Web is one of many Internet applications, some consider the Web to be synonymous with the Internet. As mentioned previously, Tim Berners-Lee of CERN is credited with developing the World Wide Web. The Web was a revolutionary advancement over previous data-sharing tools for several reasons:

- It allows many users to simultaneously access the same information.
- It provides hyperlinked information—clicking a textual link takes a user to another location.
- It combines multimedia information such as video, text, image, and sound.
- It allows access to anyone connected to the Internet from any computing platform.
- It provides searchable information.
- Anyone can develop their own information site and inexpensively make it available to millions of people.

Of course, the Web is only as useful as the software that allows users to access and search for information. Web browser software emerged shortly after the emergence of the Web. The first popular browser was Marc Andreessen's Mosaic, which was eventually named Netscape Navigator. Equally important to using the World Wide Web are search engines. Without search engines such as Google and Yahoo, finding specific information on the vast World Wide Web would be difficult.

The Web is a client/server system, as shown in Figure 14-9. It uses a standard network protocol, Hypertext Transfer Protocol (**HTTP**), to establish and maintain communications over the Internet between a computer user (client) and a Web site (server). Web interactions also require the encoding of information in a standard format called Hypertext Markup Language (**HTML**) or eXtensible Markup Language (**XML**). HTML (see Figure 14-10) allows for the encoding of static text and other types of information, including images and hyperlinks that enable the user to jump to a new Web page. The HTML code can transfer the user to another Web page through a URL. When a user wants to view a Web page located on a server, the browser sends a request over HTTP for the HTML document that represents the Web page. The server responds to the request over HTTP by sending the HTML document to the client.

Figure 14-9

Client/server Web interaction

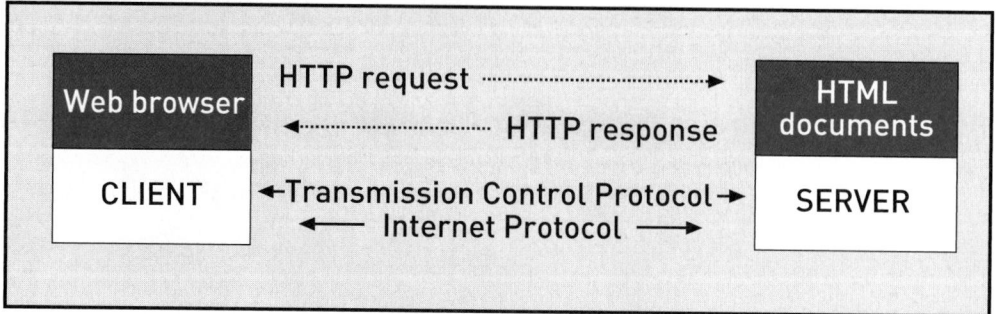

Figure 14-10

HTML code

FILE SHARING AND P2P FILE SHARING

One of the original and still-popular Internet applications is file sharing between users who operate in different computing environments. The TCP/IP suite has historically provided a specific protocol to enable file sharing over the Internet: FTP, or File Transfer Protocol. FTP is often transparent to the user because it is embedded in other applications.

As Internet technologies have grown, file sharing has expanded to include stored videos, audio files, and images. File transfer over the Internet historically used a client/server model—a client device downloaded videos, text, and other files from a shared server.

Peer-to-peer (P2P) file sharing uses a different approach. P2P file sharing entered the consciousness of Internet users primarily through music- and video-sharing systems. Rather than storing files on a server or large database management system, P2P technologies distribute files that are stored on the hard drives of individual users. Each user's hard drive acts as a server that is accessible to other users. Each distributed system involved in P2P file sharing uses common software and is connected to at least one other computer on the P2P network, as shown in Figure 14-11. When a user wants to locate a file such as a song, the computer makes a request to a connected peer, which makes a request to its peer, and so on, until the file is located.

P2P file sharing of any copyrighted information, including music and movies, is illegal though widespread. A series of well-publicized lawsuits, especially those brought by the Recording Industry Association of America (RIAA), have attempted to curtail downloading of copyrighted information. However, countless applications of P2P network technology are legal and hold great promise for efficiently sharing information.

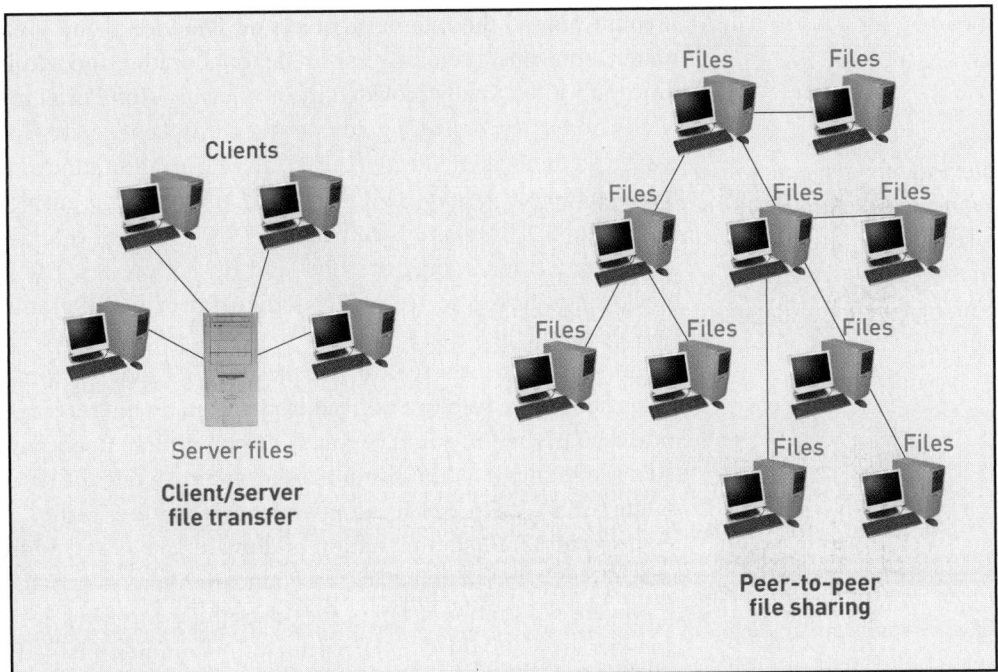

INTERNET TELEPHONY

To make telephone calls over the Internet, you use the **Voice over IP** (VoIP). VoIP is a cost-effective alternative to traditional telephone service and has quickly become a major Internet application. The main advantage of Internet telephony is that telephone calls are virtually free to users who already pay for an Internet connection. Chapter 17 discusses the technological architecture of Internet telephony as well as its advantages and disadvantages.

INTERNET BROADCASTING

For years, audio and video broadcasters have simultaneously extended their programs to the Internet. The advantage of "simulcasting" over the Internet is that the broadcast has no physical or geographical limitation. For example, the transmission range of a radio or television broadcast in the New York metropolitan area is limited by the station's antenna size and transmission power. However, the station can easily broadcast over the Internet and reach a worldwide audience at very little cost without having to contend with spectrum limitations and regulations. Cellular phones and personal devices afford mobility to Internet radio listeners and video viewers. Also, because the Internet is a multimedia environment, Internet broadcasting is much more sophisticated and flexible than traditional broadcasting. Images, video, text, message boards, and Web links can easily accompany Internet radio broadcasts and provide much greater user interactivity.

INTERNET ADMINISTRATION

Does anyone run the Internet? Who is in charge of the many administrative functions and standards setting that keep the Internet up and running? The success of the Internet as an interoperable, universal communications medium requires common, compatible standards. The Internet also requires some centralized administration to ensure that each domain name and IP address is unique. The following section discusses some important organizational and administrative considerations for the global Internet.

ADMINISTRATION OF INTERNET NAMES AND NUMBERS

If connecting to the Internet requires an IP address, and if each address must be globally unique, someone has to be responsible for allocating and administering these resources. This function was originally provided by one person—Jon Postel maintained a list of host names and IP addresses, first as a graduate student in Los Angeles and later as part of the Internet Assigned Numbers Authority (**IANA**), an organization funded by the U.S. Defense Advanced Research Projects Agency (DARPA). The IANA, under the auspices of the Internet Corporation for Assigned Names and Numbers (**ICANN**), still has centralized responsibility for the IP address space, including both IPv4 and IPv6 addresses. The IANA, in turn, allocates large blocks of addresses to regional Internet registries (RIRs) and national Internet registries (NIRs) around the world.

An individual user typically requests an IP address from an Internet service provider, which has received a block of IP addresses from an Internet registry. Internet registries may be local or may serve an entire region. For example, the American Registry for Internet Numbers (ARIN) allocates IP addresses in North America, while the Asia Pacific Network Information Centre (APNIC) allocates IP addresses in Asia and the Pacific.

The centralized administration of domain names and addresses has been a complicated issue. In 1992, the National Science Foundation hired a private company, Network Solutions, to provide registration services for the popular top-level domains of .com, .net, and .org. However, in 1998, the U.S. Department of Commerce issued a white paper calling for the creation of a private, nonprofit corporation to administer these names. This new entity became ICANN.

The process of authorizing domain names sounds simple, but it has been a controversial issue in the Internet community. The organization in charge of domain names has to set policy, and some policies enter the realm of intellectual property. For example, who should own united.com: United Airlines or United Van Lines? Other policies concern free speech. For example, can an organization assign a domain name that might be controversial or objectionable to some? Still other policies involve privacy—for example, should organizations publish the contact information of people who own a given domain name?

The greatest controversy over ICANN has involved the questions of who should make these policy decisions and have the authority to allocate IP addresses. ICANN is a nonprofit organization incorporated in California and historically overseen by the U.S. Department of Commerce. The international controversy over ICANN has involved the question of whether it is sufficiently international to make global policy decisions or whether it is too American, because of what international communities describe as U.S. unilateral control of the Internet governance functions of addressing and domain name systems. From ICANN's inception, the plan was to transition away from federal government control to a combination of private and more global governance structure. As one example of the ongoing tensions over ICANN, the

United Nations in 2005 released an Internet governance report advocating that the United States relinquish unilateral control of names and addresses to a U.N. institution. In response, the United States released a statement of principles asserting that the U.S. government would retain its responsibility and oversight of ICANN. The United States argued that the current system was working and that oversight changes might have a negative impact on the stability and security of the Internet. For years, tensions have continued over who should control centralized Internet administrative functions such as names and numbers.

INTERNET STANDARDS SETTING

Internet standards are constantly evolving, and they address literally thousands of technical specifications. As mentioned in the previous chapter, the IETF establishes common network technical specifications and standards for the Internet. Most of the IETF's efforts are handled via electronic mailing lists. The Internet standards process is complex, but in a nutshell, it involves the proposal of a draft standards specification followed by a period of iterative revision by a "working group" that anyone can join. The participants in standards setting often include network designers, engineers, and researchers affiliated with companies whose products require standards. Technical specifications step through a progressive approval process that begins with the designation of a proposed standard, evolves to a draft standard, and culminates in a standard.

Other standards organizations also establish Internet technical specifications. For example, the World Wide Web Consortium (W3C) develops Web specifications and the Institute of Electrical and Electronics Engineers (IEEE) establishes Internet-related LAN standards such as the Wi-Fi specifications.

INTERNET OPEN ISSUES

As with most technological innovation, the Internet has been accompanied by a host of economic and social policy questions. For example, should sales over the Internet be taxed? If so, should the tax be levied in the place where the user resides, where the server resides, or where the corporation doing the selling resides? Voice service over the traditional telephone system has historically been taxed. Should voice services that use Internet telephony (VoIP) be regulated and taxed, like other more traditional services?

The Internet is an integral part of the infrastructure that supports business transactions and other critical services. How might a major Internet outage or cyberterrorist attack affect nations economically?

Some countries restrict or prohibit Internet access for political or religious reasons. Countries have different laws about Internet access, freedom of speech, restrictions on information access and what can be sold online, and general approaches to censorship. Whose laws should apply, and how should they apply to a network that transcends national boundaries? Similarly, what are the trade-offs between individual privacy on the Internet, law enforcement, and antiterrorism measures? How can these concerns be balanced?

Politicians sometimes announce their candidacies on campaign Web sites, and use the sites to communicate with citizens and the media and to raise campaign funds. Bloggers use the Internet to comment on politicians. Videotapes that surface on the Internet have changed the course of politicians' career paths and their electability. Governments directly provide services to citizens via the Internet. In what other ways will the Internet intersect with politics?

The Internet has produced enormous benefits, but they have been accompanied by social challenges. For example, what are the ramifications of the international digital divide, in which some countries have widespread Internet access and computing resources and others have limited resources? Other social realities of the Internet are the problems of online predators and children being exposed to objectionable material.

Net neutrality is a phrase that has received a great deal of attention in the early twenty-first century. Net neutrality has several meanings, but it generally refers to the principle of nondiscrimination on the Internet. The basic idea is that all applications, devices, and content should be treated equally when transmitted over the Internet. Net neutrality advocacy usually addresses the Internet's "last mile" and focuses on residential Internet service such as cable company or telephone company broadband connections into homes. For example, according to the Net neutrality principle, a cable company that controls a residential broadband connection should not be able to serve as a gatekeeper that makes certain content more readily available to consumers. Those who oppose Net neutrality legislation argue that the Internet's architecture is not neutral already, because of how content is presented by search companies and because of the tiered service-level pricing offered by service providers. Opponents also note that engineering quality of service (QoS) prioritization is necessary on an application basis so that latency-sensitive applications such as video and voice are given a higher transmission priority than information that is not as time sensitive, such as data.

Finally, Internet policies have been challenged by new technological threats in such areas as spam, identity theft, online fraud, viruses, and worms. These important subjects are addressed in the next chapter.

CHAPTER SUMMARY

- The Internet is not a single network or technology, but a collection of systems that can interconnect because they use common TCP/IP and routing protocols and common architectural approaches such as packet switching.
- Different service provider networks interconnect at locations called IXPs. Peering agreements dictate how they share costs and provide acceptable performance.
- The use of IP is arguably the defining architectural characteristic of being "on the Internet."
- Devices that are connected to the Internet require an Internet address—either a 32-bit address under the IPv4 standard or a 128-bit address under the newer IPv6 standard.
- The DNS is a hierarchical, distributed database management system that performs the important task of address resolution.
- Internet applications are constantly evolving, but they fall into the broad categories of e-mail, messaging, the Web, file sharing, telephony, and Internet broadcasting.
- The Internet requires centralized administrative coordination such as managing the IP address space and establishing standards.

KEY TERMS

address resolution	network service provider
Border Gateway Protocol	P2P
Domain Name System	packet switching
dotted decimal format	peering agreement
hop	Post Office Protocol
HTML	root name server
HTTP	root zone file
IANA	router
ICANN	routing protocol
Internet backbone	routing table
Internet exchange point	SMTP
Internet Message Access Protocol	store and forward system
Internet Protocol	TCP/IP
IP address	TLD
IPv4	Uniform Resource Locator
IPv6	Voice over IP
MIME	XML
Net neutrality	

REVIEW QUESTIONS

1. In what decade was ARPANET created?
2. Is the Internet backbone owned and operated by a single entity or by numerous interconnected entities?
3. Search the Web for the names of three commercially available router products.
4. How do individual networks interconnect to form the global Internet? What are these interconnection points called?
5. What is the function of the Internet Protocol?
6. How many bits long is each IPv4 address and each IPv6 address?
7. How many IP addresses are available under the IPv4 standard?
8. Describe the primary function of the DNS.
9. What is the root zone file?
10. Name five TLDs.
11. Describe several technical differences between electronic mail and text messaging.
12. What network protocol is used to transmit information over the Internet between a Web client and a Web server?
13. Describe the difference between client/server file transfer and P2P file sharing.
14. What Internet administrative functions require centralized coordination?
15. What organization currently oversees domain name administration?
16. Name a protocol used for Internet telephony.

DISCUSSION QUESTIONS

1. What do you consider the most significant technological development in the history of the Internet? What future breakthroughs can you envision?

2. Can you devise a definition of the Internet? What problems do you encounter? What would you say is the fundamental technology that enables the Internet?

3. Name three positive and three negative social impacts of the Internet.

4. Given the international importance and widespread distribution of the Internet, who should control the centralized distribution of unique Internet names and IP addresses? What technical limitations determine how centralized or decentralized this function can be?

5. Should citizens someday be able to vote over the Internet? Develop political and technical arguments to support your position.

CASE PROJECT

Select three countries other than the United States and develop a presentation on their Internet infrastructures and policies. Your presentation should include the following information:

- *Country background*—What is the population and political system of each country? What percentage of citizens have Internet access in each country?
- *IXPs*—List the names and locations of the Internet exchange points (IXPs) in each of these countries. Can you tell who operates the IXPs (for example, the government, a nonprofit entity, or a business consortium)?
- *Service providers*—Name a major network service provider that offers Internet access in each of these countries. Provide some information about the types of services they offer.

- *Country code*—What is each nation's "country code" in domain names?
- *Public policy*—Can you find any information about these governments' policies toward Internet privacy, censorship, freedom of access and expression, taxation, and other regulatory policies?

CHAPTER 15

NETWORK SECURITY

LEARNING OBJECTIVES

In this chapter you will:

- Understand the main types of network security threats, including denial-of-service attacks, viruses, worms, identity theft, and password theft

- Examine why critical infrastructure attacks are a concern in the current economic and political context

- Understand how to significantly reduce the risk of attacks through basic security approaches such as firewalls, access control software, and encryption

- Distinguish between packet filtering, stateful inspection, and application proxy firewall approaches

- Learn about public key cryptography

- Become familiar with advanced security techniques such as digital signatures and biometric identification

INTRODUCTION

Network security is one of the most critical technical areas within IT. No one hears about network security functions when things are operating well, but when a worm, virus, or other attack spreads across the Internet, the topic grabs the attention of network users and the media. The ubiquity and accessibility of networks make them indispensable to individual users, businesses, and governments, but these same features enable hackers, organized crime, and cyberterrorists to disrupt networks. This chapter describes the major security challenges that global information networks face, including denial-of-service attacks, viruses, worms, identity theft, spam, piracy, and threats to critical infrastructure. Fortunately, many security measures can counteract these threats. For example, public key cryptography helps protect the privacy of information as it traverses a network. Authentication techniques, including digital signatures and "token-based" products, help validate the identity of network users. This chapter describes such security measures and addresses more advanced techniques that use biometric characteristics such as retinal scans, DNA, and fingerprints. Another important security function is access control—determining and monitoring who can access a network and what they can access. The chapter also discusses the primary access control mechanism for network security, the firewall.

The first highly publicized network security problem occurred in 1988, when a Cornell University student named Robert Morris unleashed an Internet worm that affected thousands of computers and disrupted the Internet for several days. The attack used self-replicating computer code and produced widespread outages at institutions that were the predominant Internet users at the time—major universities, U.S. military sites, and research facilities. The code, now referred to as the "Morris worm," was technically benign, in that it did not gain unauthorized access to information or modify or destroy any data. However, its effects were not benign; by replicating itself, the worm effectively disabled systems by consuming network bandwidth and computer processing resources. This type of attack is still one of the biggest security challenges for the Internet. Morris was ultimately convicted in a U.S. district court of unauthorized access to a federal government computer under the Computer Fraud and Abuse Act. The incident drew attention to the Internet's growing importance and the network's many underlying vulnerabilities.

Another infamous security incident also occurred in 1988, when Kevin Mitnick was sentenced to jail for breaking into the computer network of Digital Equipment Corporation (DEC). Many years later, after thousands of stolen credit card numbers and millions of dollars of lost productivity and sales, security attacks are a bigger problem than ever. Hackers have infiltrated computers at countless locations, including Microsoft, Yahoo, the Department of Defense, and the White House. Internet viruses and worms such as the "I Love You" virus, "MyDoom," and the "Blaster" worm have plagued individual users as well as corporate and governmental users.

Network security breaches are not a hypothetical or sporadic problem but a daily occurrence. Some of these security breaches are highly publicized, while others remain unreported. News of Internet worms or security holes in popular software usually reaches the public via the mainstream media. Conversely, businesses that suffer security breaches often seek to avoid the corresponding negative publicity. They don't want shareholders to think their computer networks have security vulnerabilities, and they don't want customers to worry about the security of their personal account information. The annual costs of preventing attacks and implementing reparative security measures are massive. They include expenses for software, hardware, technical staff, and policies, not to mention the cost of productivity lost during network outages. Consequently, network security is a multibillion-dollar industry.

Although no one can accurately quantify the total number of security breaches or the extent of susceptibility to computer attacks, a snapshot of these problems is captured by an organization called US-CERT. In the aftermath of the Morris worm in 1988, the U.S. government funded the establishment of the Computer Emergency Response Team (**CERT**) to spearhead responses to computer security problems, report incidents, research security technologies, and educate network users about security. CERT was launched at a research center at Carnegie Mellon University in Pittsburgh, Pennsylvania. Since that time, numerous CERTs have been formed around the world to detect, report, and respond to network security problems. In 2003, the U.S. Homeland Security Department established a federally run US-CERT (now the Computer Emergency Readiness Team) to serve as a public/private partnership that protects the U.S. Internet infrastructure and responds to security problems. US-CERT works with the CERT Coordination Center (CERT/CC) at Carnegie Mellon as well as other CERTs internationally.

US-CERT, like other CERTs around the world, tracks security incidents and serves as a technical resource for solving and preventing Internet security problems. More than 100,000 security incidents are reported to US-CERT annually, and the agency suggests that this number represents only a fraction of all security attacks. Statistical information about incidents and vulnerabilities is available at the US-CERT Web site (*http://www.us-cert.gov*).

WHO IS THE THREAT?

In general, the term **hacker** refers to people who gain unauthorized access to a computing resource either for malicious or benign purposes. In the early days of the Internet before security became a significant problem, hacking had a more positive connotation of creative programming. Today, the term has a different connotation. Some hackers infiltrate computer systems out of curiosity or sport. Other more malicious hackers might modify and damage code, unleash destructive computer viruses or Internet worms, steal resources such as credit card numbers, or engage in bandwidth piracy. The perpetrators of network or system attacks have a wide range of motives and capabilities, and they fall loosely into several categories: hackers, spammers, rogue employees, corporate or national spies, and cyberterrorists.

Some programmers also make a distinction between what they call white hat, gray hat, or black hat hackers. A white hat hacker is someone, often a security professional, who attempts to hack into systems to identify security vulnerabilities for the system owner. White hat hacking is sometimes called ethical hacking. In contrast, a black hat hacker, or cracker, breaks into systems for malicious reasons or personal financial gain. A gray hat hacker is an ambiguous term describing someone who falls between the two extremes. Yet another term is "script kiddie," a young and inexperienced hacker who maliciously attacks systems.

A serious security challenge comes from people who send enormous amounts of unwanted e-mail solicitations, or **spam**. Spam is a significant security consideration for corporate network managers and Internet service providers (ISPs) because its sheer volume can flood networks—it can constitute 50% of all e-mails on a network. Spam is not just an irritation but a serious problem, because the unsolicited e-mails waste bandwidth capacity and processing power and diminish worker productivity.

The origin of the term *spam* is unclear, but one explanation cites a skit by Monty Python, the British comedy troupe. In the skit, a waitress recites an enormous menu of dishes made with Spam, a canned meat product. A group of Vikings in the restaurant picks up on the theme and repeatedly sings a silly song about Spam. The irritating repetition of the word throughout the skit supposedly inspired a group of computer game players during the late 1970s or early 1980s, who applied the term to the equally irritating repetition of unsolicited bulk e-mail.

A more serious security problem sometimes exists within corporations and other enterprises. Disgruntled or rogue employees with legitimate network and system access present an ongoing security threat that is difficult to prevent. These employees can use their access to damage or steal computer files or damage critical infrastructure. In one famous example, a disgruntled former employee of an Australian sewage treatment plant broke into the company's computer control system and intentionally released raw sewage into the surrounding water and grounds.

A related but more external threat to business networks and systems is **corporate espionage**—rival organizations seeking competitive information or, less frequently, sabotaging systems. This type of information espionage also occurs between rival nations.

Cyberterrorism is perhaps the most serious potential threat to networks. A terrorist attack could employ a variety of tactics to disrupt or disable networks for hours, days, or even weeks. Such a disruption would present much more than an inconvenience to public network users who want to access financial systems, airline reservation systems, or other useful Web services. A coordinated attack on critical networks could disrupt stock market networks, systems of automatic teller machines, the power grid, water systems, hospital information systems, or the air traffic control system. Even a short-lived outage in critical networks could have economic or logistical consequences and an enormous psychological impact on the public. Network security breaches or disruptions do not require that the attacker be in the vicinity of network equipment or even in the same country. The ubiquitous reach of the public Internet provides global launching points for cyberterrorism.

TYPES OF ATTACKS

The main tactics for invading or disrupting networks and systems fall into the following categories:

- Viruses and worms
- Denial-of-service attacks
- Identity and password theft
- Data interception and modification
- Bandwidth piracy
- Critical infrastructure attacks

Viruses and Worms

Like biological viruses spread among people, computer viruses propagate from computer to computer. A **virus** is malicious code embedded within a seemingly legitimate program that only becomes active when the program is executed. For example, a file attached to an e-mail may actually be a virus that executes when the file downloads or the user double-clicks the link. Often, the link contains an extension such as .exe rather than .doc or .jpg, meaning it is raw code that executes once activated by a user. For example, the virus might send the seemingly legitimate program to all the e-mail addresses in the address book of the infected computer, thus propagating itself exponentially. A virus is not self-replicating or self-disseminating. Before the virus can spread, a user must activate the host program in which the virus is embedded. A computer virus can have devastating impacts, such as erasing files on a hard drive.

A virus that is cloaked in a legitimate program or file is called a **Trojan horse**, named after the ancient myth of the Greeks, who brought the Trojans a giant wooden horse as a gift. Once the gift was taken inside the city gate, Greek soldiers spilled from the hollowed insides of the horse and attacked the surprised Trojans.

The malicious potential of viruses was brought into the public consciousness in 1999 when the Melissa virus rapidly spread across the Internet and disabled major computer networks. The virus infected Microsoft Word documents attached to electronic mail messages. When a user opened the infected attachment, the virus was automatically e-mailed to 50 names in the infected computer's address book. If one of those 50 users opened the attachment, the infected attachment was sent to 50 more users, creating an exponentially increasing dissemination of virus-infested messages. CERT's description of the virus stated that one location received more than 30,000 Melissa messages in less than an hour. A U.S. Department of Justice press release said that the virus caused at least $80 million in damages. The Melissa virus was developed by an American, but many viruses also originate in other parts of the world. For example, the famous Love Bug virus originated in the Philippines.

A **worm** presents a greater threat than a virus because it is self-propagating and self-replicating, rapidly copying itself from computer to computer in a matter of hours. A worm is autonomous—once unleashed, it replicates itself without any action on the part of users. To propagate autonomously without needing a user to open a Web link or download a file, worms exploit existing vulnerabilities, or **security holes**, in operating systems, Web browsers, and other applications. Worms can modify files, launch coordinated attacks that flood a target computer with messages, or simply overwhelm a network with debilitating amounts of traffic.

Some of the most destructive worms—Nimbda, Code Red, Slammer, Blaster, and Sasser—struck early in the twenty-first century. For example, the Code Red worm in 2001 was considered the most costly worm or attack of any kind to ever hit the Internet. Code Red exploited a vulnerability in Microsoft Windows, disseminated by locating systems that contained the vulnerability, and then used these systems to launch attacks on other targeted systems. The Code Red worm was an example of a distributed denial-of-service attack, which is described in more detail in the next section. Like many worms, the effects of Code Red involved network performance degradation and overloaded servers. At the time, Code Red was considered to have cost the industry more than $1 billion in lost productivity and financial and human resources required to disarm the attack.

Researchers have speculated about the maximum theoretical speed at which future worms could disseminate over the Internet. For example, "Warhol" is the name given to worms that could exploit a protocol vulnerability and infect all systems with that vulnerability throughout the Internet in less than 15 minutes. Other studies speculate that even faster "flash worms" could exploit these system vulnerabilities throughout the Internet in less than a second.

Regardless of dissemination speed, worm attacks are effective because they can scan networks for computers that contain security holes, disseminate without human intervention, and stay one step ahead of the security software designed to stop them. Security holes in software are common; they require constant vigilance and awareness by users, and they require the periodic installation of vendor-developed **software patches** to repair problems. A software patch is a piece of code developed by a product vendor to address a software problem or to upgrade software. Unfortunately, some attacks are "zero-day attacks" that have hit networks before the software or hardware manufacturer of the affected product is aware of the vulnerability.

Even viruses and worms that are hoaxes can have a detrimental impact. Hoax viruses use **social engineering** techniques to make users take some action that simulate the actual effects of a virus. For example, the famous Good Times virus hoax warned users that opening any message with the phrase "Good Times" in the title would erase the users' hard drives. The words in a message subject line cannot really erase a hard drive, but the effect of the hoax was that thousands of users forwarded the e-mail warning. Commensurate with a real worm, e-mail propagation induced by social engineering consumes network and system resources and reduces worker productivity.

Denial-of-Service Attacks

A **denial-of-service attack** floods a targeted computer with so many requests that it cripples functionality. Such an attack is easy to perpetrate and hard to prevent. For example, if a Web site receives too many requests, it will not be available for other users who want access. Denial-of-service attacks consume bandwidth and system resources and are extremely difficult to combat. Attacking a computer in this manner does not require a hacker to gain unauthorized access, but simply to overwhelm a system with requests. The most damaging approach is a distributed denial-of-service attack (**DDoS**), as shown in Figure 15-1. This attack simultaneously floods a single system with requests from thousands of unwitting computers. As an analogy, consider the effect of thousands of simultaneous, groundless calls to a 911 dispatcher, flooding the system so that legitimate emergency calls could not connect. Denial-of-service attacks do not damage or steal information, but they create system outages. DDoS attacks have affected critical Internet root servers, prominent Internet routers, and numerous electronic commerce sites.

DDoS tools that are designed to disable systems are freely available on the Internet. The tools include a master program called a "handler" and agent programs called "zombies" or "daemons." The attacker often uses worms to scan unwitting, third-party computers for known vulnerabilities and installs agent programs on these systems. When the perpetrator wants to launch an attack, the master program issues instructions to the agent programs to begin making requests of the targeted system. The owners of these agent-infested systems usually are unaware that their computers are involved in an attack. In addition to attacking Internet root servers, this type of attack has reportedly disrupted high-profile Web sites such as *google.com*, *ebay.com*, *etrade.com*, *amazon.com*, and *cnn.com*.

The following preventive measures can reduce vulnerability to a DDoS attack:

- Patch management strategies for quickly upgrading software that has known vulnerabilities
- Monitoring tools that scan for the presence of agent software
- Distribution of traffic loads across multiple servers so that loads can be shifted to redundant systems if an attack occurs
- Using traffic monitoring tools to identify significant changes in traffic flow patterns

Figure 15-1

Distributed
denial-of-service attack

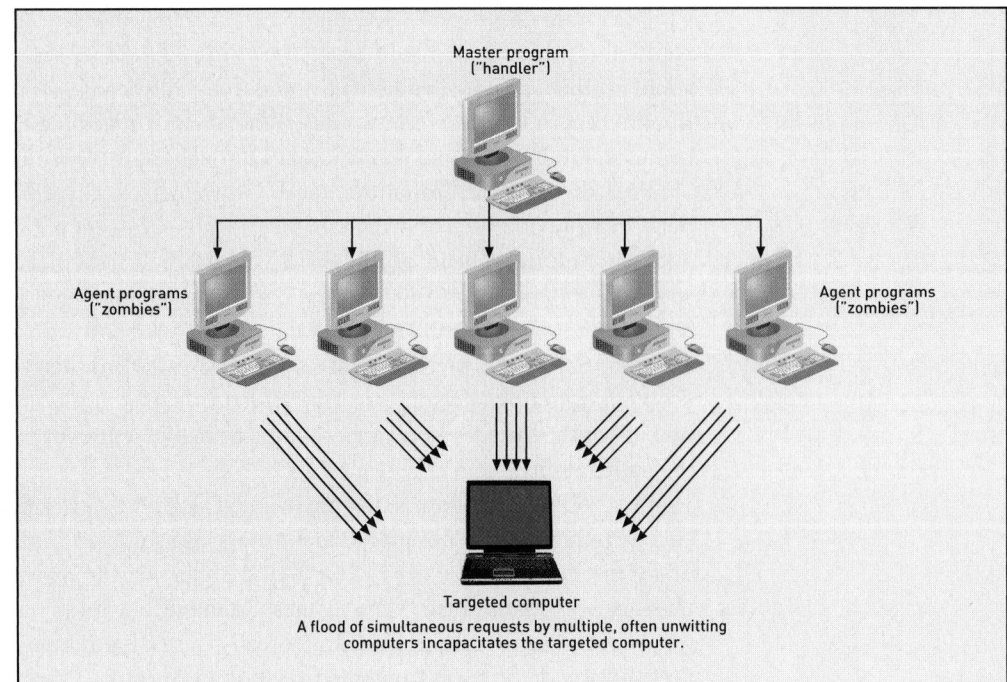

Identity and Password Theft

Another common hacker technique is to assume the identity of an authorized network user, often by obtaining a network or system password. This is easier than it sounds, especially if the hacker is in the same building as an authorized network user or somehow gains physical access to a user's networked laptop. Hackers obtain passwords in a variety of ways, including searching through office trash cans or desk drawers and peering over a user's shoulder. In other cases, users simply display their passwords on a sticky pad posted to their monitor. Hackers also can gain access by guessing a password or by soliciting one from an unsuspecting help-desk operator. Even more efficient password theft is possible with software tools that rapidly cycle through various letter and number combinations or capture passwords sent over a network (see Figure 15-2). Hackers use similar methods to steal other information that users enter online, such as names, addresses, Social Security numbers, and credit card numbers.

Figure 15-2

Password interception

Data Interception and Modification

Information transmitted across any shared network is vulnerable to interception or modification. Wire-based systems that use fiber-optic, coaxial, or twisted pair cable are susceptible in varying degrees to such attacks, but wireless networks are especially vulnerable. Unencrypted, sensitive information sent from a laptop to a wireless access point is wide open for anyone to intercept, as shown in Figure 15-3. Home wireless systems and business WLAN environments are both vulnerable to data interception and modification. For example, Wi-Fi broadcasts are sent on unlicensed and unregulated frequency bands of 2.4 GHz and 5 GHz. Anyone can buy simple equipment that receives this frequency and software that logs transmitted data, and then the person can read all transmitted information, including network addresses, personal information, and credit card numbers. Hackers have been known to intercept data by driving around neighborhoods with the appropriate equipment and detecting unsecured wireless access points, unencrypted data, and loosely encrypted data. The act of accessing unsecured wireless LAN transmissions is known as **Wi-Fi sniffing.**

Figure 15-3

Unencrypted wireless transmission is a security problem

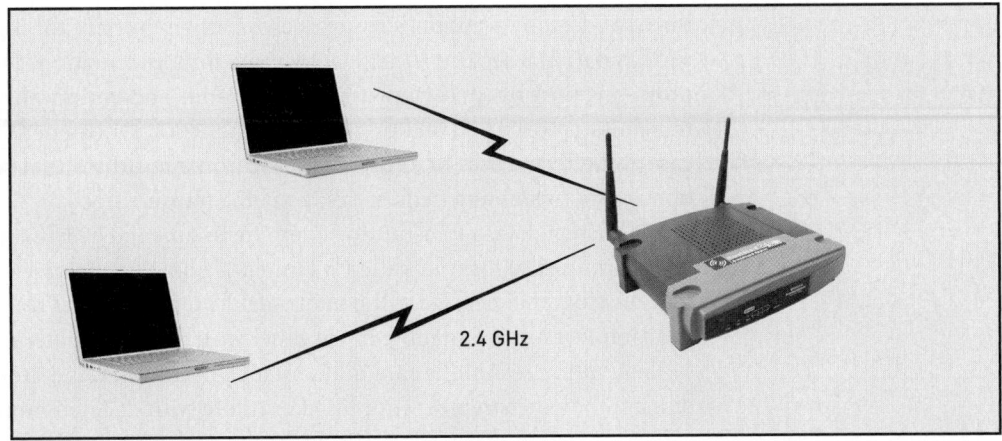

Bandwidth Piracy

Whether the network in question is the traditional telephone system, cell phone services, cable TV, satellite services, or the Internet, unauthorized network use has always occurred, but the widespread deployment of wireless Internet access has expanded the problem. Hackers use Wi-Fi sniffing equipment to intercept information and to gain free access to the Internet through wireless LANs and other unsecured networks. This type of network access theft is known as **bandwidth piracy**. Because wireless access points are so easy and inexpensive to establish, many people at a business, school, or other environment set them up outside the purview of technical administrators without being aware of the necessary security requirements. These ad hoc arrangements are known as **rogue access points**.

Critical Infrastructure Attacks

Denial-of-service attacks sometimes target the critical infrastructure of the Internet, such as DNS servers or routers in high-traffic areas of public networks. Because of the economic, social, and political importance of networks and our reliance on them for everyday functions, the threat of a terrorist attack on critical network infrastructure is a serious concern. Since the attacks of September 11, 2001, there has been a significant focus on shoring up physical network components. Many companies have strategies to protect key components that underlie mission-critical networks. For example, a network is only as reliable as its underlying power grid. When a part of the country has a major power outage, networks cease to operate. Attacks on power grids, telecommunications systems, cell phone networks, Internet infrastructure, stock market networks, and ATM networks are considered very real threats.

NETWORK SECURITY STRATEGIES

Fortunately, security strategies can significantly reduce the risk of the attacks described in the previous section. System administrators and network managers are responsible for **patch management**, the process of staying vigilant about product vulnerabilities and the availability of vendor-supplied software patches and instituting procedures for systematically upgrading systems with these patches. Sometimes, a large software patch that addresses significant security weaknesses in a product is called a **service pack** upgrade. Other strategies span areas as diverse as physical site security, access control software, and complex encryption algorithms. Most security approaches fall into one of the following three categories:

- Privacy
- Access control
- Authentication

PRIVACY

The privacy of information transmitted over a network must be safeguarded to combat the threat of data interception. Privacy is especially important for highly sensitive transmissions such as national security, financial, and medical information. The most effective method of protecting the privacy of network information is **encryption**, the scrambling of data prior to transmission over a shared or vulnerable network. One benefit of digital technology is the ease and effectiveness of applying encryption algorithms that scramble ones and zeros, as opposed to scrambling frequencies in analog transmissions.

To encrypt data, a transmitting computer mathematically manipulates data according to a predetermined algorithm called a cipher. If someone accesses this encrypted data during transmission, the message will be unreadable. Once the data reaches its destination, a receiving computer can unscramble it—in other words, the computer can decrypt the data.

Encryption techniques are mathematically complex and are implemented in the digital domain, but an extremely simplified example with decimal numbers can help illustrate the concepts. Suppose someone wants to transmit a personal phone number but wants the intended recipient to be the only person who can view the number. How could the sender encrypt it? One oversimplified cryptographic technique would be to add 2 to each digit in the number. For example, this approach would convert 703-555-1212 into the encrypted 925-777-3434. The number could then be transmitted across a network and would be unreadable to anyone who did not know the encryption technique. Presumably, the recipient would decrypt the data by subtracting 2 from each digit in the phone number (see Figure 15-4).

Figure 15-4

Simplified encryption example

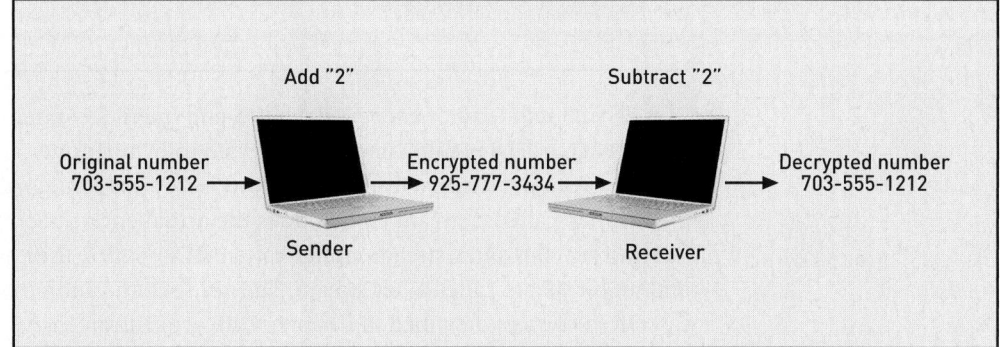

While this example is simple, and easy to crack, it does illustrate some components needed for encryption. One component is the algorithmic approach of shifting each number by a fixed amount. The second involves the actual number that is added (in this case, the number 2). This number is the **encryption key**, the cipher that encodes and decodes the encryption. For the receiving computer to unscramble the message, it must know both the algorithmic approach (shifting the number) and the key, which is essentially a long bit stream.

A fundamental characteristic of any encryption's strength is its **key length**. Key lengths of at least 128 bits are commonplace. The longer the key is, the stronger the encryption. A key length of n bits provides 2^n possible unique keys, so a key length of 128 bits provides 2^{128} or 3.4×10^{38} unique keys.

Encryption approaches take one of two forms: private key and public key. In **private key encryption**, the sender and receiver use the same key to encrypt and decrypt a message, as in the preceding example. Because the same key is used, private key encryption is also known as symmetric encryption. Private key encryption historically dominated cryptographic approaches, but it has some problems. For example, the private key must be known between two parties in advance; otherwise, it must be transmitted prior to sending the message. Transmitting the key presents obvious security risks because it might be intercepted by a third party. On the other hand, requiring every combination of sender and receiver to use a unique, private key is an unwieldy proposition. A single computer communicates with countless other computers, and should not have to maintain a unique key for every other computer with which it might exchange information.

A more recent, safer encryption approach is **public key encryption**. Using this technique, every recipient possesses two keys: a private key that no one else can access and a public key that is accessible to anyone. Sending an encrypted message requires looking up the recipient's public key and encrypting the message according to this key. The recipient then uses a private key to decrypt the message (see Figure 15-5). This approach is much more secure than private key encryption because the recipient's private decryption key is never exposed to transmission over a network. Third-party trusted entities called **certificate authorities** verify that the public keys belong to their rightful owners. Public key cryptography is vital for any type of electronic commerce over the Internet. For example, when an individual user purchases something from a Web site, the individual's browser obtains the commerce site's public key and encrypts data such as a credit card number before sending it over the network.

Figure 15-5

Public key encryption

A *de facto* standard for encrypted, end-to-end Internet communication is **Transport Layer Security** (TLS). TLS is still sometimes referred to by its former name, **Secure Sockets Layer** (SSL) encryption. Netscape Communications Corporation originally developed SSL.

A cipher, called a stream cipher, does the actual encrypting of each plain-text digit. SSL encryption relies on a stream cipher called **RC4**, which uses an algorithm to generate a random bit stream called a key stream. The key stream combines with the plain-text message via XOR (a concept described in Chapter 4) to generate an encrypted message, also known as the ciphertext (see Figure 15-6).

Figure 15-6

Generating an encrypted message

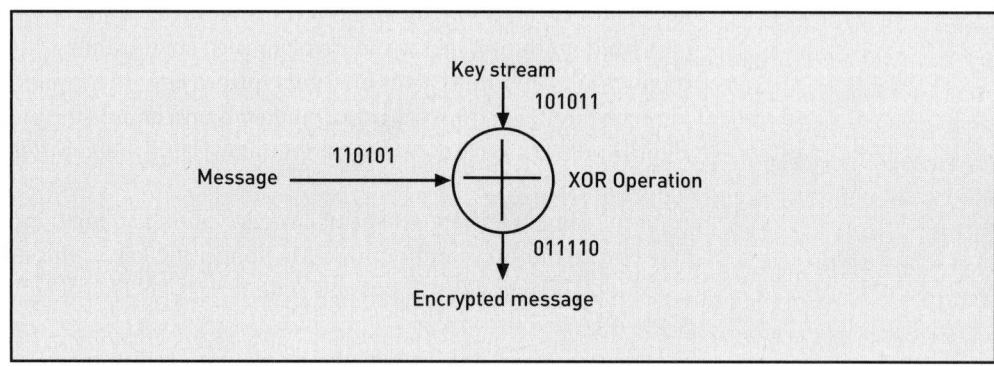

Encryption for wireless Internet access is especially important, because anyone with rudimentary tools in close proximity to a Wi-Fi link can access unencrypted wireless transmissions. One of the first 802.11 encryption approaches was based on a technique called Wired Equivalent Privacy (**WEP**). Vendors incorporated WEP into wireless cards, and the standard became the *de facto* approach for wireless LAN security. WEP's main objective was to ensure the confidentiality of data transmitted over 802.11 wireless LANs, but it also was intended to provide some access control and data integrity by preventing modification of transmitted messages. Unfortunately, WEP contained some inherent inadequacies. Several studies by cryptographic experts determined that WEP encryption was relatively easy to break. In particular, mobile devices and access points that used WEP encryption shared a secret encryption key. Researchers were able to intercept and read WEP keys transmitted over 802.11 networks using commercially available hardware and software, and by implementing an attack method originally described by Fluhrer, Martin, and Shamir. Anyone with a Wi-Fi-enabled computer positioned close enough to a WEP-secured WLAN could gain access as a falsely authenticated user by obtaining the LAN's secret key.

Understanding the immediate market need for greater security than WEP provided, an IEEE 802.11 working group began developing a standard called 802.11i. As an interim step, the Wi-Fi Protected Access (**WPA**) encryption protocol was introduced until the stronger wireless protocol, known as the **802.11i** standard, was formally completed by the IEEE.

ACCESS CONTROL

Access control mechanisms determine and enforce who or what can gain network access to computers. An important aspect of access control is physical security—ensuring that unauthorized users have no physical access to a networked computer in an office building or to a person's networked laptop. Other access controls include the systematic use of passwords and implementation of firewalls.

Firewalls

A **firewall** is one of the most important network security measures that an institution or individual user can install. A firewall is essentially an access control device. A network firewall is installed between a secure private network and a nonsecure public network to regulate access to and from the private network's resources, as shown in Figure 15-7.

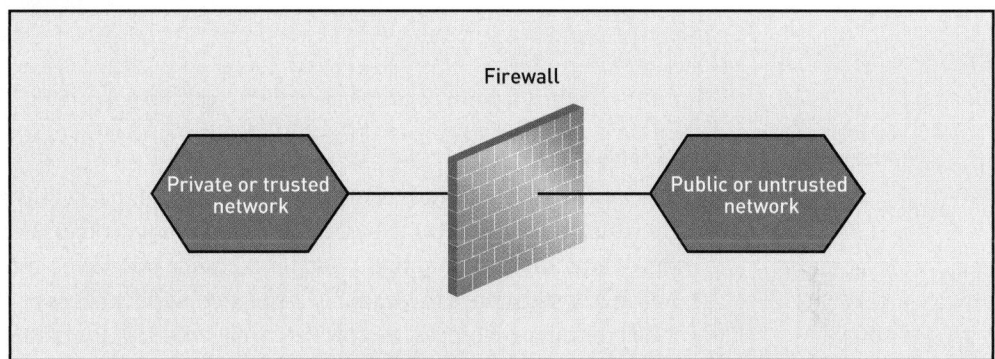

Figure 15-7

Function of a network firewall

Firewalls can be implemented in hardware or software. For example, a firewall can be a process on a router. Firewalls can also be installed on systems, or hosts, to protect access to certain applications or systems. Users can configure access control requirements that must be met before the firewall will permit access to a network or system. Users who connect to the Internet via a broadband connection usually install **personal firewall** software on their home computers. Personal firewalls detect and prevent intrusions, block hackers from seeing content on the protected computer, monitor incoming and outgoing traffic, and perform automatic updates to detect the latest security holes, worms, and viruses.

Organizations use much more complicated firewalls that provide security services for Internet applications, including Internet telephony. Firewalls can prohibit certain network activity such as the ability of outside users to log on to a protected computer or download files from it. Similarly, a firewall might enforce predetermined rules for which Internet addresses it allows and which ones it blocks.

Figure 15-8 illustrates a typical firewall configuration for an organization connected to the public Internet. A trusted network, in this case a private LAN, is connected to a firewall. Usually, the organization sets up a separate server and installs firewall software on it. A router connects the firewall to the public Internet. Inside the router, but outside the firewall, is an area called a demilitarized zone, or **DMZ**. The DMZ is less secure than the private network protected inside the firewall. Companies might place their public Web servers inside the DMZ and keep them separate from the company's internal private network.

Figure 15-8

Typical firewall
implementation

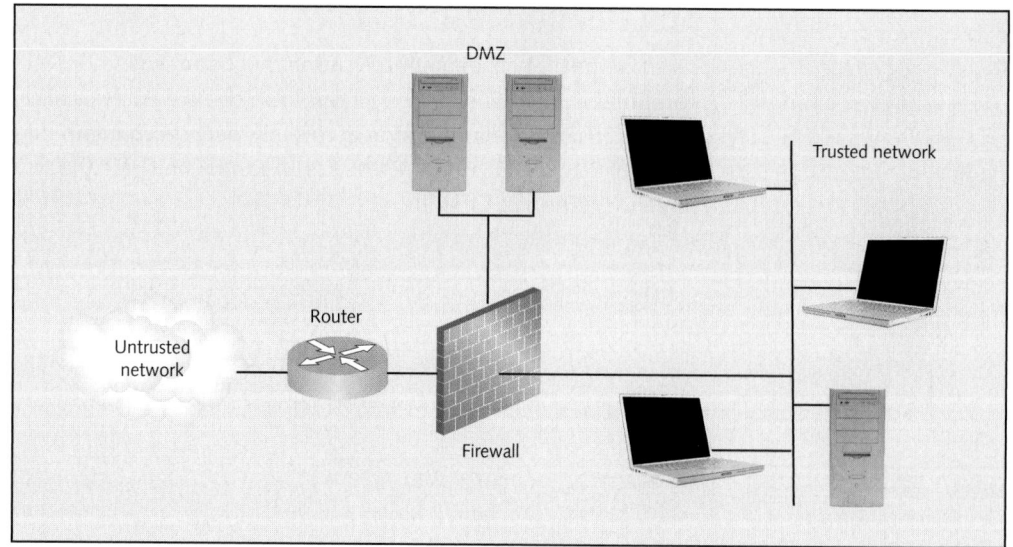

Packet Filtering

One way that firewalls can restrict access is through packet inspection. This type of **packet-filtering firewall** intercepts packets and then inspects their header contents, including the source IP address, destination IP address, source port, and destination port. A port number indicates what service the traffic is associated with, such as HTTP or FTP. The firewall then either permits or blocks the packet from entering the network. One downside of this approach is that the firewall must inspect every packet that traverses it. Also, the return traffic from any request the firewall permits is subject to the same access rules.

Stateful Packet Filtering

To overcome these disadvantages, other types of firewalls use **stateful inspection**, a more intelligent form of packet filtering that notes when an incoming response is expected after an outgoing request is made. In other words, the stateful packet-filtering firewall knows to expect traffic transmitted from a certain IP address or port, and can allow this traffic to go through. If an unexpected packet arrives and indicates that it is a response to an outgoing solicitation, the firewall knows to block this traffic if no such state exists.

Network Address Translation

Some firewall approaches use **Network Address Translation** (NAT), in which the firewall converts the IP address of every outgoing packet into a shared IP address before the traffic is sent over a network. NAT was originally developed as a solution to concerns about IP address scarcity. To conserve addresses, multiple users on a private network share a public IP address. NAT provides security primarily because it prevents bidirectional transmission. Only connections that are initiated on a local, private network are established. Any communication that originates on a public network is stopped by the NAT firewall, which automatically prevents malicious attacks such as worms from entering the protected network. This approach also has significant disadvantages. For example, because the IP address translation disrupts end-to-end transmission, it is difficult to apply end-to-end protocol features, including some encryption approaches.

Application Proxy Firewalls

The most complex type of firewall is the **application proxy firewall**, which filters information based on the application data itself. For example, rather than filtering packets based on allowing or denying HTTP traffic, an application firewall looks at the application content and distinguishes between normal and unexpected HTTP traffic.

Password Strategies

Password protection is an obvious line of defense in computer security. As shown in Figure 15-9, passwords are routinely instituted to authorize access to individual computers, networks, accounts, or applications. While passwords are an important aspect of security, they provide insufficient security on their own. It is often easy to guess passwords, intercept them over a network, or even convince users to volunteer their passwords. Security policies help instruct users about strong password strategies. For example, strong passwords should be at least eight characters long and should include a combination of uppercase letters, lowercase letters, and numbers. Furthermore, passwords that incorporate special characters such as &, $, and * provide greater security. Most policies also describe passwords to avoid. For example, a password should not be the same as the user's ID, and it should not be the user's name, birthday, address, Social Security number, or pet's name.

Figure 15-9

System login with user ID and password

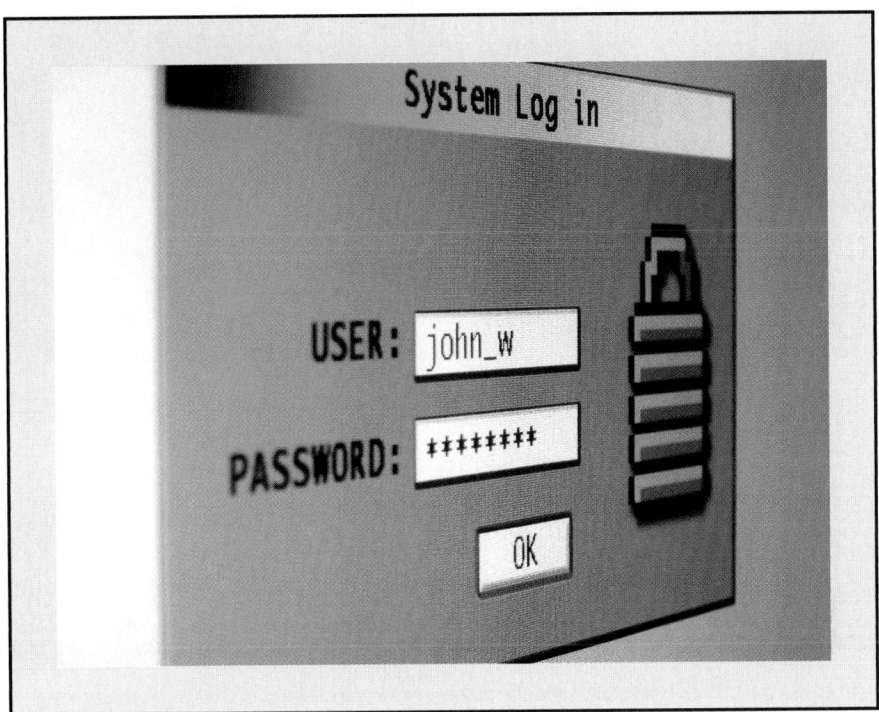

Security policies also warn network users about **phishing**, social engineering schemes that use e-mail or text messages to try to coerce users into relinquishing sensitive information such as a password, bank account number, Social Security number, or credit card number. Phishing is organized criminal activity in which the perpetrators usually disguise themselves as a legitimate entity such as eBay, PayPal, or a bank. Phishers usually try to instill fear into their targets. A typical phishing e-mail might read, "Your bank account has been temporarily suspended because of attempted identity theft. To reinstate your account, you must verify your identity at the following Web site." The phishing scheme then attempts to solicit the user's bank account number, password, and other personal information.

Organizations' security policies seek to educate network users about the dangers of phishing schemes and warn them not to succumb to solicitations for personal information without first verifying with the company that the request is legitimate. Providing users with a list of recurrent phishing schemes is also advisable, and technical measures such as spam filters and browser controls can help users to avoid such schemes.

Physical Security

Physical security is an important and sometimes overlooked form of access control, especially considering the significant percentage of security breaches that insiders perpetrate within a company, organization, university, or even a home. Routine physical safeguards include installing door locks for rooms that house servers and network equipment (including wiring closets) and providing adequate building security. Figure 15-10 shows a secure server room.

Figure 15-10

Server racks in a controlled environment

AUTHENTICATION

How can a network discern whether users are who they say they are? A correct password does not automatically indicate an authorized user. What if someone steals or guesses the password? Achieving higher levels of security requires **authentication**, the process of verifying a person's identity before allowing network access. Besides passwords and personal identification numbers, authentication methods include "token-based" authentication, biometric identification, and digital signatures.

Token-Based Authentication

One level of authentication that surpasses a memorized password is token-based authentication, which requires a computer user to physically hold a device called a token. A token is a matchbook-sized device with a liquid crystal display that provides a one-time password for gaining network access (see Figure 15-11). To log on to a network, the user first enters a conventional password. Once the network accepts this password, the user enters the access number displayed by the token. This number changes approximately every 10 seconds and is synchronized with the network. Token-based authentication provides a strong extra layer of security. Even if a hacker steals a user's password, there is no way to gain access without physically holding the token.

Figure 15-11

Token-based
authentication

Biometric Identifiers

Biometric identification is an extremely effective authentication approach, although it is sometimes controversial because of concerns about user privacy. Password authentication is not foolproof because passwords can be stolen or forgotten, and network users can lose token-based authentication products. In contrast, a person's biometric characteristics can never be lost, forgotten, or easily copied. Biometrics can identify any of a person's unique physical characteristics, including fingerprints (see Figure 15-12), facial features, voice patterns, retinal patterns, irises, or DNA.

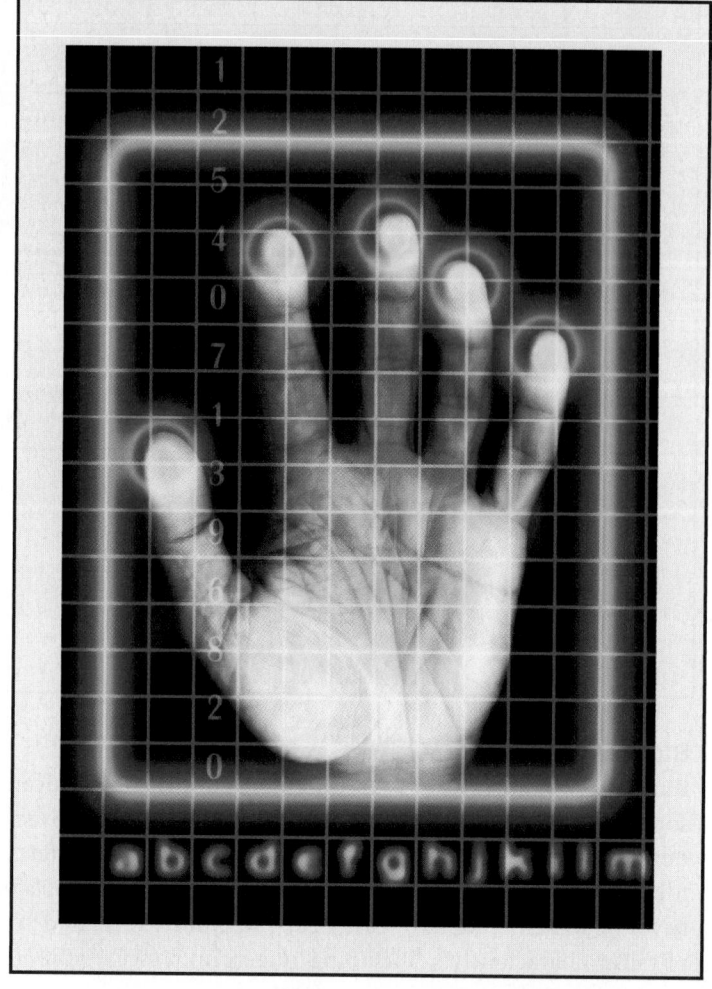

Some laptops include electronic fingerprint readers, as shown in Figure 15-13. To log in to a computer, the user must first scan a finger over the electronic sensor. If the scan matches the fingerprint stored in the computer for authentication purposes, the user can access the computer. Fingerprint scanning mouse technology is a similar authentication technique (see Figure 15-14). Electronic fingerprint sensors have exhibited some problems; for example, false negative readings can impede legitimate user authentication. Also, the sensors are vulnerable to hacking, such as gaining access using a paper or gel copy of the user's fingerprint.

Figure 15-13

Laptop with fingerprint-
scanning technology

Figure 15-14

Mouse with fingerprint-
scanning technology

In addition to routine laptop access, biometrics are also used in highly secure computing environments such as those that contain classified national intelligence or highly sensitive financial information. Biometric authentication that allows access to secure computing environments can use unique qualities of the human eye, such as characteristics of the iris or retina (see Figure 15-15). For example, the capillary (blood vessel) pattern of each person's retina is unique and generally remains fixed throughout a person's life, although some eye diseases can modify retinal patterns over time. Retinal authentication requires a person to look through an eyepiece that emits infrared light into the eye. The unique reflections from the eye can be converted into binary code and compared to the stored bit pattern. This authentication technique is extremely reliable because of the uniqueness of retinal patterns.

Figure 15-15

Unique eye patterns as biometric identification

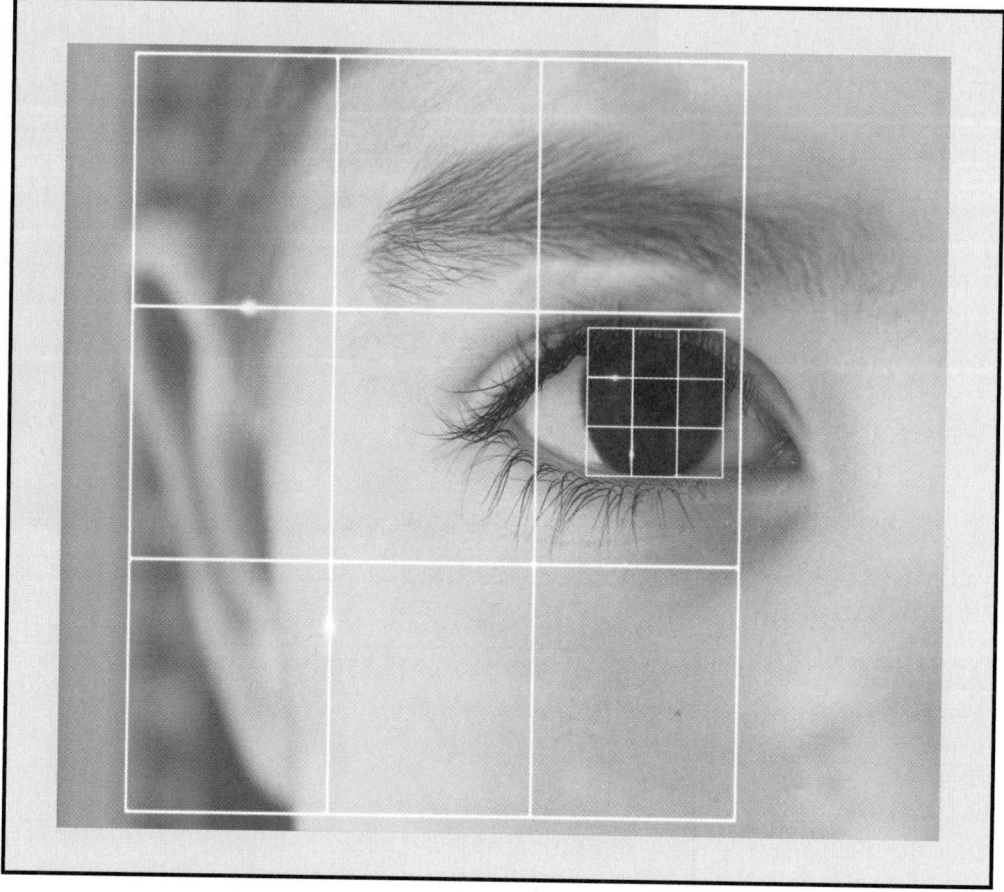

Other physical characteristics that can be used to authenticate users include scanning faces, hands, and even the way a user walks.

In some cases, biometric identification is stored in an ID card. For example, a card can hold fingerprint data and binary code representing DNA. A potential problem with this approach is that the information can be stolen. Biometric authentication is most effective when it requires the physical presence and testing of the person being authenticated. In practice, biometric approaches are typically used in conjunction with other security measures such as password authentication.

Digital Signatures

Digital signatures, which are related to public key encryption, offer another option for authentication. A sender encrypts information using its private key and transmits the information over a network to its intended destination. Once the data is received, the destination device looks up the sender's public key and uses it to decrypt the message. If this decryption is successful, the data is verified as having originated with the presumptive sender. This process is actually a reversal of public key encryption—instead of encrypting data with the recipient's public key, the sending device uses its own private key to encrypt data.

How is a public key assigned? Upon request, one of several certificate authorities (CAs) creates a public key for a site. The CA goes through a process of verifying the requester's identity, then creates the certificate and sends it to the requester while keeping a copy of the public key in the CA's database. When a site transmits a public key, the recipient contacts the CA to verify the transmitter's identity.

CHAPTER SUMMARY

- In the United States, the Department of Homeland Security tracks security incidents, publicizes security vulnerabilities, and provides information about necessary software patches and upgrades at its US-CERT Web site, *http://www.us-cert.gov.*
- The people and organizations that attack networks and systems have a variety of motives and capabilities, but they generally fall into the following categories: hackers, spammers, rogue employees, corporate spies, and cyberterrorists.
- A virus is malicious code embedded in a seemingly legitimate program. It becomes active only when a user executes the legitimate program.
- A worm is an autonomous, self-propagating, and self-replicating program that exploits existing security vulnerabilities to perpetrate attacks, such as erasing files, modifying files, or overwhelming a system.
- In a distributed denial-of-service attack, numerous computer systems—some of them unwittingly—flood a targeted computer with an overwhelming and crippling number of requests.
- Other types of common security attacks include identity and password theft, data interception and modification, bandwidth piracy, and critical infrastructure attacks.
- Three important categories of security strategies include access control, authentication, and privacy.
- Important techniques for ensuring information privacy over a network include encryption approaches such as SSL and 802.11i.
- Various types of firewalls provide access control between a public and private network, including packet-filtering firewalls, stateful packet-filtering firewalls, and application-level firewalls.
- Authentication is the process of ensuring that a person or system is who it claims to be. Authentication is accomplished via passwords, token-based authentication, digital signatures, and biometric identification.

KEY TERMS

802.11i
application proxy firewall
authentication
bandwidth piracy
biometric identification
CERT
certificate authority
corporate espionage
cyberterrorism
DDoS
denial-of-service attack
digital signature
DMZ
encryption
encryption key
firewall
hacker
key length
Network Address Translation
packet-filtering firewall
patch management

personal firewall
phishing
private key encryption
public key encryption
RC4
rogue access point
Secure Sockets Layer
security holes
service pack
social engineering
software patches
spam
stateful inspection
Transport Layer Security
Trojan horse
virus
WEP
Wi-Fi sniffing
WPA
worm

REVIEW QUESTIONS

1. What critical systems rely on secure and reliable availability of information networks?
2. Who poses a threat to information networks?
3. What is the difference between a virus and a worm?
4. What damage can a worm cause in computer systems?
5. Who issues software patches for security vulnerabilities in applications?
6. What is a distributed denial-of-service attack, and how can it be averted?
7. What type of network is most susceptible to data interception by a hacker?
8. Explain the difference between private key and public key encryption.
9. How is key length related to encryption strength?
10. What is Trojan horse software, and how does it get its name?
11. Name five physical characteristics that can be used for biometric identification.
12. Explain why digital signatures are related to public key cryptography.
13. What is the difference between authentication and access control?
14. What is the downside of packet-filtering firewall approaches?
15. Explain the difference between a packet-filtering firewall and a stateful packet-filtering firewall.
16. What is Network Address Translation?
17. What does a certificate authority do?
18. What is the primary function of RC4?
19. What is the main disadvantage of using WEP?

DISCUSSION QUESTIONS

1. In addition to being an irritation to users, why is spam a serious problem for business Internet users, network managers, and ISPs?
2. How would a 24-hour network outage disrupt your daily life? What if you suddenly lost all access to telephones, the Web, ATMs, and satellite and cable technology? What day-to-day functions would you lose, and how would this affect your day?
3. What is the relationship between information networks and the underlying power grid? What critical information networks are supported by the power grid? How reliable is the power grid? Can you find any examples of major power outages and how they affected information networks?
4. Do you think it is possible for the Internet to be disrupted by hackers for a long period of time? What architectural features of the Internet make it susceptible or resistant to this type of service outage?
5. Have you ever had a problem caused by a worm or virus? Which worm or virus was it, and what did you do about it?
6. What are some possible ethical issues related to the use of biometric identifiers?

CASE PROJECT

You have been charged with protecting your company's network from cyberattacks. One of your responsibilities is to monitor the US-CERT Web site for new security vulnerabilities and product upgrades. Visit the Web site at *http://www.us-cert.gov/* and write a report to your company's CIO that answers the following questions:

1. What is US-CERT, who runs it, and how can it help your company to prevent cyberattacks?
2. How can you receive up-to-date, real-time information about current security vulnerabilities, worms, and viruses? Does US-CERT have mailing lists or RSS (Really Simple Syndication) feeds that can help?
3. What are all the other CERTs worldwide? What is the relationship between these groups and US-CERT?
4. Find a weekly vulnerability summary on the US-CERT Web site. What are the high-vulnerability security problems for the week? Choose five current product vulnerabilities and list the following for each:
   - Vendor name
   - Product name
   - Description of security problem
   - Description of security solution or software patch
5. If your company detects a security vulnerability in a product, is subject to a cyberattack, or has learned about a new worm or virus, what corporate procedure would you institute to report the problem to US-CERT? Is there a place at the US-CERT Web site to report problems?

PART
6
TELEPHONY AND WIRELESS MULTIMEDIA

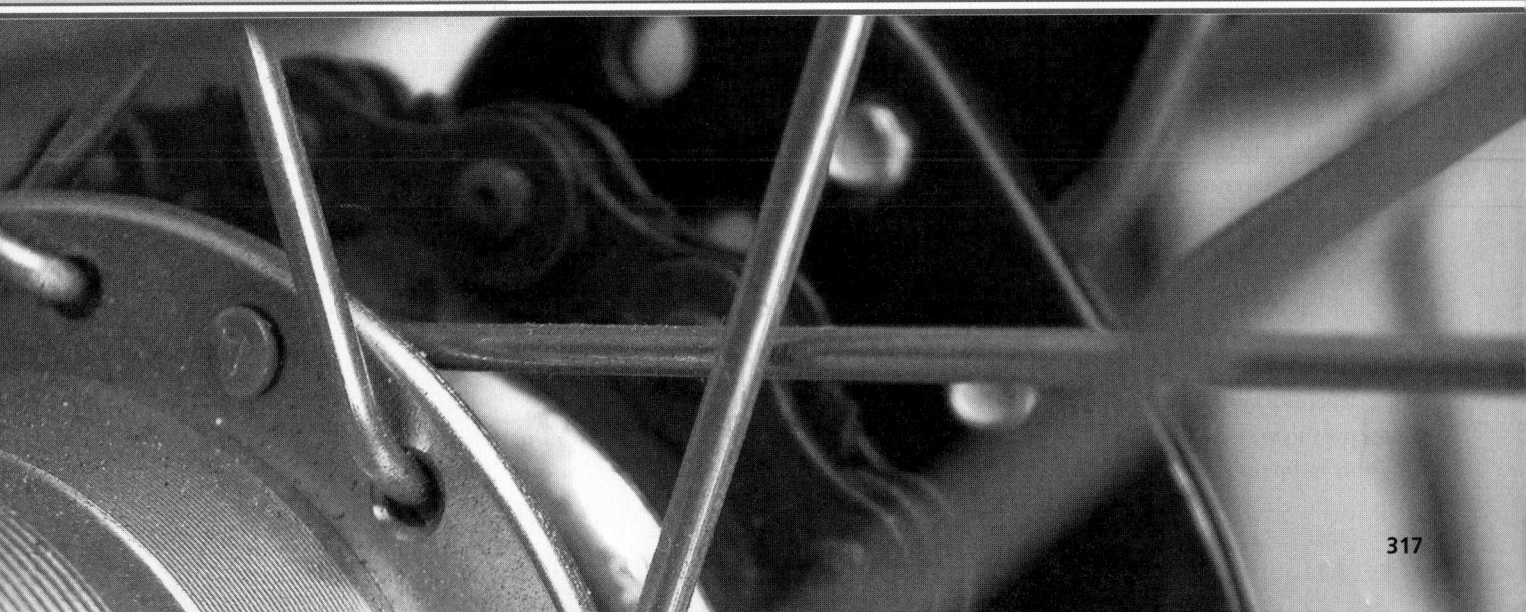

THE TELEPHONE SYSTEM

LEARNING OBJECTIVES

In this chapter you will:

- Understand the main physical components of the traditional Public Switched Telephone Network, including transmission facilities, switching equipment, and customer premises equipment

- Become familiar with the functionality of private branch exchanges (PBXs) and Centrex systems

- Understand pulse code modulation techniques for converting analog into digital signals for the digital telephone system

- Distinguish between wavelength-division, time-division, statistical time-division, and frequency-division multiplexing

- Identify the function of the T-carrier system, Synchronous Optical Network (SONET), and Synchronous Digital Hierarchy (SDH)

- Understand signaling protocols for digital telephone systems such as Signaling System 7 (SS7)

INTRODUCTION

Because of the ubiquity and longevity of the telephone system, it is taken for granted, perhaps more than any other information system. By picking up a telephone and dialing a few numbers, a caller can reach anyone almost anywhere in the world. This chapter, along with the following chapters on Voice over IP (VoIP) and multimedia wireless, introduces voice communication technologies. The chapter begins by providing some historical context on the evolution of the telephone system dating back to the late 1800s. The physical components of the telephone system are discussed next: transmission systems, switching systems, and customer premises equipment. Underlying techniques for transmitting voice information over the telephone system include multiplexing, pulse code modulation, and signaling. These technical concepts are the main subject of the chapter, and they form the foundation for understanding more advanced concepts in telecommunications such as cellular phones, other wireless telephony techniques, and Internet telephony.

The history of the telephone system is not one of linear technological development but a complex story of patents, antitrust, government regulation, and monopolies. There is not *one* history of the telephone system but *many* histories of similar systems that developed differently in various countries. The following brief history uses the development of the telephone network in the United States as an example.

Prior to the introduction of the telephone in 1876, the primary methods of communicating over long distances were the postal service and telegraph. At the time, America's Civil War had recently ended, the Wright brothers' invention of the airplane was still decades away, and European immigrants were flooding into the United States. The transmission of voice over an electrical circuit was a novel innovation compared with the communications technologies of the time. Only three decades prior to the introduction of the telephone, a telegraph line was installed between Washington, D.C. and Baltimore. Using a telegraph key and a receiver, Samuel Morse used a combination of dots (short electrical pulses) and dashes (long electrical pulses) to transmit a simple message. Telegraph systems were widely deployed in the United States to encode and transmit textual information, but they did not convey the sound of a human voice.

This changed on March 6, 1876, when Alexander Graham Bell successfully called his assistant, Thomas Watson, in another room. Bell was not the only inventor of telephone technology, but he was the first to successfully patent the invention. Bell's invention of transmitting human voice over electrical wires became the basis of telecommunications development for more than a century.

Early telephones had the same primary components of today's telephones: a switch acting as a mechanism to connect a caller to the telephone network, a speaker to hear the other caller's voice, and a microphone to detect human speech and convert these sound waves into electrical current. The molecular vibrations created by human speech can be detected and converted into electrical signals. Early telephone microphones consisted of small carbon granules that would compress and decompress in response to the sound waves they detected. This compression and decompression would then alter the resistance of the granules and therefore "modulate" or alter the electrical signal flowing through the material. Recall from Chapter 8 that a material's resistance affects the amount of charge flowing through a conductor according to Ohm's Law, which defines the relationship between voltage (V), current (I), and resistance (R) as $V = IR$.

In 1877, Bell founded the Bell Telephone Company, which would become the largest company in the world. The telephone system that emerged from the company and its later incarnations would provide the infrastructural underpinning for voice telephony, fax transmissions, data communications via modem, and eventually Internet transmissions. To demonstrate the enduring technological approach of Bell's invention, you could plug an antique phone from the early twentieth century into the current phone system and still make a call. For a century, the Bell Telephone Company—also called the Bell System or "Ma Bell," and later AT&T—served as a telecommunications monopoly in the United States.

In the early 1980s, a Department of Justice antitrust suit against AT&T ultimately resulted in the company's breakup. The **Modified Final Judgment** (MFJ) was the antitrust settlement agreement between the Department of Justice and AT&T. AT&T's local phone service was divested into seven independent regional companies called regional Bell operating companies (RBOCs). These seven RBOCs were Ameritech, Bell Atlantic, Bell South, NYNEX, Pacific Telesis, Southwestern Bell, and US West. AT&T would retain its long-distance services and manufacturing business but would not be able to compete in local services. The RBOCs would provide regional telecommunications services but were prohibited from selling long-distance services or equipment.

The MFJ resulted in competition for long-distance telephone services and in equipment manufacturing, and it regulated monopolies in regional phone services. The RBOCs were effectively regulated regional monopolies, while long-distance services were open to competition. In addition to AT&T, long-distance service providers called Interexchange Carriers (IXCs) included companies such as Sprint and MCI.

Under the terms of the MFJ, the United States was divided into regions called Local Access and Transport Areas (LATAs). These geographical areas were regulatory designations of regions within which each RBOC could provide services. Any communications originating in one LATA and terminating in another was considered inter-LATA traffic and was handled under law by an IXC.

More than a decade later, the regulatory framework of the U.S. telecommunications industry changed again with the passing of the **Telecommunications Act of 1996**. The act was the first significant overhaul of U.S. telecommunications law since the passing of the Communications Act of 1934. The objective of the 1996 act was to promote greater competition, allowing any service provider to compete in any sector of the telecommunications market.

The telecommunications service provider industry has experienced a series of mergers and acquisitions and continues to be in a state of flux. The following changes are a few examples of industry consolidation that occurred over a 10-year period: Bell Atlantic acquired NYNEX and GTE and then became Verizon, which also acquired MCI; Qwest acquired US West; and Southwestern Bell became SBC, which acquired Ameritech, Pacific Telesis, and AT&T, and then changed its name to AT&T. Today's telecommunications market in the United States is, for the most part, characterized by competition between service providers. One commonality among these competing service providers is that their offerings extend beyond traditional phone service to include wireless services, broadband Internet access, and Internet telephony.

The history of the telephone system in the United States includes Bell Labs, the Bell System's research arm, which was founded in 1925, and its manufacturing arm, Western Electric. Bell Labs, now part of Alcatel-Lucent, was responsible for many of the technologies and techniques described in this book. Their value is inestimable, but the following innovations are just a few to which Bell Labs contributed:

- The transistor
- The UNIX operating system
- The C programming language
- The LASER
- Touch-tone telephones
- DSL

PUBLIC SWITCHED TELEPHONE NETWORK

The Public Switched Telephone Network (**PSTN**), which to the public looks like Figure 16-1, is much more than just telephones. It is an intricate system of transmission lines, switching equipment, standards, and conceptual approaches that enable any telephone to connect with any other telephone in the world, regardless of service provider or location. Traditional phone service has historically been referred to as POTS, which is short for Plain Old Telephone Service.

The following main physical components of the PSTN are illustrated in Figure 16-2 and described in the following sections:

- Transmission facilities
- Switching equipment
- Customer premises equipment

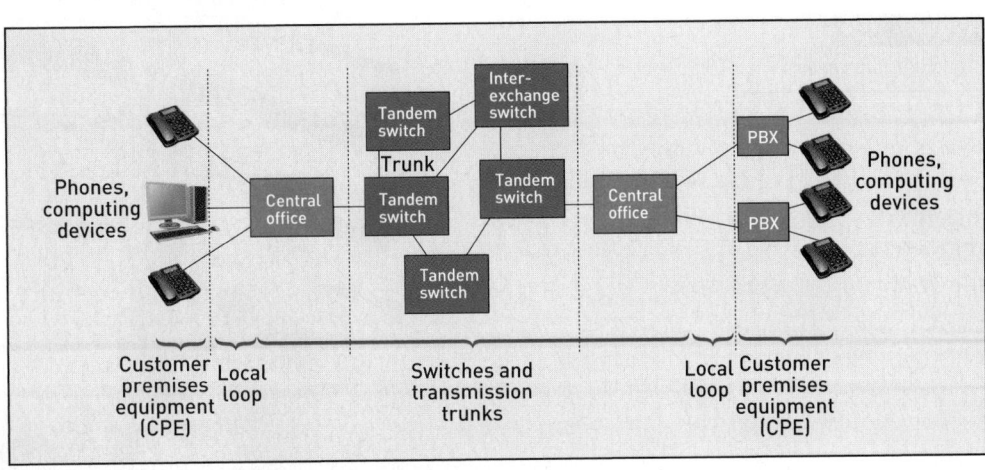

TRANSMISSION FACILITIES

The PSTN uses almost every type of transmission medium. The transmission facilities that connect homes, businesses, and institutions to a switching center called a **central office** are known as the **local loop**. The local loop spans a distance of at most three miles, extending from a demarcation point at a customer location to a local telephone company switch located at a central office. The demarcation point is where the customer's own telephone line ends and the local loop of the carrier begins. Telephones obviously connect to a wall jack, over twisted pair cable via a phone plug (see Figure 16-3). The demarcation point and the central office are usually also connected by twisted pair cable, but they are sometimes connected by fiber-optic cable, referred to as fiber to the premises (FTTP). Signals that are generated by the subscriber's telephone are digitized within the central office and transmitted digitally through the other parts of the PSTN, such as between other switching centers including tandem switches and interexchange carrier switches. Therefore, the telephone network is primarily digital with the exception, in some cases, of the local loop.

Figure 16-3

Phone plug

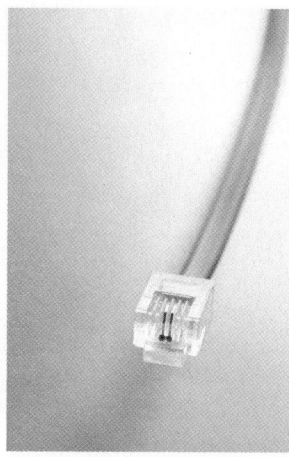

In the telecommunications industry, the term *outside plant* (see Figure 16-4) refers to telephone system facilities that are outdoors. These facilities include telephone poles, cables, microwave towers, repeaters, and any outside equipment between customer premises and switching centers.

Figure 16-4

Outside plant repairs

Longer-distance transmission facilities, such as those between switching centers of the PSTN, are called **trunks**. The telephone network uses a combination of fiber-optic cables, wireless transmission links involving microwaves and satellites, and copper cables such as twisted pair and coaxial cable. Fiber-optic cables serve as a significant portion of the long-distance trunk facilities.

In some regions, high-capacity microwave radio links also provide some of the long-haul transmission links for the telephone network. **Microwave** antennas, like all antennas, propagate electromagnetic energy into space and can carry thousands of simultaneous voice conversations.

Transmission between two microwave antennas in the telephone network is point to point. It involves a highly focused beam over an approximately straight path between antennas, in contrast to the almost 360-degree broadcasts of radio stations. Antennas typically are divided into two general classes: directional and omnidirectional. A **directional antenna** such as a microwave system radiates energy in a single direction, while an **omnidirectional antenna** can transmit or receive energy from all directions.

Microwave transmission requires a **line-of-sight** path, meaning that obstructions such as buildings or mountains between antennas disrupt transmissions. As shown in Figure 16-5, microwave antennas usually appear in the shape of a drum; they are typically mounted on the tops of buildings, mountains, or high platforms, or used in geographical areas that are relatively flat, such as the U.S. Great Plains.

Figure 16-5

Microwave antennas

Transoceanic, long-haul transmission is possible because of submarine cables and satellite communications. International calls that use satellite communications present a small but perceptible delay; the signal must reach a transponder on an orbiting satellite, regenerate, and be transmitted back to Earth, traversing a path of tens of thousands of miles. Submarine cables are special fiber-optic or coaxial cables that span the ocean floor to enable intercontinental communications. These cables have special engineering design requirements such as the ability to resist saltwater corrosion and endure the ocean's extreme temperature and pressure conditions. The cables also must be carefully placed or buried to evade anchors, construction, or fishing trawlers in shallow areas. Submarine cable systems require special materials to encase and protect the cables, and repeaters at regular intervals across the cable span to regenerate and strengthen the signal.

The telephone system uses a combination of digital and analog transmission facilities, but is mostly digital. As discussed in previous chapters, digital transmission systems work better than analog systems because they can reconstruct a degenerated or weakened signal after it has passed through a transmission medium and has been subjected to noise and interference. Regenerative repeaters can detect a signal and reconstruct the original signal (see Figure 16-6).

Figure 16-6

The role of regenerative repeaters

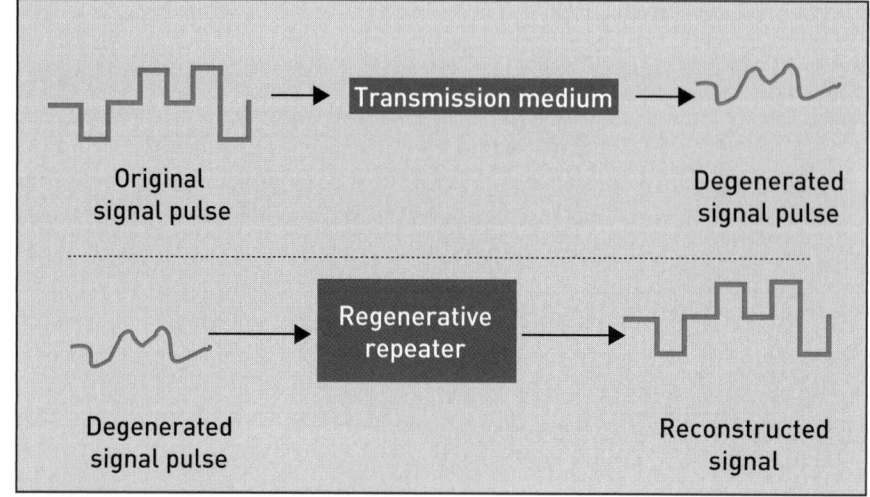

Note that in Figure 16-6, the signal pulse fluctuates from a negative to a positive value. The ones and zeros vary from a positive voltage to a negative voltage. This is an example of bipolar signaling, which is traditionally employed in voice communications. Previous discussions in this book have described unipolar signaling techniques, which involve voltage fluctuations that are always positive (or always negative). For example, the logic state representing a 0 could be 0 V, while the logic state representing a 1 could be 5 V.

As shown in Figure 16-6, the original digital signal consists of positive pulses, negative pulses, and empty spaces. After passing through a transmission medium, these pulses become degraded, losing some of their shape and energy. The repeater's first task is to amplify the signal to a level at which it can determine the signal's state at various intervals. If the signal exceeds a certain threshold, the repeater assumes the presence of a pulse, as shown in Figure 16-7.

Figure 16-7

Signal regeneration

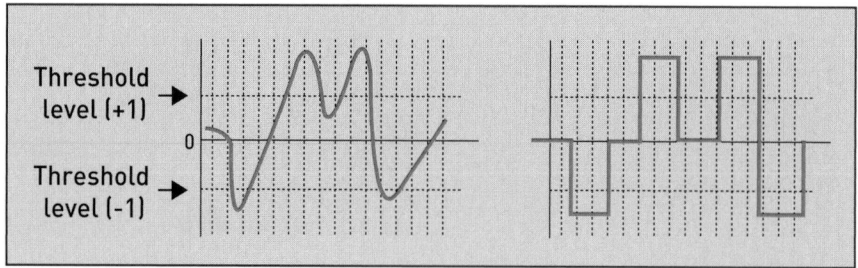

Once the repeater makes its determinations at each point in time, it regenerates the pulse. The reconstructed pulse needs to be identical to the original pulse prior to degradation, but several circumstances can cause the pulse to be reconstructed incorrectly. First, if too much noise is introduced into a signal over a transmission channel, it may produce a negative pulse where the original pulse was positive, or a positive pulse where the pulse should be negative. Each repeater has a certain error rate. Additionally, the process of detecting and regenerating a signal via a repeater requires careful timing. If the timing between pulses is not precisely reproduced, the quality of the signal becomes distorted. The telecommunications industry refers to this problem as **jitter**.

Another problem within the telephone system is **crosstalk,** a form of electrical interference between adjacent communication channels. Electromagnetic fields generated by one signal cross into another nearby signal to create the interference. In voice telephony, crosstalk presents itself in the form of noise on the signal. In some older systems, crosstalk could lead to overhearing someone else's conversation on your own private line, although the conversation might be unintelligible.

Crosstalk can be a problem when multiple pairs of twisted pair cable are distributed together over a single cable. Numerous high-frequency signals within these cables can interfere with each other. To mitigate this problem, technicians can use wires with greater shielding, create greater separation between wires, design adjacent wires with different twist characteristics, or use devices that reduce noise.

SWITCHING

Switches located throughout the telephone network play a critical role in establishing connections between two devices connected to the PSTN. Without switches, two telephones would require a fixed, hardwired link between them. Such links would soon prove untenable, because an enormous number of lines would be necessary to link every pair of phones that might connect to each other. As shown in Figure 16-8, connecting just six telephones in this manner would require 15 lines.

Mathematically, connecting *n* telephones with each other would require the number of transmission lines calculated in the following equation:

$n(n-1)/2$ where *n* is equal to the number of devices being connected

Figure 16-8

Telephone network without switching

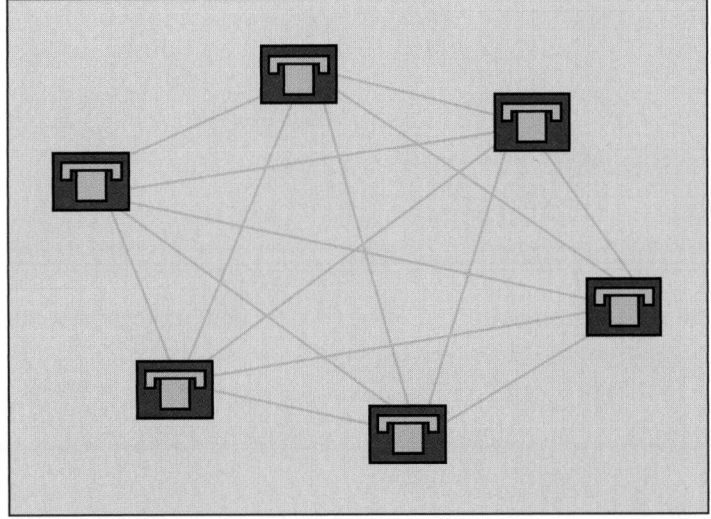

Providing a transmission line between every two devices that might communicate with each other is clearly untenable, so switching is used. Individual telephones connect through a single transmission link to a switch, which establishes a temporary path to connect two telephones for a call (see Figure 16-9).

Figure 16-9

Telephone network with switching

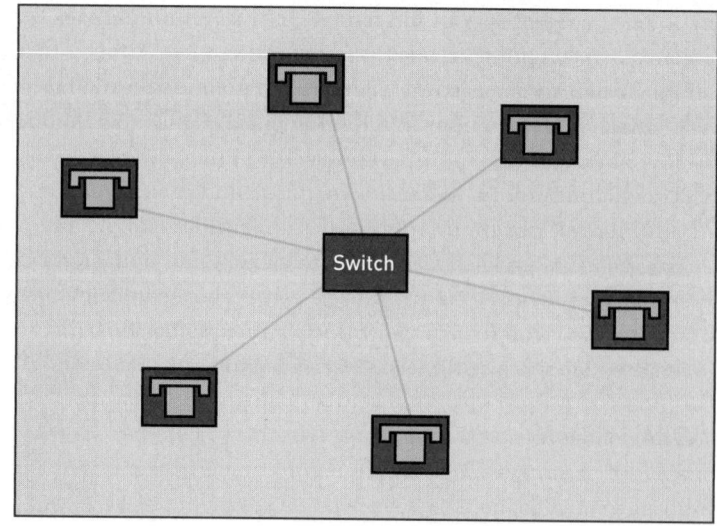

Originally, switching was provided through a switchboard (see Figure 16-10), where an operator would manually connect two telephone subscribers. Before long, this switching function was mechanized and performed automatically through electromechanical switches. Today these switches are electrical.

Figure 16-10

Antique telephone switchboard

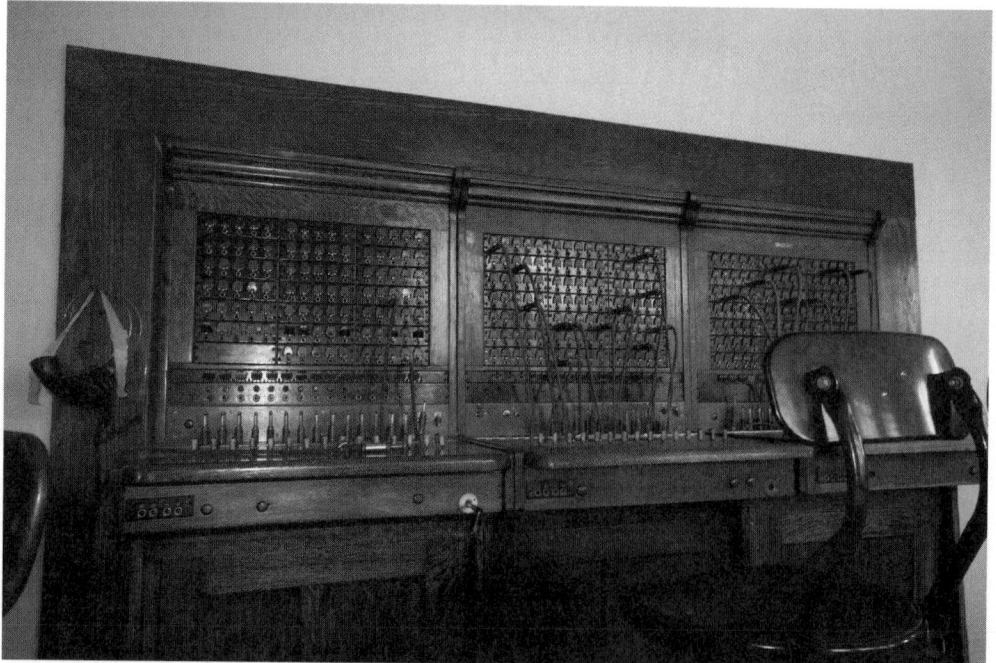

As discussed previously, the switching point to which telephones connect locally is usually called a central office. The automatic switching that connects a telephone subscriber to its destination takes place through a large, versatile, computer-controlled switch called an **electronic switching system** (ESS). An ESS is also referred to as a class-5 switch (5ESS) or a digital multiplex system (DMS), depending on the switch manufacturer. These systems can accommodate about 100,000 subscribers and perform a number of functions, including multiplexing and pulse code modulation, which are described later in this chapter. The ESS provides a variety of telephony and multimedia services that are familiar to most telephone users: messaging, VoIP, Internet services, call waiting, caller ID, call forwarding, call tracing, and multiparty calling.

The switching centers that traditionally connect central offices to each other are called tandem switching centers. Those that have the capacity to carry long-distance traffic are called interexchange carrier switches (see Figure 16-2).

The telephone system has traditionally used circuit switching, which establishes a temporary, dedicated link between two devices that remains connected for the duration of a call. This approach, as mentioned previously, contrasts with packet switching, the technique used by the Internet and other networks. Packet switching breaks transmitted information into small pieces called packets and then routes each individual packet through the network to its destination, depending on the most efficient route available at any given time.

The telephone system now uses a combination of circuit switching and packet switching, particularly as VoIP services have become increasingly popular.

CUSTOMER PREMISES EQUIPMENT

The most visible physical component of any telephone network, whether the traditional PSTN or VoIP services, is customer premises equipment (**CPE**). Everyone is familiar with a telephone set, but another important type of CPE is called a private branch exchange (**PBX**). A PBX is similar to a central office switch, except that it is privately owned and managed by a business, university, or other organization. In many large organizations, PBXs allow a sizable percentage of telephone calls to remain within the confines of their campus, building, or office complex. For example, on a university campus, many calls occur among dorm rooms, libraries, computer support organizations, and academic and administrative departments. Consider a call from an academic department to the registrar. Without a PBX, the signal carrying the call might pass through the university's network to a carrier's transmission lines and central office a few miles away, and then be switched back to the university. The signal might travel several miles before reaching its destination, a mere 100 yards from the originating caller. A PBX allows the call to be switched and routed right on campus, which is much more effective. It also allows for reduced-digit dialing (i.e., 3, 4, or 5 digits) instead of the traditional 7-digit dialing approach.

When a substantial portion of traffic originates and terminates on site, a PBX is a cost-effective technique. Most modern PBX systems also integrate multimedia Internet services and VoIP, which are discussed in detail in the following chapter.

An alternative system is called **Centrex**, which is similar to a PBX but is owned and operated by the telephone company. The advantage of Centrex services is that a business customer can use features offered by PBX without actually purchasing a PBX. By paying a monthly fee, Centrex customers can avoid the maintenance costs and equipment fees that are associated with a PBX.

Another type of CPE is the hardware and software that support **call centers**. Call centers are large offices that accommodate high volumes of incoming and outgoing telephone calls. Staffed by call center agents (see Figure 16-11), these facilities receive high volumes of inbound calls to provide such functions as catalog ordering services or customer product support services for software and hardware companies. Alternatively, call centers might support high volumes of outgoing calls such as those made by telemarketers. Call center technologies almost always provide the integration of computer systems and voice systems. For example, call center workers sit at desks with a computer terminal and a voice headset. The computer, usually tied into a corporate network, provides account and customer information and serves as a mechanism to access or input additional data about the customer interaction. One important example of call center CPE is the **automatic call distributor** (ACD), equipment that receives incoming calls and makes decisions about how to allocate these calls to available agents. ACDs make call-routing decisions based on the calling number or information they obtain through interactive voice response prompts and the skills of available agents. ACDs are typically integrated with corporate data, and they push account information to the agent simultaneously while routing the voice call.

For outbound services such as telemarketing, call center software often provides "predictive dialing," in which the software begins dialing other potential customers before the call center agent has finished a call. The software relies on historical data about average call times; it also accounts for the fact that calls take a certain amount of time to be answered and that not every call is answered.

Figure 16-11

Call center agents

Call centers can be located anywhere in the world, especially considering the low cost of phone calls. Because personnel (i.e., the agents) costs are a call center's primary operating expense, these offices have been part of the historical trend of **outsourcing** labor, especially to other countries. For example, many large companies that provide customer support services operate call centers in India, which provides a large supply of low-cost workers.

As a final note on this topic, large companies often prefer to provide chat-based customer support. This approach can be cheaper than telephone-based customer support. It also avoids having to put the telephone customer on hold, which increases customer satisfaction through faster response to customer support inquiries.

TELECOMMUNICATIONS PRINCIPLES

This section discusses several important techniques and standards for successfully transmitting information over the telephone network:

- Pulse code modulation
- Multiplexing
- T-carrier
- SONET and Synchronous Digital Hierarchy
- Signaling

PULSE CODE MODULATION (PCM)

As mentioned previously, the Public Switched Telephone Network is predominantly digital, but it still involves some analog transmissions. For example, a telephone call that originates from a home may be transmitted as an analog signal over the local loop and then be converted into a digital stream within the central office for transmission over the digital backbone portion of the network. In the telecommunications industry, the technique for converting an analog audio signal into a digital bit stream is called **pulse code modulation** (PCM). Recall from Chapter 6 that the process of digitizing an analog signal begins with taking samples of the signal at regular intervals. The amplitude values at each sample are then quantized, or rounded off to one of a finite number of possible values. Finally, a binary code is assigned for each value, as shown in Figure 16-12. As discussed previously, this process introduces quantization error, because the real amplitude value of each sample is not transmitted. Instead, the real amplitude value is quantized. How often the signal is sampled and how many discrete values the sample can hold affect the accuracy of the regenerated signal after transmission.

Figure 16-12

Digitizing voice

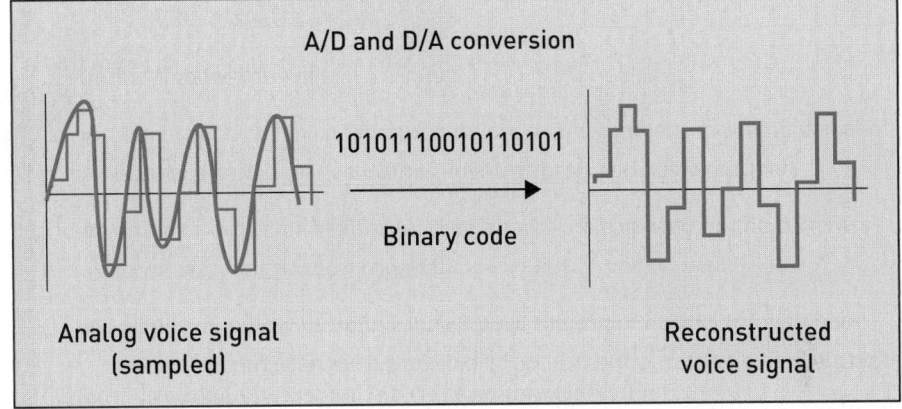

How often should an analog voice signal be sampled? Recall from Chapter 6 that analog voice signals are sampled at a rate of 8000 Hz, or 8000 samples per second. Each sample of an analog signal is a measurement of that signal's amplitude at that instant in time. This amplitude could have an infinite number of potential values, so it must be rounded off to one of 256 discrete possible values. (Over the telephone network, each sample is encoded with 8 bits, which provides 2^8, or 256, possible values.)

Because voice signals are sampled at a rate of 8000 samples per second, and because an 8-bit code represents the value of each sample, each voice transmission that uses PCM techniques to convert an analog voice signal into a digital binary stream requires a 64-Kbps transmission channel, calculated as follows.

Problem—What is the data rate of a single voice channel?

8000 samples/second $\times$ 8 bits/sample = 64,000 bits per second = 64 Kbps

Because of this requirement, the phone system is engineered using bundles of channels, each with a 64-Kbps data rate.

The signal is sampled 8000 times per second, so the time between samples is 1/8000, or 125 microseconds. The sampling period (T) is equal to the inverse of the sampling frequency, as follows:

$T = 1/f$

$T = 1/8000 = 0.000125$ seconds = 125 μs

MULTIPLEXING

As described previously, transmission media provide greater capacity than a single channel usually requires. In other words, sending a single signal over any transmission line results in wasted capacity. To optimize the available capacity of a transmission medium, it is better to simultaneously transmit multiple signals over the line, a procedure known as multiplexing. This technique is used widely within the PSTN to combine multiple channels within a central office and then to use a trunk to send the combined signals to another switching center. Multiplexing techniques provide cost-effective optimization over almost any medium, including wireless transmission facilities such as microwave links, copper cable, or fiber-optic cable. Multiplexing (see Figure 16-13) can interlace thousands of distinct signals over a single transmission line.

Figure 16-13

Multiplexing

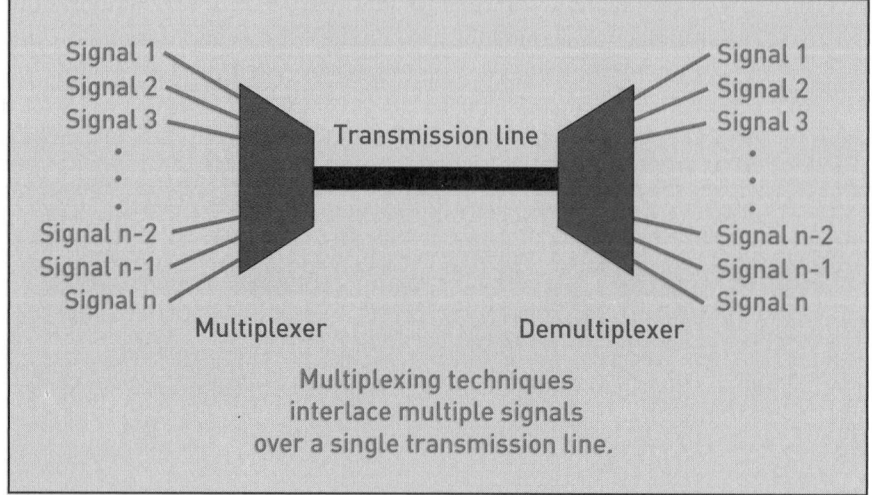

The following sections discuss several methods of multiplexing signals: an earlier method called frequency-division multiplexing, a common method called time-division multiplexing, and a fiber-optic technique called wavelength-division multiplexing.

Frequency-Division Multiplexing

Frequency-division multiplexing (FDM) served as the primary multiplexing technique when the telephone network predominantly used analog transmission.

FDM enabled the simultaneous transmission of multiple signals over a single medium by assigning a different frequency to each channel. To help understand the concept of FDM within the telephone system, consider how other communications systems use simple forms of FDM to allow many users to simultaneously share a system's capacity. For example, radio stations each use a unique frequency to transmit information over the same medium (air) within the same geographical area.

In the same way, FDM within the telephone network allowed multiple signals to transmit over a common medium without interfering with each other. Why can a single medium simultaneously support different frequencies? Any voice conversation has a predominant frequency range between 300 Hz and 3300 Hz. Even if you round up this range considerably, the maximum frequency of a voice conversation is no more than 4000 Hz, or 4 KHz. On the other hand, a copper transmission channel can easily support a frequency range of 96 KHz, or 96,000 Hz.

This range of 96,000 Hz can be divided into smaller frequency ranges of 4000 Hz, with each smaller range accommodating a unique voice transmission. For example, the range of 0–4000 Hz could be allocated to one call, the range of 4000–8000 Hz could be allocated to another call, and so on.

By assigning each voice call its own frequency band, numerous calls could be simultaneously transmitted over a single line. Figure 16-14 illustrates how three distinct voice signals from three different circuits could enter an FDM, be combined, and sent over a single transmission medium by modulating a distinct carrier frequency with each voice signal and combining the modulated signals.

Figure 16-14

Frequency-division
multiplexing

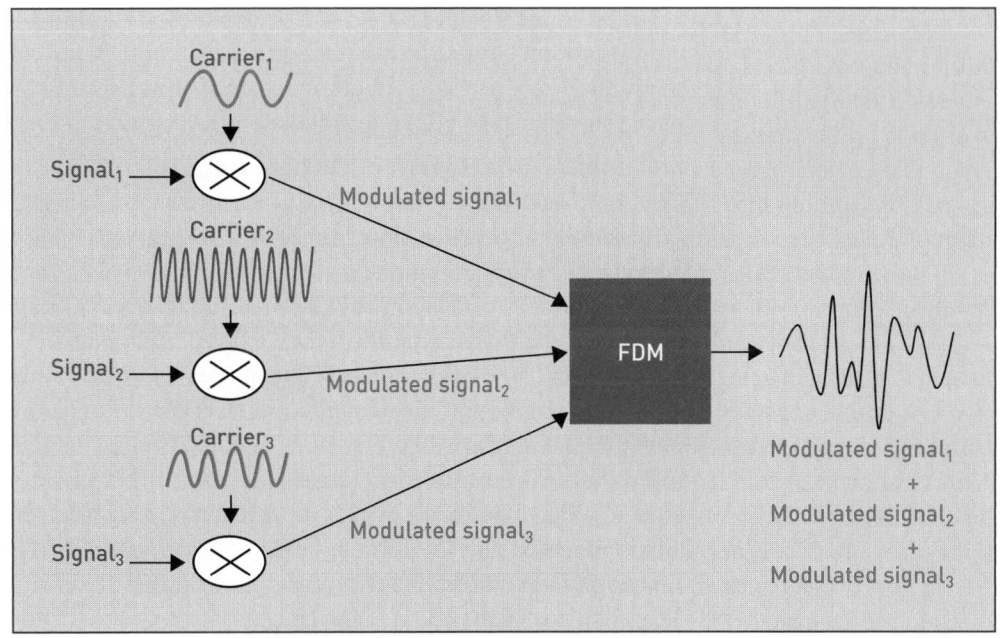

Time-Division Multiplexing

In contrast to dividing and distinguishing signals based on frequency, **time-division multiplexing** (TDM) divides signals based on time, allocating different time slots for each signal. The TDM approach, shown in Figure 16-15, combines the multiple signals entering a multiplexer by sequentially scanning and slotting them. For example, if 4 voice signals connect to a multiplexer through 4 ports, the time-division multiplexer allows the signal from port 1 to transmit in time slot 1 for a short, fixed length of time, the signal from port 2 to transmit in its fixed time slot 2, and so on. Time slot 1 is always reserved for port 1, so TDM is called a static, or fixed, configuration. If port 1 has nothing to transmit at its designated time slot, then the multiplexer transmits nothing during that slot.

Figure 16-15

Time-division
multiplexing

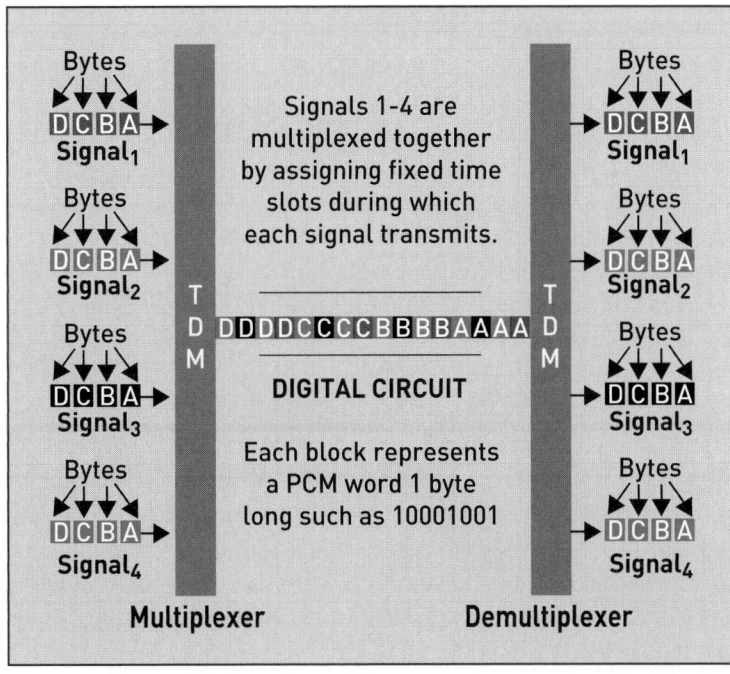

In each time slot, the TDM multiplexer places a PCM word—the 8-bit (i.e., 1 byte) sample of information—from the allotted signal. Each port "takes a turn" during its allotted time slot. If the signals entering a multiplexer are analog, they must first be converted into a digital binary stream using pulse code modulation, as discussed previously.

Statistical Time-Division Multiplexing

One significant limitation of TDM is that it wastes capacity in certain circumstances. TDM exclusively reserves time slots for each incoming signal channel. If a specific channel has nothing to transmit at a given point in time, the digital circuit transmits an empty time slot, wasting valuable bandwidth. The preferred and widely implemented version of TDM is called **statistical time-division multiplexing** (STDM). STDM devices, commonly called stat muxes, can recognize conditions of inactivity from one of their incoming signals. Rather than transmitting an empty time slot, a stat mux offers the slot to another incoming signal that has something to transmit. In other words, stat muxes allow for dynamic allocation of time slots based on the activity of incoming signals to be multiplexed. Stat muxes can also perform other intelligent functions such as recognizing priority levels of traffic and prioritizing transmissions accordingly, performing some error detection and correction functions, and even applying some compression techniques on the traffic to further optimize bandwidth.

Wavelength-Division Multiplexing

For light transmission over fiber-optic cables, **wavelength-division multiplexing** (WDM) enables several high-speed channels to work over a single fiber-optic transmission system (see Figure 16-16). If the number of channels is small, the technique is called Coarse WDM (CWDM). If the number of channels is large, the technique is referred to as Dense WDM (DWDM). WDM is somewhat like frequency-division multiplexing, which enables the coexistence of multiple frequencies over a single analog transmission system. However, FDM separates transmission into channels based on the different frequencies, and WDM makes the same type of separation based on distinct wavelengths of light. The wavelength of light is the distance light travels to complete a single cycle of its waveform.

Wavelength, as discussed previously, is directly related to frequency, as demonstrated in the following equation:

$$\lambda = c/f$$

In this equation, λ = wavelength, f = frequency, and c = the speed of light, or 3×10^8 m/s (300,000,000 meters per second).

Because wavelength and frequency are directly related, you can understand the similarities between dividing channels based on wavelength and on frequency. Likewise, you can see the similarity between FDM in analog electrical systems and DWDM in digital optical systems.

Figure 16-16

Wavelength-division multiplexing

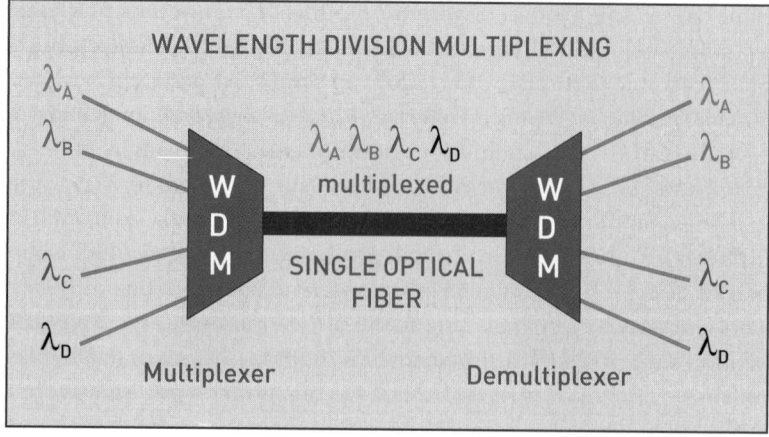

THE T-CARRIER SYSTEM

Using multiplexing, the telephone network can combine and transmit multiple signals over a single physical medium such as twisted pair or fiber-optic cable. The standard multiplexing method of the telephone network is the **T-carrier system**. The multiplexing system applies to the backbone transmission facilities that make up the Public Switched Telephone Network.

Recall that the data rate requirement for a digital voice channel is 64 Kbps, which is calculated by multiplying the 8000-Hz sampling rate by 8, the number of bits per sample. This rate of 64 Kbps serves as the base transmission rate for the T-carrier system. The 64-Kbps channel is known as Digital Signal Level 0 (DS-0). The T-carrier standard is based on guidelines issued by the American National Standards Institute (ANSI). Within the standard, a "T1" (also called DS-1) operates at 1.544 Mbps and carries 24 DS-0s (64-Kbps signals) that are multiplexed together. A T2 (DS-2) operates at 6.312 Mbps carrying 96 DS-0s; a T3 (DS-3) operates at 44.736 Mbps carrying 672 DS-0s; and a T4 (DS-4) operates at 274.176 Mbps carrying 4032 DS-0s.

To help explain these data rates, the following problem explains the transmission rate that a T1 carrier must have.

Required Transmission Rate of a T1 Line

Problem—If a T1 line carries 24 channels multiplexed together, with each signal sampled 8000 times per second and 8 bits used to encode the value of each sample, calculate the data rate of the T1 line.

We know that each channel has a data rate of 8000 samples/second $\times$ 8 bits/sample = 64,000 bps. With 24 channels, the data rate is 64,000 bps $\times$ 24 = 1,536,000 bps = 1.536 Mbps.

The data rate of a T1 line, however, is listed previously as 1.544 Mbps and not 1.536 Mbps. This discrepancy is due to the addition of an extra bit called a "framing bit" for purposes of synchronizing the transmitter and receiver and sending signaling information between central offices. The addition of the framing bit results in a slightly higher data rate of 1.544 Mbps.

SONET AND SYNCHRONOUS DIGITAL HIERARCHY

The T-carrier system is a transmission standard that specifies how to multiplex and transmit information over copper cables. Another type of standard, Synchronous Optical Network (**SONET**), serves as the transmission specification for multiplexing over fiber-optic cable in the United States and Canada. Internationally, this standard is called **Synchronous Digital Hierarchy** (SDH). Dense wavelength-division multiplexing, as discussed previously, forms the basis of the telephone system's standard approach for multiplexing light signals. SONET originated in the 1980s within Bellcore, the research arm of the "Baby Bell" telephone companies. ANSI later standardized SONET, which was also adapted into the SDH standard established by the International Telecommunications Union (ITU).

Without standards to specify telecommunications rates and formats, networks run by different telecommunications operators could not interconnect. One function of the SONET standard is to provide synchronization—it coordinates where ones and zeros are located in a transmission via a central system that does not require extra framing bits within the transmission. The SONET digital hierarchy specifies the following optical carrier (OC) levels with various operating transmission capacities: an OC-1 operates at 51.84 Mbps; an OC-3, 155.52 Mbps; an OC-12, 622.08 Mbps; an OC-48, 2.488 Gbps; an OC-192, 9.953 Gbps; and an OC-768, 40 Gbps.

SIGNALING

Signaling is another critical function that enables information exchange over the telephone network. Signaling services transfer virtually all information other than the actual content of a voice conversation, a file transfer, or a fax transmission. Most signaling encompasses control functions that indicate the status of a device. For example, a dial tone indicates that a device is available, and a busy signal indicates that a device is not available. A ringing signal indicates that a device is available and being contacted. Between devices within the telephone network, signaling communications indicate such information as circuit numbers, route availability, billing data, and network management information.

The global standard for exchanging this control, routing, and call setup information is called **Signaling System 7**, known commonly as SS7. This universal protocol defines how signals are exchanged for establishing, managing, and terminating a call over a digital network. SS7 messages are transmitted over 64-Kbps channels, called signaling channels, that take place "out-of-band"—they are not transmitted on voice channels. This out-of-band signaling actually occurs on a packet-switched network. A dedicated, end-to-end path is not established using circuit switching until SS7 determines the availability of the device. Figure 16-17 illustrates the out-of-band SS7 network that carries signaling information within the PSTN.

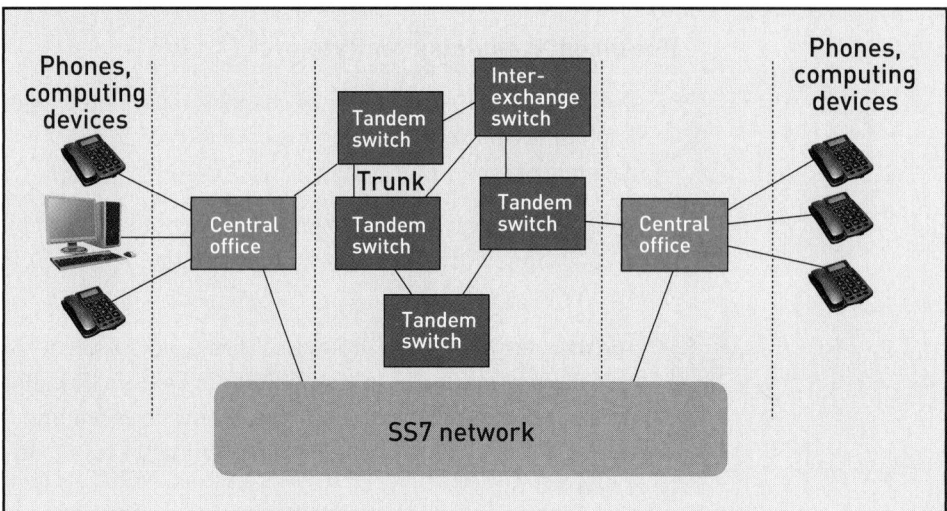

FUTURE OF THE TELEPHONE SYSTEM

Several modern trends in IT work against the long-term economic and technical viability of the traditional telephone system's architecture. For example, Internet telephony and wireless services can be less expensive than traditional phone service. (These more modern services are described in the next two chapters.)

Additionally, users expect mobility. The traditional phone system is geared toward phones that are fixed in a single location, but wireless devices provide the mobility that users want. Third, today's wireless networks and the Internet support multimedia applications such as integrated audio, video, images, and text, while the circuit-switching architecture of the telephone system focuses on voice communications. Fourth, most networks are moving toward packet-switched architecture and universal TCP/IP protocols. Some experts suggest that voice should be considered just another application over the Internet.

Despite all these trends, however, the PSTN architecture still exhibits greater reliability and quality of service. For example, the PSTN provides "five nines" of reliability—the network works properly 99.999% of the time. Cell phones and Internet telephone service have more frequent outages.

In summary, three options are available for making voice calls: the traditional telephone system, a cellular service subscription, or Internet telephony via the VoIP standard. Most service providers offer all three options for voice communications. The next two chapters discuss the technologies of Internet telephony and wireless multimedia services.

CHAPTER SUMMARY

- The telephone system dates back to Alexander Graham Bell's invention and patent of the telephone in 1876.
- For a century, the Bell System was a monopoly in the United States. This changed in 1984, when the Department of Justice ordered divestiture of AT&T.
- The PSTN uses every type of transmission medium, including microwave, twisted pair, satellite, submarine cables, and fiber.
- Electronic switching systems, including central office switches, interconnect transmission facilities and establish circuits between telephones or other devices connected to the PSTN.
- Customer premises equipment (CPE) includes devices such as telephone sets and PBXs, which are private switches owned and operated by large organizations.
- Pulse code modulation is the telecommunications technique of converting an analog signal into a digital bit stream.
- Techniques such as frequency-division multiplexing, time-division multiplexing, statistical time-division multiplexing, and wavelength-division multiplexing enable multiple signals to be transmitted concurrently over a channel.
- The T-carrier system is the standard multiplexing system for telephony in the United States.
- SONET and SDH are the standard multiplexing systems for fiber-optic transmission in the telephone system.
- Signaling System 7 is the standard that defines how network elements exchange signaling and device status information over a digital network.

KEY TERMS

automatic call distributor	omnidirectional antenna
call center	outsourcing
central office	PBX
Centrex	PSTN
CPE	pulse code modulation
crosstalk	Signaling System 7
directional antenna	SONET
electronic switching system	statistical time-division multiplexing
frequency-division multiplexing	Synchronous Digital Hierarchy
jitter	T-carrier system
line-of-sight	Telecommunications Act of 1996
local loop	time-division multiplexing
microwave	trunk
Modified Final Judgment	wavelength-division multiplexing

REVIEW QUESTIONS

1. What were the primary methods of exchanging information over long distances before the telephone was invented? When was the telephone invented?
2. List five types of transmission media used in the telephone system.
3. Which two types of transmission media enable most transoceanic communications?
4. Explain what "line of sight" indicates in the context of microwave transmissions within the telephone system.
5. Describe the purpose and function of regenerative repeaters in the telephone system.
6. Define jitter.
7. Define crosstalk.
8. How many total lines would be required to fully interconnect 10 telephones via hardwired transmission lines?
9. Explain the purpose of switching.
10. What is CPE?
11. Explain the purpose and function of a PBX. Give two real-world examples in which a PBX would make economic sense and two more examples that would require the use of Centrex services.

12. What does pulse code modulation accomplish and where does it occur within the PSTN?

13. Within the PSTN, analog signals are sampled at what rate and with what unit of measurement? Why was this sampling rate selected? Given this sampling rate, how would you calculate the associated sampling period?

14. Each sample of an analog signal on the telephone network is encoded with how many bits?

15. Explain the math that shows why digital transmission channels in the telephone system operate at a rate of 64 Kbps.

16. What is the difference between time-division multiplexing and statistical time-division multiplexing?

17. Dense wavelength-division multiplexing applies to which transmission medium?

18. List three types of control functions accomplished by signaling in the telephone system.

DISCUSSION QUESTIONS

1. If you managed a hardware or software company that fielded more than 1000 customer service calls per day, would you consider operating your call center in a country with a less expensive workforce? Name some advantages and disadvantages of this type of outsourcing. Would you provide chat-based customer support as well as voice support?

2. Do you use the traditional phone system at home, school, or work, or do you make voice calls primarily via cell phone or the Internet? What about your parents and grandparents? Is there a generational difference?

3. Comment on the future of the PSTN architecture.

4. How is the traditional telephone system being transformed in response to the popularity and ubiquity of cell phones, other handheld devices, and voice over the Internet?

5. Without looking at this chapter, can you explain the math that shows why a T1 operates at 1.544 Mbps?

6. What kind of reliability is necessary for voice calls? What kinds of applications and contexts require a network to be available 99.999% of the time?

CASE PROJECT

You have been asked to select and purchase a PBX for your company. Tomorrow afternoon, you have to present two alternative PBX products to your CIO and compare their features. Then you have to recommend one of the PBXs based on technical features, reputation of the company, and cost effectiveness.

- Search the Internet for manufacturers of PBX equipment. Find their top-line PBX products and download the list of technical features for each product. Make sure that both products you select support traditional voice, Internet service, and integrated VoIP. Try to find out what support services the vendors offer.

- Now develop a presentation for your CIO that describes at least 10 features of each product and explains why you would select one of these products.

VOICE OVER IP

LEARNING OBJECTIVES

In this chapter you will:

- Understand the concept of VoIP and how it enables the transmission of voice over the Internet

- Gain familiarity with VoIP protocols such as SIP, H.323, and RTP

- Understand options for implementing voice over the Internet

- Recognize the business drivers and advantages of using VoIP rather than traditional telephone system architectures

- Understand the security and performance challenges of VoIP and consider various solutions

- Identify the policy questions arising from the proliferation of VoIP services

INTRODUCTION

Internet telephony, also called IP telephony or Voice over Internet Protocol (**VoIP**), refers to voice communications transmitted over the Internet or other IP network, as opposed to the Public Switched Telephone Network described in the previous chapter. Many businesses, institutions, and individual users place telephone calls over the Internet rather than through a traditional telephone network, bypassing the circuit-switching approach of the telephone system. A VoIP call does not necessarily remain solely on the Internet or other IP network. A caller can use VoIP to reach someone whose phone service is provided via the traditional phone network or a cellular network, and the voice traffic is transparently converted between the various networks. This chapter describes several scenarios for using Internet telephony, discusses some of the advantages and challenges of transmitting VoIP, and describes the standard network protocols that make Internet telephony possible. The chapter concludes with a discussion of policy and regulatory challenges that have emerged with Internet telephony.

In common usage, Internet telephony is synonymous with **VoIP**, a set of standards designed for transmitting voice calls over the Internet or other IP network.

The VoIP approach converts an analog voice waveform into a digital signal format, a process discussed in Chapter 6. It then breaks the digital signal into packets for transmission over the Internet, as illustrated in Figure 17-1. One important technical difference between Internet telephony and the traditional telephone system is that Internet telephony uses packet switching, not circuit switching, to transmit voice information.

Figure 17-1

Voice communications over the Internet

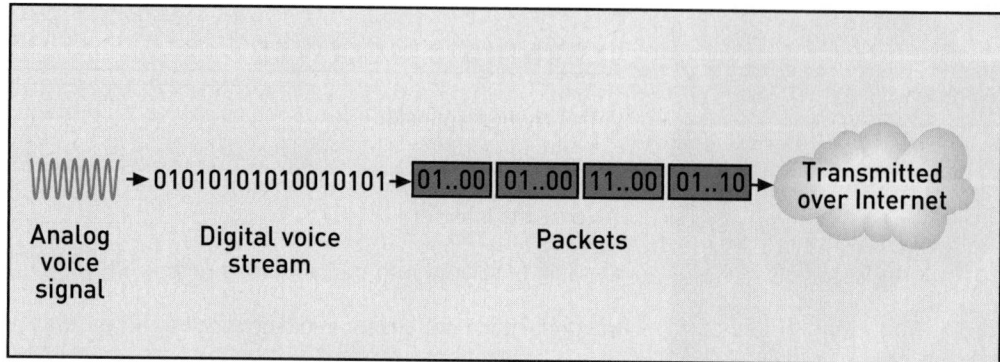

The network protocols used in VoIP are another critical distinction of voice communications over the Internet. VoIP calls have several functional components: the signaling information that establishes the telephone call, the process that digitizes the audio signal and places the information into packets, and the process of actually transmitting the packetized speech. While VoIP sounds like a single network protocol, it actually includes a variety of distinct protocols that work in combination to handle the necessary functions.

One set of standards handles the digitization of the voice signal. Another set of protocol alternatives, called **signaling protocols**, assesses user availability, rings the destination device, establishes a call, and terminates a call session. Other **transport protocols** handle the transmission of the voice call between endpoints. In some cases, proprietary (vendor-specific) protocols are used. For example, signaling protocols associated with a specific vendor's **IP PBX** (Internet Protocol private branch exchange) might be proprietary. The most popular software-based VoIP solutions, including Skype, also use proprietary protocols. Other protocols associated with VoIP are open standards developed by organizations such as the Internet Engineering Task Force (IETF) and the International Telecommunications Union (ITU).

The following sections describe two of the most well-known VoIP signaling protocols: the **Session Initiation Protocol** (SIP), developed by the IETF, and **H.323**, a family of standards developed by the ITU. We also describe the **Real-time Transport Protocol** (RTP), a popular VoIP transport protocol.

SESSION INITIATION PROTOCOL

SIP is a signaling protocol that creates, manages, and terminates multimedia sessions between end-user devices, such as two telephones. A **session** is a voice conversation or other exchange of information between users. As the IETF described in its SIP specification (RFC 3261, at *http://www.ietf.org/rfc/rfc3261*), the protocol performs five functions:

- *End-user location*—Find the end-user location with which to communicate.
- *End-user availability*—Determine whether the end user is available and willing to engage in conversation.
- *End-user capability*—Determine what type of media content will be used and what constraints might be in effect.

- *Session setup*—Establish the session between parties (also called "ringing").
- *Session management*—Handle changes to the communication session while the call is in process and terminate the session upon completion.

SIP operates at the **Application layer** of the OSI model using an information exchange that is similar to HTTP. A caller who wants to establish a voice session uses the called party's SIP identity, the SIP Uniform Resource Identifier (**URI**), to issue a SIP INVITE request. The URI looks like an e-mail address—for example, the SIP URI might be sip:name@ domainname.edu. The calling party issues a SIP INVITE request, which is received by the called party's device. The called party's device consequently rings, alerting the called party user that a SIP INVITE request has been received. The called party then issues a ringing response to the calling party notifying it that the called party device is ringing. If the called party answers, SIP sends another message, called OK, to the calling device. This message terminates the ringing tone and indicates that the call has been answered. The calling device then sends an acknowledgement, called ACK, to the called party. The devices then exchange packets that contain the digitized voice using an agreed-upon format. The session terminates when one of the devices issues a BYE signal.

This example describes the types of signaling messages exchanged between endpoints to initiate, manage, and terminate a call. Figure 17-2 illustrates some of the signaling that takes place between the two ends of the network.

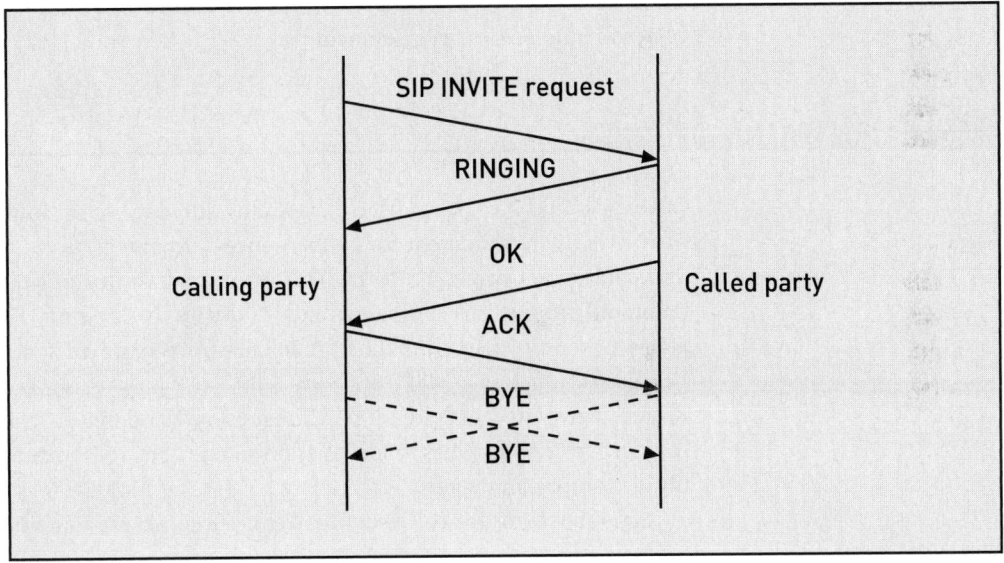

H.323

SIP is not the only signaling protocol option for VoIP. H.323 is an ITU-developed signaling standard for multimedia communications over the Internet and other IP networks. H.323 was the first widely implemented signaling protocol that supported real-time voice and video over IP networks.

Strictly speaking, H.323 is actually a series of many protocols. Some of these include H.225, a specification that describes the call signaling and control process; G.711, which describes the pulse code modulation (PCM) of voice frequencies; and Q.931, which describes the messages that establish or terminate calls. The H.323 family of protocols also includes **encoding standards**, which transform analog signals into a standard digital format. The ITU has developed encoding standards to handle this conversion, including G.711, G.721, and G.728.

REAL-TIME TRANSPORT PROTOCOL

RTP serves a much different function than H.323. RTP establishes the standard for end-to-end transmissions that carry actual digitized speech in packets. The IETF has defined the RTP protocol in RFC 1889 (*http://tools.ietf.org/html/rfc1889*). Applications such as voice and video are different from other data because they occur in real time and are much more time sensitive. To a user, a slight delay in hearing someone speak is much less acceptable than a slight delay in receiving a text message.

RTP is designed to transport voice, video, and other information that has real-time properties. As described in the IETF specification, an RTP header (group of bits) is appended to the information being transported and has four primary services: payload type, sequence numbering, timestamping, and delivery monitoring. The payload type specifies what type of information is being transmitted, such as video or audio samples. It also includes sequence numbering—the numbering of packets for transmission over the packet-switching network. When a digital signal is broken into packets, they can take different routes over the network and might arrive at their destination out of order. The sequencing function is important because the receiving device uses it to reorder received packets into their original sequence. The receiving device also uses the packet sequence numbers to detect whether any packet loss has occurred. The timestamping function indicates when the first octet in the RTP packet has been sampled, which helps to provide synchronization and good performance. The delivery monitoring function provides information that contributes to diagnostic and performance data.

Like all protocols, RTP serves a specific function and relies on other protocols to provide other functions. For example, RTP does not guarantee delivery or provide quality-of-service guarantees, but it works in conjunction with other protocols that handle such tasks. Specifically, RTP works with the **RTP Control Protocol** (RTCP), which provides some quality-of-service and session management.

IMPLEMENTATION OPTIONS

Businesses and individual users have many alternatives for implementing VoIP. At the heart of VoIP is the requirement for **convergence**—the integration of voice, music, video, images, and data over a single platform. This multimedia integration is sometimes called **unified communications**. The companies and industries that best meet user requirements for convergence will dominate, and those that fall short stand to lose considerable market share. Because so much is at stake economically, VoIP innovation is occurring at a very fast pace. Besides new VoIP entrepreneurs, almost every IT-related industry offers VoIP products and services: cable companies, traditional phone carriers, router vendors, PBX manufacturers, and phone equipment makers.

Several options for voice service over the Internet are described in the following sections.

SOFT PHONES

One simple (and free) approach to Internet telephony is to install VoIP software on a computer and use it with a headset to make a voice call, as shown in Figure 17-3. This type of implementation is often called a **soft phone,** and it requires a headset (or microphone and speakers), a sound card, freely available VoIP software, and a broadband Internet connection. Two users with a compatible VoIP configuration (see Figure 17-4) can make unlimited complimentary calls to each other. A VoIP soft phone user can also call a friend or associate who is using the Public Switched Telephone Network (PSTN) or a cell phone, but this service is not usually free. It requires a per-minute rate or monthly fee.

Figure 17-3

VoIP soft phone

Figure 17-4

Computer-to-computer
Internet telephony with
soft phones

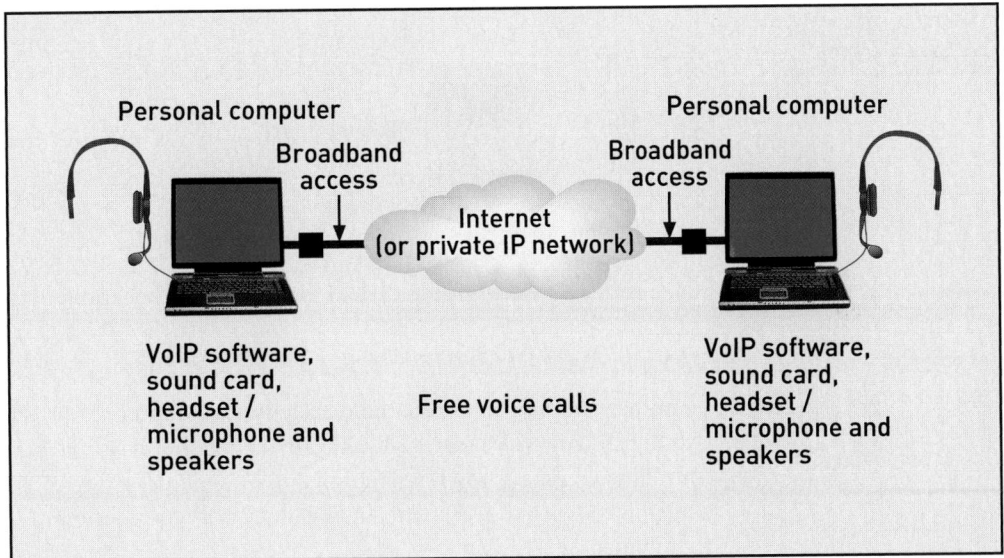

VoIP software not only allows free calls to other users of the VoIP service, it also enables conference calling and integrates other Internet applications and communications within the software, including transferring files between users and providing instant messaging.

One important feature of this type of implementation is that the VoIP software usually encrypts the voice signal prior to transmission. This security is an appealing feature for users who want to ensure privacy during a voice call, but it also raises issues for law enforcement and antiterrorism efforts. Anyone can implement Internet telephony to send encrypted voice transmissions anywhere in the world. Law enforcement and national security agencies have much more difficulty monitoring VoIP communications than accessing voice conversations over the traditional PSTN. The same complex set of trade-offs between privacy and security exists in any communications system.

Another issue arising from Skype and other free Internet telephony software is that their protocols are proprietary. They use network protocols that are closed instead of using open and publicly available VoIP standards that would promote interoperability with other vendors' products. Also, this type of Internet telephony configuration usually does not provide emergency (911) calling features, a complication discussed later in this chapter.

The same type of Internet voice calling is possible by plugging a USB-compatible telephone into the USB port of a computer. In other cases, adapters are available to plug a phone cord into a computer's USB port. Figure 17-5 depicts a voice call over the Internet using VoIP software, a high-speed broadband connection, and a cordless phone communicating with a USB-connected base station rather than a hardwired microphone or headset. This configuration allows for some mobility rather than having to stay close to the computer's microphone or headset.

Figure 17-5

Computer-to-computer Internet telephony via cordless telephone

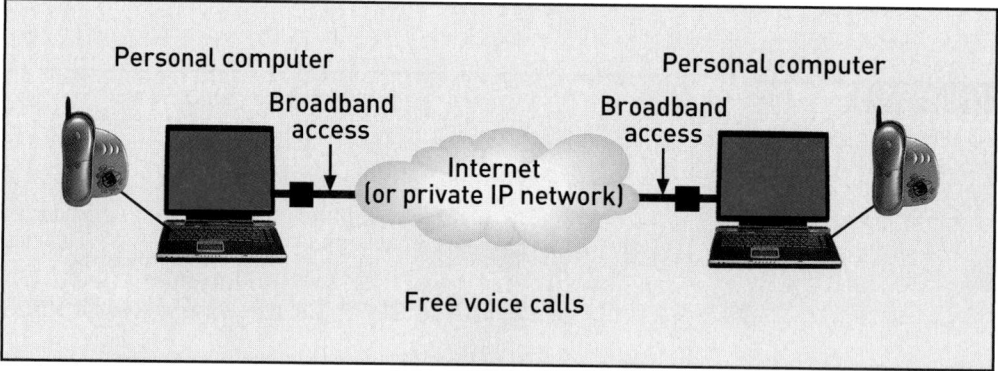

ANALOG TELEPHONY ADAPTERS

Another option is to use an **analog telephony adapter** (ATA). This device serves as an analog-to-digital converter (ADC) between an analog phone and an Ethernet network connection to an IP network, as shown in Figure 17-6. Other analog devices such as fax machines can also be plugged into the ATA. This option does not require the installation of VoIP software on a computer.

Figure 17-6

Analog telephony adapter

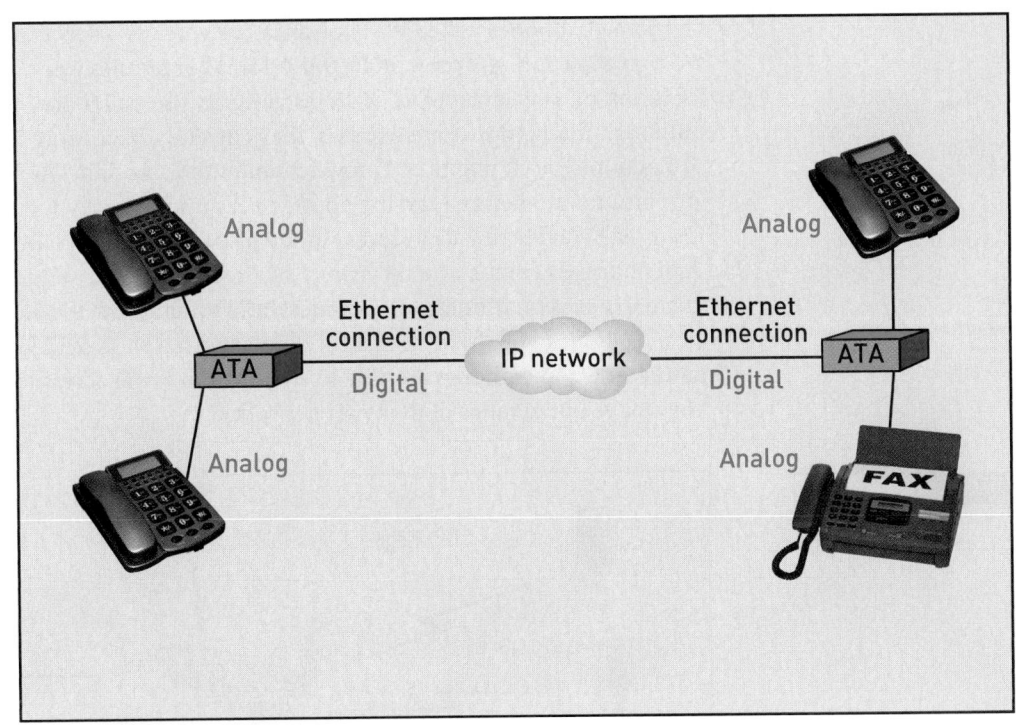

IP PHONES AND WI-FI PHONES

Alternatively, an **IP phone** (see Figure 17-7) can connect directly to an IP network without an adapter because it internally converts the telephone audio signal into digital format for transmission over the IP network. IP phones have the same features that traditional phones provide, including speed dial, voice mail, call forwarding, and conferencing. Additionally, because the IP phone is connected to a unified network that also supports non-voice information, the phone serves as a multimedia access device. IP phones have LCD displays and provide users with the opportunity to customize the information on the phone display. For example, a user can choose to display certain Web content, text messaging, a company phone and e-mail directory, business-specific data, or stock quotes. Some IP phones are also video phones.

Figure 17-7

IP phone

Some IP phones, called **Wi-Fi phones**, are wireless—they can connect to the Internet via a Wi-Fi connection, as shown in Figure 17-8. Wi-Fi phones work on unencrypted wireless connections or over encrypted Wi-Fi as long as the caller has the network encryption ID. Although these phones are wireless, they currently have more limitations than cell phones. For example, a cell phone call provides ubiquitous mobility. As a cell phone user moves from one area to another in a car, the phone call is transparently handed off from one antenna to the next. (We describe this process in the next chapter.) In contrast, Wi-Fi phone users sometimes must stay in the general vicinity of a wireless access point to which they are connected; there is not yet a ubiquitous, seamless way to hand off the call to another wireless access point. Phones with Wi-Fi capability are expected to provide greater mobility in the near future, and some phones are able to switch from Wi-Fi access to a cell phone network when they move out of range of the Wi-Fi network.

Figure 17-8

Wi-Fi phones

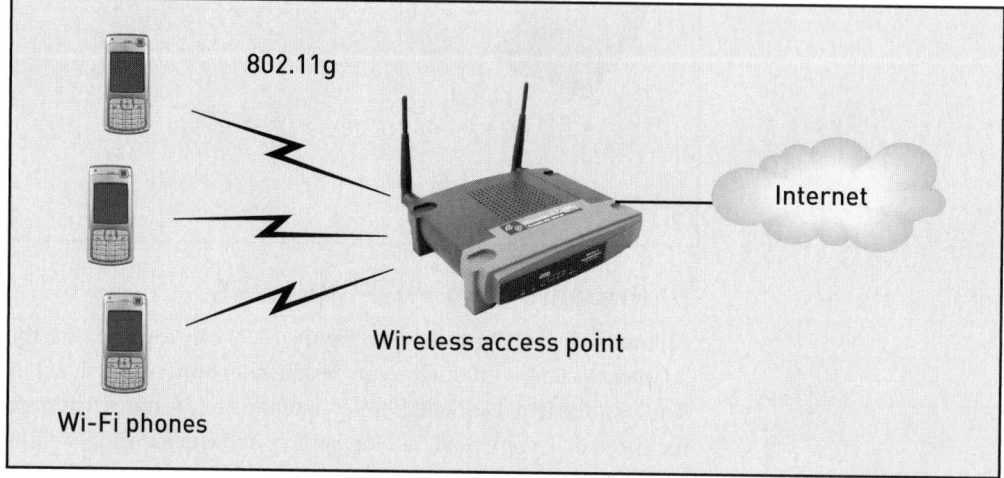

RESIDENTIAL BROADBAND VoIP SERVICES

Many residential users of voice over the Internet subscribe to a commercial VoIP service from their local cable provider or phone company. These voice Internet services work with existing phone jacks and phones and offer unlimited voice calls for a fixed monthly fee. This service requires a high-speed Internet connection, such as a cable access link via a cable modem or high-speed DSL link, and allows for simultaneous voice calls, Internet access, and television programming. These VoIP services often allow users to retain their existing phone number and provide all the advanced features of any phone service, including the following:

- Voice mail
- Caller ID
- *69 call return
- Call waiting
- Caller ID blocking
- Multiple party calling
- VIP ringing
- Call forwarding
- Anonymous call blocking

As depicted in Figure 17-9, these commercial VoIP services allow a caller to reach any other phone users, whether they use the VoIP provider's network, the PSTN, or a cell phone. Similarly, when people receive a phone call, the type of network used should be transparent to them.

BUSINESS VoIP OPTIONS

Business VoIP implementations are more complex than residential Internet telephony because they have a larger number of users and a large installed base of legacy phone systems. Institutions that have transitioned from traditional voice infrastructures to VoIP have had to weigh many trade-offs, such as the potential cost savings and integration benefits of VoIP versus the costs of migrating existing infrastructures to VoIP, training personnel to implement and manage VoIP systems, and the challenges of network security. Several benefits and challenges of VoIP technology are presented later in this chapter.

Organizations that implement VoIP can choose from one of several alternatives. They can completely replace existing phone systems with VoIP systems, or they can adopt a hybrid approach—adding some IP components in current systems or running some completely VoIP-based systems and some traditional phone systems, with a gateway between the two systems to interface them.

One option for business VoIP implementations is to use an IP PBX. The last chapter discussed how many private businesses use a local telephone switch called a PBX to connect to the PSTN. Similarly, VoIP implementations can use an IP PBX as an option for processing VoIP calls. Another option is to implement a router-based VoIP architecture. A PBX-based approach is usually a more centralized design, while the router approach is more distributed.

Regardless of the approach, many businesses use IP phones directly connected to an IP network. IP phones have their own IP addresses to identify their virtual location on the network. These VoIP devices can include integrated applications such as instant messaging and video conferencing.

Business VoIP implementations require an IP network to interconnect various VoIP components. Many companies implement their internal phone systems over a private IP network, as shown in Figure 17-10. This network usually carries both voice and data traffic. Real-time voice applications are more sensitive to latency and other performance factors, so the voice traffic is often given higher priority than other types of applications over the network. One approach to ensure adequate service for voice traffic is to prioritize voice packets over data packets by assigning them different classes of service (CoS).

As shown in Figure 17-10, calls that originate within the organization's network remain on the internal network and are routed directly to other network users. If a call's destination phone number is off the internal network, it is routed to the Internet or the traditional telephone network, depending on what the organization has implemented.

It is also common for businesses to use an IP virtual private network (VPN) for voice applications.

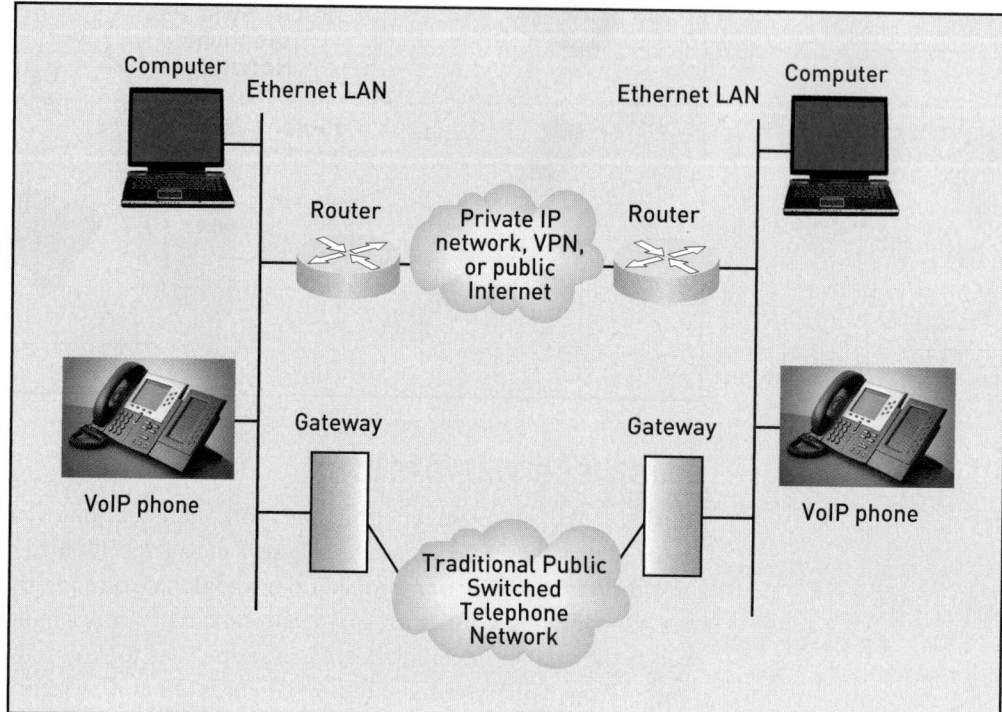

Figure 17-10

Private IP network

INTERNET TELEPHONY BENEFITS

Inevitably, voice communications will continue to be integrated with other applications over IP networks. Users who upgrade their traditional telephony equipment may ultimately have to invest in Internet telephony and services. Similarly, future innovations in voice applications are expected to be designed for use over IP networks, especially in multimedia applications geared toward integrated voice, data, and video. In other words, implementing cutting-edge multimedia applications requires an underlying VoIP infrastructure.

Using VoIP offers many advantages, but the two most significant are cost savings and network unification.

COST SAVINGS

Businesses and individual users most often select VoIP for its cost savings over traditional phone service, especially if they routinely make international or other long-distance calls or have a high monthly call volume. Rather than paying a per-minute or per-unit fee for calls, VoIP customers usually pay a fixed monthly fee (or nothing). Customers can make unlimited calls, participate in multiparty teleconferencing, and make facsimile transmissions anywhere in the world. Customers also save money by consolidating data and voice networks into a single integrated IP network.

For businesses, the consolidated network results in fewer network devices, which reduces equipment costs. For new wiring in buildings under construction, a single cable can be run to each office or unit rather than running different sets of wires for network connections and telephone connections. Also, by consolidating resources around a single infrastructure rather than two distinct network infrastructures, businesses can reduce their costs of operation, administration, personnel, security, accounting, and network management. Theoretically, savings are also produced through economies of scale and network efficiencies. Bandwidth that might have been underutilized can be optimized across the applications that share the integrated network infrastructure.

For individual users, saving money is also the primary motivation for changing from traditional telephone service to Internet telephony. A caller from a private home, for example, does not necessarily care whether a call is transmitted over a phone company's twisted pair telecommunications line, coaxial cable from a cable television company, or a high-speed wireless connection from a cable company. Rather than purchasing these different services separately, a user can subscribe to all three applications via a single broadband connection.

UNIFIED COMMUNICATIONS

Another benefit of Internet telephony is simplification. For businesses, running voice and data over the same network architecture can reduce the diversity of network equipment and standards, which in turn reduces the human expertise required and simplifies the administration, implementation, and maintenance of networks. Using a single unified network is much simpler than using two or more networks.

Similarly, individual users find it much easier to have a single, unified device for phone, Internet, and video rather than having individual devices for each application. For example, users with VoIP phone service can have their voice mail forwarded to their e-mail account.

Mobile VoIP phones, ATA, and soft phones also offer mobility and portability because they can use VoIP anywhere with the same number. The service is not tied to a particular area code.

SUMMARY OF BUSINESS VoIP DRIVERS

The following list summarizes the main factors that drive businesses to adopt VoIP technologies:

- *Cost savings*—Reduce telecommunications charges and have unlimited long-distance calling.
- *Network uniformity*—Use a single network for all information exchange rather than having a separate network infrastructure for voice and another for other applications.
- *Operational efficiency*—Reduce administrative overhead and staffing by eliminating redundant network infrastructures.
- *Application integration*—Take advantage of emerging, integrated Internet telephony applications.
- *Inevitability*—Embrace the inevitable industry direction and be positioned for innovative future applications.

INTERNET TELEPHONY CHALLENGES

The telecommunications industry has experienced a great deal of momentum toward VoIP. Nearly all equipment manufacturers and service providers are developing products geared toward voice and data integration. Large corporate users have embraced VoIP for cost savings and network simplification. However, relative to other older Internet applications such as file sharing, e-mail, and Web access, voice communications is still relatively new. Internet telephony implementations create some performance and security challenges. The following sections describe the obstacles that VoIP faces in the areas of service quality, reliability, power outages, and security.

QUALITY OF SERVICE (QoS)

Throughout most of the history of computer networking, systems for transmitting information as text and numbers have been architecturally distinct from systems for transmitting voice. One reason for this distinction has been the systems' inherently different characteristics. Every communication system has a built-in delay, though it is usually imperceptible to end users. The imperceptibility of this delay is especially important in voice conversations, which require seemingly real-time transmission—in other words, a connection with a time delay so negligible that users cannot perceive it. Anyone who has had an international conversation via satellite understands how disconcerting even a one-second delay can be. A continuous voice conversation is **synchronous**, meaning literally "with time."

In contrast, much of the data we transmit is "bursty" and **asynchronous**, meaning literally "without time." A user who accesses a Web site, sends instant messages, or reads a blog is transmitting or receiving information, but then the user might not send or receive any information for a while. No information is transmitted while the user is typing an instant message, for example. Only when the user hits the Send button is a batch of information transmitted. This type of information transmission is sometimes called "bursty" as opposed to "continuous." A half-second delay in this type of transmission is not as perceptible to users as a half-second delay in a voice conversation. The real-time, synchronous, continuous nature of voice communications presents a significant design challenge to Internet telephony.

The Internet's technical infrastructure was not designed for real-time voice communications. Transmissions over the Internet have a certain degree of packet loss, latency, and jitter, which are inherently acceptable for most bursty Internet traffic, but they present challenges for voice transmissions. **Packet loss** means that a small percentage of packets traversing the Internet never reach their destination and are ultimately retransmitted. Packet loss occurs for a variety of reasons. For example, packets are sometimes dropped if a network segment becomes overwhelmed with traffic or if a signal is degraded during transmission. Retransmitting lost voice packets is not always effective because the packets can arrive after the conversation has already advanced. VoIP users would therefore perceive **clipping**, small but noticeable losses of parts of a conversation.

Latency is another problem that affects VoIP's quality of service and that VoIP providers have to design around. Latency is the time delay between a caller speaking and the other caller hearing the speech. Miniscule time delays are introduced in several places, including the end user's location when the signal is sampled and encoded; each router a packet traverses; and the gateway between an IP network and the PSTN, if the call uses both systems.

Chapter 16 discussed the issue of **jitter** in voice communications: a degradation in call quality that results from timing problems such as variations in packet arrival time. Because each packet potentially travels a different route, and because network congestion is not uniform among all routes, packets can arrive at a destination at different times. If these time differences are too uneven, jitter can occur. The technical approach to overcoming jitter is to **buffer** the call—in other words, to introduce a slight delay to even out the timing of received packets. A jitter buffer is a space at the receiving side of a voice transmission that collects voice packets for a short period of time until the packets can be properly timed and sent to the receiving caller. The resulting voice signal has less distortion.

Manufacturers of VoIP hardware include some fixed buffering capability called **static buffering** in their products. Network managers can also program buffering rules to respond to network conditions at any time. This approach is called **dynamic buffering**. The challenge is to introduce enough delay to correct jitter but not so much that users will detect it.

POWER OUTAGES

During power outages, many users of Internet telephony will lose service unless they have a battery backup for their equipment. Personal computers, routers, and cable modems that require AC power do not work in a power outage. Outages are not a concern for telephone users with phones connected directly to PSTN services, because power is supplied by the central office. Note that power outages create the same problem for cordless phones, which do not work without power.

The inability to make a voice call during a power outage can be a major concern. Some power outages last for days, which is a long time to be without voice service and Internet service. Furthermore, power outages sometimes accompany natural disasters and other dangerous situations, during which the ability to use phone service to reach emergency responders is vital.

VoIP users have a few options to minimize the effects of a power outage. First, most VoIP users also have cell phones that can provide phone service as long as the cell phone battery lasts. Keeping an extra cell phone battery for such emergencies is advisable. Another option is to purchase an **uninterruptible power supply** (UPS), equipment that provides an alternative power source during power outages (see Figure 17-11).

RELIABILITY

Business and residential phone users have historically expected a certain level of service, features, and reliability. Users expect a dial tone when they pick up a phone. The decades-old industry metric for service reliability is "five nines," or 99.999. This means that telephone service should be available 99.999% of the time, which is all but 5 minutes and 15 seconds of a year. A technical challenge to Internet telephony systems is to align their reliability and service availability levels with those of the traditional telephone system. Upon first thought, the difference between a network availability of 99% and 99.999% may seem inconsequential. However, 99% availability means that the network is expected to be down for a total of four days per year. For business users who rely on voice service to conduct day-to-day transactions, four days would be an unacceptably high amount of time to lose service. Table 17-1 shows the downtime implications of various levels of VoIP network reliability.

Network availability	Approximate downtime
90%	36 days
92%	29 days
95%	18 days
96%	15 days
97%	11 days
98%	7 days
99%	4 days
99.9%	9 hours
99.99%	53 minutes
99.999%	5 minutes

VoIP SECURITY

Security and privacy concerns are not often associated with voice communications, but this changes with Internet telephony. Customers who are accustomed to secure voice communications over the traditional PSTN encounter new security challenges when implementing Internet telephony. Voice communications over the Internet face the same security threats as other types of information exchanged over the Internet: worms, distributed denial-of-service attacks, protocol vulnerabilities, operating system security flaws, information interception, and network intrusion.

VoIP implementations also create additional security vulnerabilities for computer data applications. Voice and other Internet applications usually share the same network in VoIP systems, so an attack on a vulnerable voice component can spread to other network devices and applications or create related performance problems.

A business VoIP infrastructure augments existing IP networks with a number of possible components, including an IP PBX, gateways, servers, and telephones. Each of these components is accessible to hackers via the IP network, just like any other computing device on the network. For example, hackers can perpetrate denial-of-service attacks on VoIP infrastructures, flooding devices with so many simultaneous requests that the network is effectively rendered unavailable.

Another issue is privacy. Private voice conversations that traverse the Internet can be intercepted and monitored. As with data communications, the solution to this problem is encryption. However, encrypting voice communications adds latency because the encryption process takes time. As discussed previously, latency is already a challenge for voice communications over the Internet.

Internet telephony implementations rely on network protocols, servers, and operating systems, and must address the security challenges that accompany them. Protocols usually have inherent security flaws and vulnerabilities that become evident over time, and VoIP protocols are no exception. They inherit the security problems of other Internet protocols and create some of their own. For example, some people are concerned about the spam problem crossing over into voice communications. This type of security breach has sometimes been dubbed SPIT, for **spam over Internet telephony**.

Spoofing, in which an unauthorized person masquerades as a legitimate VoIP caller, is another challenge to businesses and individuals that use VoIP. Caller IDs can be easily spoofed in VoIP environments. For example, some voice systems rely on caller ID for authentication, and caller ID might indicate that a bank or insurance company is calling. The person receiving the call might assume that the call is from a legitimate business and divulge confidential information such as a Social Security number or account number. This type of spoofing can present a problem for some voice mail systems that use a caller ID to authenticate the caller because the voice mailbox is linked to a specific telephone number. If this type of authentication happens from a spoofed number, the unauthorized user can access, listen to, and delete private voice mail, as well as make a number of administrative changes to the account. For this reason, VoIP systems should not rely solely on caller ID for authentication.

The US-CERT Web site (*http://www.us-cert.gov*) has published many vulnerability notes that identify security problems in VoIP protocols, equipment, and software. For example, one type of router was identified to be susceptible to denial-of-service attacks because of how it handled SIP messages. Another vulnerability note describes how products that implement SIP can be vulnerable to the execution of arbitrary code. Checking the US-CERT or other CERT Web sites for newly discovered VoIP vulnerabilities is always a good way to stay on top of potential security issues.

One security countermeasure for businesses is to run VoIP and LANs on separate networks, but this defeats the promises of economies of scale, optimal cost savings, and integration of a shared infrastructure. Another option is to use security firewalls (see Chapter 15), although firewalls introduce slight time delays, and network latency is already a challenge to VoIP. At a minimum, VoIP users should also keep software up to date, use antivirus and antispyware applications, use encryption, and be aware of other security options offered by their VoIP service provider.

PUBLIC POLICY ISSUES

Traditional phone service has been available for more than a century. During this time, regulators and public officials have grappled with public policy questions at the intersection of society and communications technology. Many of these issues involved questions of monopolies and antitrust, regulation and taxation, global economic competitiveness, public safety, and national security. The U.S. telephone industry, for example, has lived through such government actions as the Communications Act of 1934, the modified final judgment that broke up AT&T in the early 1980s, and the Telecommunications Act of 1996. Other policy decisions have addressed areas such as government access to phone records, warrantless wiretapping, and antiterrorism surveillance. Voice over the Internet is a new technological area being rapidly advanced both by new market entrants and more established companies. New questions are arising about which government institutions, if any, have regulatory jurisdiction over these new voice services and what public policy issues are emerging along with this new technological context. The following sections address two of these emerging issues: public safety, and regulatory jurisdiction and taxation.

PUBLIC SAFETY

An important feature of traditional phone service is the ability to dial 911 in an emergency and for emergency services to identify the caller's location based on the originating telephone number. The use of residential VoIP services and products presents public safety challenges. Without special battery backup, for example, VoIP services do not work during power outages. This could create a serious problem if a VoIP user had an emergency during a power outage or a performance problem with the VoIP broadband connection. In contrast, most phones in the traditional phone system still operate during power outages.

Since the inception of residential VoIP services and products, there have been problems with emergency services. For example, some VoIP services initially did not connect to 911. Others did connect to 911 but did not route to the appropriate public safety answering point (see Figure 17-12). In some cases, a caller could reach 911, but the caller's residential address was not available to the emergency services. VoIP services have evolved to include 911 access as part of their standard Internet telephony services, and they now require customers to provide their residential addresses, but this is complicated by mobile VoIP telephony. From a technical standpoint, in VoIP there is no rigid relationship between a user's telephone number and physical location. The relationship is virtual and independent of area code in the traditional sense. This independence creates complications for traditional 911 emergency services, which are accustomed to a one-for-one relationship between a phone number and a fixed geographic location. The incorporation of public safety features into Internet telephony continues to evolve.

REGULATORY JURISDICTION AND TAXATION

Public utility commissions have traditionally overseen local telephone services and have served as a body to which customers could appeal for help with service problems. Internet telephony makes voice just another Internet application. Who, if anyone, should regulate it?

The answer is not always clear. For example, some residential VoIP services operate over a broadband Internet connection in which the VoIP service and broadband connection are supplied by two different service providers. In the event of a service outage, consumers might need to deal with both providers to solve the problem. This scenario can lead to finger pointing and often leaves the burden of troubleshooting on the consumer. Even when the commercial voice service and high-speed Internet access are provided by the same company, the two services sometimes have two distinct sets of personnel to handle outages and other problems.

Similarly, federal, state, and local governments raise considerable revenue by levying taxes and fees for services offered over traditional telephone networks. The ongoing transition of voice calls from traditional phone service to VoIP raises many issues and questions. Should voice service over the Internet be subject to the same taxes and fees assessed on traditional voice service? Internet telephony that uses free computer-to-computer VoIP applications has no service fees, so what would governments tax? Voice calls are just another application over the Internet, so why should they be taxed while others (such as e-mail) are not? Would taxes on evolving services such as Internet telephony dampen rather than promote innovation? On the other hand, how would federal, state, and municipal governments make up the lost tax revenue? On what basis should traditional voice services be taxed, but not Internet voice services? These are just some of the policy challenges raised by the growth of Internet voice services.

CHAPTER SUMMARY

- Internet telephony, also called IP telephony or Voice over Internet Protocol (VoIP), refers to voice communications transmitted over the Internet or other IP network.
- VoIP products convert an analog waveform into a digital signal format, break the digital signal into packets, and transmit the packets over the Internet.
- The VoIP approach of sending voice over a packet-switching network is different from the circuit-switching approach of traditional phone networks.
- VoIP is not a single protocol, but many different standards that handle functions such as signaling, digitization, and transport.
- Two examples of VoIP signaling protocols are the H.323 protocol and Session Initiation Protocol (SIP).
- Real-time Transport Protocol is a popular protocol for VoIP transport.
- VoIP has many implementation options, including soft phones, analog telephony adapters, Wi-Fi phones, and IP phones.
- Businesses can use a private or virtual private network (VPN) to carry VoIP and other traffic.
- Compared with other types of traffic, voice transmissions have different performance characteristics that require prioritizing voice packets through class of service (CoS) designations.
- The two most significant business drivers for using VoIP are cost savings and network integration and unification.
- Some challenges presented by VoIP, such as quality of service and reliability, are related to security and performance.
- VoIP also presents some public policy questions such as its ability to reach 911 emergency services reliably and whether it should be taxed like traditional phone service.

KEY TERMS

analog telephony adapter	RTP Control Protocol
Application layer	session
asynchronous	Session Initiation Protocol
buffer	signaling protocols
clipping	soft phone
convergence	spam over Internet telephony
dynamic buffering	spoofing
encoding standard	static buffering
H.323	synchronous
IP phone	transport protocols
IP PBX	unified communications
jitter	uninterruptible power supply
latency	URI
packet loss	VoIP
Real-time Transport Protocol	Wi-Fi phones

REVIEW QUESTIONS

1. Is there a difference between Internet telephony and VoIP?
2. Is VoIP based on circuit switching or packet switching?
3. What are some distinguishing characteristics between voice traffic and other types of traffic?
4. On an integrated IP network, what would require a higher-priority designation: voice packets or packets containing text messages or e-mail? Why?
5. Define packet loss and latency.
6. What is the role of a signaling protocol?
7. Name three protocols used with VoIP implementations.
8. What equipment, network connection, and software are required to implement a soft phone?
9. What is a Wi-FI phone? To what extent does a Wi-Fi phone provide mobility? How does it compare with the mobility of a cell phone?
10. What is an analog telephony adapter?

11. Name the major incentive to use the Internet for voice communications.
12. List three technical challenges of VoIP.
13. Is VoIP subject to more or fewer security problems than other Internet applications?
14. List all of the functional features (for example, *69 and call waiting) that your primary phone service provides.
15. What is an IP PBX?

DISCUSSION QUESTIONS

1. Which of the following services do you use to make voice calls: VoIP, traditional telephone service, cellular telephony, or some other mobile handheld device? Explain your rationale for using each service.
2. When someone calls you, does it make any difference to you whether the call originates from an Internet-attached device, a pay phone, a cell phone, or other device?
3. Do you think that spam over Internet telephony (SPIT) will become a security problem like spam? Why? What could prevent it?
4. In what ways has VoIP technology raised law enforcement and public safety issues?
5. In your opinion, should Internet telephony be taxed, like traditional voice services? Provide arguments for and against taxing Internet telephony.
6. Who, if anyone, should have regulatory oversight of Internet telephony service, and why?

CASE PROJECT

Set up a soft phone on your computer. Select a popular VoIP provider of free Internet calling and download its software to your computer. Following the provider's instructions, install the software and try to call three other classmates who have implemented the same VoIP software. Write a few paragraphs in response to each of the following questions:

1. How difficult was it to select a free VoIP service, and on what basis did you make your selection?
2. Was it easy to install the software, or did you encounter problems?
3. What type of hardware did you require to make voice calls using your computer?
4. When you called your classmates, how did the call quality compare with that of a cell phone or other phone?
5. What happens if you want to call someone who doesn't use the same free VoIP service? Can you call your own cellular phone from the VoIP service? If not, how much would it cost to make a VoIP call "off the VoIP service" to reach your cell phone?

CHAPTER 18

WIRELESS MULTIMEDIA

LEARNING OBJECTIVES

In this chapter you will:

- Understand the role of multimedia access devices such as cell phones and how they have integrated previously separate applications such as voice, Internet applications, electronic commerce, and entertainment

- Understand how Bluetooth and spread spectrum technologies work

- Examine the concepts of cells, frequency reuse, and handoff in cellular networks

- Become familiar with the main architectural components of a cellular network

- Understand the role of cellular standards such as GSM and differentiate between 1G, 2G, and 3G systems

- Examine emerging wireless broadband technologies such as WiMAX and how they might transform information access

INTRODUCTION

Blackberries, iPhones, Treos, and other mobile devices are wireless multimedia platforms that integrate applications such as e-mail, text messaging, Web access, video and music players, digital photography and video recording, office software, and voice communications. These devices serve as a unified platform to provide mobile access to applications that were once supported by disparate devices: media players, telephones, laptops, and so on. An interesting feature of wireless multimedia devices is that many of them provide access to multiple network systems.

This chapter discusses the networks that enable wireless broadband multimedia. It begins with an explanation of the Bluetooth standard—the short-range, personal area network (PAN) technology that operates in the unlicensed 2.4-GHz radio frequency range and uses a technology called spread spectrum. The chapter then describes the fundamentals of cellular telephony, including the concepts of cells, frequency reuse, and multiple access methods such as TDMA and CDMA. Finally, the chapter discusses Wi-Fi mobile phones, cellular/Wi-Fi integration developments, and the prospect of WiMAX as a broadband metropolitan wireless technology.

A major transition in IT has been the integration of previously disparate functions into a more unified and mobile multimedia environment (see Figure 18-1). People can purchase several individual digital devices, including a Global Positioning System (GPS) receiver, a digital camera, a laptop, a mobile telephone, a video camera, a personal digital assistant, and an MP3 player. Consumers can also choose a single unified device that incorporates much of the functionality of these distinct devices. For individual users and business users, wireless multimedia devices such as Blackberries, iPhones, and Treos are mobile communications and information platforms that integrate various services, some of which may include:

- Mobile telephony
- Electronic mail
- GPS
- Text messaging (also known as SMS, or Short Message Service)
- Calendar

- Web browsing
- Camera
- Video player
- Music player
- Organizer
- Office applications and presentation software

These devices have improved productivity for salespeople, consultants, and other mobile business professionals. The ability to instantly check e-mail, send text messages, and access the Web "on the run" without using a laptop has greatly improved job flexibility. Businesses also tie multimedia wireless devices to enterprise servers, which provide mobile professionals with access to applications on a corporate network. Some of these services include Web-based applications, shared contact management and calendar software, and other business applications.

Some multimedia wireless devices are designed more for personal use than business use. These devices contain more entertainment-focused features such as built-in digital cameras, gaming interfaces, video and audio media players, and 8 GB or more of storage capacity.

An important feature of wireless multimedia devices is that they can often simultaneously interface with multiple wireless network technologies, including GPS, Wi-Fi, cellular networks, and Bluetooth technology.

Figure 18-1

Wireless multimedia integration

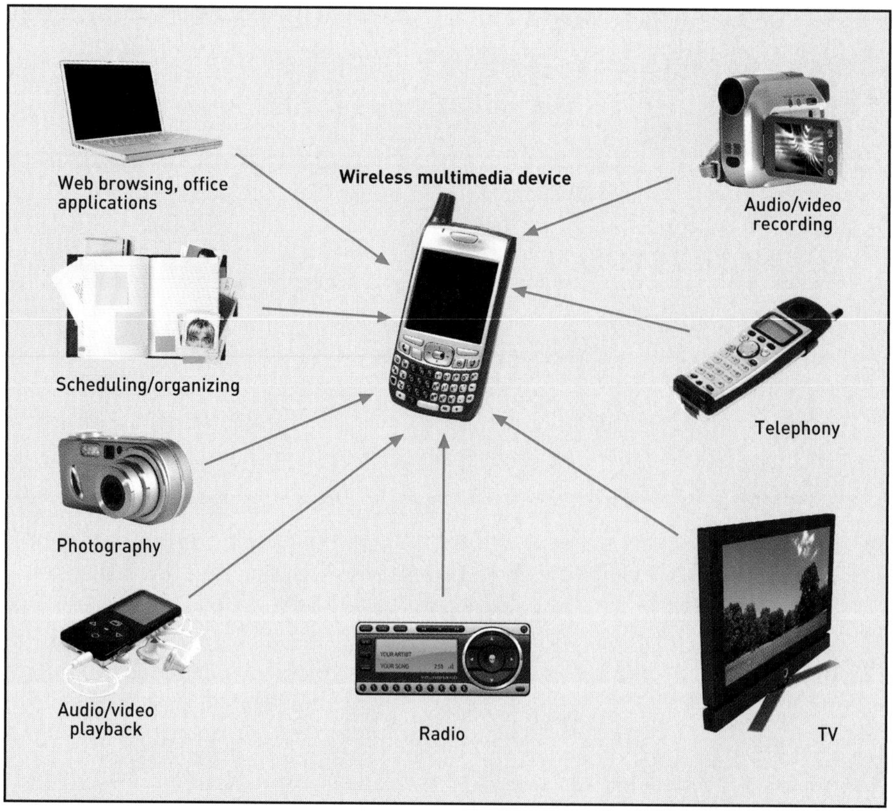

Bluetooth technology is a short-range wireless communication standard designed for geographically limited transmissions. Bluetooth allows communication between a wireless headset and a multimedia wireless device (see Figure 18-2), between a headset and a PC for Voice over IP (VoIP) communications, between a car and a remote-entry device, between a wireless mouse and a PC, and between wireless controllers for video game consoles. Another feature of Bluetooth technology is its ability to wirelessly project and control presentations from Blackberries and other multimedia handheld devices.

Also known as the **IEEE 802.15.1** standard, Bluetooth is an industry specification for encrypted radio communications. It uses an unlicensed (2.4 to 2.485 GHz) frequency band of the radio spectrum, similar to Wi-Fi frequencies. Bluetooth is an example of a **personal area network** (PAN), a network that transmits within a range of 1 to 100 meters to users. Transmitting over such short distances requires very little power and uses cost-effective communication chips embedded in Bluetooth-enabled devices.

Figure 18-2

Bluetooth transmission between wireless headset and mobile device

The Bluetooth standard specifies three classes of devices that operate at different ranges:

- Class 1 radio, which operates at up to 100 meters
- Class 2 radio, which operates in a range of approximately 10 meters
- Class 3 radio, which operates in an extremely short range of up to 1 meter

Most mobile devices that use Bluetooth are in the Class 2 category of approximately 10 meters. They have extremely low power requirements (approximately 2.5 mW) and achieve a transmission rate of around 2 to 3 Mbps.

Bluetooth uses a wireless transmission approach called **frequency hopping spread spectrum** (FHSS), which divides the allocated frequency band into smaller bands and rapidly hops between 79 frequencies at a rate of 1600 hops per second. Each frequency is used for only a fraction of a second.

Recall from previous chapters that simple radio wave communication requires the use of a carrier wave over a certain frequency for transmission, whereby the information-carrying signal is superimposed onto the carrier wave by modulating its amplitude, frequency, phase, or other property. FHSS technologies, in contrast, are more complex and rely on a large number of carrier frequencies, whereby each frequency is used for only a fraction of a second as the transmitter hops from one carrier frequency to another. In Bluetooth, for example, the transmitter hops from one carrier frequency to another at a rate of 1600 hops per second. Furthermore, if the transmitting device detects another device operating in a frequency within the unlicensed band that Bluetooth uses, it will adaptively move (or hop) to a different frequency to avoid possible interference. This **adaptive frequency hopping** approach serves to reduce interference with nearby devices operating in similar frequency ranges, such as microwaves and cordless telephones.

FHSS is actually only one type of **spread spectrum** technology while the other is **direct-sequence spread spectrum** (DSSS). The Bluetooth standard uses FHSS, the technique that transmits the signal across a transmission channel using various frequencies. The duration in which each frequency is used is extremely brief, considering that more than 1000 frequencies are used per second. DSSS techniques, in contrast, do not hop from one frequency to the next, but instead rely on *spreading* the energy of the transmission onto a large range of frequencies at any given time.

Spread spectrum was initially intended for military applications that required high security and reliability. Interestingly, however, some of the first developers of spread spectrum technology were not scientists or the military. The development of FHSS is in fact traditionally attributed to Hedy Lamarr, an actress, and George Antheil, a musician. During World War II, they developed an idea to remotely control torpedoes that later became known as FHSS.

It is inherently difficult to jam spread spectrum transmission because transmissions are distributed across a range of frequencies rather than a single frequency. This distribution makes it difficult to intercept and eavesdrop on a conversation. Spread spectrum has many other applications other than Bluetooth. It is also used within cordless telephones for transmission between the telephone and the base as well as in cellular communications and computer networking.

The sequence of frequencies that the transmitter hops in FHSS is based on a "pseudo-random" code—because the code is a binary number that is actually generated with an algorithm, it is called pseudo-random. To generate the sequence of pseudo-random numbers (i.e., codes), the algorithm relies on an initial number called the "seed" to subsequently generate the series of binary pseudo-random codes (see Figure 18-3). These pseudo-random binary codes are then used as a basis to choose the sequence of frequencies that the transmitter must use for frequency hopping. The seed may be considered an initiation number from which the pseudo-random numbers are derived. In order for a receiving device to decode the transmission, it needs to be able to generate the pseudo-random code and hence must have access to the seed; otherwise, the number is very difficult to predict. Once the code is generated at the receiving end, the receiver can extract the data from the transmission. This method of generating pseudo-random numbers is also used in encryption systems other than spread spectrum technologies.

Figure 18-3

Pseudo-random code generation

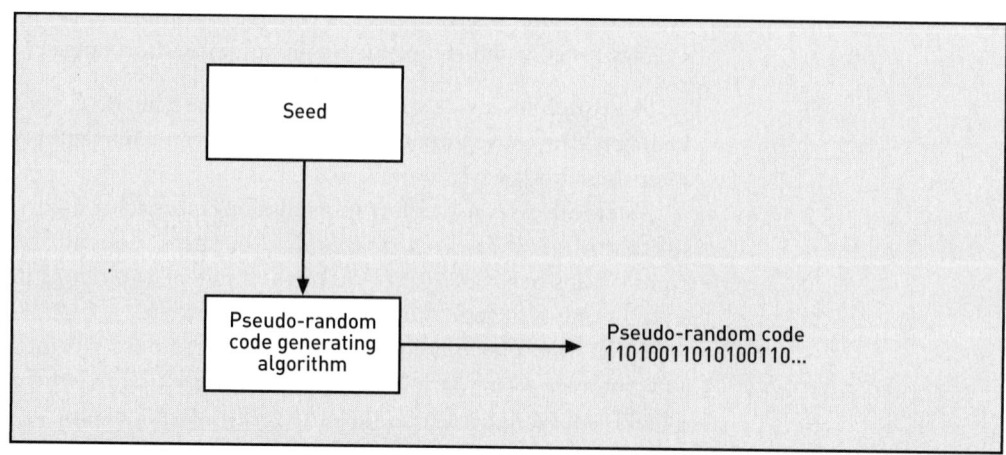

FHSS and DSSS technologies are both used in wireless local area networks, such as those that use the IEEE 802.11 standard. However, DSSS does not use frequency hopping like FHSS does, as mentioned previously. Instead, it spreads the energy of the transmission across the entire available bandwidth based on a pseudo-random code, which makes the transmission appear as noise to an eavesdropper. In fact, the receiving station also receives the transmission as a noise-like signal but has the ability to extract the information-carrying signal from it. The receiving end can generate the spreading code (i.e., pseudo-random binary code) if it has access to the seed and therefore can recover the transmitted data.

Another fundamental technical property of Bluetooth technology is that a group of Bluetooth devices, usually called a **piconet**, is synchronized both by time and by a common pattern of frequency hopping. A "master" device provides the definitive clock and hop pattern information to which the other "slave" devices synchronize (see Figure 18-4). While hopping between multiple frequencies, Bluetooth also divides transmissions into time slots. Packets are divided in time and transmitted across multiple frequencies. As a final note on Bluetooth, transmissions have a certain degree of inherent security, including 128-bit encryption. Also, Bluetooth products use a PIN code authentication the first time they connect.

Figure 18-4

A Bluetooth piconet

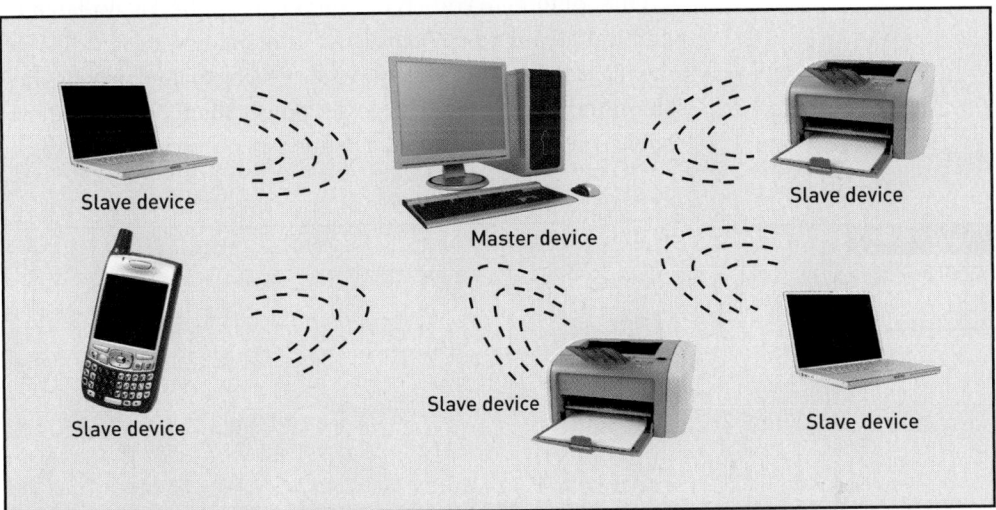

Slave device

Master device

Slave device

Slave device

Slave device

Slave device

CELLULAR TECHNOLOGY

Cell phones and most other wireless multimedia devices connect to cellular networks. Cellular systems use different frequency bands depending on whether the system is in North America, South America, Asia, Europe, or elsewhere. A government authority, such as the Federal Communications Commission (FCC) in the United States, allocates frequency spectrum for cellular services. Providers cannot use any frequencies they want.

Throughout the world, the frequency spectrum allocated for cellular transmission systems ranges from hundreds of megahertz to a few gigahertz. In the United States, the allocation is roughly from 800 MHz to 2.700 GHz. So, unlike the unlicensed band used with Bluetooth, cellular technologies use licensed frequency bands within the radio spectrum.

FREQUENCY REUSE

Consider the simple radio system illustrated in Figure 18-5, which consists of an antenna that transmits radio signals at a certain frequency. The coverage area, or reach of the signal, depends on the size, shape, and height of the antenna; the topography of the surrounding area; the transmission power of the antenna; the frequency being used; and any ambient atmospheric conditions such as weather. One option for providing radio coverage in a metropolitan area is to install a single, high-powered, large antenna such as that shown in Figure 18-5. This system would have a finite number of available channels, hence serving a limited number of users within the large geographical area. Because only a limited number of radio frequencies are made available to a service that provides wireless access, conserving frequency is a key requirement of the system design. For example, if a radio system with a single antenna had seven available frequency channels and used no sophisticated techniques to conserve channels, the system might be able to support only seven simultaneous transmissions if each channel required its own frequency. What if more subscribers wanted to use the system?

Figure 18-5

A geographical region served by a single antenna

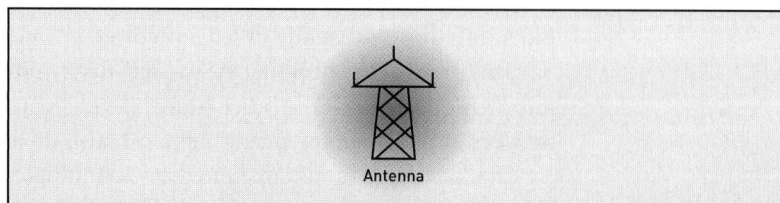

One solution to conserving a limited number of available frequencies within a large geographical area is to split the area into smaller areas called **cells** and place an individual antenna called a **base station** within each smaller area, as shown in Figure 18-6. Each base station transmits at a much lower power than a single antenna that could cover an entire region. A cell may therefore be defined as a geographical area covered by a base station.

Figure 18-6

A large geographical region divided into cells

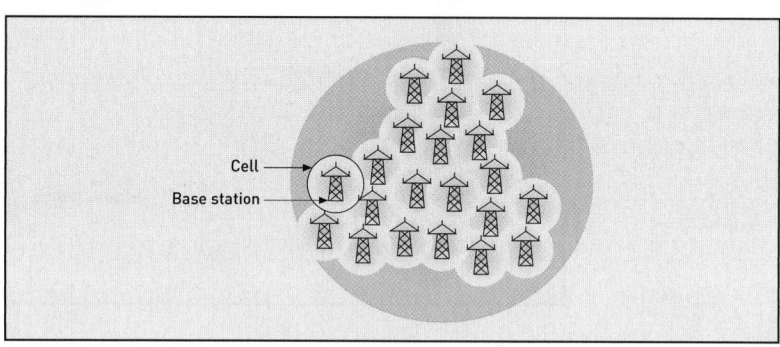

Although the shape of a cell is actually more circular, the theoretical shape of the cell is traditionally depicted as hexagonal (see Figure 18-7). A hexagon is typically used to denote a cell in cellular communications because its shape is nearly circular, and hexagons can be neatly tiled next to each other without gaps between them. In reality, however, gaps may exist between cells, and the cells may overlap because of their circular shape.

Figure 18-7

Hexagonal depiction of cells

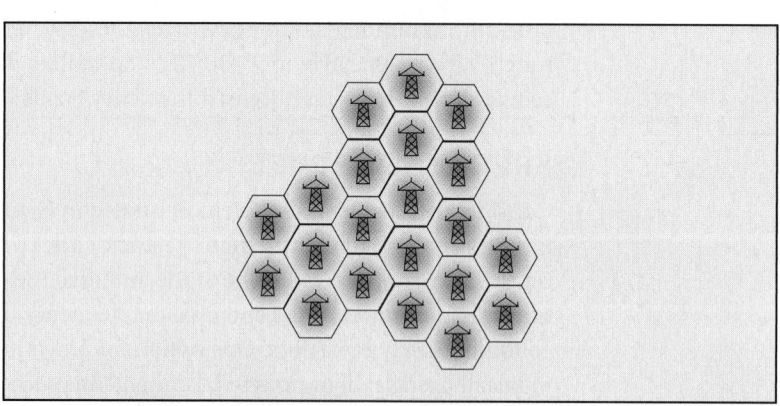

As mentioned previously, the purpose of this multi-cell configuration is frequency conservation. The same radio frequencies can be used in multiple cells as long as the base station transmitters are far enough apart and if the level of transmission power is low enough not to interfere with other cells that use the same frequency. For example, adjacent cells would not use the same frequencies because of the potential for interference.

Figure 18-8 illustrates 21 cells that cover approximately the same geographical area originally covered by a single antenna. Each 7-cell group, or **cluster**, uses the same set of frequencies, which allows the system to support many more users than a single antenna. The trade-off, of course, is the need for many more antennas, which increases costs, complexity, and power consumption. Note that the cells are color coded to indicate that a separate set of frequencies is used within the same cluster. The same sets of frequencies, however, have been reused within a different cluster, as illustrated by the repeating color patterns. For example, the red cells are geographically distant enough from each other to reuse the same set of frequencies without interference. Cellular networks can handle a large number of calls with a relatively small number of channels because of this **frequency reuse**—a single frequency can be used simultaneously in different nonadjacent cells.

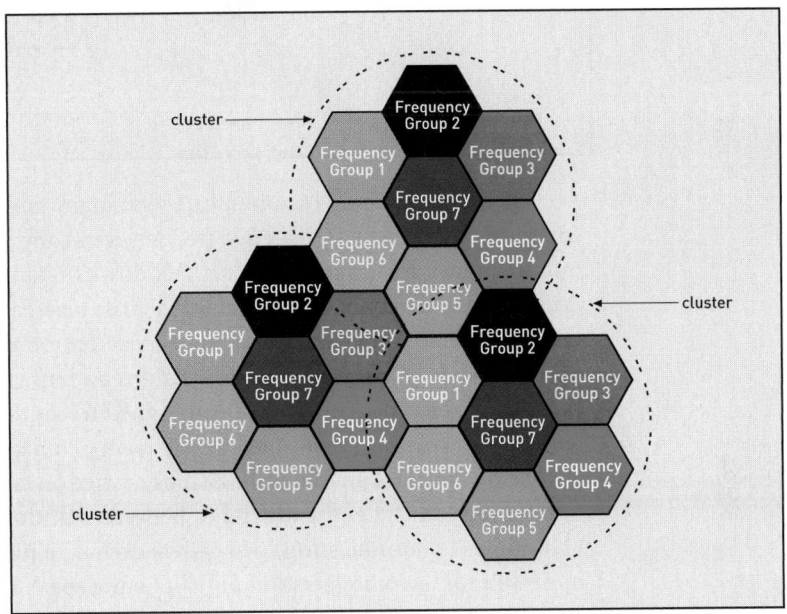

Figure 18-8

The concept of frequency reuse

The number of cells into which a larger geographical area would be divided mainly depends on the **subscriber density**, the number of users who require service in the area. There are several ways to increase the capacity of a cellular system as subscriber density increases. If some of the assigned frequency channels are reserved by the provider, they can be allocated to a cell as demand increases over time. One of the most common approaches for handling increases in subscriber density is **cell splitting**, which takes advantage of the uneven usage over a geographical area. For example, in a metropolitan area, urban centers have greater usage patterns than suburban regions outside the city. In areas of higher usage, cells can be split into smaller cells while cells in less dense areas are kept larger. This pattern is shown in Figure 18-9. To achieve these smaller sizes, power levels must be diminished in the base station antennas within each smaller cell, or **microcell**. A third technique to increase system capacity is **cell sectoring**, in which individual cells are sectored into wedge shapes via directional antennas radiating from the base station. Additionally, access mechanisms enable numerous users to share each channel, improving the capacity of cellular systems. Access mechanisms are described later in this chapter.

Figure 18-9

Cell splitting

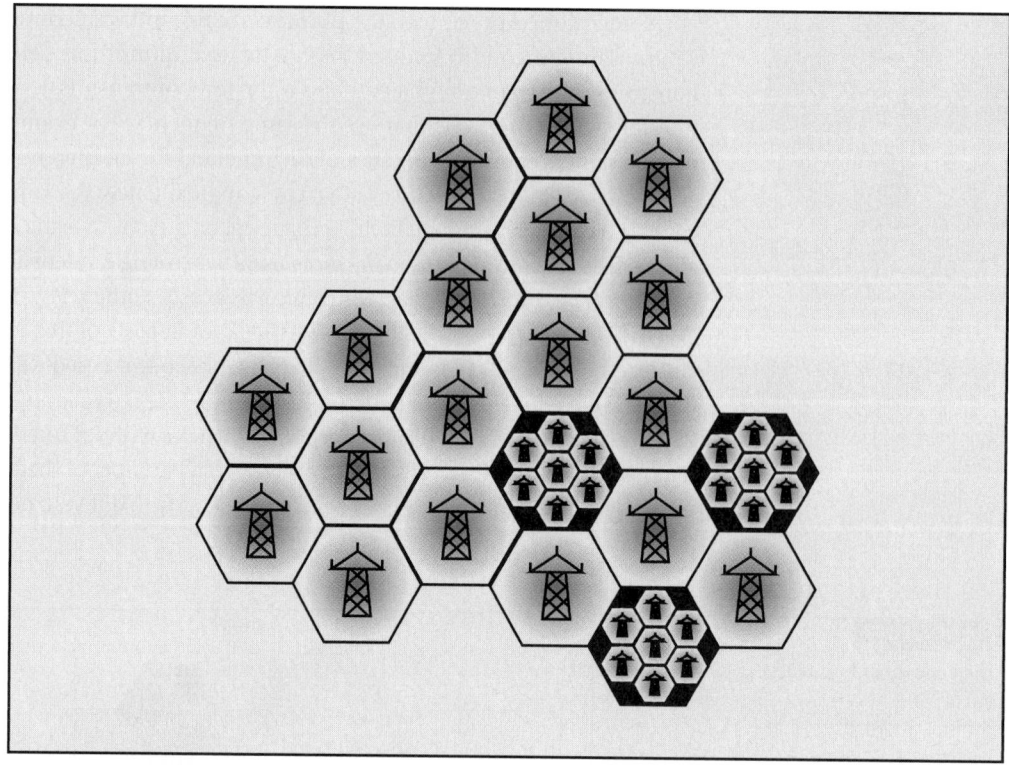

One base station can communicate with many mobile devices. Traditionally, each of these communications actually uses two frequencies in any given instant, as shown in Figure 18-10: one frequency for transmitting signals downstream from the base station to the mobile device, and one frequency for receiving signals upstream in the opposite direction. (An exception to this rule is the "walkie-talkie" service option that some cellular providers offer.) The upstream and downstream communications both use the same frequency, and only one caller can talk at one time. This system is designed to conserve frequency because it requires only half the frequencies as real-time, two-frequency communication.

The distinction between walkie-talkie service on cell phones and real-time, two-way communications illustrates the difference between **full duplex** and **half duplex** transmission. In half-duplex communication, two callers exchange information over a single frequency and must wait for the other person to finish before speaking. Full duplex, shown in Figure 18-10, allows both parties to speak at once, using one frequency to speak and one to listen.

Figure 18-10

Upstream and downstream dual frequencies per user

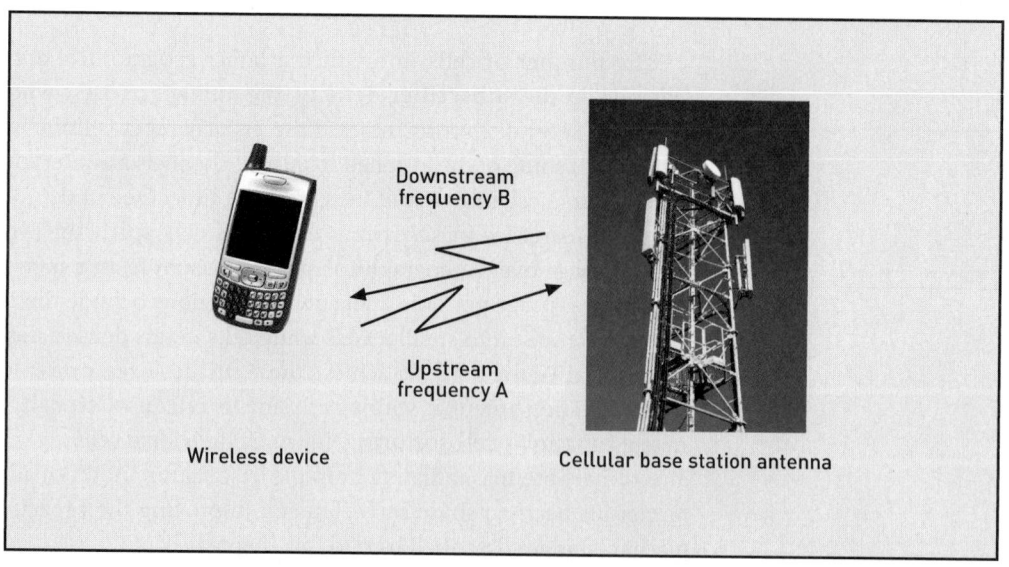

As a caller moves away from a base station, the signal becomes increasingly weak. Any adjacent area will not offer the same radio frequency channel because of the potential for interference, so there has to be a way to drop one channel and gain access to another. When signal dissipation reaches a certain threshold, the mobile device receives a control signal that requests a handoff to another base station antenna. The base station to which the mobile device is connected notes the signal dissipation, and the base station to which the caller is headed detects an increasingly strong signal from the phone. The base stations coordinate this signal strength detection through a switching center called a **mobile telephone switching office (MTSO)**. When the mobile device receives notice, it terminates one channel and changes frequencies to establish a connection with the next base station. There is a transparent transfer from the base station of one cell to the base station of another, and an associated change in the frequency channel that carries the transmission. This process is called a **handoff**, as shown in Figure 18-11.

Figure 18-11

Frequency handoff between adjacent cells

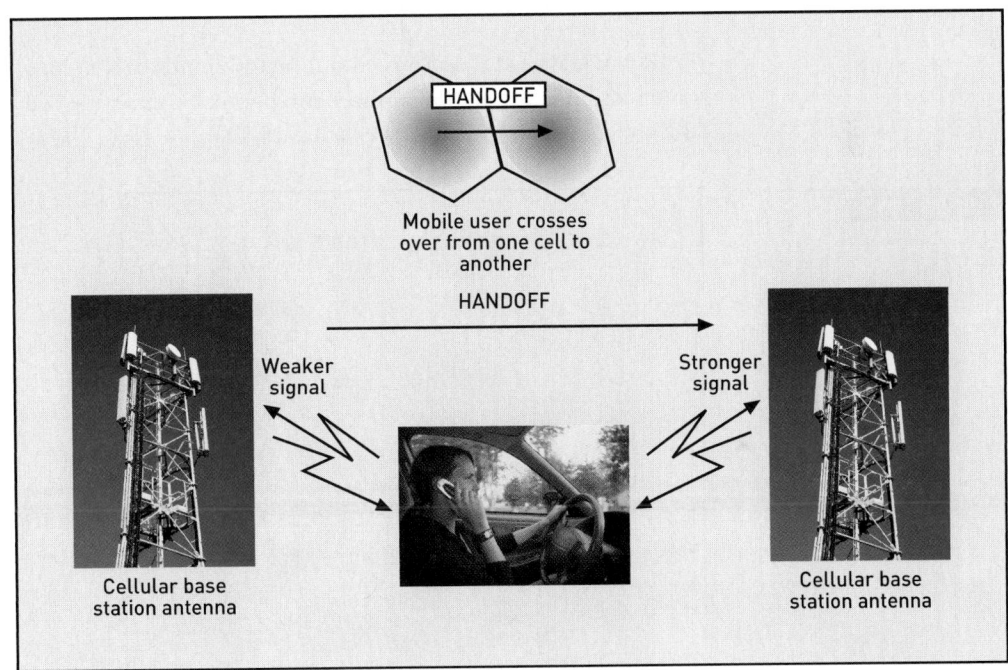

The main components of a cellular system are the base station antenna and control unit in each cell, the mobile devices that communicate with each base station, and an MTSO. An MTSO is similar to a central office of the Public Switched Telephone Network (PSTN), but it is run by the cellular provider. The three main functions of the MTSO are to:

- Interconnect surrounding base stations.
- Perform control functions such as handoffs and channel assignment.
- Serve as the interface between the base station and PSTN.

Figure 18-12 depicts an MTSO that interconnects nearby base stations and serves as the interface between the base stations to the PSTN. The connection between the MTSO and the base station is usually via land lines, so only a limited portion of most cellular transmission is actually wireless.

Figure 18-12

The mobile telephone
switching office

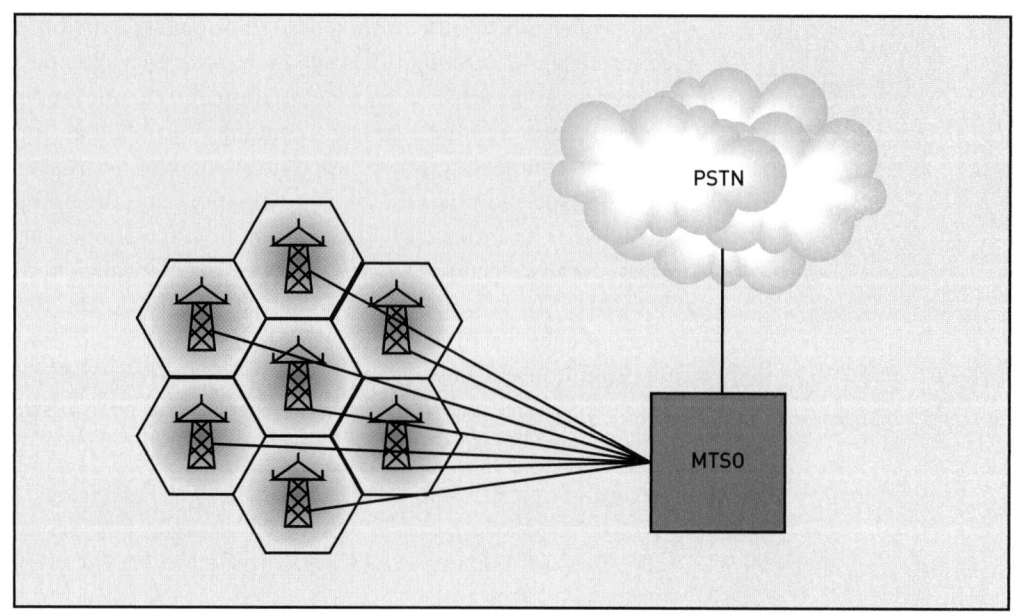

If you glance at a nearby cell phone tower, notice that it usually has two or more sets of antennas and base station equipment, because competing cell phone providers often share the same antenna structure (see Figure 18-13).

Figure 18-13

Base station with
multiple sets of antennas

When a user turns on a cell phone or other wireless device, it scans for the strongest **control channel**. This signal is sent from base stations and performs control functions such as monitoring the location of a wireless unit or establishing a call. Do not confuse the control channel with **traffic channels**, which carry the actual information of text messages, voice calls, and video streams. When a wireless device detects the strongest control channel of a nearby base station, it exchanges control information with the base station to make the network aware of its position.

MULTIPLE ACCESS METHODS

One important design characteristic of a cellular network is its **access method**, the approach that allows numerous mobile users to share a transmission medium within the same cell. In the case of wireless, the access method determines how numerous devices share the finite resource of a radio spectrum for transmission.

FDMA

Historically, the first analog cellular systems used an approach called Frequency Division Multiple Access (**FDMA**). Like other FDM techniques described in this book, FDMA divided the available spectrum into smaller frequency channels. Each device that used the system occupied one frequency channel for transmitting information and another frequency channel for receiving information. These channels were occupied for the duration of the call and were unavailable for use by any other devices on the system. As subscriber growth increased, the finite number of available frequencies made these systems unable to meet demand. Figure 18-14 illustrates the concept of FDMA for digital channels (historically, FDMA was used for analog cellular systems, but it was later used in digital versions as well). Each mobile user within the same cell is assigned a different frequency for transmitting data simultaneously with other users. Because each user is assigned a different frequency, the base station can distinguish one user's transmission from another's. Note that the frequencies assigned to each user are chosen from among the available frequencies for each cell. Recall from previous discussions that each cell within a cluster is assigned a different set of frequencies compared to a cell within the same cluster. In FDMA, each user within a particular cell is assigned a frequency from the designated frequencies for that cell, and all users may transmit simultaneously to the base station. Based on Figure 18-14, mobile user 1 transmits using frequency A, mobile user 2 transmits based on frequency C, and mobile user 3 transmits based on frequency B.

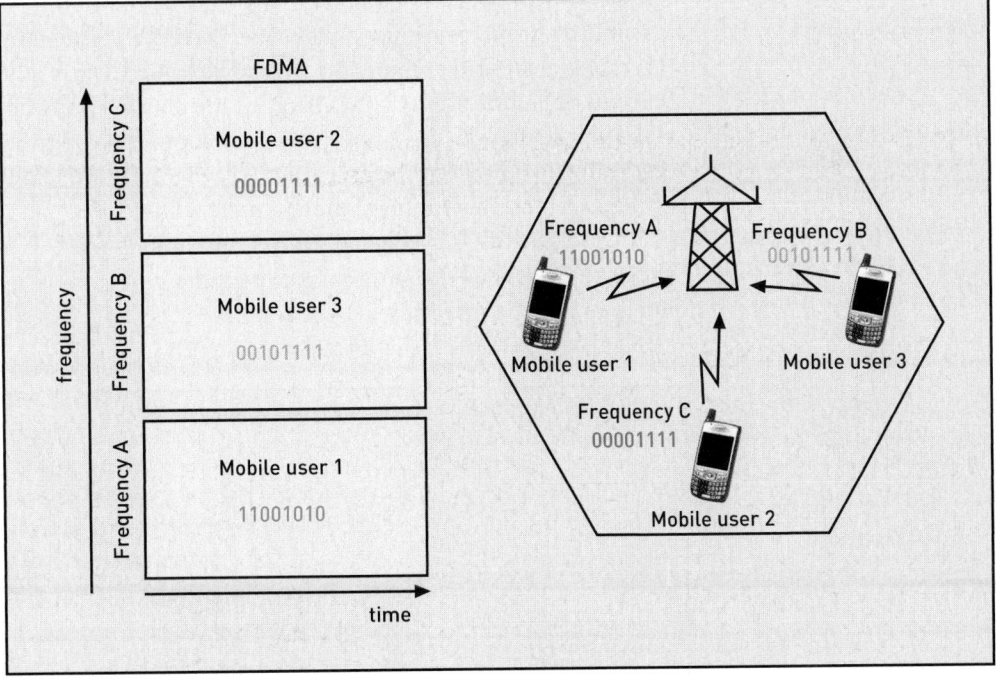

Figure 18-14

The concept of FDMA

TDMA

Digital cellular systems that convert voice into a digital bit stream improved greatly upon their analog predecessor. One example of a more efficient multiple access method is called Time Division Multiple Access (**TDMA**). Cellular systems that use TDMA divide each frequency band into time slots. In this way, multiple devices can share a single frequency (such as frequency A) by being allocated their own time slots within the band, as shown in Figure 18-15. In the figure, all mobile users transmit based on frequency A; however, mobile user 1 is assigned time slot C, mobile user 2 is assigned time slot A, and mobile user 3 is assigned time slot B. This approach greatly increases the number of users that can simultaneously access a cellular system. As with statistical time division multiplexing, which was described previously in the book, dynamic TDMA can assign time slots on demand rather than rigidly fixing time slots for each wireless device. In TDMA, users within the same cell can operate at the same frequency; however, each user is allowed to transmit only during an assigned time slot. In other words, mobile users take turns transmitting their signals. Note that this process takes place at a very fast rate within cellular networks, and mobile users are unaware that they are taking turns. This approach greatly increases the number of users that can be supported by a single base station within a cell, compared to FDMA.

Figure 18-15

The concept of TDMA

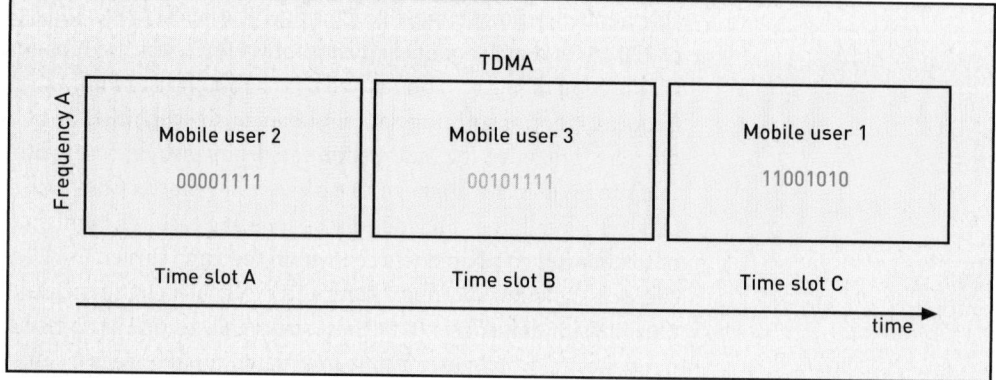

CDMA

While FDMA assigns a certain frequency to wireless devices and TDMA assigns each device time slots within a given frequency, Code Division Multiple Access (**CDMA**) assigns each transmitting device a unique digital code. Each user has a unique pseudo-random code for spreading transmissions across the frequency band. CDMA is therefore classified as a spread spectrum technique whereby the energy of the transmissions from each user within the same cell is spread out using the unique pseudo-random code of each mobile user. The codes therefore distinguish users from each other (see Figure 18-16), and the base station uses these codes to keep track of multiple mobile users within the same cell. Mobile user 1 is assigned code A, mobile user 2 is assigned code C, and mobile user 3 is assigned code B. As shown in the figure, users may transmit data at the same time and occupy the same frequency band entirely as other users within the same cell. As a result, CDMA supports more users within the same cell and provides higher data rates than FDMA and TDMA.

Figure 18-16

The concept of CDMA

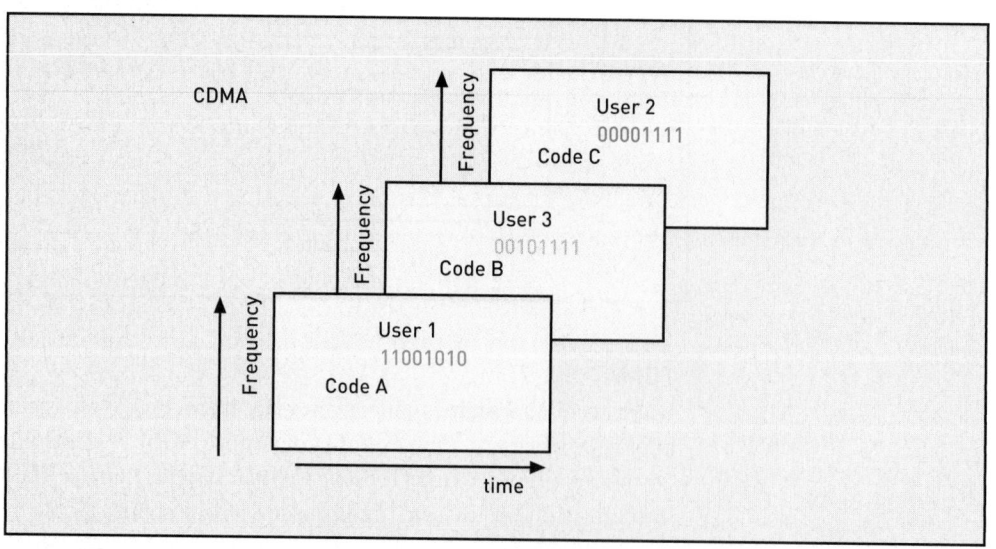

GENERATIONAL EVOLUTION OF CELLULAR STANDARDS

The IT industry uses a taxonomy of cellular network generations to categorize different systems. Like the software industry's taxonomy of first-generation through fifth-generation languages, cellular systems are usually classified into four generations (1G through 4G).

The first widespread cellular service was an analog standard called Advanced Mobile Phone Service (**AMPS**), which predominated in the United States for years. This generation of analog cellular telephony is usually called **1G** (first generation). AMPS used FDMA as its access method and required considerable bandwidth to accommodate areas of high subscriber density.

Digital cellular approaches quickly replaced analog services such as AMPS. Second-generation cellular, or **2G**, refers loosely to the first generation of digital cellular approaches. The advantages of digital over analog were described in Chapter 2; they illustrate how digital systems provided great advantages over their analog predecessors. Some of these improvements included higher transmission capacities (data rates), more reliable signals, and less susceptibility to noise. Additionally, the digitized signals were easy to encrypt, which provided greater security. It was also easy to compress the signals and apply error detection and correction mechanisms to them. Finally, digital systems introduced efficient access methods that allowed numerous users to share each channel.

A digital version of AMPS, called D-AMPS, uses TDMA as its multiple access scheme. Another example of a 2G standard is **GSM** (global system for mobile communications). GSM is a popular standard (especially in Europe) for voice and data over cellular using TDMA. Yet another example is **IS-95** (Interim standard 95), which is used in the United States and is based on CDMA.

The industry calls some modern digital systems 3G, for third-generation wireless. **3G** systems usually refer to digital approaches that have high data rates. They are optimized for today's multimedia wireless requirements to access Web applications and text messaging; download audio, video, and images; and transmit voice communications. One 3G cellular standard that uses the CDMA access method is **CDMA2000**. Another example is Universal Mobile Telecommunications System (**UMTS**), which is also primarily based on CDMA technology.

Wi-Fi, WiMAX, AND CELLULAR INTEGRATION

Manufacturers of cell phones and other handheld multimedia devices sometimes offer both Wi-Fi and cellular access from a single device, which is sometimes called a dual mode GSM/Wi-Fi device. Most of these devices integrate VoIP technology as well. For example, a Wi-Fi-only phone uses VoIP to allow calls through any Wi-Fi hot spot, but it may not allow for traditional mobile telephony via a cell phone service. Other devices incorporate Wi-Fi, cellular capability, Bluetooth capability, and sometimes GPS services.

Devices that support both cellular and Wi-Fi can use Wi-Fi connectivity when users are in proximity with a Wi-Fi network (using VoIP) or connect to a GSM network when the user is outside the range of the wireless local area network. Businesses use this capability to reduce cell phone expenses because employees can use existing wireless LAN access at little extra cost when in the office and only use cellular service when away from the office or other Wi-Fi network. The ability to seamlessly hand off calls from a Wi-Fi network to a GSM network and vice versa is an obvious requirement to which many multimedia device manufacturers are responding.

Other manufacturers are introducing GSM/Wi-Fi base stations designed for homes, which connect a user's wireless device to a broadband connection such as DSL or a cable modem. For example, a phone wirelessly connects to the access point, which connects directly to a broadband Internet connection.

Another network standard that could enable further network convergence is the emerging broadband wireless technology known as **WiMAX** (Worldwide Interoperability for Microwave Access). WiMAX is actually another name for the formal networking standard known as IEEE 802.16 (see Figure 18-17).

WiMAX would provide wireless access at high transmission rates, like Wi-Fi, but over much greater geographical ranges. The range could span several miles depending on whether the user is mobile or stationary; stationary users can connect across much greater distances than mobile users. Some areas of the developing world do not have an extensive telecommunications infrastructure, so a technology such as WiMAX has been cited as a possible way to deliver broadband access much more economically than installing land-based fiber optic and copper cables. WiMAX is different from Wi-Fi in many ways: it is designed for much greater distances, it uses a different channel access method, and it uses a licensed part of the radio frequency spectrum.

Figure 18-17

The concept of WiMAX

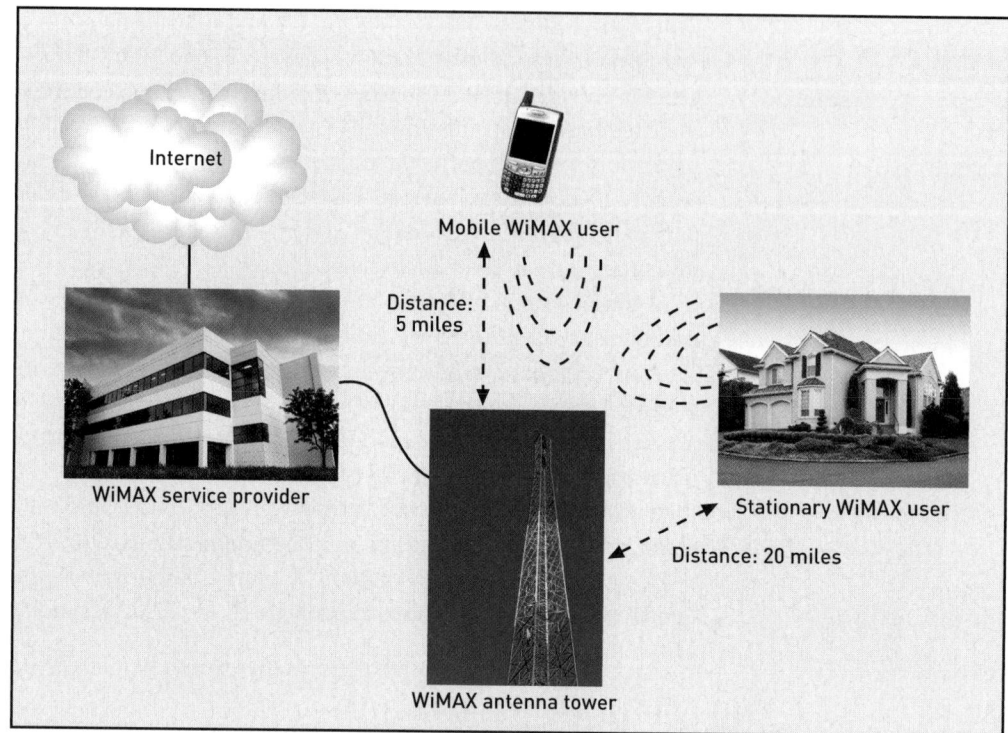

WiMAX is sometimes called a **4G** wireless system. This loose category usually indicates that the service is completely IP-based, will operate consistently at high speeds from 100 Mbps into the Gbps range, and will be oriented toward further convergence and seamless integration of previously disparate networks. Such capabilities would presumably support high-bandwidth applications such as mobile high-definition television, video chatting, and multimedia messaging. Whether a single wireless technology will rise to become the standard for seamless global connectivity remains to be seen.

CHAPTER SUMMARY

- Wireless multimedia devices are mobile communication and information platforms that provide a number of services, including mobile voice calls, electronic mail, text messaging, GPS, Web browsing, calendar applications, office applications, cameras, and media players.
- Wireless multimedia devices often have built-in capabilities to interface with multiple wireless network technologies, including cellular systems, Bluetooth technology, GPS systems, and Wi-Fi.
- Bluetooth wireless technology is a short-range digital wireless standard. It operates at an unlicensed radio frequency band and uses a transmission approach called spread spectrum technology.
- Frequency hopping spread spectrum (FHSS) divides a frequency band into small channels, with transmissions rapidly hopping between frequencies at rates of more than 1000 hops per second. This approach enables efficient spectrum usage, provides significant security, and reduces interference with nearby radio transmissions.
- Direct-sequence spread spectrum (DSSS) spreads the transmission across a range of frequencies using a pseudo-random code.
- Cellular systems are divided into smaller areas called cells. Each cell has its own base station antenna that transmits a radio frequency signal.
- Frequency reuse in cellular systems maximizes available frequencies by reusing them in nonadjacent cells.
- Most cellular networks use either Time Division Multiple Access or Code Division Multiple Access methods to facilitate channel access among numerous devices.
- The cellular industry categorizes networks into generations known as 1G (such as AMPS), 2G (such as GSM), and 3G, which includes CDMA2000 and UMTS.
- Some people refer to a fourth generation of wireless, 4G, and cite WiMAX as an example. WiMAX, also known as IEEE 802.16, is a formal networking standard developed by the IEEE for metropolitan area high-speed wireless service.

KEY TERMS

1G, 2G, 3G, and 4G	full duplex
access method	GSM
adaptive frequency hopping	half duplex
AMPS	handoff
base station	IEEE 802.15.1
Bluetooth	IS-95
CDMA	microcell
CDMA2000	mobile telephone switching office
cell sectoring	personal area network
cell splitting	piconet
cells	spread spectrum
control channel	subscriber density
cluster	TDMA
direct-sequence spread spectrum	traffic channel
FDMA	UMTS
frequency hopping spread spectrum	WiMAX
frequency reuse	

REVIEW QUESTIONS

1. What is a personal area network?
2. What part of the radio spectrum does Bluetooth use? Is this an unlicensed or licensed band?
3. Does Bluetooth require higher or lower power consumption compared to Wi-Fi?
4. Explain why adaptive frequency hopping in Bluetooth helps to avoid interference with nearby transmissions in a similar frequency band.
5. The reach of a cellular signal depends on what factors?
6. Explain the concept of frequency reuse.

7. Describe the general structure of a cellular telephone network. Draw a diagram if necessary to aid in your explanation.

8. Why do cellular providers implement microcells in urban areas?

9. What is a "handoff" in a cellular network?

10. Describe the difference between full duplex and half duplex communication.

11. Explain three ways to increase the capacity of a cell phone system.

12. Define *control channel* and *traffic channel* in the context of cellular telephony.

13. What is the general difference between 1G, 2G, and 3G wireless systems?

14. Describe the difference between FDMA, TDMA, and CDMA.

15. What is a Wi-Fi phone, and what limitations can you envision for this approach?

16. Describe some differences and similarities between Wi-Fi and WiMAX.

DISCUSSION QUESTIONS

1. Should users with dual mode GSM/Wi-Fi devices expect to be seamlessly handed off from a Wi-Fi network to a cellular network when moving out of the Wi-Fi network? How would this arrangement economically affect cellular service providers?

2. Discuss the advantages and disadvantages of having a single multimedia device that performs many different functions versus having individual devices that perform the same functions.

3. Various studies have alternately suggested and discounted a possible link between widespread wireless use and illnesses in people. Do some quick research and report your findings.

4. Wireless multimedia devices, including Bluetooth, cellular, Wi-Fi, and GPS, access many kinds of networks. What are the merits or drawbacks of a device that connects to multiple networks? Would it make more sense to have one network that interconnects wireless multimedia devices? Why or why not?

5. If a friend asked you to explain spread spectrum, how would you describe it in an understandable way? Can you think of any analogies to help describe it?

6. Will broadband metropolitan wireless services such as WiMAX live up to the hype? Will some new alternative emerge?

CASE PROJECT

Suppose you have taken a summer job with an organization to design and install a broadband wireless network in a developing country. By doing research on the Internet, try to find some initiatives that are implementing wireless solutions in developing countries. Answer the following questions:

1. What countries, regions, or towns have decided to embark upon wireless broadband projects?

2. What are the expectations and goals for what these broadband wireless networks can and cannot achieve? Is the wireless infrastructure enough? What else is necessary to achieve the project's objectives?

3. Does the area have a telecommunications infrastructure to which the broadband wireless network will connect, or will it be an entirely wireless solution?

4. Did you find any regulatory or legal impediments to the project? What are the potential legal impediments to wireless implementations?

5. What technology was implemented? Wi-Fi? GSM? Something else?

6. Was the initiative run by communities, governments, private industry, or nonprofit organizations? In your opinion, what are the possible advantages and disadvantages for the involvement of these different groups?

1G — First-generation analog cellular system such as the advanced mobile phone system (AMPS).

2's complement notation — A notation used to express negative decimal integers in binary form.

2G — Second-generation cellular; refers loosely to the first digital cellular approaches.

3G — Third-generation cellular; refers to digital cellular approaches with higher data rates than 2G systems.

4G — Fourth-generation cellular; a loose category indicating that the service is IP-based, operates at high data rates, and is oriented toward multimedia convergence and interoperability between heterogeneous networks.

802.11i — An IEEE-developed encryption standard for wireless LANs.

A

AAC — Short for Advanced Audio Coding; a lossy audio compression format.

absorption — The effect whereby a radio wave loses its energy due to certain materials in the atmosphere that act to absorb the energy of the radio wave.

access control mechanism — A procedure that specifies and enforces rules for when each device may transmit or receive information over a network.

access method — The approach for sharing a channel among numerous devices.

active tag — A radio frequency identification device that can emit radio waves through its antenna using energy from the battery when prompted by a tag reader.

adaptive frequency hopping — A wireless transmission approach used by the Bluetooth standard that avoids interference with other wireless devices.

address resolution — Domain Name System translation between alphanumeric domain names and the associated IP addresses required for routing information across the Internet.

agile software development — An approach based on developing software in multiple teams with a strong sense of interaction between them. The agile approach provides the customer with a limited version of the software product in a short amount of time.

AIFF — Short for Audio Interchange File Format; an uncompressed audio format associated with Apple computers.

aliasing — Occurs when a signal is sampled at a rate lower than the Nyquist sampling frequency during analog-to-digital conversion. The signal is said to be aliased into a new form when it is reconstructed.

alpha testing — The initial software testing phase; normally conducted by the software development team when only a limited version of the product is made available to the customer.

amperes — A unit that indicates the magnitude of electrical current.

amplitude — The magnitude of a signal at a given instant of time.

amplitude modulation — Also called AM; the amplitude of the carrier wave is varied in proportion to the amplitude of the analog information signal, resulting in the modulated waveform/signal.

amplitude shift keying — Also called ASK; binary amplitude shift keying varies the amplitude of the carrier wave according to the binary amplitude of the digital information signal.

AMPS — Short for advanced mobile phone system; a first-generation analog cellular standard.

analog device — A device that processes information in analog form.

analog information — Information that is presented in analog form.

analog signal — A signal that varies continuously over time and takes on an infinite number of amplitude values.

analog telephony adapter — Also called ATA; in VoIP implementations, serves as an analog-to-digital converter between an analog phone and an Ethernet network connection to an IP network.

analog-to-digital converter — A device used to convert signals in analog form to signals in digital form.

ANSI — The American National Standards Institute; a U.S. standard-setting institution.

Application layer — Layer 7 of the OSI reference model.

application proxy firewall — A firewall that filters information based on the application data itself, such as allowing HTTP traffic.

ASCII — Short for American Standard Code for Information Interchange; a standard used to express alphanumeric characters in binary.

ASF — Short for Advanced Systems Format; a Microsoft streaming media file format.

aspect ratio — The ratio of width to height in a display technology.

assembler — A specialized program used to convert assembly language into machine language.

assembly language — A low-level programming language whose instructions are the human-readable version of machine language.

asymmetric digital subscriber line — Also called ADSL; a type of digital subscriber line network access, provided over twisted pair cable, that supports asymmetrical data rates.

asynchronous — A term describing communications such as text messaging that do not require the synchronization between transmitter and receiver with respect to time.

Asynchronous Transfer Mode — Commonly called ATM; a wide area network alternative that formats information into 53-byte, fixed-length packets called cells.

attenuation — The loss of a signal's energy while traveling over a transmission medium.

audio signal — The physical signal that carries audio information.

authentication — The process of verifying a person's identity before granting network access.

automatic call distributor — Also called an ACD; telecommunications equipment that receives incoming calls and makes decisions about how to allocate the calls to available agents, or that initiates calls and assigns them to agents.

AVI — Short for Audio Video Interleaved; Microsoft's format for multimedia applications in the Windows operating environment.

B

bandwidth piracy — Refers primarily to the unauthorized access and use of unsecured wireless LANs.

base 2 — Also called the binary numbering system; a system comprised of only two symbols. Any binary number is expressed as a combination of these two symbols.

base 8 — Also called the octal numbering system; a system comprised of only eight symbols. Any octal number is expressed as a combination of these eight symbols.

base 10 — Also called the decimal numbering system; a system comprised of only 10 symbols. Any decimal number is expressed as a combination of these 10 symbols.

base 16 — Also called the hexadecimal numbering system; a system comprised of only 16 symbols. Any hexadecimal number is expressed as a combination of these 16 symbols.

base station — In wireless communications, the transmitting/receiving antenna and associated equipment that communicates with wireless devices in its geographical vicinity.

beta testing — A software testing phase in which the complete version of the software is supplied to the customer so it can be used for a specific purpose. The customer then uses the software and reports any bugs to the software development team. The customer essentially tests the software during the beta testing phase.

beta version — The version of the software supplied to the customer during beta testing.

binary code — A code made up of two symbols (1 and 0) called bits. Combinations of bits can be used to represent any form of information, including numbers, text, images, and sound.

binary coded decimal — An alternative standard used to express decimal numbers in binary. Each decimal number is encoded as a distinct 3-bit binary number.

binary digits — Also called bits. Bits are symbols (1 and 0) of the binary code.

binary language — The language of digital devices that is based on representing information in binary.

binary numbering system — A numbering system comprised of only two symbols: 0 and 1.

binary point — A symbol used within the binary equivalent of fractional decimal numbers. It may be thought of as the binary equivalent of the decimal point.

binary signaling — A method for sending digital information across a communication system; corresponds to sending data one bit at a time and switching between two discrete signal values, such as between two distinct voltages or light levels.

binary symbols — Same as binary digits or bits.

biometric identification — Authentication based on a person's unique physical characteristics such as fingerprints, facial features, voice patterns, retinal patterns, iris recognition, or DNA.

bit depth — In digital imaging, the number of bits allocated to represent the color information of each pixel.

bitmap image — A digital image representation in which each binary code corresponds to a pixel and can be mapped onto a specific part of the image.

bits — Same as binary digits or binary symbols.

block codes — A class of error-control codes that are based on adding redundancy to fixed lengths or blocks of data.

Bluetooth — Also known as the IEEE 802.15.1 standard; a wireless standard that uses an unlicensed (2.4 to 2.485 GHz) frequency range for geographically limited transmissions, such as those between a wireless headset and a cell phone.

BMP — Short for bitmap picture; one of Microsoft's image file formats; stores images in the form of a bitmap.

Border Gateway Protocol — A routing protocol usually known as BGP.

brightness resolution — In digital imaging, relates to bit depth, and consequently the number of different colors that are inherently present in the image.

broadband — In everyday terminology, the term indicates that a transmission medium has a high information-carrying capacity.

buffer — A term that relates to introducing a slight delay to even out the timing of received packets.

bug — An error in software.

byte — A grouping of 8 bits.

C

call center — A large facility staffed by agents who receive or initiate high volumes of incoming or outgoing telephone calls such as for catalog ordering, customer support, or telemarketing.

cathode ray tube — A tube in some computer monitors and traditional television sets that emits three electron beams whose strength and direction of travel is controlled electromagnetically by a beam deflection mechanism to follow a raster scan format. The electron beams paint a picture across the screen of the CRT display.

CCD array — Short for Charge-Coupled Device array; a CCD array is comprised of rows and columns of minute sensors that detect light and produce electricity proportional to the intensity of the light impinging on them.

CDMA — Short for Code Division Multiple Access; a spread spectrum-based cellular access technique that assigns each transmitting device a unique digital code.

CDMA2000 — A 3G cellular standard that uses the CDMA access method.

cell sectoring — A technique to increase cellular system capacity by using directional antennas radiating from the base station to sector individual cells into wedge shapes.

cell splitting — The process of dividing cells into smaller areas to increase the capacity of a cellular system.

cells — Cellular systems are divided into small geographical areas called cells, which are serviced by base stations.

central office — In the Public Switched Telephone Network, a service provider-owned and operated building that contains a large telephone switch and associated equipment, and that terminates local loop transmission facilities.

Centrex — In telecommunications, a service that provides businesses with functionality similar to a PBX, except the switch is owned and operated by the telephone company rather than residing at a customer location.

CERT — In the United States, the US-CERT is the federally run Computer Emergency Readiness Team, which protects Internet infrastructure and responds to security problems. Many CERT organizations exist around the world.

certificate authority — A third-party company that assigns public encryption keys and verifies online identities.

circuit switching — An architectural approach that establishes a physical, dedicated, end-to-end path through the network between a caller and receiver, and maintains the path for the entirety of the call.

cladding — In fiber-optic cable, the glass/plastic material surrounding the core.

clipping — Usually in voice communications, the small but noticeable loss of parts of a conversation.

coarse WDM — Short for coarse wavelength-division multiplexing; this approach multiplexes a small number of channels onto a fiber-optic cable.

coating — In fiber-optic cable, it surrounds the cladding to protect the fiber from its surrounding environment and to provide rigidity.

coaxial cable — A transmission medium consisting of a central, usually solid cylindrical core conductor made of copper and an insulating material surrounding the core called a dielectric. The structure is further encapsulated in braided metal strands that act as a shield. A tough cable jacket encloses the complete cable structure.

color filter array — A component used within a digital camera for capturing a color image that comprises small sections of red, green, or blue filters positioned directly on top of each sensor in a special pattern, usually called a Bayer pattern.

color gamut — The spectrum of colors a digital camera or scanner can capture, similar to an artist's palette.

command-line interface — A type of interface usually associated with operating systems that requires the user to enter commands to perform specific functions related to the operating system, such as deleting files, copying files, and so on.

compiler — A program used to convert high-level programming language instructions into machine language.

complex sound — A type of sound that is comprised of two or more frequency components.

compression — The use of mathematical algorithms and digital conservation approaches to decrease file sizes for more efficient storage or transmission.

computer forensics — A law enforcement specialization concerned with recovering computer evidence and detecting, preventing, and prosecuting crimes that have been committed using computers.

configuration management — An administrative responsibility for tracking and managing all the hardware and software associated with a network, and for managing any changes that occur to these architectural elements.

connectionless — In networking, a term indicating that a dedicated end-to-end connection is not established for the duration of transmission.

control channel — In cellular telephony, a channel reserved for performing control functions such as monitoring the location of a wireless device or establishing a call.

convergence — The integration of voice, music, video, images, and data on a single electronic platform.

convolutional coding — A powerful error detection and correction approach; instead of grouping data into blocks and applying redundancy to each block as in block coding, convolutional encoders take in a continuous stream of bits at their input and produce a continuous stream of encoded information at the output.

core — In fiber-optic cable, a cylindrical material made of glass or specialized plastic that constitutes the central portion of the cable. Light remains trapped within the core during fiber-optic communications.

corporate espionage — A term in network security that refers to the possibility of rival organizations seeking competitive information or, less frequently, sabotaging other systems.

coupler — In the context of optical communications, couplers are placed strategically between the light source and the fiber input and between the light detector and the fiber output. These devices efficiently couple light into and out of the cable so that light can be inserted at appropriate angles (greater than the critical angle) and to minimize insertion loss.

CPE — Short for customer premises equipment; the term historically refers to voice devices such as telephones or PBXs that are owned and controlled by a customer.

critical angle — In optical communications, if a light ray strikes the core/cladding boundary at an angle that is greater than the critical angle, the ray undergoes total internal reflection, where it is totally reflected from the boundary back into the core.

crosstalk — A form of interference between adjacent communication channels. The signal from one communication channel may interfere with a signal on another communication channel and hence create crosstalk if the signals are in close proximity to each other.

CSMA/CD — Short for Carrier Sense Multiple Access with Collision Detection; a nondeterministic LAN access control mechanism.

CSU/DSU — Short for Channel Service Unit/Data Service Unit; network equipment that serves as a physical interface to a digital transmission circuit and performs important functions such as signal formatting, timing, and some error control.

cyberterrorism — A network security threat that a terrorist attack could disrupt or disable a major network such as the Internet.

cyclic redundancy checking — A popular error detection technique based on performing a simple mathematical operation on the original bit stream to append a stream of bits to the end of a data block. The receiver can then apply the same mathematical operation to check for errors based on the stream of bits at the end of the data block.

D

dark fiber — A common term in the telecommunications and IT industry to describe a fiber-optic infrastructure that has been established but is not yet being used.

Data Link layer — Layer 2 of the OSI reference model.

data rate — The number of bits per second transmitted over a transmission line.

database management system — A system that manages the organization of information and allows for queries in response to requests from users or other software programs.

databases — A compilation of large amounts of information that can be organized, classified, searched, and retrieved in various ways.

DDoS — Short for Distributed Denial of Service, an attack that floods a single system with requests from thousands or more unwitting computers.

de facto standard — A standard that becomes dominant not through a formal process of collaborative effort, but gains popularity over time.

de jure standard — A standard developed through a formal, premeditated process, such as in a standards-setting organization.

decimal numbering system — A numbering system comprised of 10 symbols: 0, 1, 2, 3, 4, 5, 6, 7, 8, and 9.

decimal point — The symbol used within fractional numbers to segregate the fractional part of a decimal number from its integer part.

decompiler — A program used to convert machine language into a high-level programming language.

demodulation — The process of extracting information-carrying signals from modulated waves.

denial-of-service attack — A network attack flooding a targeted computer with so many requests that it cripples its functionality.

dense WDM — Short for dense wavelength-division multiplexing; this approach multiplexes a large number of channels onto a fiber-optic cable.

deterministic — An access method, such as in a LAN, that provides each network node with a predetermined and orderly opportunity to transmit information. The method is called deterministic because you can determine the maximum amount of time it takes information sent from any node to reach its destination.

device driver — Software that provides an interface between a computer and its peripheral components, such as printers and mice.

DGPS — Short for differential global positioning system; enables near-perfect calculations by including alternative measurements taken by designated earth stations.

digital audio broadcasting — A term associated with digital radio; offers high-fidelity music that is close to CD quality.

digital device — A device that processes information in digital form.

digital information — Information that varies discretely and is represented as a combination of bits.

digital light processing — Also called DLP; a projection technology based on Micro Electro Mechanical Systems (MEMS). DLP projectors employ chips called DMDs that comprise an array of tiny mirrors.

digital micromirror device — A typical DLP projector uses a chip called the digital micromirror device (DMD) that is comprised of an array of very tiny mirrors.

digital signal — A signal that carries digital information and varies discretely.

digital signature — User authentication via public key encryption.

digital subscriber line — Also called DSL; a WAN access alternative that connects a user's DSL router to the twisted pair cables installed as part of the traditional telephone network.

digital-to-analog converter — A device that can convert a digital signal into an analog signal.

direct sequence spread spectrum (DSSS) — A type of spread spectrum transmission approach that does not hop from one frequency to the next but spreads the energy of the transmission onto a large range of frequencies at any given time.

directional antenna — An antenna, such as in a microwave system, that radiates electromagnetic energy in a specific direction.

disassembler — A program that converts machine language instructions into assembly language instructions.

discrete — A term used to refer to the variation of a signal based on a finite number of values.

DMZ — Short for demilitarized zone. For network security reasons, companies sometimes establish a DMZ—a network segment outside a firewall-protected, internal, private network—to house networked resources such as a public Web server.

Domain Name System — Also called DNS; an enormous globally distributed database system that provides Internet address resolution.

dots per inch — Usually abbreviated as dpi; within the context of digital scanners, dots per inch essentially corresponds to the number of pixels per inch produced by a scanner in capturing an image.

dotted decimal format — A shorthand convention for representing 32-bit Internet addresses.

drop — In local area networking, a cable connecting a network node to a device such as a central switch or hub.

dynamic buffering — Buffering that introduces a delay in real time based on prevailing network conditions, in contrast to static buffering, which introduces a fixed time delay to even out the timing of packets.

E

EBCDIC — Short for Extended Binary Coded Decimal Interchange Code; an IBM standard used to express alphanumeric characters in binary.

edge router — A router that sits at the edge of a network and interconnects one or more LANs at a customer location to a wide area network.

EIA — The Electronic Industries Alliance; a trade organization that sets technology standards.

electrical insulators — Materials such as glass or plastic with tightly attached electrons within their atoms so that no electron flow occurs even if a voltage is applied to them.

electrical signal — A signal that is physically in the form of a varying electrical entity.

electromagnetic energy — A form of energy that wireless technologies employ for carrying information. Electromagnetic (EM) energy travels in the form of EM waves, such as radio waves, light waves (infrared, visible light, ultraviolet), X-rays, and Gamma rays.

electromagnetic spectrum — Types of EM waves, such as radio waves, light waves, X-rays, and Gamma rays, that vary within a range of frequencies. These waves are ordered according to their wavelength/frequency, resulting in a spectrum of waves called the electromagnetic spectrum. Frequencies within the electromagnetic spectrum extend from approximately 3 Hz to more than 1×10^{20} Hz.

electronic storage — A form of storage that is used to store bits electronically by storing varying levels of electrical charge corresponding to bit values.

electronic switching system — Known as an ESS; a large PSTN switch capable of connecting large numbers of subscribers and performing functions including multiplexing, pulse code modulation, messaging, VoIP, call waiting, caller ID, call forwarding, call tracing, and multiparty calling.

encoding standard — In the context of VoIP, protocols that transform analog signals into standard digital formats.

encryption — The algorithmic manipulation of data designed to make information unreadable to unauthorized entities.

encryption key — A number, or cipher, that is used to encode and decode information during encryption and decryption.

enterprise networking — Communications geared toward business information exchange among companies, customers, and suppliers.

error — A term used to refer to the incorrect reception of bits within transmission and storage systems. An error is said to occur if a 1 is received as a 0 or vice versa.

error control — The process of detecting errors and in some cases correcting them within transmission and storage systems.

error control coding — Techniques to encode a bit stream prior to transmission or storage for the purposes of error detection and/or error correction.

Ethernet — The most prevalent local area network standard.

Ethernet switching — In contrast to a shared Ethernet LAN, Ethernet switching assigns each device or group of networked devices a dedicated segment that is directly connected into an Ethernet switch, which then switches traffic to the receiving computer based on address.

EUI — Extended Unique Identifier network address space.

F

fault management — A network management function responsible for detecting, preventing, and solving a network outage or performance problem when it occurs.

FCC — The U.S. Federal Communications Commission.

FDDI — Short for Fiber Distributed Data Interface; a type of local area network configured in a dual-ring topology, using the token passing access method, and operating over fiber-optic cable.

FDM — Short for frequency-division multiplexing; a technique in which each channel is assigned a specific carrier frequency for transmission.

FDMA — Short for frequency division multiple access; an access method that divides available frequency into smaller frequency channels and assigns each device one frequency channel for transmitting information and one frequency channel for receiving information.

fiber connector — A device that connects one fiber segment to another with minimal loss of light.

fiber optics — The transmission of information via light over a glass or specialized plastic medium.

fidelity — The degree to which an audio signal that has undergone processing resembles its original form.

firewall — An access control device installed between a secure private network and a nonsecure public network to regulate access to and from the private network's resources.

flash card — A type of flash memory device that is shaped in the form of a small card.

flash drive — The drive used to read and write to flash memory devices such as flash cards.

flash memory — A type of electronic storage medium that may be in the form of a flash card or disk. Flash memory devices are highly popular for portable storage.

frame rate — The rate at which frames are displayed in a video.

frame relay — A connection-oriented wide area network service alternative.

frames — Within the context of networking, frames are segments of information represented by a string of ones and zeros that contain various components, including the actual information content, source and destination addresses, and other components necessary for correct transmission. Within the context of video technology, a frame refers to an individual image in a video.

frequency — The number of cycles a wave completes in one second.

frequency bands — Refers to the division of the radio spectrum into regions, or bands, such as very high frequency (VHF) and ultra high frequency (UHF).

frequency component — Within the context of sound, the term refers to each component that is found within complex sound.

frequency hopping spread spectrum (FHSS) — A wireless transmission approach that divides the allocated frequency band into smaller bands and rapidly hops between frequencies during transmission.

frequency modulation — Also called FM; the frequency of the carrier wave varies in proportion to the amplitude of the analog information signal.

frequency reuse — An approach in cellular telephony whereby a single frequency can be used simultaneously in different nonadjacent cells.

frequency shift keying — Also called FSK; binary frequency shift keying varies the frequency of the carrier wave according to the binary amplitude of the information signal.

frequency spectrum — A representation, usually in the form of a graph, that gives information on the number of frequency components and the frequency of each component within a signal.

full duplex — A communication approach that allows two devices to simultaneously transmit information in both directions.

G

GEO — Geostationary earth orbit satellite.

GIF — Short for Graphics Interchange Format; an image file format.

GPS — Short for global positioning system; a network of satellites and numerous earth stations that constantly monitor these satellites; provides positional information in three dimensions to GPS receivers.

graded index fiber — A fiber-optic cable whose core refractive index is nonuniform and varies gradually. The refractive index toward the core is usually the greatest, and it gradually diminishes toward the cladding.

graphical user interface — A type of interface usually associated with operating systems that does not require the user to enter commands to perform specific functions. Instead, graphical structures such as icons represent files and other entities; the user can manipulate these icons to copy files, delete files, and so on.

grayscale — An alternate term for black and white images.

GSM — Short for global system for mobile communications; an example of a second-generation mobile communication standard.

H

H.323 — A family of VoIP signaling protocols developed by the International Telecommunication Union (ITU).

hacker — In modern IT terminology, a person who gains unauthorized access to a computing resource, modifies or damages code, unleashes computer viruses or Internet worms, steals resources such as credit card numbers, or engages in bandwidth piracy.

half duplex — In half-duplex communications, two devices can exchange information over a single transmission medium in both directions, but each device must wait for the other device to finish transmitting before using the medium to transmit.

ham radio — Amateur radio that uses specifically assigned frequencies from the EM spectrum to connect people from all over the world. Ham radio operators must have a license.

handoff — In cellular systems, the transparent transfer of a call from the base station of one cell to the base station of another.

HD radio — Short for high definition radio; a radio transmission approach that supports very high-quality audio broadcasts.

HDTV — Short for high definition television; refers to a digital broadcasting approach that provides much higher resolution than traditional broadcasting formats.

hexadecimal — A term used to refer to the base 16 numbering system.

high data rate digital subscriber line — Also called HDSL; a type of digital subscriber line network access service provided over twisted pair cable.

high-definition television — Also called HDTV; digital television broadcasting with superior picture and sound quality.

high-level programming language — A programming language whose instructions are more abstract compared to low-level languages such as assembly language.

hop — In the context of a communications network, the number of hops measures the number of times a packet traverses various routers.

HTML — Short for Hypertext Markup Language, which is used to encode Web information in a standard format.

HTTP — An acronym for Hypertext Transfer Protocol, a standard used to establish and maintain communications over the Internet between a Web server and a client.

hybrid fiber coaxial — Also called HFC; an arrangement frequently used by cable service providers to convert signals from coaxial cable into optical signals at the curb to be carried across fiber-optic cable to the service provider's facilities.

I

IANA — The Internet Assigned Numbers Authority; the organization with centralized responsibility for the IP address space, including both IPv4 and IPv6 addresses.

ICANN — The Internet Corporation for Assigned Names and Numbers; an Internet governance organization responsible for administering Internet domain names and numbers.

IEEE — The Institute of Electrical and Electronics Engineers; an organization involved in standards setting and other tasks.

IEEE 802.15.1 — The formal name of the Bluetooth short-range wireless communication standard.

IETF — The Internet Engineering Task Force; an organization that establishes Internet standards.

information — A fact or series of facts that carry meaning.

information technology — A system of hardware and/or software that captures, processes, exchanges, stores, and/or presents information, using electrical, magnetic, and/or electromagnetic energy.

Inmarsat — International Maritime Satellite Organization.

integrated circuit — An electronic device comprised of transistors, interconnections, and other devices packaged together; also referred to as chips.

Internet backbone — The global collection of high-speed transmission lines that carry the bulk of Internet traffic.

Internet exchange point — Also called IXPs; the switching locations that interconnect Internet service provider networks.

Internet Message Access Protocol — Usually abbreviated as IMAP; an electronic mail retrieval protocol.

Internet Protocol — Also called IP; an essential Internet standard that specifies the addressing and routing of information from a source to a destination over the Internet or other IP network.

interoperability — The ability of heterogeneous equipment and software made by different manufacturers to work together.

interpreter — A program that converts each line of source code into object code, so that each line of object code can be executed before going on to the next instruction.

IP address — Short for Internet Protocol address; a globally unique 32- or 128-bit network number required for Internet connectivity.

IP PBX — An IP-based Private Branch Exchange.

IP phone — In the context of VoIP, a phone that can directly connect to an IP network without an adapter because it internally converts the telephone audio signal into a digital format for network transmission.

IPv4 — Short for Internet Protocol version 4; a prevailing Internet standard that specifies 32-bit Internet addresses.

IPv6 — Short for Internet Protocol version 6; a new Internet standard that specifies 128-bit Internet addresses.

IS-95 — Interim Standard 95; a 2G cellular standard based on CDMA.

ISO — The International Organization for Standardization; an international standards-setting organization made up of national standards bodies.

iterative model — A software development model that relies on breaking down the development process into smaller segments and developing each segment in a limited amount of time. The segments may not necessarily be completed in a specific order. The results are combined to fine-tune the final software product.

ITU — The International Telecommunication Union; an international standards-setting organization.

J

jacket — In fiber-optic cable, it surrounds the cladding to protect the fiber from its surrounding environment and to provide rigidity.

Java — A popular high-level programming language used to develop software.

Java applet — A small program that is frequently used to create interactive Web pages.

jitter — Degradation in call quality that results from timing problems, such as variations in packet arrival time.

JPEG — Short for Joint Photographic Experts Group; a family of compression standards and one of the most predominant image file formats.

K

key length — In encryption, the number of bits in the encryption key.

L

LAN — Short for local area network; a network that spans a confined geographical distance, such as a building or home.

LAN operating system — Software that manages and controls networked access to LAN resources such as printers, files, applications, and messaging services; it also provides some network management and provides security by managing user directories, monitoring remote LAN access, and incorporating encryption and other security features.

latency — Within the context of computer networking, a term that relates to the time delay between the transmission and reception of packets over a network.

layers — In the OSI reference model, conceptual categories that group protocols by function.

LD — Short for Laser Diode; a device that emits light with the application of electricity. It is usually employed within optical communication systems.

LED — Short for light emitting diode; similar to small light bulbs that emit light at an intensity based on the electricity applied to them.

LEO — Short for low earth orbit satellite.

line-of-sight — Refers to the wireless transmission requirement for a clear, unobstructed path between two microwave antennas.

Linux — A type of operating system with open source code that may be freely downloaded from the Web.

liquid crystal display — Also called LCD; display technology that uses liquid crystal materials whose properties can be varied by electrical signals to display an image.

local loop — In the Public Switched Telephone Network, the transmission facilities between a demarcation point at a customer location and a local telephone company switch at a central office.

logical topology — Dictates how packets of information logically flow among nodes of a local area network.

lossless compression — A compression technique that does not result in loss of fidelity.

lossy compression — A compression technique that results in loss of fidelity.

low-level programming language — A programming language whose instructions are primitive compared to high-level programming languages.

M

M4A — An audio-only variation of the MPEG standard.

MAC — Media Access Control network address space.

machine language — A programming language whose instructions are solely in the form of ones and zeros; also referred to as machine code.

magnetic storage — A form of storage used to store bits magnetically by varying magnetic field orientations corresponding to bit values on a magnetic medium.

magneto-optical storage — A form of storage used to store and retrieve bits using a combination of magnetic and optical energy.

markup language — A language comprised of special codes used to define how any document or resource is displayed to a user.

M-ary signaling — A form of signaling in which M different signal levels may be used to transmit $\log_2 M$ groups of bits with each signal level.

mechanical storage — A form of storage used to store bits mechanically by creating physical effects such as indentations or holes on a medium corresponding to bit values.

mechanical waves — Waves formed due to transfer of mechanical disturbances of particles such as air or water molecules.

media-free information — A term used to describe the storage of information across computer networks that can be accessed online from anywhere in the world.

megapixel — A term used to refer to approximately one million pixels.

memory module — A printed circuit board that carries memory chips such as RAM chips.

memory slot — A slot residing on the motherboard into which memory modules are connected.

MEMS — Micro Electro Mechanical Systems; technology that involves the manufacture of extremely small components such as gears, mirrors, and other parts.

MEO — Medium earth orbit satellite.

metallic conductor — A material such as a copper wire that comprises atoms with loosely attached electrons or negatively charged particles around their nuclei.

microelectromechanical system — See *MEMS*.

microprocessor — Also called the central processing unit; the chip that is considered the main processing chip of a computer.

mnemonic — The human-readable version of each machine language instruction.

mobile telephone switching office (MTSO) — In cellular systems, the central office that interconnects surrounding base stations, performs control functions such as channel assignments, and provides an interface between base stations and the Public Switched Telephone Network.

mobility — An increasingly important trend in IT that provides users with the capability of roaming around and being "unplugged."

Modified Final Judgment — Also called MFJ; the 1980s antitrust settlement agreement between the U.S. Department of Justice and AT&T.

modulation — The process of superimposing information signals onto a carrier wave such as a radio wave or a light wave.

molecular storage — A form of storage that relies on storing information at the molecular level.

Moore's Law — An observation that the number of devices on an integrated circuit will double every 18 months.

motherboard — The main printed circuit board of a computer that carries the microprocessor, memory chips, slots, and other devices.

mouse port — An external port used to interface a mouse to a computer.

MOV — A video format developed by Apple Computer and primarily associated with Apple's QuickTime media player.

MP3 — A lossy digital compression and formatting standard commonly used to reduce the size of audio files.

MPEG — Abbreviation for Moving Pictures Experts Group; a family of lossy compression standards for moving images and associated audio.

MPLS — Short for multiprotocol label switching; MPLS is designed to simultaneously support many types of WAN traffic and can transport IP packets, frame relay frames, and ATM traffic.

multicore computer — A computer comprised of a microprocessor with multiple cores.

multimedia — A term used to describe the integration of information in the form of text, numbers, sound, images, and video.

multi-mode fiber — Optical fiber with a relatively large core diameter, ranging from tens to hundreds of micrometers.

multi-mode graded index — Multi-mode fiber-optic cable whose core refractive index is nonuniform and varies gradually.

multi-mode step index — Multi-mode fiber-optic cable with a uniform refractive index throughout its core.

multiplexing — A technique to enable a single transmission line to simultaneously transmit multiple information-carrying signals.

multiprocessor computer — A computer comprised of multiple microprocessors.

N

NAND gate — A type of logic gate that performs the Not AND operation.

nanotechnology — Technological devices so small they are measured in nanometers.

narrowband — In everyday terminology, a term that indicates a transmission medium has a low information-carrying capacity.

Net neutrality — The general advocacy of technical nondiscrimination on the Internet.

network address translation — The conversion of a private IP address into a shared public IP address before the information is sent over the Internet.

network interface controller — Also called a NIC; a small card (printed circuit board) that provides the physical interface to a network medium or wireless LAN and that supports an addressing system critical to the LAN's operation.

Network layer — Layer 3 of the OSI reference model.

network operating system — An operating system that can support computer networking.

network port — A port used to interface a computer to a network.

network protocol — Specifications and rules that enable the exchange of information over a network.

network security — The protection of computing resources from hackers, viruses, worms, spam, protocol vulnerabilities, terrorism, natural disasters, and other potential threats.

network service provider — A company that provides networking services such as Internet access.

node — A device connected to a network.

noise — An effect that disrupts information; an unwanted occurrence that interferes with a signal and consequently the information carried by the signal.

noisy signal — A signal that includes unwanted noise.

nondeterministic — An access method, such as in local area networking, that allows any device to transmit a packet at any given moment, provided that no packets are already being transmitted over the network.

nonvolatile memory — A type of memory that retains its contents even if power to the computer is switched off; also called permanent memory.

NOR gate — A type of logic gate that performs the Not OR operation.

NOT gate — A type of logic gate that produces the inverse of its input.

notebook — A type of portable computer that is also called a laptop.

Nyquist rate — The minimum rate at which an analog signal should be sampled during the analog-to-digital conversion process so that it can be perfectly reconstructed back into its original form.

Nyquist sampling theorem — The theorem that the minimum number of samples per second required to perfectly reconstruct an analog signal should equal at least twice the value of the difference between the signal's highest and lowest frequency component.

O

object code — A program in a form that is directly executable by the computer.

obsolescence — An information technology trend in which equipment or technology becomes outdated over time and is no longer used.

octal — A term used to refer to the base 8 numbering system.

Ohm's Law — Defines the relationship between voltage (V), current (I), and resistance (R) as $V = IR$.

ohms — A measure of electrical resistance.

omnidirectional antenna — An antenna that can transmit or receive electromagnetic energy in all directions.

open source code — A term used to describe software that anyone can access, view, and modify.

open standard — A standard that allows diverse computing environments to interoperate by providing published guidelines for how to format and exchange information; sometimes also defined as one that is developed in a forum that allows open participation or that is available to license under reasonable and nondiscriminatory (RAND) licensing terms.

operating system — A piece of software that performs various functions, including providing a user-friendly interface to the computer, supporting computer networking, managing memory, hardware, and directories, and so on.

optical communication — Refers to the technology that is concerned with the transmission of light via fiber-optic cable or other optical medium to carry information from one point to another.

optical computing — A form of computing that relies on the use of light for processing and transmission of information.

optical interconnect — An approach that relies on the use of light for interconnecting devices such as chips or boards to each other instead of metallic wires.

optical medium — A physical medium that is capable of transmitting light. Examples include transparent glass or plastic and water.

optical signal — A signal that is physically in the form of a varying optical entity.

optical storage — A form of storage that is used to store bits optically by varying surface reflectivity corresponding to bit values on an optical medium.

OR gate — A type of logic gate that performs the OR operation.

OSI reference model — A conceptual framework to understand protocol functions and help organize the large number of IT protocols.

output device — A device that is used to obtain an output from a computer. Examples include printers and monitors.

outsourcing — The transfer of labor, such as programming, call center telemarketing, or IT support functions, to an external party. In the context of information technology, outsourcing often involves "offshoring" to workers in other countries such as India and China.

overhead — Within the context of computer networking, overhead includes everything transmitted over a network that is superfluous to the bytes of actual content. For example, start and stop bits, network addresses, and control information would be considered overhead.

P

P2P — Short for peer-to-peer; refers to file-sharing architectures in which files are distributed on multiple peer devices rather than stored in a centralized system.

packet — A small segment of information routed over a packet-switched network.

packet-filtering firewall — A firewall that intercepts packets and makes access control decisions based on their header contents, including the source IP address, destination IP address, source port, and destination port.

packet loss — The phenomenon of packets being dropped on a network that is overwhelmed with traffic or as part of a signal that is degraded during transmission. A dropped packet never reaches its destination and must be retransmitted.

packet switching — A network switching approach that breaks information into smaller segments called packets, sequences them, and routes each packet individually through the network, usually over the most efficient or most available path. Each individual packet may reach its destination across a different route and at a different time compared to the other packets. The packets are then reassembled at the destination in their correct sequence.

padding — Extra bits added to an Ethernet frame or other frame format to bring the bit length of the frame to the minimal required length.

parallel ATA — Short for parallel Advanced Technology Attachment; a standard for attaching devices such as hard drives over parallel connections to the computer via special connectors on the motherboard.

parallel bus — A bus comprised of multiple parallel lines capable of carrying bits in parallel form from one point to the other, such as between the CPU and memory chips.

parallel port — A port used to interface a device that requires a parallel connection to the computer. Such a device may include a printer.

parallel wire — A transmission medium constructed by grouping together several wires in parallel, each surrounded by an electrical insulating material such as plastic. Another insulating material, called the jacket, usually holds the insulated wires together.

parity bit — A bit added to the end of a data block to perform error detection.

passive tag — A radio frequency identification device that emits radio waves only when induced by a radio frequency identification reader.

patch management — In network security, the process of learning about product vulnerabilities and the availability of vendor-supplied software patches, and instituting procedures for systematically upgrading systems with these patches.

PBX — Short for private branch exchange; a telecommunications switch that is similar to a central office switch, except that it is privately owned and managed by a business, university, or other organization.

PCB — Short for Printed Circuit Board; a board made of resin or other materials that carries the microprocessor and many other chips, including their interconnections.

PCI bus — Short for Peripheral Component Interconnect bus; a bus that carries data between devices such as network adapters, sound cards, and other parts of a computer.

PCI-E bus — Short for Peripheral Component Interconnect-Express bus; a faster version of the PCI bus.

PCI/PCI-E slot — A slot on the motherboard used for interconnecting peripheral devices such as network adapters and sound cards to a computer.

peering agreement — An arrangement between service providers about sharing costs and guaranteeing performance at shared network switching points.

perceptual coding — The process of compressing information in consideration of the limitations of human senses.

performance management — An administrative function responsible for ensuring that a network is performing adequately for the applications and users it supports.

period — The time a wave takes to complete a single cycle.

peripheral — A hardware component that is not vital to a computer's operation, but is attached to the computer to enhance its capabilities.

permanent memory — Also called nonvolatile memory; a type of memory that retains its contents even if power to the computer is switched off.

personal area network — Also called PAN; a network that transmits within a range of 1 to 100 meters and sometimes provides services for a single user.

personal firewall — Access control software installed on a computer to detect and prevent intrusions, block hackers from accessing content, monitor incoming and outgoing traffic, and perform automatic updates to address the latest security holes, worms, and viruses.

phase difference — A term that describes the alignment of two waves with respect to each other in time.

phase modulation — A modulation process whereby the phase of a carrier wave is varied in proportion to the amplitude of the analog information signal.

phase shift keying — Also called PSK; binary phase shift keying varies the phase of the carrier wave according to the binary amplitude of the digital information signal.

phishing — A social engineering technique that uses e-mail or text messages to coerce users into relinquishing sensitive information such as a password or bank account number.

phonograph — An early analog audio recording and playback device.

photodiode — A device that converts light into electricity. The strength of the electrical signal it generates is proportional to the intensity of light impinging on it.

photophone — A nineteenth-century device constructed by Alexander Graham Bell that demonstrated how sound could be transmitted via an optical beam traveling through free space.

phototransistor — A device that converts light into electricity.

Physical layer — Layer 1 of the OSI reference model.

physical topology — A term referring to how multiple devices (often called network nodes) are physically connected to each other over a local area network such as in a ring, star, or bus.

Piconet — A group of Bluetooth connected devices.

pit — A structure found on the surface of a compact disc that is used to store a single bit of information.

pitch — A measure of the tonal quality of sound.

pixel — Short for picture element; the elementary unit of a digital image.

pixels per inch — Abbreviated to ppi; essentially corresponds to the number of pixels per inch usually produced by a scanner or digital camera in capturing an image; another term for dpi.

pixel pitch — Also called dot pitch; an indicator of display quality relating to the space between pixels on a digital display device.

plasma display — A high-quality display technology that uses an ionized gas trapped inside cells between two glass plates.

PNG — Short for portable network graphics; a lossless compressed image file format that stores images in the form of a bitmap.

port — A component used to interface peripherals to a computer.

Post Office Protocol — An electronic mail retrieval standard usually abbreviated as POP.

Presentation layer — Layer 6 of the OSI reference model.

private key encryption — An encryption approach whereby the sender and receiver use the same key to encrypt and decrypt information.

programmable device — A device that can be directly programmed using software.

programming language — A language used by programmers to write software.

projector — A display device that projects images onto a screen using various optical and electronic components.

proprietary protocol — A closed standard not available for other product manufacturers to create competing products.

protocol suite — A group of protocols such as TCP/IP.

protocols — Agreed-upon rules that specify how information is formatted, exchanged, and interpreted between nodes over a network.

PSTN — The Public Switched Telephone Network; the traditional system of transmission lines, switching equipment, standards, and conceptual approaches that enable global telephone interoperability.

public key encryption — An encryption technique in which users possess two keys: a private key that no one else can access and a public key that is accessible to anyone. Sending an encrypted message requires looking up the recipient's public key and encrypting the message according to this key. Then the recipient uses a private key to decrypt the message.

pulse code modulation — Also called PCM; the three-step process used to convert analog audio signals into digital.

pure sound — A sound that has a constant pitch and that varies sinusoidally when captured and plotted in the form of a graph.

Q

QoS — Short for quality-of-service; a network performance metric measuring characteristics such as latency, dropped packets, and network availability.

quantization error — A term used within the context of the analog-to-digital conversion process to indicate the difference between the actual value of a sample and its rounded-off value.

quantum computing — A form of computing that is based on quantum physics.

R

RADAR — Short for RAdio Detection And Ranging; a radio wave remote sensing technique used in navigation, weather forecasting, mapping, imaging, space exploration, and a variety of military and research applications.

radio frequency signal — A signal that is physically in the form of a radio frequency wave.

radio spectrum — The frequency range of the EM spectrum between approximately 3 Hz and 300 GHz.

RAM — Short for Random Access Memory; RAM is classified as volatile memory because its contents are lost when power to a computer is terminated.

random access — An access method used to retrieve information from any part of a storage medium without having to serially access the contents of the medium from start to finish.

raster scan — In a CRT display, a procedure that forms an image on the screen one small area at a time by scanning the screen from side to side and top to bottom/bottom to top.

RC4 — A cipher used in SSL encryption.

Real-time Transport Protocol (RTP) — A VoIP transport protocol defined by the Internet Engineering Task Force (IETF).

Red Book standard — A standard format used to store files such as music on a compact disc.

refraction — The bending of light when it crosses the boundary from one optical medium to another with different refractive indices.

refractive index — Also called the index of refraction (IOR); a property of an optical medium that affects the speed of light traveling through that medium. The refractive index also affects the degree to which a light ray will be bent as it passes from one optical medium to another.

registers — Small areas usually within the CPU for temporarily storing information such as instructions and data.

repeaters — Electronic devices placed at certain intervals over a transmission line to amplify weak signals and relay them along the transmission line. Repeaters enable long-distance communication and are commonly employed in transoceanic communication systems.

resistance — The extent of a material's inherent property of opposing the flow of current through the material; measured in Ohms.

RFC — The "Request for Comments" series documenting Internet standards.

RFID tags — Short for radio frequency identification (RFID) tags; a small radio frequency emitting device used for identifying and tracking shipping containers, pets, and sometimes even people.

RGB additive color model — Short for Red Green Blue additive color model; specifies that any color of light can be created by combining various proportions of the primary colors red, green, and blue.

RISC — Short for Reduced Instruction Set Computer; an approach used to design microprocessors with reduced complexity for performing a particular function using a small number of instructions.

rogue access point — A wireless LAN access point established in a business in an ad hoc manner outside the purview of network administrators.

ROM — Short for Read Only Memory; a type of memory classified as nonvolatile because its contents are not lost when power to the computer is terminated.

root name server — Part of the Internet's Domain Name System (DNS), root servers maintain the master file tracking the top of the hierarchical DNS namespace.

root zone file — The master file listing the names and IP addresses of the official DNS servers for all top-level domains.

router — An intelligent switching device that determines how to direct (or route) a packet across a network, based on the packet's destination address and network conditions.

routing protocol — Protocols, such as Border Gateway Protocol, that enable routers to share network changes that are reflected in updates to router tables.

routing table — A database that a router uses to "look up" information necessary to direct a packet in the most efficient way toward its destination.

RTP Control Protocol — In VoIP implementations, a standard that provides some quality-of-service and session management.

S

samples per inch — Abbreviated to spi; essentially corresponds to the number of samples (such as pixels) per inch produced by a scanner in capturing an image.

SAR — Short for Synthetic Aperture RADAR; a technology primarily used to map the surface of the earth.

SATA — Short for Serial Advanced Technology Attachment; a standard for attaching devices such as hard drives over serial connections to the computer via special connectors on the motherboard.

satellite Internet — A technology that enables Internet access via satellite.

satellite radio — The transmission of digital audio signals via satellite to terrestrial listeners.

satellite television — The transmission of digital television programming via satellite to a subscriber's dish antenna.

satmodem — A modem used for communicating directly with a satellite, usually for Internet access.

scattering — Occurs when a radio wave encounters an object that is close to the dimensions of its wavelength, whereby the energy of the radio wave is scattered across different directions.

script — A short program that other applications can interpret and execute.

scripting language — A language used to develop scripts.

SCSI — Short for small computer system interface; a high-speed standard for attaching devices such as hard drives to a computer.

sector — A part of a hard disk that is formed as a result of formatting the hard disk. Information on a hard disk is stored in an organized manner within sectors.

Secure Sockets Layer — Usually referred to as SSL; an encryption protocol; also called Transport Layer Security (TLS).

security hole — An inherent software vulnerability, usually in an operating system, Web browser, or other application.

security management — A vital IT function responsible for network access control, user authentication, information assurance, firewall management, and critical infrastructure protection.

sequential access — A form of access method that is used to retrieve information from a storage medium by having to serially access the contents of the medium. The entire medium is scanned until the desired information is reached with serial access.

serial bus — A bus comprised of serial lines capable of carrying bits in serial form from one point to the other, such as between the CPU and the keyboard.

serial port — A port used to interface a device that requires a serial connection to the computer. Such a device may include a keyboard.

server — A type of computer that supplies resources including files and various services to other computers called clients.

service pack — A large software patch that addresses significant security weaknesses in a product.

session — A term used to denote the process of information exchange between devices.

Session Initiation Protocol — Known by the acronym SIP; a VoIP signaling protocol developed by the IETF.

Session layer — Layer 5 of the OSI reference model.

Shannon-Hartley capacity theorem — In digital systems, the maximum number of bits per second (channel capacity, or C) that can reliably be carried over a channel depends on the bandwidth B (expressed in Hz) of the channel and a unitless ratio called the signal-to-noise ratio (SNR) as follows: $C = B \log_2(1+SNR)$, where SNR is the ratio of signal power to noise power in the channel.

shielded twisted pair — A cable that uses the same fundamental wire configuration as UTP, but adds a shield surrounding the twisted pairs to further reduce the amount of EMI on the signal.

signal — A term used to refer to the physical variation of a form of energy such as electricity or light.

signal bandwidth — The difference between the highest and lowest frequency components of a signal.

signal dynamic range — Within the context of audio digitization, the range of amplitudes between which an analog audio signal varies.

signaling protocols — In VoIP, a set of protocols that perform signaling functions such as assessing user availability, ringing the destination device, establishing a call, and terminating a call session.

Signaling System 7 — Also called SS7; a global signaling protocol for establishing, managing, and terminating a call over a digital network.

signal-to-noise ratio — The ratio of signal power to noise power in a channel.

simultaneous masking — A perceptual phenomenon of filtering out less prominent information.

single-mode fiber — An optical fiber with a very small core diameter, on the order of a few micrometers (μm).

single-mode graded index — Single-mode fiber-optic cable whose core refractive index is nonuniform and varies gradually.

single-mode step index — Single-mode fiber-optic cable with a uniform refractive index throughout its core.

sinusoidal — A term used to describe the variation of a signal according to the sine function.

SMTP — An electronic mail standard; also called Simple Mail Transfer Protocol.

social engineering — Techniques to make users take some action that simulates the actual effects of a virus, such as forwarding a hoax e-mail that consumes resources and reduces worker productivity.

soft phone — An end user's VoIP device usually consisting of a headset and VoIP software installed on a computer.

software — Traditionally defined as a series of instructions written by computer programmers in a language that people can understand and that the computer can translate into binary.

software development process — The process used to develop software.

software patch — A piece of code usually developed by a product vendor to address a software problem or to upgrade software.

SONET — Synchronous Optical Network; a North American transmission standard for multiplexing signals over fiber-optic cable.

sound port — A type of port that connects peripherals including microphones, headphones, and speakers to a computer, allowing users to listen to music and other audio.

sound wave — A mechanical wave of energy produced due to disturbances of air molecules.

source code — The instructions of a program.

spam — A term used to refer to unsolicited, unwanted e-mails that flood Internet mailboxes.

spam over Internet telephony — The problem of unwanted solicitations over VoIP systems; abbreviated as SPIT.

spatial resolution — In digital imaging, an indicator of the number of pixels per unit length (such as inches or centimeters) in an image.

spectrum analyzer — A device used to produce the frequency spectrum of a signal.

spinning disk filter — A circular rotating filter positioned on a CCD array.

spiral model — A software development model that is heavily concerned with risk and tries to minimize it during every stage. The development team tries to find alternatives to reduce risk before proceeding with the next phase.

splicing — The bringing together, or fusing, of two cables.

spoofing — The process of an unauthorized user or system masquerading as a legitimate user, system, or IP address.

spread spectrum — A wireless technique that distributes a transmission across a range of frequencies rather than a single frequency. Types of spread spectrum include direct-sequence spread spectrum (DSSS) and frequency hopping spread spectrum (FHSS).

Sputnik — The first successfully launched satellite; launched in 1957 by the Soviet Union.

stack — Refers to a protocol implementation; vendor-specific software or hardware that adheres to protocol specifications.

stateful inspection — A packet-filtering approach that makes access control decisions based on expected traffic patterns, such as allowing an incoming response from an IP address after an outgoing request was made to that address.

static buffering — The introduction of a fixed time delay to even out the timing of received packets.

static RAM — A type of RAM whose contents do not have to be periodically refreshed. The contents of static RAM are retained as long as power is supplied to the computer.

statistical time-division multiplexing — A form of time-division multiplexing that allows for dynamic allocation of time slots based on the activity of incoming signals to be multiplexed.

step index fiber — Fiber-optic cable with a uniform refractive index throughout its core.

step size — A term that is associated with the analog-to-digital conversion process; its value is directly affected by the range of values that the analog signal varies between and the number of bits assigned per quantized sample during the third step of the analog-to-digital conversion process. In essence, step size = $\text{Range}/2^{\text{number of bits}}$.

storage media — Physical media used to store information using various forms of energy.

store and forward system — A network approach, such as those used by e-mail systems, that does not require the simultaneous presence of senders and receivers because information is stored on a server.

streaming multimedia — Refers to the ability to view multimedia information in near real time as it is transmitted over a network, in contrast to being able to view the content only after downloading the entire multimedia information.

streaming video — Refers to the ability to view video in near real time as it is transmitted over a network, in contrast to being able to view the content only after downloading the entire video.

subscriber density — In cellular systems, the number of users requiring service in a given geographical area.

supercomputer — A type of computer that is extremely fast and expensive and that can process vast amounts of information within a very short amount of time.

superparamagnetic effect — An effect that causes random fluctuations in the magnetic field if bits are packed too closely to each other on a magnetic storage medium.

surge protection system — A system that is used to protect a computer against sudden surges in electricity supplied from a public electrical power outlet.

S-video port — A special type of fast port that connects a computer to a display device such as a monitor.

symmetric digital subscriber line — Also called SDSL; a type of digital subscriber line network access service provided over twisted pair cable that supports symmetric data rates.

Synchronous Digital Hierarchy — Also called SDH; an international transmission standard for multiplexing over fiber-optic cable.

synchronous — A term describing communications that require the synchronization between transmitter and receiver with respect to time.

system bus — An important parallel bus on the motherboard that connects the CPU and main memory.

system software — Software such as the operating system that supports the various operations of the computer. Compilers, assemblers, and device drivers are also traditionally classified as system software.

T

tags — Special codes traditionally used with markup languages to encapsulate text for the purposes of formatting it and controlling how it is displayed.

T-carrier system — The standard method for multiplexing over the copper cables of the traditional telephone network.

TCP/IP — Acronym for Transmission Control Protocol/Internet Protocol; a family of protocols used for Internet connectivity.

TDMA — Short for time division multiple access; an access mechanism that allows multiple devices to share a channel by dividing a single frequency into time slots and allocating time slots to each transmitting device.

Telecommunications Act of 1996 — An overhaul of U.S. telecommunications law designed to promote greater competition, allowing any service provider to compete in any sector of the telecommunications market.

telephone port — A type of port that connects a telephone line to a computer for dial-up connections via a modem.

temporary memory — Also called volatile memory; a type of memory that loses its contents after power to the computer is switched off.

thermal noise — A type of noise that arises from heat-induced agitation of electrons in a conductor material.

thin client — A type of computer that does not have the regular hardware typically found on other computers, such as a hard disk or other disk drives; it also has a scaled-down version of an operating system stored on a special memory chip.

TIFF — Short for tagged image file format; a digital image file storage format.

time-division multiplexing — Also called TDM; an approach that divides a channel based on time slots, allocating different time slots for each signal.

TLD — Short for top-level domain; a suffix (e.g., .com, .edu, .org) at the top of the Internet's DNS hierarchy.

token — A specific bit pattern transmitted from one device to another in sequential order in a token ring network.

token passing — A deterministic LAN access mechanism in which each network node is given a predetermined, orderly, and sequential opportunity to transmit information when it possesses a token.

token ring — A LAN standard, defined in the IEEE's 802.5 specification, that uses the token passing access method and is configured in a ring logical topology.

total internal reflection — In optical communications, it occurs when a ray of light traveling through one medium hits another medium of lower refractive index and reflects back into the original medium if the angle of the incident ray is greater than the critical angle.

track — The physical path on the disk where bits are stored. Tracks may be circular in form as on a hard disk or spiral as on a compact disc.

traffic channel — In cellular telephony, the channel that carries the actual information of text messages, voice calls, or video streams.

transducer — A device that converts one form of energy to another, such as light into electricity, electricity into sound, and so on.

transistor — An electronic switch that is the elementary component of all digital devices. A transistor can be controlled electronically to be switched on or off.

translator — A program used to translate programs written in one programming language to another.

Transmission Control Protocol — Also called TCP; part of the TCP/IP protocols underlying the Internet.

Transmission Control Protocol/Internet Protocol — Usually referred to as TCP/IP, a family of networking protocols that enable interoperability between heterogeneous devices over the Internet.

transmission media — Physical media through which signals pass.

Transport layer — Layer 4 of the OSI reference model.

Transport Layer Security — Commonly called TLS; a common standard for encrypted, end-to-end Internet communications.

transport protocols — In VoIP systems, they handle the transmission of a voice call between endpoints.

triangulation — The basis of deriving a location via GPS; positional information can be obtained by means of triangulation if the distance from at least three reference points is known.

Trojan horse — A virus cloaked in a legitimate program or file.

true color — In digital imaging, a bit depth of 24 is usually referred to as true color.

trunk — In the Public Switched Telephone Network, longer-distance transmission facilities such as those between switching centers.

truth table — A table that conforms to Boolean logic and conveys the values of all the inputs and outputs of a particular logic gate.

tunable laser — A device that can be adjusted to emit at a variety of wavelengths of light.

tunneling — A technique that creates a virtual, secure information link within a shared network, primarily through the use of encryption.

Turbo codes — A class of error-control codes, based on the idea of convolutional encoding, that encodes digital information so it can be retrieved with a very small amount of errors, even under the highest attenuation and noisy environments.

twisted pair — A common type of transmission medium in which two insulated copper wires are twisted around each other to reduce crosstalk.

U

UMTS — Universal Mobile Telecommunications System; a 3G cellular standard primarily based on CDMA.

Unicode — A standard used to express alphanumeric characters in all major written languages in binary.

unified communications — Multimedia integration of voice, music, video, images, and data over a single platform.

Uniform Resource Locator — Also called a URL; a string of characters associated with a specific information resource on the Internet.

uninterruptible power supply (UPS) — Equipment that provides an alternative source of power to devices such as computers in the event of power outages.

UNIX — A type of operating system that is considered robust and secure.

URI — In the context of VoIP, a caller who wants to establish a voice session uses the called party's SIP Uniform Resource Identity (URI) to issue a SIP invite request.

USB port — Short for Universal Serial Bus port; used to interface peripherals that have USB connectors to the computer.

V

vacuum tube — An electronic switch that is considered to be the precursor to the transistor.

vector graphics — A digital approach in which images can be represented in the form of mathematical expressions called vectors.

video port — A type of port that connects a computer to a device such as a monitor, projector, or other display device.

virtual private network — Also called VPN; using security techniques and performance guarantees, a VPN is a private network arrangement that runs over a public telecommunications network, usually the Internet.

virus — Malicious code embedded within a seemingly legitimate program that only becomes active when the program is executed.

visual persistence — A property of eyesight relating to the number of distinct images we can detect before they begin to blur with each other.

Viterbi decoding algorithm — An algorithm employed at the receiving end of a communications system that receives and decodes a bit stream encoded at the transmitting end with an error detection and correction mechanism such as convolutional coding.

VoIP — Short for Voice over the Internet Protocol; a set of communication standards that enable the transmission of voice over the Internet or other IP network.

volatile memory — A type of memory that loses its contents when power to the computer is switched off; also called temporary memory.

voltage — An electrical term that refers to the potential difference applied between the two ends of a conductor.

volts — A measure of voltage.

VSAT — Short for very small aperture terminal; a small satellite used primarily for the commercial application of wirelessly transmitting credit card information between points of sale and banks or credit card companies.

W

W3C — The World Wide Web Consortium; an organization that sets Web standards.

WAN — Short for wide area network; a network that spans a large geographical area such as a city, nation, or the world at large.

WAP — Short for wireless access point; a device that connects wireless computers to a wired network to enable high-speed Internet access and other services.

waterfall model — A software development model that consists of several critical phases, each of which must be fully completed before the next stage can commence.

WAV — Short for Waveform Audio Format; an audio format that is essentially uncompressed.

wavelength — The physical length between two peaks or two crests of an electromagnetic wave; wavelength (λ) can be calculated using the expression $\lambda = c/f$, where c is the speed of EM waves (the speed of light) in a vacuum measured in meters per second (m/s), and f is the frequency of the EM wave in hertz (Hz).

wavelength-division multiplexing — Also called WDM; an approach used within the context of fiber-optic transmission systems that divides a channel based on distinct wavelengths of light to concurrently send multiple channels through a single fiber.

Web server — A server that supplies Web pages to client computers.

WEP — Short for Wired Equivalent Privacy; a wireless LAN encryption standard.

wide area augmentation system — Also called WAAS; used to provide highly precise positioning information, such as for aircraft, by employing a combination of ground-based stations and satellites rotating directly above the equator.

Wi-Fi — Short for wireless fidelity; a popular designation indicating that a product complies with the IEEE's 802.11 wireless Ethernet specifications.

Wi-Fi phones — A phone capable of connecting to the Internet via a Wi-Fi connection.

Wi-Fi sniffing — Detection and unauthorized use of an unsecured wireless LAN.

WiMAX — Short for Worldwide Interoperability for Microwave Access; an emerging broadband metropolitan wireless technology; another name for IEEE 802.16, a formal set of networking standards for wireless metropolitan area networks developed by the Institute of Electrical and Electronics Engineers (IEEE).

WLAN — Wireless local area network.

WMA — Short for Windows Media Audio; an audio format that achieves lossy compression and is commonly associated with the Microsoft Windows Media Player software.

WMV — Short for Windows Media Video; one of Microsoft's video formats; prevalent because it can be played on Windows Media Player.

word length — Corresponds to the maximum number of bits of information that a computer can process at one time. Word length is expressed in terms of bits.

worm — Self-propagating and self-replicating code that threatens network security because it can modify files, launch coordinated attacks that flood a target computer with messages, or overwhelm a network with debilitating amounts of traffic.

WPA — Short for Wi-Fi Protected Access; a wireless LAN encryption protocol.

X

X.25 — A packet-switching wide area network standard popular in the 1980s and 1990s.

XML — Short for eXtensible Markup Language; used to encode Web information in a standard format.

XNOR gate — A type of logic gate that performs the Exclusive Not OR operation.

XOR gate — A type of logic gate that performs the Exclusive OR operation.

INDEX

Numbers

A